Describing the Dynamics of "Free" Material Components in Higher-Dimensions

Dr. Martin Concoyle

Order this book online at www.trafford.com
or email orders@trafford.com

Most Trafford titles are also available at major online book retailers.

Printed in the United States of America.

ISBN: 978-1-4907-2370-9 (sc)
ISBN: 978-1-4907-2373-0 (e)

Trafford rev. 01/14/2014

 www.trafford.com

North America & international
toll-free: 1 888 232 4444 (USA & Canada)
fax: 812 355 4082

Copyrights

These new ideas put existence into a new context, a context for both manipulating and adjusting material properties in new ways, but also a context in which life and creativity (practical creativity, ie intentionally adjusting the properties of existence) are not confined to the traditional context of "material existence," and material manipulations, where materialism has traditionally defined the containment of material-existence in either 3-space or within space-time.

Thus, since copyrights are supposed to give the author of the ideas the rights over the relation of the new ideas to creativity [whereas copyrights have traditionally been about the relation that the owners of society have to the new ideas of others, and the culture itself, namely, the right of the owners to steal these ideas for themselves, often by payment to the "wage-slave authors," so as to gain selfish advantages from the new ideas, for they themselves, the owners, in a society where the economics (flow of money, and the definition of social value) serves the power which the owners of society, unjustly, possess within society].

Thus the relation of these new ideas to creativity is (are) as follows:

These ideas cannot be used to make things (material or otherwise) which destroy or harm the earth or other lives.

These new ideas cannot be used to make things for a person's selfish advantage, ie only a 1% or 2% profit in relation to costs and sales (revenues).

These new ideas can only be used to create helpful, non-destructive things, for both the earth and society, eg resources cannot be exploited to make material things whose creation depends on the use of these new ideas, and the things which are made, based on these new ideas, must be done in a social context of selflessness, wherein people are equal creators, and the condition of either wage-slavery, or oppressive intellectual authority, does not exist, but their creations cannot be used in destructive, or selfish, ways.

Alternative title:

The Unbounded Hyperbolic Shapes, and the Self-Oscillating, Energy-Generating Constructs

This book is dedicated to my wife M. B. and to my mom and dad

It could be said that these new ideas about math descriptive context are so simple that the main ideas presented in this book are presented by the handful of diagrams about these simple shapes and how they are folded which are provided at the end of the book.

This book is pieced together from over 100 essays. There may be repetitions (sorry) but they are titled differently, and separated from one another, so the repetitions are hard to detect.

Note: Because these chapters were first posted on free-speech websites, occasionally, in a few sentences in the chapters, the (sometimes) unseen subject to whom there is a reference, is a person who left a comment on my free-speech post, and then my reply-or-comment was added to the chapter.

Alternative title for this book:
The Unbounded Discrete Hyperbolic Stable Shapes, and the Self-Oscillating Energy-Generating Stable Shapes

This leads to a very deep mystery relating math to physics and, in turn, forming new relations to religion

Mystery

There are two fundamental questions [in regard to the alternative title] concerning "an unbounded descriptive context," for systems which can be "energy-generating."

I. Is the stable unbounded shape still "too-big of a set" to be a part of a "descriptive context" which is both a logically consistent context, and a measurable context?

Note: A measurable "descriptive context" is defined by:
     {A precise description being contained within a high-dimensional set, composed of two types of metric-spaces, hyperbolic and Euclidean, where the hyperbolic metric-space contains within itself stable discrete hyperbolic shapes, in turn, defined by their spectral properties, so that the highest dimension "hyperbolic metric-space" (in the containment-set) possesses its own fixed "finite spectral set?"}
Note: It is only in this type of a context of measurability, eg descriptions within metric-spaces, that both (1) stable patterns can be described (and, thus, descriptive information is reliable and can be used in a practical manner), and (2) measuring is reliable (and, thus, a system can be built).

II. Do such "large sets," ie the unbounded shapes, allow there to exist "arbitrary" discrete jumps…, in regard to a system's containment set…, which can occur, so that the…, apparently, "arbitrary"… discrete jump is both (1) intended, by the system, and realized by the system (by means of the system's capability to cause and control, or to allow, internal energy changes), and (2) so that the discrete jump of the system's context occurs between two extremely different containment contexts, eg each containment context is defined by a different spectral set, but where the two containment-set contexts are both measurable contexts.

To provide a different way, in regard to a creation mythology, in which to consider these, seemingly, mathematical questions, put into a religious context, one might also ask:
Is this the context of the ancient origin and creation myths, quite often concerning the actions of gods, but where, in fact, the gods are human-beings (or equivalent highly capable living systems), possessing a knowledge which is more consistent with the true properties of existence, where human-systems are both unbounded shapes and energy-generating shapes?
Or

Is this the ancient context of humans traveling (between different places [eg between Africa and Australia] or between different contexts of existence) by using the mythical "rainbow-bridge?"

That is, math mysteries have a surprising relation to the true properties of the containing-space of existence, and this, in turn, has a surprising relation to mankind's understanding of a spiritual-world, where a many-dimensional descriptive construct takes one's mental construct of existence into a new spiritual context.

Contents

PART IV
(MORE EMPHASIS ON TECHNICAL, MEASURABLE, DESCRIPTION OF EXISTENCE)

PART V
NEW MATERIAL

Foreword

S tability and set-containment need to be considered in new ways. (also see the last page of this book)

One's model of existence needs to extend beyond (or transcend) the idea of materialism.

The open-closed topologies of metric-spaces which possess shapes is one of the sources for the idea of materialism, but it also fits into the new constructs for the containment sets for existence. The metric-spaces have shapes and they fit, as shapes, into a higher-dimensional containment-set, so that each subspace of each dimensional level is modeled as a stable shape [except for the highest-dimensional level which is an 11-dimensional hyperbolic metric-space, where a hyperbolic metric-space is equivalent to a space-time metric-space]. These shapes (especially, of different dimensions) are discontinuously related to one another, and each possess an open-closed topology, ie the properties of higher-dimensional-shapes are difficult to encounter, and each shape can contain lower-dimensional material components (except a 1-dimensional shape), and differential equations can be defined, in regard to each independent shape's material-components, which can exist within each of these metric-space shapes.

A stable-system which is modeled as a "3-dimensional hyperbolic metric-space shape" is contained in a 4-dimensional containing metric-space (a space which contains, amongst other things, eg ourselves), this 4-dimensional metric-space is our external structure . . . , the idea about which we view an external-set of properties is in a context of materialism . . . . , and the internal-structures of our being, are not so well divided from ourselves "as distinctive material objects," which is the sense we have in our 3-dimensional material containing (and what is often believed to be) an all-inclusive containing space.

Furthermore, since we are contained in a metric-space, which also contains "the solar-system as a stable 3-dimensional-shape," we, ourselves, since we are a higher-dimension than a 3-dimensional shape, and we are a part of the solar-system, then we would be bigger than the solar-system (if we are represented as a higher than 3-dimensional stable shape) and thus, we our 3-space-shape in the 4-space . . . , which represents our 4-dimensional subspace shape . . . , is quite large. However, this would also mean that we are related to the same mythology of the, so called, "giant Gods" and their associated mythological "descent to earth" (or singular, God, if one wants)" so that, in this case (ie since our 3-shape contained in 4-space would be very large, ie bigger than the solar-system) the sense of our "correct-size" would be (is) in opposition to our idea of materialism, and this is because we pay-attention to "our 2-shape size," which is contained in 3-space, and we think of as being a material object on the earth's surface.

Our sense of being, as well as our sense of perceiving an "external existence" is made difficult in a 3-dimensional metric-space, which is the shape which defines the earth's stable planetary-orbit . . . , {ie the metric-space where our notions of materialism are formed, and the space where an external existence (ie external to-ourselves) is well defined} . . . , so that the idea of materialism is difficult to transcend, within pour own minds, because of the open-closed topology of the 3-dimensional metric-space, where we identify spatial position and relative size of material components contained in 3-space, and within which we define the differential equations of our material-based ideas about physics.

Thus, within the context of materialism, and this materialism being associated to an open-closed topology, it is not possible to identify the source of stability for the observed (stable) properties of material systems based on an open-closed topology related to a model of material interactions based on (partial) differential equations.

That is, the stable properties of material systems come from the stable circle-space shapes, which, in turn, can have many different dimensional-values, as well as being shapes which have many different sizes associated to themselves, so that existence is contained in a many-dimensional context and is associated with many different set-containment constructs.

Note: Since we do not think of ourselves as having a shape which is 4-dimensional, then "How can one discount these (above) ideas?" One can only discount them if one has complete faith in the idea of materialism, but materialism is not a logical necessity, and it fails to describe the stable properties which (it is observed that) material-systems possess.

This is not about opinions . . . other than . . . interpretations of math patterns which are associated to the idea of measuring reliability and stable (math) patterns.

Measuring reliably requires that a stable uniform unit of measurement be defined and maintained, that is a unit of measuring must be identified with very stable patterns (or very stable properties). Since only the line and the circle can be made quantitatively consistent with one another, as has been demonstrated within the complex-numbers. Thus, the stable shapes are the circle-spaces, eg the tori (or doughnut-shapes) and shapes which are composed of toral-components (eg an n-holed-doughnut, ie an n-toral-component shape).

Furthermore, in regard to physical properties, they are identified with distinctly different metric-spaces, thus a description of systems which have several properties associated to themselves eg they possess the properties of position in space and the systems have stable energies associated to themselves, must be contained in several different metric-spaces, whose spatial-subspaces have the same dimension, at the same time, or a collection of different metric-space types are needed to contain the different properties which are associated to the system.

Note: These are the stable shapes identified by Thurston-Perlman in their geometrization theorem, but geometrization is only required by the math-community, since the above sentence proves geometrization based on elementary considerations about quantity and stable shapes.

Preface

The new math context is much simpler and logically more consistent than today's vision of physical description, and yet the new description opens-up a descriptive context which is much more diverse, eg capable of describing living systems based on a (higher-dimensional) unifying-form (shape) associated with the living system.

To present the new math context of the new descriptive construct in few words (here it is in about 1½ pages) and given in math words which are used within the range of their technical meaning (Also go to the figures provided in the back of the book):

In this new descriptive context the stable shapes (or stable patterns) . . . , {in a context where measuring is reliable (ie both linear and metric-invariant as well as continuously commutative almost everywhere (or except for one-point), [where continuously commutative everywhere, is a property which characterizes the circle-spaces, or the discrete isometry shapes which are of non-positive constant-curvature])}, . . . , are based on the discrete isometry (and associated unitary) subgroups (of various dimension metric-spaces with various signature metric-functions) so that the associated unitary groups (which are associated to isometry groups) are based on the pairs of opposite metric-space states, and where the metric-functions have constant coefficients, and the descriptive context is expanded to a set of properties of existence, ie physical properties . . . , {where new physical properties, which begin as 1-dimensional shapes (of either inertia-displacement or charge-energy properties), depend on both changing dimensional levels, and changing the metric-function signature, ie the various signature metric-functions of the various dimensional metric-spaces are associated to specific physical properties} . . . , which are associated to each of the different metric-spaces (different spatial-dimensions and different metric-function signatures, especially, when the metric-spaces have the same spatial-dimension, then the different signatures represent different physical properties, all contained within the same spatial-dimension subspace, where these math properties (eg spatial-position or stable-(energy)-pattern) represent the possible physical properties of a system-defining shape, which is contained within the spatial subspace of the given dimension).

A physical system is a stable metric-space shape, which is in resonance with a finite spectral set (see below), and its (physical) properties require that its description by contained in "a mixture of different metric-spaces," which are related to either the same, or adjacent, spatial-dimension subspaces, where the relations are based on material interactions.

These physical properties are essentially defined by the physical symmetries identified by E Noether, eg invariant spatial-displacements are associated to inertia, invariant temporal-displacements are associated with energy, etc.

The changes between dimensional levels can define discontinuous discrete changes in open-closed topological metric-spaces, so the (local) operators which define the properties of local measuring (defined on metric-spaces) act in a discrete and discontinuous manner, and so that action-at-a-distance can be defined in the Euclidean part of the system's (interaction) properties (assuming that action-at-a-distance is the essence of Bell's non-locality property, a property which A Aspect measured, to confirm the property of non-locality in physical systems) in an inter-dimensional model of material interactions.

Furthermore, this context is contained in (or organized around) an 11-dimensional hyperbolic metric-space, due to properties about discrete hyperbolic shapes uncovered by D Coxeter, where the 11-dimensional hyperbolic metric-space, which is partitioned by discrete hyperbolic shapes . . . . , {so that in each subspace of each dimensional level, there is a largest discrete hyperbolic shape, which is a part of the finite partition}, . . . , so that all discrete hyperbolic shapes in the 11-dimensional hyperbolic metric-space must be in resonance with (at least) one of these largest shapes (largest for each different dimensional level and defined for each such subspace of each given dimension) in the partition.

This causes the 11-dimensional hyperbolic metric-space to define a finite spectral set, to which all components of discrete hyperbolic shapes [and associated discrete Euclidean shapes which possess various set-containment relations, as well as discrete shapes of other metric-function signatures of other metric-spaces] are in resonance (ie with this finite spectral set).

In fact, the finite spectral set is defined on the set of 1-dimensional to 5-dimensional, bounded, discrete hyperbolic shapes which compose the partition, since, according to Coxeter, the 6-dimensional and higher-dimensional discrete hyperbolic shapes are all unbounded.

Furthermore, discrete Weyl-angles can be used to fold the discrete hyperbolic shapes in a natural way, where Weyl-angles are a part of properties which distinguish the different conjugation classes of the maximal tori of the fiber Lie groups, ie the classical Lie groups, eg the finite dimensional isometry and unitary groups. There is a finite set of discretely defined Weyl-angles. That is, any element of a Lie group can be conjugated to be in one of the Lie group's maximal tori, where the matrix elements of a maximal torus are all diagonal, ie linear and always (or continuously) locally orthogonal, as long as (or when) the local coordinates transformations (defined by the Lie group elements) stay within the given maximal tori.

A word about containment: If there are a set of 3-cubes of various sizes, each defining a different 3-subspace, then each 3-cube has a choice of being in any of (11-3) = 8 different 4-cubes (or 4-subspaces), but some of these 4-subspaces will be excluded due to size restrictions required on the size of the 4-subspace cubes, so as to be able to contain the size of the 3-cubes. This construct allows for trees-of-containment which can have various types of branching structures.

Partial differential equations and material interactions are re-defined as discrete, discontinuous operators, defined for small discrete time-intervals, in turn, the time-intervals are defined by (time) periods of spin-rotations of metric-space states. These new models of material interactions are similar to the classical model of local measuring of physical properties, and second-order 2-body classical problems are used to adjust the (orbital) properties of the stable spectral-orbital constructs, where these very stable constructs for material systems are the main result of this new description.

(preface continued)

One must always identify the exact problem which physics and physical science is trying to deal. This is because the media and journalists are indoctrinated to believe that the truth is difficult to identify, but really the truth, which the media represents, is adjusted to fit the interests of those who control creative actions within society by their investments, and that only a few people have the capability to discern truth, namely, those scientists indoctrinated by . . . and competing within . . . an institutional vision of truth and its relation to creative actions (ie adjusting the complicated instruments owned and controlled by the very wealthy) . . . and to deal with the range of language which is needed to find the truth, where it is this limited and narrowly defined "truth," which the journalists find from the experts, and then the journalists present this "expert truth" to the public.

Thus, the journalist is all about the institutional truths of the age, and this truth must be associated with the documentation and quotes from the experts, so as to prove the point about the truth which the journalist is making within the media.

This is all nonsense, truth is about the language developed around the principles upon which one is claiming to base:

1. society's organization, and behavior within society, or
2. thoughts, in regard to discerning a truth, and in regard to a context for creative actions, about evidence and its interpretation, and subsequently organized in a descriptive context associated to a simple principle (simple stable pattern) which allows one to find hidden information which is both accurate (to sufficient precision) and is practically useful, either as information or in relation to controlling the properties which exist in a practically creative context of action, which, in turn, exists in a context of reliable and repeatable measured properties, which are within the context of one's stable thought patterns, in regard to creative efforts.

There are many ways in which this can be done, and given the failures of the current way in which this is being done now (2013), and presented in very narrow contexts by the propaganda system, this should be a prevalent activity for rational thought, which is trying to identify a "precisely described truth" which is accurate and possesses practically useful capabilities.

While journalists are all about elitism and inequality, and narrow expressions of truth.

Publication controlled by journalists and editors (or based on peer-review, remember peer-review cannot review new ideas) is all about indoctrinated people who express the institutional elitism, which is defined by the powerful, where elitism defined within the narrow context determined by the blinders which they (ie those serving the powerful as wage-slaves) have had fixed to their eyes, or to their vision, which requires that they see a narrow institutional truth.

Though C Hedges is one of the more independent journalists, yet, he cannot see the failing of society as a failing of the experts, ie a failing due to the narrowness of the indoctrination and reward system. He essentially believes the "security argument," which is an argument for tradition, and established authority, and fixed ways of doing things, where if one does not follow such a conservative path then one will end-up with chaos and unnecessary destruction, a conservative path which acquiesces to wage-slavery and the attachment of blinders so as to keep our focus narrow and safe. That is, the great intellectual achievements of our society must be made secure.

Even seemingly courageous journalists, and, apparently, whistle-blowers too, are so well indoctrinated, so as to serve the failing (so called) "institutional truths" where these "unquestionable truths" are the, so called, "truths" expressed by the illusional-icons of the intellectual illuminati, eg Einstein, ie the indoctrination (of those who serve the moneyed-interests) is a deep belief in inequality and a belief-in a rigged word-game, which is supposed to rigorously identify an absolute truth . . . , and one can be assured (as the propaganda system assures us all) it will be an authoritative truth.

There is a vague questioning of economics, politics, and the justice system but the illuminati, who express our deepest cultural truth, can not have their dogmatic truths (or their integrity) or their intellectual capacities questioned.

Thus, the journalist is not assertive of their belief, rather they are authoritative in that they researched the established authorities so as to ascertain a truth about, which the sole voice of an authoritative truth of the media, is allowed to express, where that sole voice of truth is the propaganda system.

It is a "catch" in the endlessly repeating, and circularly-referential, propaganda system, which is allowed to trap what is supposed to be the voice of rationality . . . , but the well meaning journalist who seeks to provide truth . . . , instead espouses the greatest of illusions, where one of these illusions is a religiously-faithful belief in the authority of science.

Saying these words results in the readers "buttons being pushed," (this is a result of the deep control that the propaganda system has on everyone's ability to think) and as a result it needs to be stated that "this skepticism of science's authority is mainly because a better alternative 'mathematical model' is being presented," and it is not motivated so as to support either skepticism of global warming (the CO2 global-warming issue should have been settled back in 1900 with renewable energy sources) or to support creationism or creative-design ideas which are supported by groups of people who do not supply a valid alternative math model. Nonetheless Darwin's probability based model of evolution needs to be strongly criticized, since (1) the origins of life cannot be random, since life formed on earth almost as soon as the earth cooled, and (2) DNA cannot be put-together as hypothesized, apparently 90% of a living system's physical structure emerges (during embryonic development) from the properties of the epi-genome.

On the other hand these new math models provide a new context within which to view life and how the epi-genome might work.

The, so called, "truthful" journalist only supports (or reports on) institutional truths.

But institutions always (or also) provide, in hidden ways, red-herrings (deceptions) about the nature of their (the institutions'), so called, truths, so the focus becomes about the red-herrings, and not about the very questionable validity of institutional truths, which are provided by university departments of, so called, public institutions, whereas a university-department's "truth" serves the interests of the ruling-class, not the creative interests (or the creative capacities) of the public (where creative capacity is expanded by, equal free-inquiry).

Similarly, all ideas, associated to organizations, which are, identified as being against the interests of the ruling-class, are attacked by the justice department, and subsequently these attacks are legalized by the political structure, ie or equivalently the propaganda structure (that is, the social function of the political system is to serve as a well paid-cog in the propaganda system, and the function of journalism seems to be that of uncovering injustices so the politicians can re-write the law so as to make these injustices legal).

One cannot live in a sustainable manner where society is based on property rights and minority rule, this leads to selfishness, violence, and destruction

Rather

Law must be based on equality, where the context is each person is an equal creator.

That is, human-value is about creativity.

That is, value is not to be based on the ownership of material property, nor based on scarce material-types.

The material of the earth must be used in ways which are harmonious with the earth's structure.

The market is about the gifts which human creativity provides to society, and to the earth.

The market needs constraints in regard to advertising, and the large-scale of the creative efforts, and the subsequent over-use of certain types of material, ie it needs constraints based on curtailing domination and control. Etc.

But essentially "anything can be done to the public within today's society," as it is constantly demonstrated in the news, provided that one:

uses extreme violence, and
the media announces the changes and

the violent institutions of society enforce what is to be done (as it was announced, in a way which is consistent with the communication-political-propaganda channels of society).

However:

In science there is the need to develop (new) ideas in an intuitive manner (not in an overly authoritative, and overly formal axiomatic structure), and one wants many different ideas to be expressed, so that they are different, and they challenge the accepted dogma about science expressions, ie expressions which also have great limitations as to what patterns, the (given) language being used, can express.

That is, one wants new language to be built based on new: assumptions, contexts, interpretations, purposes, ways of organizing language, and new ways to identify a containment-set.

The versatility of language needs to be exploited in regard to seeking new realms of creative actions, and this exists in a context of identifying:

What is one trying to do (what purpose)?
In what context?
and
How is that context to be contained and organized (by words and relationships)?
What observed property is to be re-interpreted?
Etc etc.

In physics one sees the containment set defined in either In a very narrow vision such as Einstein who believes materialism, and one must define material and its containing space, ie the material is defining the containing space, and all of physics properties are to be contained in this metric-space with its open-closed topology and its metric-function signature being related to R(3,1), where 3 is the dimensional of the spatial subspace, and 1 is the dimension of the temporal subspace, so that R(3,1) is a (3 + 1) = 4, ie a 4-dimensional metric-space. Furthermore, along with the measures for material quantities all the physical properties are related to local measures of the given property so as to be locally measured in the function's domain space, where the physical-property of the material system is represented as a function, and this is assumed to be able to identify and determine all of the system's physical properties if the local measured properties satisfy the physical laws, because the system is contained in the domain space where the domain space is a metric-space. Furthermore, Einstein wants to unify all the forces associated to the different materials into one expression for force so that this is based on covariant invariance for arbitrary changes in coordinate motions so that all force fields can be related to the distortion of the shape of space and the subsequent geodesic structures on this shape geodesics which are supposed to define inertial properties.

or

In a very broad vision about physics, expressed in quantum-physics and particle-physics, where the system is basically random (so it is modeled as a function-space of harmonic-functions oscillating about a geometric structure of a system's energy, ie the potential energy term) but the system defines a set of discrete spectral-values, where these values can be related to particular operators, and so that the system's properties are represented as a set of operators (a complete set of commutative Hermitian operators) which act on the quantum system's function-space of harmonic-functions, where the function-space is complex-valued, and "the" Hermitian-form-operator (defined on the function-space) is invariant to unitary operators, where for Hermitian operator, H, then e^iHt is a unitary operator, and where (so that) the Hermitian-form, along with commutativity, is used to separate (or identify) the spectral functions of the system's function-space, and thus, the unitary operators preserves the spectral-structure of the system's function-space, where completeness allows convergence to all spectral (or energy) states.

But then the quantum system (or any material system) is assumed to reduce to a finite-set of elementary-particles, which have internal particle-states associated to different energy-properties, so that wave-function is provided with an internal particle-state structure, and all quantum-material-interactions are modeled as particle-collisions, where the internal particle-states of the colliding particles are always changing, ie locally the particle-states are moving between different (particle) energy-values, apparently, due to the collisions. This is supposed to perturb the wave-function, which is a sum of the set of spectral-functions, where the particle-state operators (which change the internal-particle-states) are (non-linear) connections (or derivatives), which act on the particle-states of the quantum-system's wave-function, so that in the perturbation the wave-function is summed over all such particle-state changes, ie apparently for a lot of particle-collisions as a part of the material interactions.

But such a containment set has

1. many different local vector-space structures associated to the physical system in regard to the interaction fiber factor-group, U(1) x SU(2) x SU(3), and
2. the wave-function is no longer defined on a smooth containing metric-space, but space breaks-down around particle-collisions (apparently, allowing for all the different particle-collision adjustments in the perturbation sum) ie the continuum (upon which the wave-function's smooth structures depend) breaks-down, and
3. it is a random basis for the local particle-spectral events so the particle-collision geometry also breaks-down, eg due to an uncertainty principle which is always associated to a random basis for description.

That is there are several local space structures, not simply the metric-space, which is defined by materialism, and furthermore, the continuum loses its properties, so convergence loses its defining context, the geometry of a particle-collision cannot fit into a random context used for the descriptive basis, and there is not any notion that the interacting material quantum-system has any valid structure for containment, rather it is a set of different math constructs applied for the convenience of identifying certain math patterns (or to use certain math structures) and then abandoning these structures (in another stage or) in the next step of (in) the interaction descriptive sequence.

That is, in quantum description, there is a set of separate-islands of math structures, which apply when one is within any of the (different) particular island, but which are not relevant on the next island, in the, apparent, island-jumping sequence of descriptive constructs, which are associated to describing the quantum-model of material interactions.

Thus, it is incomprehensible.

But, apparently, the cohesion of this descriptive structure is claimed to be resulting from the "axiomatic formalism of mathematics," and thus, the patterns are not stable, eg non-linear changes of particle-states, and the quantitative structures are not consistent eg the smooth wave-function and the break-down of the continuum for point-particle collisions, but, it is claimed that this is OK, within each formalized context of math.

Furthermore, this is a perturbation process used to adjust, slightly, the energy-values of the system's spectrum. But, in regard to the wide array of many different quantum systems, the wave-function for general, many-but-few-component quantum systems cannot be found. Thus, this is all about completely indefinable randomness, and it simply does not work for a wide enough range of stable systems, which are observed.

That is, consider the many fundamental and very stable quantum systems: nuclei, general atoms, molecules, molecular shapes, crystals, and then the macroscopic systems of classical physics or of general relativity, namely, the stable solar system, and then there are the very stable and highly controllable living systems. The properties of these systems are not being described to sufficient generality and with sufficient precision and to a level where this information has practically useful value, so as to be based on the, so called, laws of physics, as these, so called, laws are now being expressed in the propaganda based "set of institutional truths."

Perhaps, all material systems do not reduce to elementary-particles, and perhaps randomness is not the basic property upon which to base a description of material systems which possess stable properties.

Perhaps materialism is also wrong.

Perhaps the geometry of material systems defined within the assumed to be material containing metric-space is not capable of describing the observed order of the many-but-few-body solar-system, or the order and control which are possessed by living systems.

Perhaps it is not sets of operators which represent physical properties but rather it is sets of different dimension and different metric-function signature metric-spaces which represent the physical properties as well as the materials which exist and which are the parts of the properties of a stable system composed of different materials.

The containment space for existence is the main structure upon which the observed stable order of the most fundamental physical systems depend.

These fundamental stable systems are: charges, nuclei, general atoms, molecules, crystals, closed thermal systems solar-systems and apparently the closed stable motions of material within galaxies and the motions of entire galaxies.

The new containment set is characterized by:

1. its high-dimension, relative to the dimension defined by the idea of materialism, and
2. by the different dimensional levels being determined by stable hyperbolic metric-space shapes (see below).

The observed order of material systems are derived from the set of stable shapes, which are a part of the containment space, and it is also based on these stable shapes being contained in a many-dimensional context.

The idea of materialism defines a topologically open-closed material containing metric-space (or metric-spaces), so as to be based on only one metric-space with a metric-function which has a particular signature, with one fixed dimension . . . , though there is some latitude in regard to the two metric-spaces R(3,0) 3-dimensional Euclidean space and R(3,1) where R(3,1) is a 4-dimensional space-time metric-space . . . , so that there are either force-fields based on material geometry or randomness modeled as complex-valued wave-functions, where these functions have a space-time domain space, ie materialism is maintained, so that the definite spectral-values for quantum systems are to be related to sets of operators, specifically the energy-operator . . . , where the assumption of randomness was a result of seeing the random events of particle-spectral values identified at single-points of such random events found in space-time . . . , then it was believed that all of material systems could be reduced to elementary-particles.

Then, when trying to identify material interactions for quantum systems, based on the reduction of material to elementary-particles, the idea of "hidden" particle-state was developed, to be used to describe material interactions for quantum systems . . . , where it is assumed that the basic wave-function for the system can be found, ie harmonic waves oscillating about the system's locally measurable average energy-operator (or oscillating about the system's potential energy term).

Unfortunately, this is far from being true.

That is, these models of quantum interactions are defined in the random context of quantum descriptions, ie the uncertainty principle applies, but the model depends on point-particle collision geometry, a model which is incompatible with randomness.

Nonetheless, with each collision there is an associated set of changes of each particle's internal particle-state, which reflects the range of energy of the system and the energy of the particle-collisions, so that different internal particle-states are activated by different energy-ranges.

This is a model . . . , in which neither the geometry of particle-collisions, nor the discrete random energy changes of the internal particle-states, are consistent with the quantum system's smooth wave-function, which represents the random basis for quantum description, and an associated set of discrete energy (or spectral) values associated to the function-space's spectral decomposition by the set of operators, where these random, discrete energy-particle events are observed for the quantum system being described, . . . , in which the notion of the containment of such a quantum system in a metric-space, which possesses the properties of a continuum, which, in turn, is needed to define its smooth wave-function, are obliterated.

That is, either

(1) the construct of "randomness with the geometry of particle-collisions"
or
(2) the churning-energy-changes associated to changes of internal particle-states, and an apparently ever-present set of virtual particle-collisions associated to these changes in internal particle-states, . . . , are two "math" constructs, which obliterate the continuum upon which the smooth structure of a quantum—system's wave-function depends. But, furthermore, it is the structure of the smooth wave-function upon which the assumed wave-function's internal particle-states act in a manner which is supposedly a set of perturbing agents.

Thus, it is very difficult to comprehend, or to see, any valid content in such a description of particle-physics, or likewise, it is difficult to conceive that any physical construct, which is derived from particle-physics, can possess any content . . . , such as string theory, etc.

Where are the rigorous descriptions, based on the laws of physics, which describe to sufficient precision the spectra, of the entire range of all quantum systems in the following short list of: nuclei, general atoms, molecules etc?

The idea of defining a material-system's observed stable measurable patterns based on material interactions so as to be defined, either in a metric-space which possesses the properties of a continuum (either the wave-function, or classical material geometries), or (following quantum description) acting on energy-levels based on the collisions of elementary-particles, . . . , is a description of particle-properties which defies attempts to make such an idea fit into a mathematical context.

Both of these math models of physical systems are ideas which "have not worked," based on either geometry or the pair of associated ideas of randomness and reduction to elementary-particle collisions, but where internal particle-states must conform to materialism, by requiring that particle-events stay inside an "assumed to exist" system-containing metric-space.

That is, it "has not worked," since the stable precise spectral-orbital properties of material-systems cannot be described (or identified) in a context of either geometry or random-waves (which have been reduced to particle-collision-determined particle-states).

Try some new ideas (let us have a scientific revolution, the very type of thinking which the public is indoctrinated against pursuing)

On the other hand partitioning an 11-dimensional hyperbolic metric-space by a dimensional-subspace-size determined partition, which is composed of stable shapes, where each dimensional level and each subspace (of any given dimensional level) has a largest shape, so that all the other shapes must be in resonance with the spectral values of some largest hyperbolic metric-space shape defined in the partition. In such a context there is, thus, a set of set-containment-trees, defined in the 11-dimensional hyperbolic metric-space, where these trees depend on the dimension-subspace-size properties of the shapes of the partition, which are associated to the spectra of the partition.

The order of material systems is either due to condensed material following the stable orbit structures defined by the shape of the material containing metric-space shape within which the condensed material is contained, but these orbits can be perturbed by the second order dynamics defined on the condensed material, which is contained in the metric-space, ie perturbed by the usual material-interaction models of classical physics (or by general relativity, but the context of general relativity is now contained in a linear, solvable shape).

or

are a result of component-collisions defined by second order dynamic equations, in turn, defined in the metric-space, but so that during such component collisions (which are also defined within a compatible energy-range, so that) there is defined a resonance between the colliding component-complex and the finite spectral-set defined by the spectrally partitioned 11-dimensional over-all containing metric-space, so that the colliding component complex re-forms into a new stable material component, which is in resonance with the finite-spectra of the over-all containing space.

The new model of material and of material interactions depends on new types of discretely defined operators, so as to be defined on discrete time-intervals (which separate opposite metric-space states) and discretely defined between different metric-space shapes, and between different dimensions (and/or dimensional levels), so this discrete-operator descriptive context has a mixture of:

1. metric-space states,
2. shapes,
3. discrete action-at-a-distance interaction geometric structures, and
4. there are also new physical properties and associated new materials,

where the basic physical properties are not operators, as in quantum physics, but rather the physical properties are determined by the presence of metric-space types, which, in turn, are determined by the metric-space's spatial-subspace dimension and the metric-space's metric-function signature, where the full identification of each different metric-space also determines different physical properties which can be associated to both physical systems and their interactions. Furthermore, the sets of opposite metric-space states (each defined by a metric-space's associated physical property) determine discrete time-intervals associated to the (new) discrete math structures of material interactions.

Note: Describing physical stability: The differential equation vs. New containment constructs

It needs to be noted: That the truth of this new math construct cannot be judged by today's experts, since that is like having the authorities of the Ptolemaic system judge the truth of Copernicus's system in that age.

That is, the truth of the new math-physics construct will be found based on both (1) its descriptions of new properties, and descriptions of new contexts within a new interpretive format, and (2) the practically useful instruments, and inventions, as well as new creative contexts, which emerge from this new math-physics descriptive structure.

The propagandists will claim that "its truth" must be determined by "the intellectually superior people," those, so called, superior people, which the media has cataloged in the worship of personality-cult, ie the religion which the media develops and maintains (and propagates), and the personal history (eg the CIA, NSA records) of "m concoyle" does not show him to be either a genius, or at all that capable, so this description he gives cannot be true, based on the risk assessment, which is a mathematics based on indefinable randomness, ie meaningless mathematics, which, apparently, (eg the Risk Management company which went bust in, about, 1998) is now (still) being used to determine investment risks, eg Chase's about 8-billion dollar loss on its bets about a year ago (now it is 2013).

Note: apparently, Now intelligence is a behavior, which can be learned (or improved) through a scheme of behavior modification, ie PBS 8-9-13 About the Brain (?). Intelligence is (was) measured as achievement, in regard to "an assumed set of educational development levels," ie it was a measure of high-cultural advancement, where "high-culture" is, effectively, determined by the owners of society.

But this is the way the propaganda system operates (ie to serve the interests of the investment-class).

That is, the journalists are essentially mindless brown-nosers, and the experts are pathetically indoctrinated obsessive types, who are also mindless, ie without a capacity to judge the value of a new idea by themselves, the experts have always been guided (and motivated) by the propaganda-education system.

So one sees the circular-ness of the meaningless logic, which is the basis of modern rationality, a rationality which is based on both propaganda which guides the experts, and experts, who simply compete within the narrow rules which are provided to them (provided for them by the propaganda-education system, ie brought to them by the investment-class), but these competitive people have not judged "truth" for themselves. The expressions of "ideas" of both the media and the wage-slave experts, are based on this circular-rationality, because these efforts (of the media and the experts) are all about serving the interests of the owners of society.

Neither the math nor physicist asks: "Why all this pointless complication?" [Where the word, "pointless," is a reference to the failed attempts, by the experts, at using the, so called, "laws of physics" to accurately describe the measurable properties of the very common types of very stable physical systems.]
"Why are the very stable properties of a wide range of general and very fundamental physical systems not being described based on the laws of physics and mathematics?"

They seem to accept the ridiculous claim that "these systems are too complicated to describe," even though they are stable systems, and can be distinguished from one another based on their very individualistic, stable and precise properties. But the existence of these properties, of stability and precisely measurable properties, imply a linear, stable, metric-invariant, and continuously commutative (almost) everywhere, set-structure, of both physical properties, as well as these math properties being related to such consistently measurable stable physical systems.

"Who is there who can judge fundamental truths?"
"Who is there to judge the value of someone's creative efforts?"
The truth of an idea, or of a construct, is related to the range of understanding, that the viewpoint provides, and the range of practically creative things, and the new range of "the creative context," which determine such "value" to be associated to new ideas and new constructs.

The bankers and the Roman-Emperors were both incorrectly attributed to possess (within themselves) such capability of judgment about both value and (to be able to judge) truth, eg investment risks. But investments, and directing the development of already "known technologies," are not about developing new ideas, rather they are about exploiting traditions and exploiting animosities and competitions within a narrow context.

We need to re-kindle the idea that the law of society is to be based on equality, in relation to knowledge and creativity, a viewpoint which might be attributed to Socrates.

Part I
Book VII 3,
Describing physical stability:
The differential equation vs.
New containment constructs

1.

Education and the free expression of ideas

Society's institutions of education need to consider:

The current structure in which knowledge is related to creativity is related in a very narrow way so as to result in the destruction of: knowledge, education, and creativity within society, because knowledge is only being related to the narrow interests of big businesses, eg military, oil, and banking.

For example, there are fundamental, stable, definitive physical systems which exist at all size-scales which do not have valid descriptions if one tries to apply the, so called, physical laws in regard to identifying (calculating) the well defined properties of these physical systems.

There is a new descriptive context based on partitioning an 11-dimensional hyperbolic metric-space by means of very stable circle-spaces, or discrete hyperbolic shapes, where these stable shapes model a metric-space associated to particular subspaces, in turn, associated with the different dimensional levels, and by this model the stable properties of fundamental physical systems can be identified (calculated), as well as developing relatively simple ideas about life-forms, so as to also model the control which life possess over itself.

Yet, this type of creative effort is marginalized, due to intellectual arrogance and dogmatic authority whose basis is from the social forces of monopolistic socio-economic institutions (the military, oil, and banking) which limit both knowledge and creativity within society, so as to strangle the society itself.

The entire culture must serve the narrow interests of these monopolies.

Education (or knowledge and knowledge's relation to creativity) should not be about focusing on a handful of very technical (or very secretive) instruments for the purpose of upholding the very selfish monopolistic and socially domineering business interests of the society.

Today's math and science university departments focus on descriptive structures based on indefinable randomness (where elementary events are unstable) and non-linearity (a quantitatively inconsistent descriptive context) so as to relate these poorly defined math patterns to feedback systems with limited range of purpose (so as to guide a drone or a missile to be used in war), or to design a nuclear bomb, also one must mention the "risk fiasco" where apparently, so called, "deep MIT PhD mathematicians" were used to calculate "investment risks" (where the failures of these calculations were directly related to the 2008 economic collapse). That is, these risk calculations were not true, because these descriptive structures are not capable of identifying a pattern which is stable, or valid, so as to result in the lose of trillions of dollars.

The reason for this failure is:

A risk calculation, where the elementary events are not stable, cannot hope to be all that realistic of a model.

A distinguishable property is not a valid basis for the definition of probabilities (eg the events of a valid elementary event space must be stable and well defined, not simply identifiable).

Furthermore, randomness and indefinable randomness cannot be used to describe the stable properties of fundamental physical systems, as is now being attempted.

On the other hand:

Creativity results from people who are equal, and who are seeking changes, most notably changes in knowledge and its relation to new contexts for creativity. Whereas, fixed authority is related to holding social patterns fixed so as to sustain monopolistic, socially domineering business interests.

Equality is for creativity, while inequality is about fixed structures and extreme violence is needed to maintain a state of inequality.

Learners need to be equal free-inquirers, but when one lives in a society where the propaganda system is the solitary voice of authority, and that single voice supports fixed monopolistic socio-economic interests, then the learning institutions become based on dogmatic authority and are characterized by arrogant intellectualism, which leads to social failure and to the failures of knowledge.

Creativity cannot be defined in terms of strumming a guitar, manipulating society by a psychological controlled use of icons of social value (allowed by the single-authoritative-voice model of the media, ie propaganda system), or working on math patterns which cannot describe a stable set of observed patterns.

That is, the military oil banking business interests should not be the primary (only) social forces defining the single voice of authority (of the media that they own) for a free and equal society. Human creativity is much more diverse.

The correct interpretation of Godel's incompleteness theorem is that if one cannot create "what one wants" by using the existing precise language, then invent a new language, within which new contexts of creativity can emerge.

Education should not be about defining psychological types such as "autism with capacity to use language" ie obsessive personality types who are competitive but not reflective nor are they particularly curious, so as to work on a small set of complicated instruments, which are used by the ruling-class (ie military, oil, banking interests) in a fixed way, associated to maintaining society in a fixed way of organizing a society so that the society is required to use the products produced by the monopolies.

People are fundamentally both equal "seekers of knowledge" and equal creators, so education is about (descriptive, measurable, and build-able) knowledge and the relation that knowledge has to creativity and a creative context.

Creativity is best related to a precise language which is closely related to the elementary language structure of assumption, axiom, context, interpretations, and organization of a containment set. Stating elementary properties for a new language allows creativity to be accessible to many people.

The best example might be Faraday who developed the language of electromagnetism and while he invented this language he also developed technology and the use of the ideas of electromagnetism.

In regard to Faraday's creative actions it should be noted that the working aspect of today's, so called, highly technical society is all centered around the development of the ideas related to electromagnetism, while the undefined randomness associated to quantum description has not led to any technical development based on the information gained from quantum description, eg the crystals were doped with impurities based on thermal techniques, the properties of transistors were determined experimentally, and then coupled to classical systems (electric circuits). Particle-physics is only related to rates of nuclear reactions. This is another example of how big business directs knowledge's relation to creativity within our society, where erroneously calculating the "investment risks" for banks (where these calculations were based on institutionally determined truths about math, which are related to quantum physics math techniques) is, yet, another example.

Why should randomness (or limited contexts of feedback) be a basis for stable patterns?

The observed stable properties of physical systems: nuclei, general atoms, molecules, crystals, the solar system, dark matter (motions of stars in galaxies) etc. all go without a valid description based on what is supposed to be physical law.

The descriptions of these stable systems require stable math patterns to both describe these stable systems and to allow for the stable measurable context within which these systems are observed. But such a simple idea, which makes so much sense, is carefully ignored by society and the institutions, which so scrupulously serve the narrow interests of big business

Today's education institutions should take a broader view of knowledge and creativity so as to not define knowledge as a process of fine-tuning the instruments which are of most importance to the ruling class.

That is, the society needs to be out-side the oppression of wage-slavery, and monopolistic domination negated, and this can be accomplished (or remedied) by practical technical creativity.

Teaching needs to be based on equal free-inquiry and not be based on narrow dogmatic authority. Authority is not truth, yet, today, "educational institutions" are not distinguishing between authority and truth.

Since the reader must also be confused, let me inform you; truth needs to be related to practical development, and remember the ideas of Ptolemy were measurably verified (measurable verification can often be irrelevant, in regard to determining truth).

If educational institutions want to become valid institutions of knowledge, which is relevant to widely diverse viewpoint concerning creativity, then consider new ideas about math and physics.

Consider research about how to further develop the use of stable circle-spaces in a many dimensional context to describe the observed stable properties of physical (eg nuclei) and living systems.

2.

Further Comments

The social comments are derived from many attempts to express new ideas, the point of expressing ideas in "a society of elites" is that the person expressing the idea needs to be considered, by the public, to be an elite member. Here are some science and math comments, too.

There is no reason for such an idea, especially, if one simply tries to reason on one's own.

In other words it is an example of people believing the propaganda system and subsequently discounting themselves, ie letting others decide "what the truth is" for them, apparently, based on the emotional need for people to look-up to the elites (the main point which the propaganda system is expressing, ie only the superior people should have a say in things), apparently, an emotional need, which people are taught to possess, is their need to look-up to the elites.

The technical claim within the propaganda system, which expresses a deep belief in elitism, is that "the ideas are too complicated for people to understand." This is all nonsense, but to successfully oppose-it it requires that there actually be honest people explaining the true nature of social power to the people, amongst many such topics, ie the very point of an education system.

But the managers of the education system insist on mis-representations, not truth. Since WW II this has been about the total militarization of the US society, where the management class is often pulled from a military background.

In regard to science and math, such precise language trying to express some aspect of truth, is the language of creativity, and the example of Copernicus is that the experts, or the authorities, are floating on a context of assumption and interpretations, about which anyone and everyone can have an idea.

Furthermore, the complicated methods of the experts need to be analyzed based on what they can actually describe, and in regard to, what they are trying to describe, with their precise language.

What they are trying to describe in physics is: The stable and uniformly measurable spectral-orbital properties of all of the most fundamental physical systems, which are related to our experiences; nuclei, general atoms molecules crystals and the stable solar system; all of which cannot be calculated based on the accepted laws of physics. Math is very much about supporting this task.

That is, the methods of physics and math are failing.

This means that their assumptions, interpretations, contexts, vision of containment, etc are wrong, and at such a simple elementary level of language (ie assumptions, interpretations, contexts, vision of containment, etc), everyone can try to hypothesize about the nature of these systems, and to be a scientist or mathematician.

That is, the imbalance is that the experts have no business being considered experts by others, since their methods fail at what they are trying to describe.

Thus, a person should not "buy-into" the idea that only elites can grasp and use language, so as to describe the observed properties of fundamental physical systems, since the experts are failing so miserably.

How can anyone take the idea of an expert (who possesses a superior truth) seriously?

One cannot fail any worse than the "considered exerts" are now failing.

If one reads Einstein, one sees ideas being expressed at an elementary level of language, but the criticism of Einstein is not that he is considering the idea of moving frames all that incorrectly, the issue is that frames are not that important, Einstein has the wrong containment-set structure, the laws of physics, the focus of Einstein's concerns, are exactly what has failed to describe the stable spectral-orbital properties of material systems, and central to the containment ideas of Einstein is the idea of materialism. Nonetheless that inertial mass does equal gravitational mass does seem to require that the shape of space be a source of inertial causes (or a cause for observed inertial properties). Furthermore, it is also clear that Einstein was wrong about action-at-a-distance, etc as Aspect's experiment has shown, and his considerations about frames and relative motions are mostly irrelevant.

There are many wonderful things about Einstein, he uses intuitive language rather than axiomatic formalization, where axiomatic formalization assumes a much broader reach, in regard to application, than math can actually be applied, without adjusting how the math language is organized.

Einstein likes ideas, and considers many different ideas.

The typical expert is absorbed in their own correctness, and in their superior intellects, and their social position, which requires that they always be correct, in regard to their authoritative dogmas, if they want to get (or remain) hired, etc.

It is very easy for math to be incorrectly applied, particle-physics is an exercise in incorrect application of math, and it should be the most heavily criticized set of ideas in society, it is incomprehensible baloney, there is nothing about particle-physics which is defined in a valid manner.

As soon as one breaks free of the idea of materialism, so as to re-organize math structures so as to describe the physical world based on considering higher-dimensions, so that one models those higher-dimensions with valid mathematics, then one is also uniting science and religion.

This is about the nature of existence and the creative relation that living systems have to the creation of existence, try to do that (ie create existence) and all the other baloney concerning good and bad boils down to selfish acts vs. selfless creativity.

The media-education system is reluctant to actually provide the public with ideas different from those ideas which suit the interests of the owners of society, so mostly one does not encounter that many ideas.

Furthermore:

What is being stated is that:

The equations and subsequent (solved) formulas, which are associated to the accepted laws of physics, are the wrong patterns to use, in regard to trying to describing the (observed) stable properties of material systems, and thus they are ideas and constructs which are incapable of describing the observed stable order of some of the most fundamental physical systems (which are built from many-but-few-components), ie wrong in regard to understanding the stable structures of all existence.

Rather the correct and sufficiently precise descriptions of stable properties of existence are to be based on stable shapes of discrete isometry groups, essentially the circle-spaces . . . , where an example of a circle-space is a torus, or doughnut, these circle-space shapes are stable since the circle and line can be made quantitatively compatible, eg this is done in the complex numbers, and these circle-spaces are related to cubical shapes which make-up a checkerboard of a (rectangular) coordinate space, ... , so that in an over-all containment-space composed of many-dimensions can be organized around these stable circle-space shapes, where each dimensional-level can be related to several different metric-spaces (each distinguished by its metric-function signature, and associated [different] physical properties)

These checker-board "squares" can have definite sizes, depending on the different subspaces (of the same dimension) so that branches of the "trees of containment spaces," and their subspaces, can be determined based on a finite set of spectral-geometric values, and containment trees can be related to increasing sequences of the set of subspace's dimensional-values, where the containment-tree is about fitting smaller-boxes into bigger-boxes as the dimension increases.

(ie the spectral-values of a stable shape will be associated to the sizes of the squares in the checkerboard)

so that

(1) the branches of the containment tree
(2) the finite spectral set, and
(3) the possible angular-momentum relations (or connections) between the various sets of circle-shapes . . . , ie tori and shapes with connected toral-components, which exist both within the branches and between dimensional levels, . . . , together define the range of control over geometry and space which living systems can possess, where this could be, geometrically, quite vast.

That is, mathematics, is all about geometry, not partial differential equations, and the containment set, identified by (1), (2), and (3) is "in all likelihood" accessible to us, because we are higher-dimensional, relatively-stable, metric-space shapes.

We probably have either memory, or perception, of all of this, (especially, if we perceive the world as it really is) and we can use our capabilities of accessing both various sizes and dimensions, by using our intent, which along with correct knowledge, can, together, be used to create a further aspect . . . , an extension . . . , of existence itself.

This range of math concerning geometries and containment set-structures, can also be relatable to many other different (such high-dimension) over-all containment-sets . . . , ie containment-sets of the same types of structures but with different finite spectral-sets and different containment-trees . . . . , so that our (possible) control over this context allows us to be able to move between many such different types of spectral-geometric experiences.

That is, we are the perfect instrument with which to do these things, within the true structure of existence, and to be creative, in regard to all of existence, so that differential equations and associated solution formulas are not necessary, yet differential equations could be (or are) applicable to this containment context, because it is a measurable context based on stable patterns.

Note:

This anti-differential-equation belief, so as to, instead, favor geometric models of existence, is often ridiculed by the peer-reviewed experts, as not really being math. This is because of the fact that Newton invented the differential equation and it has never been fully understood:

Is the derivative a model of local measuring?

(as Newton saw it) [but does it really have the correct structure in regard to measuring material interactions?]

or

Is it an operator on a function-space?

An expert can be defined as "a charlatan who acts as if they totally understand Newton's differential equation."

This attitude of investigating other math patterns, which are central to the use of the differential equation, is often ridiculed by the peer-reviewed experts, as not really being math.

This is what happened to W Thurston and his geometrization, which, essentially, says that the discrete isometry shapes are the stable patterns of mathematics, and it also happened to D Coxeter, who identified all of these checkerboard patterns in the context of hyperbolic space, ie most stable shapes of geometrization are the discrete hyperbolic shapes.

These two people described, in words, the central properties about which math can deal, especially, if it stays fixed in the idea that measurable descriptions must be based on a context where measuring is reliable and the patterns either used or described are stable.

Differential equations mostly move outside of this context, but can be used to adjust the descriptions of spectral-orbital properties, which exist for material contained within a metric-space shape.

However, one does not need the, somewhat, fancy theory of geometrization, . . . if . . . . one stays true to the idea about "what math is about" . . . , (in regard to, math being about accurate descriptions of stable patterns, accurate to sufficient precision, and its relation to practical creativity)

Namely, math is fundamentally about . . . , measurable descriptions, which must be based on a context where measuring is reliable and the patterns either used or described are stable,

. . . in which case one . . . then one . . . looks at math patterns which are: linear, metric-invariant, and continuously commutative (almost) everywhere (eg except at one point) contexts . . . , [concerning patterns which can be described and which are subsequently of great practical creative value], . . . , and then one is automatically put into the context of geometrization, and the set of circle-space shapes, in turn, associated to the squares of a checker-board partition of the classical metric-invariant metric-spaces of non-positive constant curvature.

Part II (reviews)

3.

Review of book

Determining the medias pronouncements concerning:

institutional scientific truths (eg L Halle's book, "Out of Chaos."),
And, considerations about:

the intuitive voice of Einstein, and
a review of Einstein's main ideas

Determining either the correct coordinate frames or the correct containment-sets

Identifying the correct containment set is the main issue, of physical descriptions, in regard to being able to describe the stable spectral-orbital properties of physical systems of all size-scales, where this is a needed construct because the application of the laws of physics cannot be used to identify these properties based on both the laws and the associated rigorous calculations needed for any such stable systems, such as: general nuclei, general atoms, molecules, or the stable solar-system.

The public seem to be most ignorant about the nature of language in regard to its great limitations.

The positivist philosophical group focuses exclusively on the structure and properties of language to determine content and meaning concerning the expression of (philosophical) ideas.

There is a "described truth" and a "perceived truth" which depends on the constraints of a person's perception, where the religious command (of the Buddhists) is "to see the world as it really is."

Western culture is all about "described truths" and, in science, these need to be repeat-ably measurable properties, or patterns, of existence (or, in the west, of a material existence), but they also need to be truths which are related to practical creative activities.

Godel's incompleteness theorem is about the limitations of a quantitative language, where the quantities are the integers, in regard to the range of patterns it can describe. Thus, any quantitative structure which is consistent with the integers, eg the rational numbers and the real numbers, will also possess these great limitations as to what measurable patterns are describable within a idea about a particular use of the measured values, or a particular context of a quantitative language.

Furthermore, if a quantitative structure does not satisfy the properties of the integers then it cannot be quantitatively consistent. This means that chaotic quantitative patterns, which emerge from non-linear partial differential equations, will be a part of these quantitatively inconsistent constructs, where in chaotic quantitative patterns "from time to time" there exist multiple-values for the same property, eg bifurcations, and thus, such a multi-valued property cannot be a part of a measured property of a quantitatively consistent pattern, ie math methods cannot be applied to these types of patterns in any reliable manner.

These ideas about the limitations of language are best exemplified by the . . . , Copernican system vs. the Ptolemaic systems, . . . , where the beginning assumptions concerning the content of the descriptive patterns affect the patterns which can be described (or predicted), eg predicting planet positions in the heavens.

Then there is the very important relation, identified by a PBS TV show (perhaps the show was Frontline, around 2005, concerning the LA-Times newspaper, and the Chandler family, whom owned the LA-Times), that, because this family controlled about 65% of the circulation of local newspapers in LA, this allowed them to make investments, and then convince the public, through carefully edited newspaper reporting, to support laws which can cause these investments to become very profitable, eg changing zoning laws, etc, (ie one who can control "what is said in the newspaper" can control the thoughts of the people, especially if the people view the media as the sole voice, which represents a carefully-determined, and thus, an authoritative truth, to be accepted by the entire culture as being a singular expression of truth.

Thus, there are all these limitations concerning the nature and limitations of a "described truth," but nonetheless, the public considers the propaganda system to be the sole voice of a protected vision of truth, so that the propaganda system is supplying the society with an absolute authoritative truth which is verified by both skepticism and rigor. This is essentially the claim of the journalists, and the education system, and business interests, and the politicians; and so that this truth, so responsibly provided to the public by the propaganda system, is only accessible to superior intellectuals, ie truth itself demands inequality. Inequality is always the main message of the propaganda system.

This is the baloney which the western public actually takes seriously.

If one considers the US society's propaganda system, then one finds it continually repeats the same basic ideas, which support the social power structures, where, effectively, only the powerful social institutions are allowed to express ideas, concerning their social and/or knowledgeable and creative interests, within the media, where the powerful social institutions are those institutions about which society has come to be organized.

The main message of the propaganda system is that people are not equal, and that success comes from competition defined within a small set of "narrowly focused" categories of thought, about which one is allowed to consider and express ideas.

Thus, the narrowly based viewpoint of business interests becomes, through the media, the dominant belief structure, and defines the focus of discussion, within the public (and throughout all of society).

The main idea presented, by the media, is that of inequality and the need for competition.

Subsequently, scientific truth is defined as a narrowly defined competition devised for the interests of society's elite intellects.

But in such a scenario, science, becomes an authoritative dogma, which is adhered to by competitive obsessive personality-types, whose realities have become their memorized symbolic constructs, where these intellectual categories represent the categories of creativity of society, which are related to the investments of big businesses. That is, the scientific community is a set of wage-slaves competing for the best jobs.

That is, it is very easy to dominate the nature of a society, as well as the structure of its knowledge, by controlling its language, and by controlling the personality types which are placed in particular social positions of both management and research.

In reality, as the example of Copernicus shows that, science is supposed to be a discipline where the authority of language is to always be challenged.

But if "what a scientist can say" is required to be based on a narrow authoritative viewpoint, which has come to be a narrow dogma because of the control of scientific publishing by business interests, ie peer-review, then science and math will come to be a technical language related to servicing the complicated instruments, whose functioning results in power being given to the ruling class, (eg the owners of the publishing instruments of society), but in such a context of expression (ie for science and math) a person is not allowed to challenge the authority (or the use) of the language used within the context of the authoritative publications.

Such a challenge is simply identified as being the result of an incompetent person and a result of a non-competitive person.

One can see how the notion of an elite belief structure of an authoritative journal is used to create a competition, which becomes the authority of science and math, and the basis for a who is to be hired as a scientist in a society organized around wage-slavery.

The dogmatic authoritative expressions, provided by peer-review journals, (most of the authorities demand peer-review so as to ensure that [their] truth is protected, a demand not surprising for a set of people defined to be authorities due to their peer-review publishing) are not science, but rather it is an expression of a religion, which worships the god-like figures of society's very rich, whom according to Calvinistic Puritanism are the moral-souls chosen by God to lead society, ie the effective Gods on earth.

Considering science outside of the hype

There can be two types of measurably descriptive languages or ways in which to apply measurable patterns to physical systems:

1. The use formalized axiomatic where the assumptions and interpretations and contexts and containment constructs for descriptions are pretty-much fixed and formal and authoritative, this requires that the descriptive patterns keep getting ever more complicated . . . as well as ever more irrelevant to useful information,
2. The intuitive approach, where the assumptions, contexts, interpretations, and containment sets are all defined by the new way in which a quantitative structure and stable math patterns are being applied to a description of a physical system.

In reality, there are a few constraints which do, really, apply to expressing "actual scientific" ideas (rather than the scientific constraints of the business interests on scientists where the investment-class wants their instruments maintained and adjusted, so the ruling-class designs both peer-review, and axiomatic formalism, so as to ensure qualified competitors, who are to compete for the few positions of instrument maintenance, eg particle-accelerators [associated to nuclear weapons] in which the ruling-class invests).

The simple constraints (of a reliably measurable and practically useful scientific descriptive language):

One wants a reliably measurable, and verifiable, set of stable patterns (or descriptions), which are associated to the observed patterns within the world, so that relatively simple laws and principles form the basis for the calculations (or determinations) of the stable, measurable properties of the observed system's being described, so that these principles apply to a wide range of physical contexts, so as to be solvable (or calculable, or determinable), to the point of providing quantitative values of the observed patterns of the system, which hold to sufficient precision (for a wide range of general systems) and so that the information provided is (accurate and is) practically useful [information] concerning the observed properties of the system (or about the properties of existence). Ideally, one wants to be able to control the physical system.

This has, so far (2013), been based on the idea of materialism which identifies set-containment, and the subject matter of objects and geometries or events, whose relatively rigid properties (or inter-relationships), or stable properties, are measurable properties.

It has also been based on a need to model measuring in a local context, in regard to properties of systems and such local measuring is related to their containment sets.

And there is a dichotomy (paradox) concerning whether the main attribute of the descriptive patterns are geometric patterns, ie general relativity, or random patterns, ie quantum- and particle-physics, but with the notion of a random pattern winning-out, but still, based on the idea of materialism, the randomness identifies random particle-spectral events in space and time (or in space-time, ie the Dirac operator) but the system is modeled as a function-space, upon which internal particle-states are also defined.

Thus there is a focus on

1. coordinate frames,
2. structures for measuring, eg differential-forms and partial differential equations, definitions of material, and material systems (partial differential equations), and their measurable properties, ie the laws of physics, or
3. for allowing quantitative structures to be associated with space and time, ie assumptions about containment-sets, where these assumptions are based on the idea of materialism, within their containing coordinate (metric-space) frames (of reference),
4. laws of local measuring, either certain partial differential equations or the quantum-physics' rule of:
    "finding a complete set of commuting Hermitian operators for a quantum system" and the relation of these operators to either geometries, or spectra (or contexts of adherence, such as a system's energy), and within a closed system.

The point of such a quantitative description is to find accurate (to sufficient precision) and useful information concerning a system's measured, and (if possible) controllable properties, where the controllable properties are the most practically useful descriptive contexts. That is, finding to sufficient precision the observed spectral-orbital properties of the most fundamental physical systems of: nuclei, general atoms, molecules, crystals and the stability of the solar system ie to sufficient precision solving the many-but-few-component systems, which exist at all size-scales, and which possess stable spectral-orbital properties.

These problems, which are such a very fundamental aspect of the physical world, have never been solved.

However, the descriptions of physical systems, which are solved, are most often related to closed thermal systems in equilibrium, and linear circuits whose voltage signals are related to a given range of spectra.

Feedback systems, for systems which possess a non-linear context, such as controlling satellite motions (in the solar system), can also be important. However, the control of non-linear systems is always unreliable. That is, controlling these systems by the use of feedback is always tentative, such systems can never be identified as being precisely controlled, as the switching circuits in computers can be much more precisely controlled.

There is a set of fundamental physical systems built from charged material components, whose system-identifying numbers can be identified by integers, eg atomic numbers or atomic weights, etc, whose observed properties are stable, to the point where their spectral properties are precisely identifiable, and they are systems which are (mostly) neutrally charged. Furthermore, they are systems which are (mostly) inert until excited, where they get excited by being stimulated by the correct frequencies, where the correct frequencies induce these (mostly) inert systems to radiate, but to stay stable, unless interactions take place at very high-energies, which can cause either changes or decomposition of the system.

The stable, neutral, and unexcited properties of these physical structures, whose very stable properties are based on numbers of charged and neutral components, cannot be described based on general principles of (what are now considered to be) the "laws of physics," and the main statement about these systems is that "based on physical law 'these systems are too complicated to describe.'"

But the stable and precisely identifiable properties of these systems suggests that they are systems which possess a controllable geometric structure which has an open-closed topology, where topology is about the set structure of a continuum, and where a continuum is the model of the system-containing coordinate-space, wherein the system's properties depend on models of local linear measuring constructs (which are math processes, which are supposed to depend on a continuum).

An open-closed topology would be the context of a system whose charged fields are being contained by, or restricted to, a region of space, or to a frame.

Galileo, Newton, and Einstein discussed these main issues about the definition of inertial material, a material system's measurable properties, and their relation to local measuring and to containment frames.

Is a frame to be based on the fixed stars and the relative different sets of frames defined by different constant motions in regard to the fixed stars? (Newton) or (Einstein) Is inertia more like charged systems whose defining spatial relations in space-time, relate the electromagnetic force-fields, associated to charged materials, where the electromagnetic force-fields are determined by a non-homogeneous hyperbolic wave-equation, but which also possesses inertial properties (in space and time), which are best modeled by the Lorentz-inertial-force, but where for an inertial-frame the field (of Einstein) is the metric-function itself, so that, in Einstein's case, force is "an expression of inertia defined on geodesic paths in space," where the geodesics are found at a point in space so as to be related to a (local) metric-function?

Whereas in Newton's inertial structure, the gravitational-field depends on action-at-a-distance, it has been found that small charged-related inertial systems, ie quantum systems, are "non-local," which means that they also satisfy the idea of action-at-a-distance, eg Aspect's experiment.

Furthermore, the feedback systems for satellites in the solar-system are governed in a context of action-at-a-distance.

Thus, Newton's frames seem to win-out over Einstein's frames, and thus, there are, very real, differences between the frames for inertial-material and charged-material which have to do with the signature of the metric-function inertia is associated to R(3,0) while charged systems are associated to R(3,1).

Apparently, gravitational material and inertial material are the same value, because they have the same discrete Euclidean shapes, and are defined in a Euclidean containment metric-space.

Can there be an inertial field?

Yes, apparently in the context that all material is charged material (though it may be neutral), which, in turn, possesses inertial properties, where inertia is associated to the property of position-in-space relative to the fixed stars, as so identified by the symmetries of E Noether, but, nonetheless, a material's position can be affected by the shape of a metric-space, which has (non-positive) constant curvature, and, thus, the neutrally-charged material component's spatial position also relates to the entire constant-curvature shape of which a neutral charged-material component is a part (or within which it is contained. In other words, there are two frames to which the material's inertia can be related, (1) spatial position, and in relation to (2) a charged material's containment space, but the material-shape being moved by the shape of the material-shape's containing metric-space (but the inertial affects of a constant-curvature space, may, in some instances, be quite slight [eg if the mass-quantity is small]).

When there is a mix of physical properties associated to a system then there is a mixture of different metric-space types which are a part of the description of these properties.

The problem with an inertial-field, ie defined as a general metric-function (as described by Einstein), is that it is mostly non-linear, and thus, it is both quantitatively inconsistent and it is not a stable math pattern. Whereas, in physical description, measuring is reliable and the observed patterns are very stable, where the observation of stable material patterns also implies that measuring is "uniform" and "reliable in its uniformity," ie to be based on a metric-function with constant coefficients.

Furthermore, the stability of pattern also suggests a geometric and controllable descriptive context for such stable systems.

Thus, there is both the spatial-geometric relations between material objects in Euclidean space, and there is also a "shape of space" which also provides an inertial push on the inertial-material (toral) shape, due to the material-containing hyperbolic metric-space's shape, where the shape either pushes a toral-object toward a geodesic, and/or once the toral-object is on a geodesic path, that geodesic defines an inertial path, so that if the geodesic is a circle then the push of the inertial property of this circle-geodesic is tangent to the circle, ie not normal to the circle as the objects in 3-Euclidean-space push in the radial direction between the masses (or charges), ie normal to the circle (or towards a focus of an ellipse when in an elliptical orbit).

This extra inertial geodesic force defines an envelop of orbital stability for material orbits.

Propaganda vs. an intuitive description

Consider two authors, Einstein and L Halle, where L Halle is a general scholar (or journalist, ie propagandist), who represents the stereotypical rationality of the age, a rationality as defined by the propaganda system, and Halle is also a fairly good writer.

Einstein writes, not so much in the formalized axiomatic context, but rather in an intuitive context. He is always referencing the definition of inertia, and its relation to coordinate frames and their motions.

Einstein believed that the laws of physics, represented as the smooth math structures associated with the general principles about local measuring, which determine a physical system's interaction-determining force-field (or energy-averages) dependent partial differential equations, are central to physical description,

But rather, (according to the new alternative math context for physical description)

It is the properties of the coordinate metric-spaces, and the shapes of these metric-spaces, and the different dimensional context of containment, which are the central issues in regard to the descriptions of observed physical properties of material systems. The stable properties of material systems cannot be described in a context in which material-interactions are defined within an open-closed metric-space, as is now the basis for physical law, which is locked into an unnecessary requirement that materialism be satisfied by a descriptive structure.

That is, force-fields are higher dimensional structures associated to inter-relationships between different dimension metric-space structures, and the discrete temporal-geometric constructs which spatial properties of change require, ie changes of an object's (or a metric-space's) position.

[What should be called physical law, is about how (local) measuring is related to the various properties (metric-space states, and spin-rotation of metric-space states), and geometric structures which possess different dimension-values, and relative containment constructs, of different metric-spaces]

Science and social manipulation

Einstein's writing style is very confusing, because he appears to be so simple and intuitive, but he skips writing phrases, which would clarify his expressions, so he reads more like a riddle.

This feature of his writing, and his embracing a context of non-linearity abstractions (where embracing non-linearity, actually, contradicts his opposition to "basing physical description on the idea of randomness"), apparently, was very pleasing to elite structures within society, and he was made (by the propaganda system) into an enigmatic illuminati, ie a central cog of the aristocracy of intellect.

Although frames have importance, they are not central, because they are, in fact, specific and separated so that each of the separate metric-spaces is given metric-function structures, and containment is not general and unified within a single metric-space, as Einstein assumed that metric-space containment should be a unified construct.

That is, there are inertial frames, and electromagnetic frames, and there are also frames for other higher-dimensional materials too.

Halle's writing is taken from his book "Out of Chaos."

The blurb of this book says he presents a vision of cultural knowledge . . . . and its relation to the capacities of mankind.

But this is the capacities of mankind in the limited context of creativity as defined by the owners of society (where creativity within society is defined by their investment), and expressed by the propagandists who express this limited context, as Halle does in his book "Out of Chaos."

The main idea of the book is that "order is imposed upon a deeper and more fundamental chaos."

The fundamental chaos is identified, by the book, to be the random collisions of elementary-particles, which is supposed to be the basis for quantum (or material) interactions, and then there is the randomness of being in a particle-state (which is always changing), where the change of particle-state takes place at the time of a random particle-collision, so that the stable spectral properties of the nuclei are all supposed to emerge from this random particle-collision structure as well as the random changes of particle-state structures, as well as the spectral properties of atoms, molecules, found in regard to solutions to the wave-equations of these systems etc etc. [but the observed stable properties cannot be described within this math context, in general, the wave-equations cannot be formulated and they cannot be solved.]

But Halle never questions these constructs, he never states where there does "actually exist" in the technical (peer-reviewed) literature, a calculation of the spectral properties of a general nucleus (composed on many-but-few-components) based on the equations of quantum interactions, etc or of a general atoms composed of more that five charged components, eg a nucleus and other electrons (and neutrinos) etc, or how a molecule enters into its stable spectral-structure, which is a structure which often also possesses a relatively stable shape (along with fixed ratios of atom-types), etc . . . .

Halle states, p84, with " . . . the conception that the geometry of space-time is determined by the distribution of mass he (Einstein) implicitly established a unifying relationship between space-time and matter-energy."

This is the myth, but since these ideas of general relativity have virtually no practical value, eg they will probably never be used to identify the stable properties of the solar system, etc.

Thus, there is not much content to the above statement, other than the literary content ie used for propaganda so as to identify an elite intellectual class.

"Concerning existence, when viewed in regard to scientific knowledge when inquiry starts; space, time, energy, and elementary-particles, already exist . . . , and this inquiry concerns itself with their properties, their relationships, and the consequences of their existence."

"Life is one such consequence." (?, this is quite a jump, or a "leap in faith," a deep belief in the elite intellectual class)

"If there were no nuclear interactions that bound the elementary particles to one another, there would be no atomic nuclei;" (where is this demonstrated so as to be based on actual calculations, which in turn, are based on the laws of particle-physics? Nowhere.)
"if there were no electromagnetic interactions that draw electrons in association with atomic nuclei, there would be no atoms," (where is this demonstrated for general atoms composed of more that five charged components, eg a nucleus and other electrons, so as to be based on actual calculations, which in turn, are based on the laws of quantum-physics? Nowhere.) and "if there were no electromagnetic interactions that bound the atoms to one another, there would be no molecules" (where is this demonstrated so as to be based on actual calculations, which in turn, are based on the laws of quantum-physics? Nowhere.)

[Furthermore, the shapes of molecules are found faster and more accurately using an expert program (filled with much information, eg the computer which beat the Jeopardy champs), than using the methods of quantum physics. This implies that quantum physics has no value as a method, and that it is un-able to determine the observed properties of the quantum systems which exist in the world.];

"if the molecules were not so constituted as to combine in a variety of aggregations distinguished by varied geometric forms, and if some of the forms did not have the faculties of . . . self replication . . . there would be no life." (where is this demonstrated so as to be based on actual calculations, which in turn, are based on the laws of quantum-physics? Nowhere.)

In other words, this is a description of a fantasy, it is not about the valid descriptive capabilities of science, as the laws of physics are so construed as presented in the axiomatic formalization for peer-review physics journals today (2013).

Halle continues:
"Remaining within the bounds of science, we can base our definition of life exclusively on the properties and potential of the self replicating complex of molecules . . ."
(this description has never been within the actual bounds of valid, calculable science, rather it is within the structure of propaganda, yet it is claimed to be an authoritative, and even an absolute truth, which is protected by peer-review).

He continues:
"The universe is the same everywhere so as to be based on space time energy and elementary-particles"

(this cannot possibly be known, there are many ways in which "far away" spatial or energetic structures could be quite different from such structures around the sun, or around our galaxy, furthermore, spectra can be changed as it passes through different regions of space and thus such spectra would be difficult to interpret, etc) . . . .

"Resulting in the elements and the ways in which elements fit together . . . . to form the basis of cosmic order . . ." p108.

Reply to Halle:

There is an order which is associated to much of the material structures which are observed, but the scenario, provided by Halle, concerning the origins of this order, is far from being demonstrated, in the technical peer-reviewed literature, in particular, in regard to the most fundamental of material systems (nuclei, general atoms, molecules etc) which are observed to possess a relatively stable order associated to themselves, ie there exists no rigorous formulation nor solution to these systems based on rigorously applied laws of either particle-physics or quantum physics.

The basis for Mr. Halle's "rational 'scientifically based'" viewpoint about the world, is a carefully maintained illusion, in which the main idea is that all material properties can be reduced to random particle-collisions, but the precise development of the stable order which is observed to exist for each of the different types of material systems within the set of: nuclei, general atoms, molecules etc, so that the precise descriptions of the stable spectral properties for each of these systems, so as to be based on this main idea of particle-collision determined material-interactions, does not exist in the technical, peer-reviewed, literature.

Thus, the claim "out of chaos comes order," is not rigorously demonstrated in any convincing "all inclusive way." Nonetheless, this viewpoint is maintained in the propaganda system, so as to keep the focus of university physics departments on a "probability of collision" model (or basis) for nuclear weapons (random collisions of material components which takes place during a chaotic transition process, which, in turn, occurs between pairs of relatively stable states of material, which occurs due to the triggering of an unstable material condition for one of the stable states of material, eg triggering a critical mass of a radioactive element to come into existence).

Once such an authoritative description, based on a dogmatic truth, is placed within the propaganda system, then this viewpoint is taken to be an established absolute truth, so that to contribute to such an assumed truthful description of existence, one must compete within the rules identified by the propaganda system and education system, so as to, subsequently, become a peer-reviewed and authoritative viewpoint.

How is this different from the Ptolemaic authorities excluding any form of language which is not based on the set of assumptions associated to the Ptolemaic viewpoint?

In the education system, such a narrow (and delusional) idea about the properties of material can be maintained by having rationality based on "the memorization of disconnected rules," which is the natural operating structure of the US education system, where the study of knowledge is about separately developed and separately memorized sequences of language strings, where strings of symbols are memorized as authoritative truths, but not thought about, or the rules are memorized at a point in one's thinking where rational considerations about the descriptive structures is not possible (and one's intellectual capabilities are not sufficiently developed to deal rationally with the subject matter, but nonetheless "what are considered to be absolute truths," . . . but whose validation has not ever been made explicit . . . , are memorized).

Because of both the influence of propaganda, and the style by which learning is promoted, as a sequence of memorizations of bits and pieces of knowledge, not completely developed, yet symbolically complicated, and presented as separate strings of language pieces, which are claimed to represent an absolute and authoritative truth, so as to be represented as "high cultural truths," which are expressed in the propaganda system.

Reconsider the descriptions of the order of the material world, based on mathematics and the laws of physical description, and not based on propaganda and authoritative traditions

Formalism vs. the intuitive

Formalized axiomatic language structures vs. language structures of quantitative representations of existing properties based on intuition, different assumptions, new contexts, new interpretations and new ideas about organizing the containment structure of existence, eg where containment is based on ideas which are different from the idea of materialism.

Consider Einstein's literary endeavors, concerning physical description, he focuses on the most elementary of the basic assumptions of the development of thought within the language of physical description.

Yet, the simplicity of such an elementary language is so seldom to be considered within modern physics. Thus, it is difficult to follow the content of modern physics.

Yet the language of physical description has a simple content, materialism and the local measuring of physical properties, where containment is based on the idea of materialism (in either a geometric or a random context), which can branch-out in many different ways, based on new ideas about containment, and how material is defined, especially, in the new context, where each way (each new decision) concerning a "branching-out linguistic process" can be considered as being a different way in which to base physical description.

Consider some of Einstein's ideas as expressed in one of his essays "On the Theory of Relativity."

He points out that gravitational mass and inertial mass are the same value, but one is about the resistance to motion-changes, [and is the basis for measuring gravitational mass (in a constant gravitational field at the earth's surface)] where gravitational mass is a measured physical constant which also depends on the geometrically motivated factor (1/r^2) in the gravitational-force's mass-geometry based formula.

This (1) equivalence, of inertial and gravitational mass, along with the idea that a differential equation which is to determine (2) a law of physics . . . , ie a description of material interactions, but material interactions are not descriptive structures which can be used to describe the observed stable spectral-orbital properties of material systems . . . , is covariantly invariant to any type of relative motions which can exist between coordinate frames . . . . , ie a locally invertible diffeomorphism fiber group (to be associated with a general metric-function) and not an isometry fiber group, . . . , so that these two ideas, together, are (claimed by Einstein) to be interpreted to mean that all force is inertia, so as to be defined on (or by) the geometry of a curved coordinate system.

But then one needs to account for the electromagnetic forces as deforming the shape of space, ie the (few) charges (in a mostly neutral material context) are deforming space, but this cannot be identified (or interpreted) unless one considers charged-material (as well as space-time space itself) as defining closed metric-space shapes, so that these shapes can identify the needed geometric curvature of the inertial structure, which is associated to electromagnetic forces-fields, ie instead of electromagnetic force-fields and the subsequent application of the Lorentz-force to determine inertial forces.

But such a model of "material associated to metric-spaces" can determine the stable spectral-orbital properties of material systems directly, without a need to model material interactions. In fact, material interactions are really only adjustments to be applied to the stable spectral-orbital envelopes (or stable metric-space shapes), but it is the spectral-orbital envelops, themselves, which are the real point about the stable properties of the physical world which physical description is trying to (precisely) describe, but material interactions (and subsequently, inertial properties) do not bind material together, but they can "perturb and adjust" stable spectral-orbital properties.

So why is resistance to motion-changes the same value as gravitational mass, unless the gravitational force is . . . . , [caused by a particular type of discrete Euclidean shape so as to always be relatable to spatial displacements, which, in turn, are associated to resistance to changes in motion, which, in turn, relates one inertial shape to another inertial shape, in Euclidean space, for a force relation defined between two inertial bodies so as to be caused by the other body, equal and opposite forces characterized by proportionally different accelerations with the proportions being the ratio of the two inertial masses] . . . , the only type of force-field, which a "spatial-changes relation to motion-changes" can be so related within a containing Euclidean space, ie the electromagnetic force-fields are seen as an effective gravitational-field to inertial properties of material components in Euclidean space, R(3,0).

That is, force-fields need to be re-interpreted in each different type of containment space, or in the various types of discrete isometry shapes, namely, discrete Euclidean shapes are related to inertial properties, which are associated to physical properties, but where the natural properties of space-time, R(3,1), are energy properties, but whose coordinates are related to motions, but "motions of what?"

Answer: The natural coordinate relation of 3-dimensional hyperbolic space is to velocity, but the velocity seems to be about a constant velocity, as in the case of light, or (perhaps) of the constant flow of energy within a spectral-flow of a (2-dimensional) discrete hyperbolic shape, and not so much about a charged objects velocities.

But if a charged object's motion was modeled as an attached discrete hyperbolic shape associated to a relative velocity then . . . [such a model of velocity would require the existence of a discrete hyperbolic shape associated to a charged-component (which is also a discrete hyperbolic shape) so that] . . . . this velocity-shape is associated to the energy properties, ie two shapes where one is identified with velocity the other (discrete Euclidean shape) associated to the charge's relative position to other material, ie the other charge with which it is interacting, and this pair of energy shapes can change by small jumps of energy-value for each different shape, ie small changes in velocity as well as small changes in relative position, which is allowed to change between discrete time intervals (which are associated to the new model of local measuring, as a discrete operator).

If there is no such pair of interacting charges then the energy remains constant (within the relative frame in which the motion is defined) so as to be related to the charge's constant velocity.

But Einstein is trying to assume that containment is to be based on one type of space, which contains all the physical properties of physical existence.

That is, he wants local space-time coordinates to be the containment space for both electromagnetism, and energy, as well as the containment space for: inertia, and spatial position, and time.

But this might not be the correct containment structure for physical properties.

It should be noted that, particle-physics has (or uses) many different containment-space structures for the different particle-types, and for different representations of the Poincare group, etc.

That is, the idea of different and separated metric-spaces to be used as a part of the same description of the physical properties of a single system, especially, in a (new) model where the distinct physical properties, actually, come from different metric-spaces, is already a part of an "established" descriptive context.

However, that context, ie particle-physics, is a completely useless form of language for physical description, and should have never been a part of a carefully considered attempt to describe the properties of the physical world, in a way which is considered to actually be related to mathematics.

But then math itself allows absurdities associated to generalizations, which are unnecessary, and which seem to have their validation dependent upon an arbitrary application of math . . . . defined in a language based on formalized and fixed sets of assumptions . . . . , to arbitrary contexts, which appear to be quantitative.

Furthermore, the issue of "a need" for stable math patterns, within the language of mathematics, seems to never be brought-up by the, so called, authorities of (a dogmatic and formalized) math language.

"One containment set"

Within the viewpoint of only "one containment space" for all physical properties, Einstein sees accelerating frames as being equivalent frames (apparently, for uniformly accelerating frames) so he assumes local space-time frames need to allow for forms of physical law, ie partial differential equations, to remain invariant.

Since the metric-function is symmetric, as a 2-tensor (or as a matrix), which means that the local structure can be diagonal (for some orientation of the local coordinates). But the algebra of all (most other) local properties are non-commutative, as one immediately moves (or transforms) away from the coordinate position of the diagonal representation of the coordinate's metric-function, where the coordinates are to be associated to measuring local physical properties, but this means a quantitatively inconsistent set of relations between coordinates and measured properties, which are represented as functions, and which have the coordinates as their domain spaces.

Einstein is motivated to create this math context based on both

(1) a fixed idea about uniform containment of all physical properties within one coordinate-system, a coordinate-system defined by the idea of materialism, and

(2) the fact that local measuring breaks down when general containment frames, ie (apparently uniformly) accelerating frames, are considered, except for a very small region about a general diagonal metric-function defined on local space-time coordinates, but, in

fact, this small region does not allow for quantitatively consistent functional models of physical properties to exist.

Thus, the math ideas of reliable measuring and stable patterns are better guides to understanding the stable structures of physical systems.

That is, Einstein's assumptions about containment seem to be the main flaw of his picture for physical description, and that many spaces with different metric-functions and associated to different physical properties, and the actual containment needs to be tough about as being within a many-dimensional context where local linear operators are discrete, when described in the physical measuring processes of which the operators are a part (or which these operators are trying to describe).

Thus, the properties of relative positions and relative motions are properties which exist in different metric-space structures, each associated to different metric-functions, and with different metric-function signatures, and associated with different physical properties, and in turn, associated to separate and different material-types.

Einstein is trying to say accelerations (associated to inertia) are equivalent to gravitational fields, but more generally when the force-field is an electromagnetic force-field (associated to inertial changes of material) the idea would be that these force fields need to be associated to some energy-structure, which, in turn, allows an inertial property (a deformation of the general metric-function) . . . associated to either an electromagnetic field or to apparent forces (which exist in an inertial coordinate system which is rotating relative to the fixed stars) . . . . , to be identifiable.

That is, the centrifugal forces on a rotating object are also supposed to possess an energy-shape which also provides the patterns of those same centrifugal forces.

But this energy-shape cannot be identified, unless it is the shape of the hyperbolic shapes of an electromagnetic material system, but then this would separate the space of inertia and the space of energy for a rotating frame (unless, perhaps the shape of the envelope of stability of the solar system, a shape which is a discrete hyperbolic shape, the shape of an energy-space which might allow for the equivalence of an inertial-space and an energy-space for a rotating frame).

However, Einstein is insisting on placing these properties in the same metric-space containment-set structure.

Furthermore, geometry is concerned with spatial relations between (apparently small) material components, so space-time (or hyperbolic space) is not the correct space within which to identify material geometries, rather one needs the properties of positions of material in space, ie one actually needs Euclidean space (this is a math pattern identified by E Noether, ie spatial displacements are associated to inertia, while time displacements are associated to energy).

The property of inertia belongs in Euclidean space and it is associated with a discrete Euclidean shape, so that whether inertial or gravitational if there is a same-sized shape then it will have the same value for mass (or inertia) in any inertial context. But the property of (relative) velocity belongs to hyperbolic space.

The interaction is a property of measuring in relation to a force-field which is a geometrically measurable relation, which exists between material positions and measurable motion-changes, in turn, associated to both inertial shapes and velocity shapes.

However, material interactions are not fundamental in regard to either the properties of stable material systems, or the existence of stable material systems, rather stable material systems result from stable discrete hyperbolic shapes, as well as resulting from other constant negative curvature discrete isometry (or unitary) shapes [or discrete isometry subgroups], which exist in higher-dimensional spaces than are allowed by the idea of materialism.

The, so called, proof that the general principle of relativity is valid . . . , p 59 of the book "Essays in Science" by Einstein . . . , only has the motions of the planet mercury's axis (and gravitational affects on light), but this is not really a truthful verification since the mass of Mercury in this "verifying model" is zero, and if Mercury had a non-zero mass, the equation would not be solvable, so it is an illusion upon which general relativity is claiming its verification.

The math structures through which existence gains its stable order is from containment sets which are discrete isometry shapes.

That is, the general properties upon which physical descriptions depend are not related to a local-measuring of the process of material-interaction, and a subsequent focus on "the invariant form of a partial differential equation, which is defining a material interaction, in relation to a general metric-space, in an arbitrary frame of motion," rather the general quantitative properties which are needed for practically useful information to come from quantitative descriptions of patterns are the main properties of (1) reliable measuring and (2) the existence of stable shapes, and subsequently, this is about basing quantitative descriptions on stable shapes . . . , which exist in metric-spaces where local-measuring is definable . . . , so as to be able to describe the observed stable measurable properties of the wide array of spectral-orbital systems which exist at all size-scales, and in a wide range of different dimensions, with various metric-function signatures, and with a wide array of set-containment trees which influence the type and size of the material-component structures which a particular type of physical property (signature), defined at a particular dimensional-level in a subspace which is related to a particular size metric-space, which can contain the material-component (or shape), and where containment also depends on the existence of resonances between the component and the over-all high-dimensional containing space's set of finite spectral values.

A second order partial differential equations, which can be a part of a model of the measurable properties of a material interaction, do affect the dynamics and energy of the interacting material components, but the stable properties exist outside of the metric-space, so as to exist in a many-dimensional set containment structure where the sets contained are mostly the discrete isometry shapes which possess resonant spectral properties with the over-all containment space's finite spectral set, and different sized containment structures (containment trees) allow for the containment of the material component, ie the material-component has a smaller size than its containment metric-space's size. The second-order dynamic differential equations could be a math structure which can perturb and/or adjust the spectral-orbital properties of the material components inside the metric-space.

The structure of quantum physics, and subsequently particle-physics, seems to ultimately reduce to discrete cubical partitions of complex-values . . . , associated to mixtures of pairs of opposite metric-space states, for the various metric-space related physical properties which are a part of a physical system and its containment context . . . , to which it is assumed that all material reduces to elementary-particles [modeled as material-point-particles, which do not have mass unless there is hypothesized a scalar math pattern, symmetric about zero, modeled as a scalar-particle associated to all space, which causes the particles to choose either a positive or negative mass-value for the particle which collides with such a scalar-particle] . . . , where these point-particles can be modeled as being the distinguished-points of relatively stable sets of discrete isometry (and related unitary) shapes.

Thus, in the new descriptive context, point-particles are really related to cubical partitions of complex-coordinates.

But this type of a description is not a part of particle-physics.

Whereas, particle-physics assumes both random collisions (where the energy-level determines the types of particles in the collision process a process of mixing internal particle-states) and a set of non-linear patterns, which are associated to the math process of randomly mixing particle-states, and where each particle-type, ie the set of internal particle-states, is given its own containing-space structure with its own connection-type (or connection-dimension). These types of random math structures (associated with elementary-particles) are supposed to adjust the, assumed to be smooth, wave-function of a quantum system.

However, the containment sets of this construct do not make any sense, eg renormalization implies that there can exist various disconnected types of containment sets which can be associated to a quantitative description so that one can select from . . . , any of the different containment sets . . . , the value which best fit's the data. But such a disconnected and arbitrary, viewpoint of containment and "calculating" (or selecting) a property's value from any of several models of containment sets (within which the "calculation" is supposed to take place) is not a valid property of mathematics.

However, it is the random context of particle-collisions which most relates to the chaotic and indefinably random state of material which is transitioning between material states of relative stability . . . where it must be noted that it is the stable (and most often neutral) properties of material which go without valid descriptions.

It is the structures of chaos and indefinable randomness (random-collisions and random changes of internal particle-states of point-particles) to which string theory adapts its geometric model of particle-physics, but the assumption is that, in order to preserve the randomness and structure of particle-physics, the geometry (of string theory) must be made small.

There is no valid reason for such a decision, since higher-dimensional geometry can be modeled to be both macroscopic and very difficult to perceive, because, once one is focusing one's attention within a particular dimensional level where one is attending to the material structures of that particular dimensional-level.

Furthermore, the geometry of string-theory is modeled so as to possess an arbitrary partition between the small shape which are supposed to model internal particle-states and the subspace where materialism reigns and where the particle-collision processes are being modeled.

String theory is all about combining the symmetry points of a spherical symmetry geometry (which general relativity is supposed to model) with point-particle collision processes of particle-physics, so as to be able to relate a model of a gravitational explosion associated to particle-collisions, ie it is a model of a powerful nuclear weapon, and has no relation to determining practical developments at the level of macroscopic material, even if it were statistical in its nature.

Nonetheless a set of unitary partitions of C(3,0), ie with a fiber group of SU(3), could be related to the statistical model of partitioning a thermal system into quantum cubical components, but in such a case one needs a model of discrete shapes, which, in turn, relate to a controllable material process.

It is the descriptive context of discrete isometry (and associated unitary) shapes which provides the context of stable linear metric-invariant and controllable shapes, and to which the fiber groups structures, U(1) x SU(2) x SU(3), of particle-physics can be related, where these different "factor groups" would be related to the resonant models of inertial-masses, in turn, associated (also by resonance) to the set of stable discrete hyperbolic shapes, where SU(3) would be related to real spaces, and to both "SO(3)- and SO(3,1)-related discrete shapes" but these are shapes which would be contained in either, R(4,0), or R(4,1) and R(4,2), metric-spaces (ie different signature metric-functions), where, in turn, these discrete hyperbolic shapes, give order to both (1) the stable material components (or systems) and (2) the stable orbital properties, of material systems of various size scales.

These stable "discrete hyperbolic shapes" can be linked, by angular momentum properties, in a stable shape of a molecule (or a crystal), which is (or can be) modeled to have a very high genus (ie number of holes in its shape, eg where a 2-torus, or doughnut, has one hole ie a genus of one), where angular momentum could be defined as rotational motions (energy-flows) about the various holes in the shape. Such a shape could be similar to a controllable model of an electric circuit, so that different atomic or molecular component in such a high-genus crystal could act like different circuit components but now in a context of the flow of energy in a context of the system's over-all angular momentum properties.

Indeed these energy flows of angular momentum within a discrete hyperbolic shape are similar to the notion of travel through worm-holes, but the energy-flow would be in a discrete shape which has much greater size, ie defined between stars or between galaxies.

Note: Worm-holes, as they are now construed, are meaningless constructs since they are non-linear, and thus, quantitatively inconsistent and they are not stable shapes. That is, stars and galaxies can also have associated to themselves a very intricate and "geometrically wide-ranging" discrete hyperbolic shapes of very high-dimension, and with an intricate hole-structure, which can also be related to angular-momentum properties.

These same, stable "discrete hyperbolic shapes" can be linked, by angular momentum properties, in a stable shape of a crystal, or linked to a closed thermal system's partition of energy, which is (or can be) modeled to have a very high genus (ie number of holes in its shape), where angular momentum could be defined as rotational motions (energy-flows) about the various holes in the shape.

Back to statistical physics, in regard to practical-physics (as it so construed now (2013)), the partition of a closed thermal system in equilibrium into little separate energy regions is the closest thing, associated to valid math descriptions, which the SU(2) or the SU(3) fiber factor groups of particle-physics might be related.

The small regions of a thermal energy partition can relate to either the SU(2) or the SU(3) fiber factor groups of particle-physics, but the particle-physics model is so distant from identifying the stable properties of material systems, that such a relation to an energy partition of a thermal system is far from being relate-able to a controllable model in a context of physical model.

But these fiber groups associated to actual discrete isometry (and related unitary) shapes could be of significance.

4.

Re-introduction
[A re-statement of 1, on education]

Social comment, especially concerning math and physics

A re there stable patterns within existence?
　　or
Does everything "swim" in a context of indefinable randomness, where intent is upheld by extreme violence and intellectual arrogance?

The current state of education seems to show that the personal of educational institutions, either as administrators of . . . , or as authorities within . . . . , these people in educational institutions are failing in regard to the knowledge and creativity of society as a whole (because these institutions serve the forces of monopolistic domination over society, and these monopolies are being supported by educational institutions).

The current structure in which knowledge is related to (very narrowly defined) creativity has destroyed: knowledge, education, and creativity within society, because knowledge is only being related to the narrow interests of big businesses, eg military, oil, and banking.

There are fundamental, stable, definitive physical systems which exist at all size scales which do not have valid descriptions resulting from applying, so called, physical law in regard to using this physical law to identify (calculate) these well defined, and stable, properties of material systems.

There is a new descriptive context based on partitioning an 11-dimensional hyperbolic metric-space by means of very stable circle-spaces, or discrete hyperbolic shapes, where these stable shapes model a metric-space associated to particular subspaces associated with the different dimensional levels, and by this model the stable properties of fundamental physical systems can be identified, as well as developing relatively simple ideas about life-forms and to model the system organization which allows the description of the control that life possess over itself.

Yet this type of creative effort is marginalized, due to intellectual arrogance and dogmatic authority whose basis emanates from the social forces of monopolistic socio-economic institutions (the military, oil, and banking) which limit both knowledge and creativity within society, so as to strangle (and destroy) the society itself. [The desire to serve the monopolies that dominate society emanates from the social position of people as wage-slaves seeking high salaries and seeking the prestige of a highly-valued institution.] The entire culture must serve the narrow interests of these monopolies, and this is effectively enforced by the justice system (wage-slavery is enforced by the department of justice based on property rights).

Education (or knowledge and knowledge's relation to creativity) should not be about focusing on a handful of very technical (or very secretive) instruments for the purpose of using these instruments to uphold the very selfish monopolistic and socially domineering business interests.

Today's math and science university departments focus on descriptive structures based on indefinable randomness (where elementary events are unstable) and non-linearity (a quantitatively inconsistent descriptive context) as well as a continuum (which is simply "too big" of a set, so that words and patterns lose their meaning) so as to relate these poorly defined math patterns to feedback systems with limited range of purpose (so as to guide a drone or a missile to be used in war), or to design a nuclear bomb, also one must mention the "risk fiasco" where apparently "so called, deep MIT mathematicians" were used to calculate investment risks (where the failures of these calculations were directly related to the 2008 economic collapse). That is, these (believed to be, rigorous) risk calculations were not true, so as to result in the lose of trillions of dollars.

A risk calculation, where the elementary events are not stable, cannot hope to be all that realistic of a model (since the question is, "Is counting well-defined?").

A distinguishable property is not a valid basis for the definition of probabilities (eg the events of a valid elementary event space must be stable and well defined, not simply identifiable).

Randomness and indefinable randomness cannot be used to describe the stable properties of fundamental physical systems, as is now being attempted.

In a continuum, if one can define a value by a convergence process then it will fit into a continuum, ie a continuum allows too many types of quantities defined upon itself, eg randomness and geometry can co-exist in the same probabilistic descriptive context within a continuum, eg particle-physics.

Patterns do not exist in an abstract condition, they depend on a context and within that context patterns need to be stable and well-defined. Sets which are "too big" allow incompatible patterns, supposedly, within the same context.

This is an issue about whether descriptive and precise words have meaning, where meaning is related to stable well-defined patterns in a context, where the context is relatable to practical creative actions.

Creativity results from people who are equal, and who are seeking changes, most notably changes in knowledge and its relation to new contexts for creativity.

Whereas, fixed authority is about holding social patterns fixed so as to sustain monopolistic, socially domineering business interests.

Equality is for creativity while inequality is about fixed structures and the extreme violence which is needed to maintain such inequality.

Learners need to be equal free-inquirers, and when one lives in a society where the propaganda system is the sole voice of authority and that single voice supports fixed monopolistic socio-economic interests, then the learning institutions become based on dogmatic authority and are characterized by arrogant intellectualism, which leads to social failure and to the failures of knowledge.

Creativity cannot be defined in terms of strumming a guitar, manipulating society by a controlled use of icons of social value (allowed by the single-authoritative-voice model of the media, ie propaganda system), or working on math patterns which are not capable of describing a stable set of patterns.

The military-oil-banking business interests should not be the primary (only) social forces defining the single voice of authority (of the media that they own) for a free and equal society. Human creativity is much more diverse.

The correct interpretation of Godel's incompleteness theorem is that if one cannot create "what one wants" by using the existing precise language, then invent a new language, within which new contexts of creativity can emerge.

Education should not be about defining psychological types such as "autism with capacity to use language" ie obsessive personality types who are competitive but not reflective nor are they particularly curious, so as to work on a small set of complicated instruments which are used by the ruling class (ie military, oil, banking interests) in a fixed way, associated to maintaining society in a fixed way of organizing a society so that the society is required to use the products produced by the monopolies.

People are fundamentally both equal "seekers of knowledge" and equal creators, so education is about (descriptive, measurable, and build-able) knowledge and the relation that knowledge has to creativity (and the creative context).

Creativity is best related to a precise language which is closely related to the elementary language structure of assumption, axiom context interpretations and organization of a containment set. Stating elementary properties for a new language allows creativity to be accessible to many people.

The best example might be Faraday who developed the language of electromagnetism and while he invented this language he also developed technology and the use of the ideas of electromagnetism.

In regard to Faraday's creative actions it should be noted that the working aspect of today's, so called, highly technical society is all centered around the development of the ideas related to electromagnetism, while the undefined randomness associated to quantum description has not led to any technical development based on the information gained from quantum description, eg the properties of transistors were determined experimentally and then coupled to classical systems. Particle-physics is only related to rates of nuclear reactions. This is another example of how big business directs knowledge's relation to creativity within our society, where erroneously calculating the "investment risks" for banks (where these calculations were based on institutionally determined truths about math) is, yet, another example.

That is, this is about the law of the US society, in fact, being based on equality (because of the historic significance of the Declaration of Independence, which proclaimed American law was to be based on equality) and law not being based on property rights and minority rule.

Does equality imply that the public have access to the resources of the earth as they need to survive and create (at least put-forth plans for creativity, since resources are not unlimited), where they create for the selfless purpose of helping the public, as opposed to the idea that property can be taken (stolen) and "property rights enforced" by society with extreme violence, primarily for the benefit of the few (destroying resources and the earth), and this property is used by its owners (of society) for narrow selfish purposes?

Note: Only when everyone is considered an equal creator (not a wage-slave) can there be a "free market."

That is, Are humans wage-slaves? or Are they equal creators?

If one believes that we are unequal then "who, or what, is allowed to define the context in which value is determined?" It "should be" science and math, but where science and math are defined by their relationship to creating precise languages at elementary levels (of assumption and defining contexts for interpretations), and more importantly, whether a descriptive language is related to practical creativity, but this essentially means the existence of stable patterns and "in math" patterns are defined as geometries, while numbers are (mostly) about measuring.

Why should one believe that randomness (or limited contexts of feedback) be a basis for stable patterns?

The observed stable properties of physical systems: nuclei, general atoms, molecules, crystals, the solar system, [dark matter (motions of stars in galaxies)] etc. all go without a valid description based on what is supposed to be physical law.

The descriptions of these stable systems require stable math patterns to both describe these stable systems and to allow for the stable measurable context within which these systems are observed. But such a simple idea, which makes so much sense, is carefully ignored by society and the institutions, which so scrupulously serve the narrow interests of big business.

Today's education institutions should take a broader view of knowledge and creativity so as to not define knowledge as a process of fine-tuning the instruments which are of most importance to the ruling class.

That is, the society needs to be out-side the oppression of wage-slavery, and monopolistic domination negated, and this can be accomplished (or remedied) by practical technical creativity.

Teaching needs to be based on equal free-inquiry and not be based on narrow dogmatic authority. Authority is not truth, yet, today, "educational institutions" are not distinguishing between authority and truth. Since the reader must also be confused, let me inform you; truth needs to be related to practical development, and remember the ideas of Ptolemy were measurably verified (measurable verification can often be irrelevant, in regard to determining truth).

If an education institution wants to become a valid institution of knowledge which is relevant to widely diverse viewpoint concerning creativity, then perhaps such an education institution should try to get independent creative types to be their bosses.

To further develop the use of stable circle-spaces in a many dimensional context to describe the observed stable properties of physical (eg nuclei) and living systems.

On the other hand the extreme violence of our culture, where this violence emanates from the protection of the interests of the owners of society, has allowed a propaganda system to mold society, where the propaganda system only allows the interests of the monopolistic big business interests to be expressed. The affect of this, is that, both social organization has become very stable and fixed but also the language (which the public uses in its institutions) is also stable and fixed, leading to a context of stability upon which (basically, ill-defined) statistics can be applied, as long as the stable social context stays fixed and the propaganda system continues to use language in an "ever repeating fashion" (always the same message; People are not equal, where property and money determine value).

The social context is held fixed by extreme violence, and the terrorism of the counterinsurgency tactics which the (highly militarized) justice system applies to the public, in order to defend the national security interests of maintaining the power of the powerful, and the fixed social structure upon which the power of the powerful is based, eg requiring that people continue to use oil and not use thermal solar energy.

In other words, independent creativity and the development of knowledge, different from the knowledge used by the monopolistic business interests, is not to be allowed. This is the basis for the ludicrous demand for "peer review," in what is supposed to be an equal society.

A review of assumptions and contexts so that these basic ideas can identify the creative or knowledge intent of a person expressing ideas should be done, but then publication should be organized around the different types of assumptions and contexts and interpretations.

Thus, the creationists could publish, but placed into their narrow category for the people who are interested in those types of ideas.

That is, science and math publishing should be open, but placed in categories of assumptions and contexts and creative intent, but math patterns seem to mainly be the patterns of geometry, and geometrization seems to say, geometric patterns should focus on the very stable discrete hyperbolic shapes, if one wants to actually be expressing valid math science, measurable, stable (or actual) patterns, which might have practical applicability.

Part III
(mostly social comment, but in regard to the detached language of math and physics)

Social comments (in regard to the, supposedly, detached language of math and physics, but big business monopolies turn dispassionate [detached] viewpoints into ideas attached to specific instruments, in turn, associated to categories of their commercial, or investment, interests)

5.

"About what?" is Godel's incompleteness theorem, concerned?

The simplest statement of Godel's incompleteness theorem is that there exist measurable patterns which are true, but which a system of axioms concerning quantities, cannot be used to deduce these true math patterns.

So what to do? does one add axioms, or is this a much more significant problem, where the ideas about assumptions and contexts . . . , eg geometry must be contained within 3-dimensions, ie materialism . . . , possess a deeper relation to the statements and the contexts upon which measuring and practical building processes can be applied. Since math is about quantity and shape and randomness, and randomness depends on a condition of stability, in regard to the random events whose probability one is trying to determine (by simple counting processes) which is usually not satisfied when statistics is used, eg economic stability exists until one day everyone panics, the main issue concerning the stability of math patterns, deals with geometry, ie shape, where Thurston-Perelman's geometrization suggests that stable shapes are fairly limited. That is, a measurable pattern needs a certain amount of stability for one to be able to make meaningful statements about something, eg about a shape, as a true pattern, whose description can be used in a context of practical creativity.

The point about the media, the highly controlled publishing world (the voice of authority), is that the idea of "very rare high-intellects," where this is a myth which the media wants to develop and maintain, since the media is all about expressing the idea that people are not equal. Those who run the society, the owners of society, do not want the simple conclusion of Godel's incompleteness theorem to be expressed, since it is a statement which is equivalent to the ideas expressed by Socrates, namely, that people need to be equal free-inquirers, and the structure of language requires that language be built "close to the level of assumption and creating contexts and providing interpretations and re-organizing the ideas about containment-sets" which can be done by anyone.

The usual context of discussions about Godel's theorems is about the axioms about "what a quantity is," often discussing Z-W axioms etc, but a quantity is something built from a counting process based on defining a stable uniform unit. That is, the discussions which the academics have about Godel's theorem are all about providing the public with red-herrings.

The thing about the physical world is that it is built upon a set of fairly stable fundamental systems: nuclei, atoms, molecules, crystals, and stable solar-systems, but there are no valid descriptions of "why these systems possess their properties of being relative stability for so long.

Propaganda-education

The main problem facing the US society is the propaganda system, which is the society's sole authoritative, reliably-truthful vehicle of social expression (but which only promotes monopolistic interests) where it continuously spews-out mis-information which, when placed in the "context" of honest reporting, the public always interprets to be an absolute truth.

Thus, when monopolistic, unregulated, but almost completely controlled "market-place" fails, in a complete collapse, due to criminal fraud aided by the justice system and the congress the propaganda system continues to "sing the praises" of (and need for) the "magical" unregulated "market-place", where it should be clear that "unregulated", now especially, means license to steal along with a complicit justice system and political system, since these institutions have been manipulated by the propaganda system so as to simply to have become a part of the propaganda system, themselves.

The authorities (or technical experts) which serve this system by adjusting the complicated instruments for the monopolistic ruling interests (the interests of the owners of society) are also controlled by the absolute-truth espoused (in the context of "honest" reporting) by the media related to "technical-development" (but really only small adjustments to fixed traditional technologies, since the context of math and science is not allowed to seek new creative contexts within the context of stable math patterns, ie the monopolies depend on society continuing to use products and resources in a fixed way which allows the monopolies to continue to make money based on their products), where for the intellectual class, the academics, the authoritative experts, are provided with a set of "prized" problems whose context is:

1. Indefinable randomness
2. Non-linearity

3. Non-commutativeness, or
4. At best locally commutative, in a context of a general metric, and thus, non-linear in regard to the containing coordinates, eg geodesic coordinates, or the set of functions.

So that the propaganda system, validated by certain narrowly interested experts insists in "big bangs" particle-physics, string-theory geometry, ie theories which are needed to understand the singular points of a black hole's gravitational field, and it is claimed "by some unknown possibly" that charges can be related to black holes, in regard to singularities within nuclei, so as to realize a "grand-dream" (truly a pipe-dream) of the control of worm-holes in space, or control over a "many-world" context of existence, though the math structures through which these ideas are to be described, based on probability and non-linearity, does not allow any control of such constructs, even if they are mathematically modeled (as it is so claimed that quantum systems are correctly mathematically modeled) due to the properties of the math constructs and due to the over-whelming complexity of such models these models (of controlling a "worm-hole") cannot possibly be controlled, as the propaganda system is suggesting that they can be controlled. Consider, if the current descriptive context cannot describe the stable, definitive fundamental systems, systems so stable that it implies that these systems form in a linear controlled context, then how can their exotic models of physical systems (changing between worlds, in a many-world model) ever be realized, if they are models which are not consistent with the actual structure of the (external) world? or of these ideas

That is, these prized problems are delusional-ly based, yet the propaganda system promotes them as do the experts themselves so they become the basis for identifying an elite in-crowd of "knowledgeable" experts.

This, in-crowd, of so called superior intellectual elites, obsessed with complicated math patterns (one must note the autistic connection, the manipulation of personality types by institutional managers, used for the purpose of deceiving the public) who are led to believe, by the media itself, that they are zeroing-in on the wonderful goal of their great intellectual prowess and intellectual creativity.

Yet it is clearly a failed intellectual exercise since there are basic stable definitive physical systems which exist at all size scales but which go without valid description (based on physical law).

Nonetheless (however) when these stable systems are actually solved the intellectual community and society has a great mental inertia (almost entirely caused by the propaganda system, it unwillingness to publish the new solutions, since they define a completely new math and physics context) to realize just what they have heard, but nonetheless assured that it is they the intellectual elites who will be the ones to forge new roads into new technical landscapes and thus the elites will not (will refuse to) listen to the new ideas which must originate from an inferior mentality and thus must be wrong and surely wrong within the dogmatic authority which defines their truth for them. The elites only find these new ideas somewhat interesting but from an intellect inferior to their own since they have memorized their contexts and their always correct interpretations and their always intelligent evaluations of the state of knowledge, they are not tricked by prized problems, no not them.

That is, the propaganda system is a communication system (vehicle) which expresses a dogmatic authority (an absolute truth) which is followed by the experts which in fact determines the faith of the high-valued institutions which serve the ruling class and the experts have an absolute faith in the authoritative truth of that dogma. The propaganda system defines the true religion of society and it is a religion of personality-cult (not unlike Roman emperors or Egyptian pharaohs) and a deep belief in inequality and a manufactured property of a society dominated by selfish monopolistic interests, ie it is a society opposed to life and opposed to adapting to change

Wake-up you popes of the religion of science and math, eg general relativity, particle-physics, string-theory, indefinable randomness, non-linearity non-commutativity or only locally commutative.

Gödel's incompleteness theorem has a simple interpretation:
Because precise language has sever limitations the axiomatic basis, the containment set, the organization of this containment set, the context of the description, and the interpretations of the observed properties (of the observed measurements must be fully considered and when an alternative, well defined example of new ways in which to present axioms and contexts is provided, especially if it solves the most difficult problems in math and physics then the math physics community of experts should listen and take it seriously.

A "grand-dream" (truly a pipe-dream) of the control of worm-holes in space, or control over a "many-world" context of existence, the math structures through which these ideas are to be described, based on probability and non-linearity, does not allow any control, since there are not any stable patterns in this (such a) descriptive context.
Nonetheless these math structures do apply to the business interests:

Is math staying too: traditional, fixed, formal, complicated, irrelevant?
Is math only about serving business interests in regard to:

1. Unstable patterns to be used in fleeting contexts (feedback systems),

2. Chaotic transitory systems, eg nuclear reactions, chaos transiting between two stable states,
3. Manufacturing complications, eg formulating security codes, etc,
4. Pulling the wool over the eyes of the public, so as to provide irrelevant and inadequate descriptive structures for physical systems?

The crux of the problem with knowledge and education in society is its capture by corporate and private interests capture by the owners of society even though it is most often "public" education institutions, nonetheless the professors are trying to adjust the complicated instruments for the corporate and private interests and they are not concerned with descriptive knowledge in its most general and most powerful sense where new creative context get developed by new contexts for descriptive knowledge which result when assumptions, contexts, containment, organization of pattern use are considered at their most elementary levels, as Faraday developed the language of electromagnetic description while he also developed a new context in relation to the instruments related to electromagnetic properties. Though such a dramatic chain of developing events is not a necessary attribute for developing a new descriptive context at its most elementary level it means that new languages can be related to new creative contexts.

Since the professors of public universities have been captured by corporate and private interests through the mechanism of funding research and identifying prize problems in math so as to keep math traditional and under the control of peer review the talks at conferences, such as at the joint math meeting in San Diego 2013, are either about developing even more complicated theories and more complicated formal professional math language where these formal math languages have very limited if any relation to practical development they may only be marginally related to corporate interests, essentially related to bomb technology, or

About applying technical complicated math so as to be able to adjust rather complicated instruments of interests to corporate and private interests, eg feedback systems, imaging systems, recognition systems and improperly defined statistical constructs which often appear to be valid, since the propaganda system is capable of making all of society to continue to use language and product in very certain narrowly defined ways so as to place an artificial stability on the statistics where such stability of the statistical context does not really exist.

Propaganda, in regard to science and math is provided in the context of great breakthroughs and, supposedly, new things but which has little bearing on the uselessness of the descriptive language except in regard to weapons technology

Science and math are often about making adjustments to systems which will reduce labor if things remain in their fixed social context in regard to the corporation's products

That is developing new knowledge in regard to new contexts and new ways in which to organize descriptive language, in regard to solving fundamental mysteries, is effectively stopped by peer review prize problems traditional authority and mostly by the process of funding which is controlled by the corporations and private businesses.

Nonetheless, there are many unsolved fundamental problems in physics which are ignored due to dogmatic authority of math and funding traditions within public educational institutions an authority which is essentially related to military development and banking investment interests

The sad thing is that now (2013) these fundamental problems have been solved but the structure of both propaganda and the "knowledge institutions" keeps-out ideas which are different from the inter-related interests of business and traditional academic authority.

Academic science was invented by the mercantile class in the 1600's (after Newton appeared) to support their investment and productively-creative interests, in the 1500's science development based on measuring was centered around schools and about literate people, so that investment in science was centered around the schools (or universities).

However, public schools should take notice of Godel's incompleteness theorem and the logical positivists who proclaimed to limitations of precise language (or the limitations of measurable descriptions) and to consider the example of Faraday wherein he both invented a new math language to describe electric and magnetic properties but he also created the instrumental context through which these properties could be used and controlled
That is, descriptive knowledge best leads to new contexts for creativity at the elementary level of language assumption context interpretation etc not at the complicated formal level, eg no one can use the principles of particle-physics to describe the stable spectral properties of general nuclei.

That is, one wants new contexts for creativity to come from precise descriptive languages one does not simply want adjustments to complicated instruments since instruments also have limitations as to their capabilities or if there is not a better instrument to do the same thing, though digital electronic seems to be able to deal with arithmetic and math patterns well but this has led to an attempt to deal with non-linear systems but non-linear systems can only be related to feedback systems and only for limited ranges of time or distance in regard to arithmetically determined solutions.

There are many mysteries in regard to fundamental physical systems: why do there exist stable physical systems with definitive measurable properties, eg nuclei, general atoms, molecules and their shapes, crystals, life stable solar systems, there are many galaxies with planar spiral structures, etc.

These fundamental physical systems go without valid descriptions based on what is considered to be "physical law" instead one hears about all the hyped-up ideas through the propaganda system and the professional mathematicians and physicists about "big bangs," Black holes, wormholes, Higg's particles, transforming neutrinos, all related to general relativity, particle-physics grand unification and string-theory etc, are all expressions which are wild speculations if they cannot describe the stable properties of fundamental systems whose stability implies that they come into being in a controlled context, while believing the wild speculations is mostly driven by (or caused by) the propaganda system and a traditional context of math and physics authority and a personality cult which forms around these academics who mostly interfere with the development of new contexts for knowledge and creativity but the monopolistic business interests do not want new contexts for creativity. Where it might be noted that general relativity was shown to be untrue in regard to the non-local properties of material interactions in Euclidean space since non-localness was demonstrated by A Aspect's experiments.

The correct answer as to why there are stable fundamental systems requires that the context of containment be changed in a drastic manner from 3-space and time (quantum, Newton) or space-time (electromagnetism, particle-physics (?)) and materialism where the measurable descriptions are either Classical, often leading to non-linearity or Quantum, and indefinable randomness, spectra supposedly derived from 1/r potentials, function spaces usually non-commutative and Lie groups all used in the context of a continuum and a loose idea about convergence to this continuum, eg renormalization (something Dirac rejected) but apparently personality cult and propaganda was able to establish as an authoritative technique.

Instead consider a new context:
An 11-dimensional hyperbolic metric-space is partitioned . . . into its different dimensional levels and the different subspaces of the same dimensions . . . by discrete hyperbolic shapes which exist up to hyperbolic dimension-10. This can be used to define a finite spectra on the over-all 11-dimensional hyperbolic metric-space.

The set of all resonating discrete hyperbolic shapes which are contained within this high-dimension containing space form the bounding stable structures of the more usual physical description of material defined as one-lower dimensional shapes in each dimensional level, where the usual metric-invariant, second-order elliptic, parabolic, hyperbolic, differential equations . . . associated to material interactions or material properties . . . are defined, but the stable elliptic structures are defined on the discrete hyperbolic shapes.

The stable physical systems are the discrete hyperbolic shapes of the various dimensional levels which are in resonance with the finite spectra of the over-all 11-dimensional hyperbolic metric-space which has been so partitioned.

Each n-dimensional level "sees" the bounding geometry of the (n-1)-dimensional material "surfaces" but the open-closed topology of these shapes allows light to be observed from outside the metric-space shape's fundamental domain out to the unbounded lattice, this is especially true for metric-spaces whose fundamental domains are large eg as large as the solar system, where it should be noted that lower-dimensional shapes than (n-1)-dimension tend to condense onto the (n-1)-material's shape.

That is, the shapes imply the discontinuity of a metric-space experience between dimensional levels.

Interactions between micro-components imply Brownian motions which implies (due to E Nelson 1967) quantum randomness. Furthermore the distinguished points on discrete hyperbolic shapes implies that the interactions appear point-like, but, nonetheless, mediated by discrete shapes.

It might also be noted that this descriptive context provides a definitive spectral relation between different 11-dimensional containment sets, ie there are many different worlds where each world is well defined by a definitive spectral set, and the best instrument to realize transitions between these world might very well be a (human) life-form.

6.

Simple

A reasoned voice is drowned-out by a bamboozled public, who because of the media, support those ideas which oppress them, the public, (eg they place themselves in wage-slavery) and the indy-media's are part of the problem, all, but one, actively supports opposing free-speech, assuming the commentators to my posts are the Portland Indy's editors (which might not be true, rather they may be packs of organized groups which oppose the new ideas, which the bankers also oppose, and they oppose equal expression. The public is not to be blamed, rather the system which destroys its own public is to be blamed.

The media confuses authority with truth.

This is an expression about truth, and the relation of truth to (practical) creativity.

In regard to a measurable description of observed stable physical patterns there is a partition of both distinct subspaces and the dimensional levels of an 11-dimensional hyperbolic metric-space, so that all such components of these discrete hyperbolic shapes within the over-all 11-dimensional containing space are contained in the different-sized discrete hyperbolic shapes . . . of the proper dimension (one-dimension more than the components) . . . . of the (given dimensional-subspace of the) partition (which) where these "smaller components" are the "material" components, which are in resonance with the stable finite spectra which the partition's finite-spectral-set defines. They are the material components (or metric-spaces) which can exist within such a containment construct, and this construct allows stable patterns to exist and to be described.

Order and stability of fundamental material systems comes from (or exists within) non-decreasing spectral sequences defined on subspace-dimensional levels, which are non-decreasing sequences (in size) of metric-spaces defined as the dimension of these metric-spaces increases.

This non-decreasing size-sequence allows orbits to be defined, by means of small (material) components which exist within the higher-dimension (ie larger) discrete hyperbolic shapes, in the spectral-dimensional sequence defined on some subset of subspaces within the partition.

This newly considered space is discontinuous in regard to higher dimensions when the measurable properties of the (smaller) components are observed inside any one of the discrete hyperbolic shapes of the partition. That is, the properties of higher-dimensions are difficult to detect. Note: Frames of reference are best defined in relation to these stable components.

The current basis for measurable descriptions of physical systems is mostly concerned with wild speculations within a descriptive context wherein no stable patterns can be described (within the authoritative descriptive contexts now used) which the experts and their doting-media pass-off (these speculations) as high intellectual value (one must remember the descriptions of Ptolemy were measurably verified, ie measurable verification is a weak basis for distinguishing science as opposed to dogmatic authority [or authoritative religion]).

These speculations based on dogmas "as is everything about the propaganda system" are a fraud presented to the public and backed-up by the compliant (and true believing) authoritative experts, who are people who are willing to compete within the narrow context of authoritative dogma, so as to get-to work on the owners of society's complicated instruments.

This is a manipulation of personality by the administrative managers of these institutions (mostly controlled by the forces of wage-slavery, as well as a propaganda system which is the sole voice of authority for all of society) thus turning science back into an authoritative religion, but now serving the narrow creative interests of the monopolistic big-businesses.

Descriptive truth is best related to precise measurable sets of many different such languages, which are defined by their assumptions and interpretations etc. so that the different languages are to be associated to practical creativity, while socially everyone is an equal creator.

The issue (whose fundamental context is "What is existence?") is really about the "true" range and capacity for the human life-form to create, as opposed to creating and managing an elite, dominant social-class, based on control of material and property, wherein knowledge and creativity are subservient to this dominating social structure.

The most extreme vision concerning descriptive truth and its relation to creativity, (the vision which might well be the correct vision concerning life) is that human life is about creating existence itself, but this is not domination but humbly accepting the actual structure of existence, as opposed to the adherence to the idea of materialism, adjusting to the scientific context of dealing with the true structure of existence.

One presents this idea which contrasts the basic distinction between science and religion, or materialism and "seeing the world as it really is," where (to state the obvious) a many dimensional construct transcends materialism, yet it contains materialism as a subset, and it solves the most basic problem facing physics and math today, namely, being able to describe the very stable patterns of material systems which exist at all size scales: nuclei general atoms, the stable solar systems etc and it is a context which is consistent with a locally-dimensional idea about materialism, and all of the randomness and non-linearity associated to the idea of materialism.

In the new descriptive context there are two new types of ways by which these fundamentally stable systems come into being: (1) is a local resonance which is locally related to the material containing metric-space and the other (2) in regard to collisions of material components whose dimension is one-dimension less than the dimension of their containing metric-space where the energy-collision structure becomes resonant with the global spectral set of the over-all containing space.

Furthermore, the descriptive structure also shows randomness and non-linearity to be the prevalent context for the material contained in each dimensional level.

That is, it is a descriptive context which is both measurable, verifiable, and geometric, thus, very useful in regard to practical creativity (the true test of a valid scientific descriptive language), yet it places the context of existence again within a mysterious context of "What is existence?" where there is now a new map (or context of inquiry) leading into a higher-dimensional context, which is modeled with the simplest of geometric shapes and in a (stable, or limited) quantitative structure which can be defined on a (very large, but) finite set.

That is religious discovery and scientific discovery are back in the news.

To use arrogance and authority so as to confuse "authority (and social standing) with truth," seems to be the main media strategy, which business interests employ, as they have turned the media into an advertisement for themselves (or their selfish interests), where within the propaganda system (the media) the monopolistic business interests only allow the advertisements which support their business interests to be heard by the public.

The reason this new math construct seems so "simple" (its context is by no means simple) is that the very stable math patterns, upon which it (as a descriptive language) depends, are very simple.

[In math one often partitions regions in order to measure (or limit) them, but this partition allows the quantitative sets to be defined by a finite set of values, and it defines bounds for metric-spaces (smaller material components) and it defines bounds (or boundaries) for metric-spaces, which (nonetheless) are difficult to see within a metric-space.]

If the stable math patterns used in this descriptive context were identified by their math contexts (or math properties) and if these properties are represented by means of (partial) differential equations (or operators applied to function spaces) they would be linear, metric-invariant (ie related to the classical Lie groups), (geometrically) separable, commutative everywhere (globally commutative), ie the locally measurable properties (identified by derivatives defined on functions which represent system properties) of coordinates and shapes in the domain space are orthogonal (or independent) at each point in the containing domain space. That is they would be the solvable math patterns which can be controlled by controlling boundary or initial conditions of the system.

[Furthermore, the functions in the function spaces would also be these same types of stable discrete shapes.]

The simple stable shapes (or coordinate structures) whose fundamental domains would be relatable to "rectangular simplexes" ie thus they are either the tori or the shapes built from toral components (as well as the possibility of "cylindrical shapes")

These are the discrete (discontinuous) subgroups which are associated to the classical Lie groups SO and SU, including the SO(s,t) or SU(s,t) Lie groups, as well as the symplectic Lie groups.

The exotic Lie groups may have some importance but this would be a distant idea, since the real geometry and the relation between shape and interaction need to be worked-out in the low dimension (five and less) metric-invariant geometric case, first, including the determination of an actual finite spectral set for an (or for our) existing world.

This is basically a descriptive context about real geometries which possess local opposite-states, so that global real geometry, but also Hermitian geometry, ie the geometry of the complex variables, seems to be the correct path to explore. Note: It is difficult to envision the shapes of things whose dimension is more than three, yet we may be 5-dimensional entities. That is, it is a true mystery in which everyone is invited to explore, since the basic math constructs are simple and geometric.

The only valid criticism of these ideas must provide Either a complete answer as to the structure (or cause) of physical stability, since the new construct accounts for both randomness and point-like interactions, and it provides a new descriptive context within which answers concerning stable physical systems are given, or criticisms must identify logical or containment errors within the new language (that is, Copernicus cannot be criticized because he did not begin with the assumptions of Ptolemy [these new ideas cannot be criticized because they do not agree with currently accepted dogmas, and this is because the current scientific dogmas and this new language are built on different sets of assumptions]) or provide an alternative complete answer concerning physical stability

That is one cannot claim the new context is wrong because of the authority of the old (or currently accepted) contexts, since these old constructs have not provided any answer as to "why stable systems possess the property of stability."

That is, the old (or current) construct cannot answer the fundamental questions about the stability of the fundamental physical systems.

We are at a time during which our social institutions are failing, so what are some alternative models of society?

Now the social structure is based on the law of: property rights and minority rule, where contracts (as well as fortunes) of those in the higher social-classes will be upheld (or maintained), but the contracts of the people in the lower social-classes usually are not upheld by the justice system.

There is the other social construct of basing law on equality (this is what the US Declaration of Independence proclaims, but it has never been upheld by the ruling upper-social-classes).

This is often placed, by the media, in a context of

Competition vs. equality

Where competition is defined by very limiting sets of authoritative laws. However, the upper social-classes have used laws to support their monopolistic businesses and to destroy their competitors, where implicit in this sentence is the idea that "the politicians are easily controlled in a society whose laws are based on minority-rule."

The distinction between current social laws and "new" laws (rather laws of equality which are begging to be acknowledged and enforced) can also be represented as:

Narrow dogmas vs. build many (measurable) languages at the level of assumption

Where the intellectual elites can be defined by narrow authoritative dogmas within a context of competition and in highly managed educational institutions, where these intellectual elites have been managed to serve the interests of business monopolies (apparently strong minds but weak back-bones).

Or

Dogmatic elitists tied to a fixed social order (defined economically, or monopolistically) and a fixed relation between the knowledge of the culture and creativity which only serves monopolistic business interests, eg oil military, banking.

vs.

Free-inquiry based on equality (each person being an equal creator) where knowledge is tied to practical creativity and a truly equal free-market

Or

Elitist, wage-slaves
vs.
Equal people who possess a wide range of practically creative possibilities

The media is all about fraud, it lies and mis-represents information to serve its few paymasters. Even the alternative out-lets depend for funding, in order to have a voice, so the main idea expressed by all of the media is that "people are not equal," and "the learned ones with superior intellects are to guide the masses,"

But the masses are in such a bad way (easily led, and quite confused) because of the media.

For example, in the media science is an authoritatively based form of personality-cult.

Furthermore, the media (ever) continues its incessant repetition about inequality, and a very restricted idea about identifying high-value, (high-value is about the creative capabilities of each individual of a society, it is not to be based on an extreme violence needed to maintain inequality based on property rights, ie based on materialism)

The essential model of truth which is presented by the media is that "authority is the same as truth."

This, of course, is also the basis for repressive religions.

However, the scientific authorities have failed to provide answers to the most fundamental of physical questions "why is physical stability so very prevalent."

But the public has been bamboozled, by the media, into supporting the narrow authority of science as if they are protecting high-intellectual and cultural value the value they are protecting is narrow and directed at maintaining inequality and a fraudulent idea about value.

All of the US institutions, except the military, which are based on high-valued intellectual knowledge and intellectual capacity are failing, and this is because the basis for the fundamental knowledge of our culture . . . . , where the military is not failing because all of science is based on developing military instruments (and with all the money the politicians give to the military monopolies, "How can they fail?") . . . . , has become too narrowly defined and far too controlled.

It seems that the main social failure is supporting:

inequality over equality,
narrowness vs. wide creative range,
extreme violence vs. equality,

but alas, the elitists have the capacity to scare the public, unless "we, the people," re-define law, to be based on equality (as the US really claims law should be so-based).

7.

Authority or truth?

Since the media confuses authority with truth, ie the public has no valid idea about the determination of truth, scientific truth is not simply a "measurably verifiable" precise description, if this were so then the ideas of Ptolemy would still be our scientific truth, it has to do with practical usefulness, and often a better description has been associated with the idea of simplicity of the description, the ideas of Copernicus were simpler than those of Ptolemy etc.

Peer review of new ideas within professional publishing excludes new ideas.
This is because:
The authority of the "knowledge basis" for big business (and how it uses knowledge in relation to its creative concerns) is quite carefully guarded.

When someone advocates for a social change and "they are allowed on the media" they always support their claims by saying this is an idea which has been peer-reviewed . . . . , ... , this is sort-of-funny, since the media is all about misrepresenting and lying about the nature of truth. Thus, claiming that the science (and math) . . . , which form the basis for big business's narrowly defined creative production interests . . . . , is also about identifying a truth, which the entire public can believe-in, is a naïve vision concerning the tenuous nature of a "precise descriptive truth," and a set of high-paid (wage-slave) guardians of that "truth."

In particular, one should be suspicious of a claim of belief in an "absolutely authoritative truth" in regard to science and math, since this is what "Gödel's incompleteness theorem" warns against, and it is the conclusion of the Copernicus vs. Ptolemy controversy, but that controversy has been spun as a "science vs. religion" conflict, but it really is an "absolutely authoritative dogma" vs. "new ways in which to organize a precise measurable description" conflict, where newness and equality, in regard to free-inquiry, should be challenging a failed authority, and since a described truth always possesses limitations, authority (or its basis in a precise language) should always be challenged.

eg climate change where there should be no conflict, since all possible identifications of harm to public health or public interest should be a basis for social and business changes. That is, society must be adaptable, since this attribute of "being adaptable" has always been applied to the poorest of the citizens.

The society is allowing the inertia of big business to control society which most opposes new development, or opposes new contexts for creativity.

Business profits should be quite subservient to public interest (consent of governed, where everyone owns property equally) since it is the businesses which should be the ones most adaptable to changes, not the public adapting and conforming to business interests, and the (social) inertia associated to business monopolies which come to dominate and destroy the society for the selfish interests of the business monopoly.

It is sad but one finds web-sites which very courageously express news events, along with some of the social structures which are driving these occurrences, but who nonetheless believe the dogmas of their fellow intellectuals, who have been deceived by the overly authoritative dogmas upon which their academic positions depend (where academic positions are often about the careful use of language), for example the knowledge associated to science and math support weapons and communications instruments, whereas economics, as is also the case for evolutionary biology, are languages which appears to be based on measurable patterns, but they are based on indefinable randomness [ie the elementary event spaces of these constructs are not valid, ie they are not composed of stable and calculable events], and thus they cannot describe stable patterns, they are languages which cannot express any meaningful content, ie stable patterns cannot be determined, at best, they express the current ways in which the measurable properties upon which the description is based are organized (by other academics or by big-business), but this is an arbitrary set-up, especially since there are very big-players who can change the game (re-organize the set-up) within a time interval of hours.

Geometrization and the mathematical context for the solution of physical stability: eg. Nuclei, general atoms, molecules, and the solar system etc.

This relatively new (since 2002), and relatively simple, context of math containment provides the setting for a solution to the problem of finding the math structure for the observed stable material systems which are so fundamental and so prevalent. It also provides a basis for a quantitative structure which is defined on a finite set.

However, most stable physical systems nuclei, general atoms, molecules, crystals, solar-system stability, etc now (2013) go without any valid math structure (for these systems), in a currently accepted math context of indefinable randomness (eg improperly defined elementary event spaces), non-linearity (quantitative inconsistency and chaos), (global) non-commutativity, or only locally commutative, (eg quantitative inconsistency, eg chaos), where all of these constructs are defined by a contrived descriptive structure of convergence and divergence onto a continuum.

Where these math structures are together used to explain (or identify) the (stable) properties of physical systems. But they cannot do this.

Rather, such a math context really only applies to physical systems in a chaotic transitioning process (eg reactions in weapons) and for feedback systems (eg guided missiles) whose range of applicability is difficult to define (thus the lack of precision, whereas the media claims perfect precision for these systems), and it is a context which applies to quantitative complexity (eg secret codes).

But these complicated math structures, which fail to be able to describe the very stable fundamental physical systems (mentioned above), are also used by the media to create an illusion of expert "mastery" and "expert complexity."

Though, in fact, the academic sciences are (have been) organized by the ruling-class to provide personnel who can adjust the complex instruments needed (or used) by the big business interests.

Though many difficult problems now have solutions (due to the new math context): nuclei, general atoms, molecules, a new way in which to analyze crystals, and the stable solar system, due to these new ideas (expressed in other papers), this means that these relatively new ideas should be dominating the attention of the professional mathematicians and physicists, but they are not. Note: The media can easily deceive the experts, since the experts have had their attention diverted from the main problem of describing the basis for physical stability.

Technical note: Apparently, waves which possess physical properties can be successfully related to solutions by function-space techniques.

Apparently there are stronger social forces involved in an inability of a public, or of an expert-class, "to discern truth." The strongest social force in the US society is its privately owned and controlled propaganda system, the politicians are an arm of the propaganda system.

The US propaganda system is the sole authoritative voice for all of society, and it is the propaganda system which directs the attention of the expert researchers. These researchers are dependent on a funding process.

However, these same researchers arrogantly claim to be the personifications of the highest cultural attainments in the society (where this arrogance is based on their important role in the propaganda system), nonetheless they have social positions of being both wage-slaves and society's, so called, top intellects in regard to a religious personality-cult, expressed through the media, so public-worship consolidates their belief in their "far too authoritative" mathematics and physics dogmas, dogmas which have failed to solve the problem of "the cause of physical stability" for nearly 100 years (ie it is a failed dogma), ie the media turns "top intellectualism" and the dogmas upon which such a "measure" of intellectual-talent rests (the "intellectual winners" of the competition whose rules, in the education system, are defined by an, essentially, absolute authority) into a religion, where this is a deep "religious belief" in what the media labels as science [Copernicus would have a more difficult time persuading others to consider an alternative way in which to organize and fashion language within such a current religion (2013) of expert authority, than the difficulties he had in regard to the "authoritative religion" of his time].

The professionals are following their "deep beliefs" as dictated to them by the propaganda system.

Apparently these professionals can rigorously prove properties which are contained in a world of illusion, eg where a description based on randomness also possesses well defined geometric properties, eg particle-collisions (this is an absurdity).

It should be noted that the best interpretation of the Godel's incompleteness theorem is that precise languages can be very changeable when reduced to the elementary levels of assumption, context, containment, organization, interpretation, etc. Yet the failure to describe the stable underpinnings of physical existence has not been seen as a "crisis of the knowledge" which is being derived from the currently accepted authoritative dogmas of math and physical description.

The inability of the current scientific authority to describe the stable properties of fundamental physical systems is, in fact, a complete and total failure of that authority.

There are other social organizational properties which manage society, and with which one must deal, there is a vast social organization in regard to management of the math and physics (or science) communities, eg managing personality types, similar to the management of personality types in politics and the justice system.

In the new context of containment one uses the most prevalent of the stable geometric patterns identified in the Thurston-Perelman geometrization, namely, the discrete hyperbolic shapes, and the properties that these shapes possess, as identified by Coxeter.

Furthermore the ability to "surround" a "hole" by a closed shape, so that a continuous deformation is limited, ie the "holes" introduce stable properties into the context of the continuity of shape.

The discrete hyperbolic shapes [with component interactions mediated by discrete Euclidean shapes (tori)] are also very rigid shapes with very stable spectral properties.

That the solar system is stable is evidence, which can be interpreted, to prove this new context for mathematical descriptions of the physical world is true, especially, since the professionals have no valid model of stability for any of these many fundamental stable systems.

Note: If one criticizes these ideas based on one's deep religious belief in the current paradigm, so that the criticism is that the new ideas need to account for (or be consistent with) the current belief structure, then this is analogous to requiring that Copernicus begin with the assumptions of Ptolemy and then prove that Copernicus is correct. This cannot happen, due to the structure of language.

That is, the new ideas are built on a new language based on:

new interpretations of data,
new contexts,
new ways in which to organize the containing set,
new ways in which to define a derivative operator (as a discrete, locally-linear operator),
Etc . . . . etc . . . etc . . .

And it solves the central fundamental problems which the current paradigm has not been able to solve, yet the experts are oblivious . . . , oblivious, ... in an analogous manner as the propaganda system continues to express belief in de-regulated free-markets.

Furthermore, there is a need for a valid model for living-systems, living systems are relatively stable and they can demonstrate a capacity for very precise control over their own systems, again this is a stable structure and the new way of organizing math supplies a coherent model for such a system, one would like to say a simple model, but it is a model whose inertial properties exist in at least a 4-dimensional Euclidean metric-space, eg where the system's stable components exist in a 4-dimensional hyperbolic metric-space (eg 5-space-time), where the model provides a method of "spectral-energy control" between dimensional levels.

8.

Both questions

By partitioning an 11-dimensional hyperbolic metric-space on the subset structures of the subspaces of the different dimensional levels by means of the discrete hyperbolic shapes, where these discrete hyperbolic shapes are easiest to consider in regard to discrete hyperbolic shapes of uniform shapes, whose sizes, which change between dimensional levels, are changed by means of multiplying the subspaces of the different dimensional levels (or these shapes which are used in the partition) by constants, so that with both the discrete Euclidean shapes and an organization by means of other math constructs a model for material-component interactions between the different open-closed metric-space components . . . . of the various dimensions and containment contexts "which the different size-containment subspace structures allow," ie metric-space components which are resonant (with the finite spectra that the partition defines) are allowed, ... can be determined.

It is this type of math construct which allows for the stability of math patterns which exist within a quantitative set to have their quantitative structure be based on a finite set.

It is a math construct which solves both of the most fundamental problems in regard to the cause of "the stability of fundamental material systems" which are observed to exist, in a context of materialism, and it also identifies the structure of, the true nature of existence, and subsequently, it identifies the nature of existence in regard to perceiving "the world as it really is."

It answers . . . , in a complicated many-dimensional and many-subspace context . . . , the questions about existence that were posed by D Juan, in C Castaneda's classic anthropological (or religious) expressions concerning the cognitive (or perceptive) and physical structure of the world "perceived by" and "acted upon" by the advanced practitioners of "reason and exploration" who were (or are) in the native cultures of the Americas.

A world which Castaneda was provided entry, apparently, mostly because of his close relation and "interaction with" the cognitive center "of the being of D Juan."

A world of apparent existence, wherein the same questions are asked, as are asked in regard to the physical world, but now in regard to a wider awareness of perception, as to "what exists" and "how it is put-together," or questions about, "on what is (that more widely perceived) existence based" and "how is it organized?"

That is, the new math context allows for both an understanding of "from whence physically stable patterns come" and on what structure are the highly complex experiences, which are a part of D Juan's experience, built? That is, the command of religious experience, to "perceive the world as it really is," is (in fact) about forming the same basis in perception as science uses perceptions to determine the properties of existence.

Furthermore, in biology a higher-dimensional and rather simple model of life can be identified, a model which is closely identified with simple physical properties, and a model which allows the organization and control over a living-system can be determinable, but within a higher-dimensional context wherein a hierarchical (and highly organized) organization of a living-system can be identified (it is a many-dimensional model for a living-system so it cannot be all that simple, yet, it is a system (or model) which can be described in a simple manner).

That is, this is a metric-space structure which is consistent with materialism, but it allows for the containment and existence of a set of very stable, yet very simple, math patterns (or stable math-material patterns) within the context of "what appears to be" an ordinary metric-space containment set, but it is an 11-dimensional hyperbolic metric-space. It is a model which is consistent with materialism as a subset, but which transcends the construct of materialism.

It is a description which also allows for (and can be used to guide the interpretations within) the much more varied experience of existence which was reported to be experienced by C Castaneda when he was being mentored by D Juan, but they seem to be experiences which D Juan could not interpret to his own satisfaction, since he seemed to be trapped within this . . . , apparently, real, but far from usual . . . , experience, where such a sense of being trapped was reported by M Tunnneshende (ie who reported D Juan's being trapped).

Despite all of these answers to some of the most fundamental questions facing human culture, in the broadest way in which to consider culture, none of these factions:

physics,
math,
biology,
religion (perception of "the world as it really is"),

. . . , seem to be willing to accept such answers, (apparently because) since within their (bounded) culture, they (these different factions) are, apparently concerned that they are not receiving a message (from the media) of there being a new description of the truth about existence.

That is, these arrogant and authoritative social factions are not being called by that "solitary voice of authority" to which they have all been trained to listen.

Namely, they do not hear the "OK" given to them by the propaganda system, a propaganda system which is controlled by the few owners of society who also define, through their money (as the paymasters for the wage-slaves), the value of things and ideas within the culture, the arbitrary value of things and ideas (in society) is expressed to the world through the propaganda system.

That is, it is unfortunate that the separate groups of (disciplined and accomplished) people . . . , eg mathematicians, physicists, biologists, and religious adepts who have experienced the world in a deeper form as to its real nature . . . , are held isolated in their own experience, due to their arrogance and an incapacity to determine truth in regard to any form of a new information source, ie new information which is given to them by an outside source, but a new source of "descriptive reason," but which is outside of their own group, and the propaganda system, whose main purpose is to control information has not blessed the new truth with its OK (and, of course, the media would never bless such a new truth, since it is a set of ideas which undermines the creative efforts of the very rich, ie the idea that human creativity is mostly found outside the confines of the material world is something which upsets the factions of: business, science, math, and religion, where these groups are narrowed by their own assumptions, traditions, interpretations, and goals, (or their arbitrary judgments concerning what has high-value)).

But each of these different categories of types of people (math, physics, religion, etc) relies on the authority of the other group's truths.

That is, they are arrogant in their own disciplines and confuse authority with truth, when they look to other groups for interpretations concerning knowledge and in regard to considerations concerning into "what precise language can a set of very diverse set of observed patterns of relative stability be interpreted?"

That is, the human being is being destroyed by the propaganda system which serves the selfish interests of the few owners of society, but which separates and marginalizes various groups which are assigned to different separate categories of awareness and action, and various ways in which to use precise descriptions of relatively stable patterns so as to interpret the wide range of various experiences of very disciplined people.

This is (in itself) a good enough reason to abandon the idea that law be based on "property rights and minority rule," as expressed by the main body of the US constitution (while the Bill of Rights has consistently been un-enforced) where the claim (or relation) to reason which the owners of society use to defend and justify themselves is that "using resources is best done on a large-scale and organized around wage-slavery (and to support their selfish interests)."

The alternative, in law, is that the law be based on equality.

It should be noted that, it has always been that, "a long period of relative equality for the people which exist within a society, has often resulted in the subsequent building of a strong society" but once the society is full of riches, those on the top of the society have turned to, and arranged for, extreme-violence and inequality to be the society's basis for rule.

The US Declaration of Independence was to end that aspect of both western thought and western actions, ie ending the extreme violence by ending the inequality.

Law in the new US nation was to be based on equality.

Unfortunately, (or even though) the corruption of the western culture had already infiltrated the US culture . . . , what was essentially an equal society for white males (except in egalitarian Pennsylvania) . . . , and this corruption had already created a barbaric culture which was already involved in the extermination of the native peoples, and the extermination of the noble cultures of the native peoples (much more noble than the western culture), when the Declaration of Independence was written, so the early US society was easily led astray by a rich ruling minority.

This identification of a set of elites is always about those on the inside in regard to activity which is taking place within the culture at the moment when equality transitions to inequality, or new ideas (whose origin is the space of equal thought) are accepted as being a part of a long authoritative tradition. That is, the development of knowledge is about considering a wide range of assumptions contexts interpretations etc so as to try to develop new precise descriptive languages which can be useful in regard to some form of creative attempt, but in society, the social context of wage-slavery causes the learned peoples to be manipulated by the few very rich by a funding and filtering social process to limit the nature of language so as to define a dogmatically authoritative language of the learned, where the knowledge of "the few learned" is the knowledge which best suit's the "productive" interests of the rich. This corralling of those who use language in the sense of "being learned" within society, causes their to be authority and an "in crowd" within a learned community, and this "in crowd" identifies a set of bullying intellectuals who are stuffy and traditional in their outlook in regard to a precisely described truth.

What exists in the US society is that law and the justice system is a bullying structure associated and in collusion with narrow selfish interests which have come to dominate and destroy society the monopolistic business institutions of: oil, military, and banking which control the basic resources associated to the living conditions of the public, but the control of these resources should be classified as part of the public welfare not part of a selfish scheme to dominate society. This violent assistance which the justice system provides to these domineering private institutions leads to ever greater violence and social collapse since it opposes sets of descriptions about (or a description of) the true nature of existence, and it opposes the creative nature of mankind and instead supporting a traditional way of acting whose creative capacities have failed and now these attempts are selfish attempts to control markets and destroy the DNA of plants because the knowledge used is only a partial knowledge about life, eg the life-sciences need a valid model of life not simply identifying chemical structures and then trying to identify correlations on very poorly understood and very large sets of chemical data, ie the models of life are simply wild speculations where some of these speculations are identified to be a part of a "scientific" tradition associated to authority and allowed to bully other challenges to these somewhat arbitrary authorities. However, fitting a precise language with observed patterns is necessary (in order to be science), but then why is it that neither physics nor math are required to describe the stable properties of material systems (nuclei general atoms molecules the solar system etc) in order to maintain their authority?

That is, in the US society; what resources are used and who uses them and how they are used is not based on issues concerning reason or public welfare but are rather based on a deep belief in inequality and the process of bullying and corralling wage-slave groups of accomplished people, where this social organization based on bullying is done for the selfish interests of a very few: oil, the military, and banking. Thus ideas which are given authority are those expressed by the people most compliant and obedient to the bullying forces of society. These obedient authorities are not capable of discerning truth, other than the truth which supports their own social positions. Thus, new ideas cannot be expressed anywhere within the society. [thus the biggest threat to new ideas {as Copernicus had new ideas} is not simply the authority of the rationalism of the age but rather the structure of violence and bullying which is related to the social structure which is effectively owned by a few and the violence and bullying which actually upholds these authoritative expressions]

But the main issue of science is "how does one precisely describe the stable properties of fundamental physical systems which exist at all size scales: nuclei to solar systems and beyond?" and the ideas which are expressed in the papers . . . . , of m concoyle, who tries to express these ideas on web-sites where the ideal of equal free-speech is honored . . . , in fact, do solve these problems.

But the bullied and obedient (as well as manipulative) science and math authorities of our age have had their attentions diverted from the main issue of science (or they understand their opportunities, to not solve the most fundamental problems of physics): describe fundamental patterns which are observed, eg the stability of fundamental material systems whose properties of stability imply they form in a controlled mathematical context (eg in a context of linear geometry), and then use these descriptive forms so as to be able to create over a wide range of contexts and in a very practical manner.

That is, the newly provided many-dimensional metric-spaces partitioned by stable discrete hyperbolic shapes, allows one to represent existence in a new context, which, though apparently a simple structure (because stable patterns are simple math patterns), it is, in fact, quite a complicated construct which has many-dimensions of stable shapes . . . , which the authorities of society cannot consider, where this is due to the authority of their own professional dogmas (which apply to military instruments), but these are dogmas which best serve the business monopolies and not serve the public welfare and the need for people to be creative in the context of practicality (as opposed to the limited creative context of literature), but also, our (own) material based viewpoint tries to exclude from our considerations new constructs for existence, in order (for western religion) to support the material-based construct of a society based on "control by extreme violence." (The true religion of western thought is materialism, and using language to mislead and misrepresent both knowledge, and the social structure, and social power of money and all of its inter-connections within society.)

(Western religion consistently pushes its people to the viewpoint of materialism, and not toward trying to perceive the world as it actually exists, although such an idea is hidden in its well-intentioned symbolic claims.)

Nonetheless, the complex nature of the new math-or-a-measurable context within which to interpret the world which we experience (where we should interpret the world in this context since it answers the unanswered questions of the various categories of (pigeon-holed) thought) it is a context which also allows one to understand the widely diverse viewpoint of a person such as D Juan, and apparently he himself (D Juan) could have benefited from such a clear vision of: existence, pattern, action, and thought.

Moving between "different worlds" could be interpreted to be about experiencing an 11-dimensional hyperbolic metric-space whose subspace-dimensional-level partition defines a finite spectral set which is different from the finite spectral set of the 11-dimensional hyperbolic metric-space within which our experience is contained a finite spectral set upon which metric-space-material components depends (by resonances) for their existence.

This new math and/or physical model within which existence is contained accounts for (or is consistent with) materialism, randomness, and the point-like nature of micro-material (material component) interactions, but it also allows for stable patterns of material-component systems to exist. It also allows for a quantitative set to be built from a finite set of values which determine the finite set of the types of material-component (or metric-space component) systems which can exist in a particular 11-dimensional hyperbolic metric-space. Thus, within such a containment set it is sufficient to model a continuum by (or within) a set of rational numbers.

It is a containment set which essentially allows the geometry within a metric-space to be identified by the material components which the metric-space contains, and once a person is taught to view existence in a way consistent with a subspace of a particular dimension and in a particular subspace then it is difficult to perceive an experience which is outside of that particular dimension, ie it is difficult to perceive the full set of dimensional levels which actually compose existence.

70

However, it is a model which is consistent with the idea of religion which claims that the command of religion is to "try to perceive the world as it really is" and the idea of materialism which is attached to a particular subspace of a particular dimension is the proper definition upon which a measurable description of existence is to be based, ie this defines science as being related to materialism and randomness and non-linearity in regard to locally measurable relations between coordinates and measurable properties of physical systems, ie functions defined on the system-containing domain space. However, in order to understand stability (eg stability of fundamental physical systems eg the many-bodied lead nucleus or the many-bodied solar system) then one must exist within a many-dimensional existence which is partitioned by stable shapes eg an 11-dimensional hyperbolic metric-space partitioned by the very stable discrete hyperbolic shapes into a subspace and dimensional-level partition.

Thus "perceiving the world as it really is" is now considered to be a type of religious statement concerning a religious experience, but in reality it is an idea which points to the true context within which to organize and to re-categorize both science and math as well as the idea of a measurable existence, wherein stable structures can and do exist, and these stable components are fundamental to the nature of existence.

What should be considered to be?

To all of you ditto-head worshippers of the very rich, or to all you stuffy, elite "liberals" who worship the intellectuals, both of these groups, which support the "status quo," identify high-value in a narrow but socially acceptable manner. They are the palace guards who have been placed at the temple of high-social-value (or to guard the arbitrary whims of the very rich and powerful, who actually define high-social-value).

To be a serious critic of these new ideas then one must first explain physical stability, eg describe the laws of physics from which the stable nature of the lead nucleus (ie Pb-nucleus) can be derived, or from which the stable solar-system can be derived, etc. If you cannot do this, then read and learn.

You should note that, as a society fails, or its institutions have failed, truth becomes something which exists outside the traditional categories of the society (in fact, this always defines truth). This also identifies the social context in which Copernicus put-forth his ideas. That is, new ideas are not allowed into the intellectuals' tight set of dogmas and (all of) the news outlets (who all practice some form of ditto-head-ness, since they all need to be funded) also will not allow such (intellectually revolutionary) information to be distributed, this includes alternative sources.

The point of free speech is that a society based on equality is to allow all expressions of ideas. But ideas are about knowledge's relation to practical creativity, not simply someone's opinion or belief where the belief is not relatable to practical creativity.

This is not about a narrow definition of how a society functions, it is about actual "precise descriptive knowledge" and its relation to creativity, ie it is about all people regard-less of how society treats its people.

These new ideas are both news and it is new knowledge, which is applicable to the equal freedom of all people, to create at a practical level, as opposed to creating at a literary level (where current math and physics apply only to literary creativity, and has no relation to practical creativity), or as opposed to the narrow channels of creativity defined by the rich and their investments.

The problem is, that people are allowed (and encouraged) to bully each other. This defines the managerial-class of helpers for the rich and socially-powerful.

Is the force of the language surrounding "bullying" based on a precise descriptive truth and the relation of that truth to creativity (this would be a good definition of either science or math), or is it based on violence and selfishness?

The bullying is often based on ways (or categories) of how language is used.

Thus, the core of an agenda of bullying is about the authoritative use of language in certain categories of thought and actions.

Because language and categories of thought are closely associated with bullying, where authoritative language has become equivalent with a self-righteous truth (a truth upon which one's social value depends) which one violently protects, or others are used to violently protect someone's self-righteous truth in a society organized around wage-slavery.

Thus for authorities to acknowledge their failures effectively negates their social capacity to bully others. That is, even though these new ideas solve fundamental problems in physics and math they are outside the categories which grant the expressions of authoritative abilities, ie competitions, to bully others and fit into the very narrow creative interests of big businesses.

The propaganda system is all about what gets bullied and how this is done.

It is clear that the state must be the main player in public welfare, while free markets based on the public being equal free inquirers and equal creators and it is this idea which should determine the creative products which might enter into public welfare, eg transportation, medicine etc. ultimately the use of resources is about public welfare but selfishness and bullying need to be carefully controlled within society.

One wants a wide range (or a large set) of possible precise languages whose assumptions, interpretations, contexts, etc. are well identified and distinguishable so as to relate practically creative contexts to the observed stable patterns of existence. This gives more equality and more freedom to be practically creative and it allows for a greater range in which to express ideas. If one is "religious" (belief in existence beyond materialism so that one's actions are a part of this context which exists beyond materialism, ie lip-service is not enough, one needs to talk-it and walk-it) then one seek a math structure beyond a structure based on materialism, but which contains materialism as a subset, would be a believable descriptive construct of existence.

The idea of the western culture is a belief in materialism and the right of superior individuals to bully and to perpetuate violence upon those judged to be inferior specifically if the bullies support some form of creative production which is organized so as to socially support the bullies, *eg neo-liberal capitalism.

Is human value to be defined in a context of knowledge and its relation to "practical" creativity? This is a social analysis based on man's relation to the world (or to society) either as a bully or as one who develops knowledge and creates in a practical context.

This is a very positive and creative message, but it comes-up against very negative-nellies, in the form of the, supposedly, (self-proclaimed) very-positive, media, which, in fact, spurns true creativity, and supports traditional outlets for creative behavior, namely, to serve the interests of the ruling-class. The supposedly very creative intellectual-class . . . . , but the intellectual-class only relate to literary creativity self-righteously proclaiming that they are the gate-way to practical creativity but they . . . , have not been able to describe the stability of the lead-nucleus based on physical law.

In other words, what is called positive-ness is the cynical expressions of the upper-class elites acting as the guards of the narrowly defined creativity which is controlled by the very rich and powerful.

9.

Clear

New ideas are not allowed into the media especially in the authoritative peer-review journals which are based on authoritative dogmas similar to how the Catholic church excluded the heretical ideas of Copernicus, since they were ideas which challenged the rational authority of their day, of which the church was the gate-keeper.

Peer-review is equivalent to the authority of the Catholic church.

It is only free speech which allows ideas to be expressed, but most often the indy-medias do not sustain their claim to support free speech, and they waver based on the subtle notion of an expression of ideas must be from a "competent source," ie free speech is placed into the letter of the law, and only certain funded people have been shown to be "competent sources," ie sources which are easy to manipulate. However, the media itself is mostly about expressing lies and mis-information, so "why is it the gate-keeper?"

(only one indy-media actually supports free expression of ideas.

It is the expression of new ideas which the paymasters [of the wage-slaves] fear most, since the normal operation of the media is to supply mis-information and lies, it does not pass the test of "competent source," {so why believe the science of these paymasters, when there are many good alternative ideas about measurable descriptive constructs, for one such valid alternative construct of science read-on} yet the main stream media is the designated gate-keeper (designated by the owners of society).

It is a rigged game of logical-circles which support the rich and entrap the poor, ie where the poor are those who do not adequately help the rich get ever more rich.

It should be made clear, that, if one tries to use the laws of physics to describe the (relatively) stable definitive observed properties of: nuclei, general atoms, molecules, crystals, as well as the solar system, and beyond (eg dark matter) the stable definitive spectral-orbital properties of these systems, properties which can be used to uniquely identify these systems, cannot be calculated.

That is, these most fundamental physical properties have not been adequately identified by using, what passes for, the laws of physics today (2013). [This can only be interpreted to mean that today's physical laws are failures as a model for a measurable and practically useful descriptive construct.]

Furthermore, this spectral-orbital stability does not imply randomness as a basis for a description of these properties, but rather: linearity, geometry and causality are the patterns which cause the formation of these very stable physical systems.

Thus, if one interprets the data from particle-collision experiments to mean that existence is many-dimensional and (locally) unitary, then one can hypothesize that existence is based on a high-dimensional discrete lattice structure, upon which the stable discrete shapes are based (the stable shapes identified by Thurston-Perelman in geometrization).

This simple construct allows the description to enter the context of being: linear, geometric, and (subsequently) causal, but the causality really about the resonances of "contained metric-space or material components" with the spectral properties of the high-dimension containment space, whose subspaces of the different dimensional levels have been partitioned by the lattice structure associated to discrete hyperbolic shapes. It is an apparently very simple construct and it accounts for materialism, randomness and point-like local-interactions and assigning math-physical properties to the metric-spaces eg position and stability of patterns, allows for there to be metric-space states which require unitary rotations of these metric-space states, these metric-space states are also part of the interaction construct.

However, it is really a very complicated and a very rich context in which to explore both math and physical patterns to be used in physical descriptions based on geometry, as well as to be used to describe the patterns concerning living-systems.

The geometric base is very significant since it allows the description to be very useful, especially, if all the dimensional and subspace as well as containment properties can be determined, but again it is geometric so this mystery seems to be solvable. This also provides a math model of different sets of spectral structures on other high-dimension metric-spaces (this has always been religion, but now it is (there is) a mathematical model of an existence which is not based on materialism), and the accessibility of these other spectral structures or these other-worlds might be best traversed by the high-dimension "instruments" of life.

It provides the modeling basis for the stability of: nuclei, general atoms, solar systems etc (though Bohr's model with Somerfeld's orbital corrections might be the better model to have in mind for the stable [small] systems, but Weyl-folds (associated to the Lie group properties of Maximal tori) of the discrete hyperbolic shapes would be part of the description).

One way of putting-it is to say that the radial-equation (for the separable differential equations of these systems) can be solved in this new context, where the bounding properties of the containing space are the stable discrete shapes associated to the properties of the containing space, which is a lattice. Whereas for free systems (free material components) in the metric-space both non-linearity and randomness are common properties (these properties are now contained in a newly identified stable "bounding" construct, the bounding subset constructs of metric-spaces), ie it is a construct which is consistent with the observed data, but that observed data is, presently, interpreted to mean that classical physics and quantum physics are both correct, with quantum physics being more correct, but quantum physics is based on indefinable randomness, which means that the stable spectral properties of general quantum systems cannot be identified by means of calculations based on function spaces and randomness. This can only be interpreted to be a major failure of the math construct. Dirac held-on to quantum physics, though he suspected it did not work, and he helped model particle-physics (which is really about the spin-rotation of opposite metric-space states) but Dirac's conclusion, though only expressed as a slightly negative statement, was that the intellectual programs of both quantum and particle-physics are failures.

The reason the failed theories of quantum physics and particle-physics are still "front and center" is about the relation of the randomness of particle-physics to reactions in weapons, and the relation of non-linearity to feedback systems used in guided-flight (but feedback has an indeterminable structure so it can never be actually relied upon). That is, the physics and math communities are highly manipulated and highly managed sets of certain types of personalities, where the case of Oppenheimer is a clear example of an intellectual who possesses partial information which he is manipulated, so as to turn that information into horrific weapons (there is the similar story with GMO's), although many of his cohorts were very willing to supply . . . , what is essentially still "the Roman empire" (the European model of social inequality based on extreme violence) . . . , with horrific weapons, whose purpose serves the very foolish concern, or sport, (the blind competitive purpose) as to determine "what few personalities" will own, and thus run, the world, ie its main purpose is to serve the idea (or construct) of social inequality (an idea very prominent in the Roman empire, and subsequently taken over by the banking institutions associated to very narrow and highly controlled markets, that is, markets sustain power, and the politics becomes an easy commodity to control, especially in a social context of wage-slavery, note: equality means that "an equal general welfare" must be the priority of politics, free-markets can only exist if everyone in society is considered to be an equal creator).

It does not take much to realize that peer-review is really a bullying construct associated to the competitions based on authoritative dogmas (where authority is confused with truth, by the competing intellects) and it is used to filter people into a proper-class of, so called, high-ranking intellects, but these people are easy to manipulate, similar to Obama (or to politicians in general, divided into either those who support the owners of society or those who support the intellectuals, where, in turn, the elite intellectuals are hired to support the owners of society).

The US got strong enough to rebel from its ruling country due to equality (not equality of opportunity, rather equality in the sense that we are all equal creators) but Hamilton and the other thoughtless, or hypocritical, economic elites of the US converted it back into, being a "part of the banking empire," but this is not surprising since the, so called, highly moral Puritans were building their power on the extreme violence of exterminating the native peoples, the "moral model" of manifest destiny, and subsequently, a banking empire based on violence (Does it sound like Rome?).

Why are the following ideas (about gun control) never expressed in the highly controlled publishing community?

Whereas it is clearly the gun-industry is the "party of the discussion" which has a screw-loose (since they are promoting their profits and do not care about the community), and since the second amendment is about "not having a standing army," so as Orwell stated, the second amendment is either about "arming the citizens with nukes (otherwise the public cannot meaningfully resist an oppressive government)" or dismantle the military and let the citizenry keep their guns . . . , but it is really new ideas which the paymasters fear . . . , not guns) Note: Free-speech has never been upheld by the justice system, ie a justice system which does not follow the law, since the Declaration of Independence states that US law is to be based on equality, and politics is supposed to be about caring for the general welfare of an equal public (this is the true spirit of the US law).

10.

Dichotomy

People tend to reiterate what they have heard in the media, and the media has far too much authority, especially, since physics has not answered many fundamental questions about the stable properties of physical systems. But furthermore, their physical laws are not contributing in any significant way to practical creative development.

These are issues about using properties in a classical context vs. using properties in "the new context of the description." In regard to technology (eg associated to quantum physics) this is about quantum properties being used in classical systems, not using quantum properties in a quantum context (only the laser fits this criterion).

Consider the assumptions and contexts upon which a descriptive language is based, where the huge influence of where one begins in one's thoughts is easily recognized in relation to Copernicus and his new way in which to contain (or organize) the solar system, where the new ideas (A New Copernican Revolution) posit containment in a space, which has a lot of geometric structure (though they are natural properties of metric-spaces) while the main structure of conventional physics is a very complicated idea of materialism and the very complicated idea "that order emerges from a random structure," but they cannot actually show that their assumption is correct. That is, they cannot calculate the ordered states of material systems, so perhaps one should posit different assumptions.

The need for an example in regard to such a new set of assumptions, is what the new descriptions provide.

The very interesting thing for human thinking is that the new assumptions are many-dimensional, which means it is a description which goes beyond the idea of materialism, and the proof that this is needed, is that we have a stable solar system, all the models of the many-body solar system depend on the idea of materialism and these models are all chaotic systems.

However, many-dimensions allows for a new context within which to consider religion, but there are also odd-dimension shapes (3-, 5-, 7-, 9-dimensions) which have an odd-number of holes (3,4,5, . . .) in their shapes, where a doughnut (or torus) has a single hole in its shape, so that, if all the natural stable-flows . . . , (ie the system's natural spectral [or vibrating] properties) . . . , which exist on the shape are occupied (with charges) then the shape would naturally have a charge imbalance which would cause it to start oscillating and generating its own energy.

This could be a simple model of life, but it is a shape which is contained in a higher-dimension than 3-dimensions, ie it transcends the idea of materialism, where it is assumed (by materialism) that all material is contained in 3-spatial dimensions if one is assuming the idea of materialism.

It is this idea, of transcending materialism, which people will either shrink from, or they will welcome the idea, where one would expect the more religious people to want to transcend the idea of materialism. Unfortunately, there are intellectual turfs which feed off one another, in a back and forth way, where the religious people will defer to the material based science and then stay away from the new ideas even though the new ideas provide a map which leads into the higher-dimensions by analogy.

That is, the set of assumptions upon which a descriptive language depends determine what type of properties which the description will have, that is, in your language these different language would "discriminate" in different ways. Godel's incompleteness theorem is really an invitation into new ways in which to organize a language and identify assumptions for a language, but it is good to base the language on measuring and on stable shapes (quantity and shape) so that the descriptions can be both verified by measuring and they are practically useful descriptions (geometric descriptions are (or can be) practically useful).

Main part

Either

There is many-dimensions partitioned by stable shapes with existence dependent on resonances between metric-space-shapes and the finite spectral set of the partitioning stable shapes (where the stable metric-space shapes [contained in a greater sized and one-higher-dimension {containing} metric-space] are the smaller material-components which are contained in a metric-space whose dimension is one-dimension-greater than the dimension of the material components and whose size (of the material containing metric-space shape) is greater than the size of the stable metric-space shapes, or material components, it contains). This structure allows for the many-body systems, (systems with relatively few bodies) which possess the properties of stability to be solved.

Or

There is materialism, which is simply given (or assumed), and randomness (to fit data) and non-linear models of (material) interactions, but with no valid models for stable many-body material systems (though, many-body material systems, with only a few bodies).

Randomness is simply a data-fitting technique associated to a set of data which, in turn, is associated with random-event properties (in space). It is a data-fitting technique which appears to be more sophisticated . . . . , based on operator and function space math techniques which nonetheless gives no valid model which measurably describes the stability of many-bodied systems based on randomness . . . , than the data-fitting epicycle structures of Ptolemy, but, in fact, both (data-fitting) constructs allow one to say that their models are measurably verified and precise descriptions.

More generally there is: existence and creativity, the opposite poles within which life (experience) is defined.

If one tries to make current physics analogous to these universal opposites there would be material and force-fields, or the basic Fermions (hadrons and leptons) and the (Bosonic) field-particles of (defined on) metric-spaces (defined by local unitary transformations of the, so called, Fermionic particle states), but nothing is actually described in this old context.

Whereas, in the new descriptive structure it would be the stable discrete hyperbolic shapes and the force-field structure of discrete Euclidean shapes, which, in turn, relate a metric-space geometry (related to material components, ie shapes) to the local coordinate transformations found within the fiber group, and related to the local interaction geometry, and where the local coordinate transformations are placed in a unitary context, but it is an interaction structure which allows for stable properties to emerge (or might emerge), ultimately determined by resonances.

The many-body systems (with relatively few bodies) which possess the properties of stability have been solved.

This over-turns the math-science context of the current science-math authority associated to materialism, randomness, and non-linearity (and irrelevant data-fitting).

The new ideas change the context of "materialism vs. a context beyond materialism" since material is shown to be a subset of the new math containment construct, where the new math construct goes beyond the idea of materialism, eg it is many-dimensional [contained in an 11-dimensional hyperbolic metric-space, or equivalently contained in a 12-dimensional space-time metric-space]. Furthermore, in the new context:

1. material is defined, and its usual properties of
2. prevalent randomness, and

3. prevalent non-linearity,

are derived (within the new descriptive context).

In the new containment-set one must assume a natural construct for containment which implies that stable patterns can be identified, and where these patterns can remain stable, and creativity requires knowledge, arrangement, and organization so as to cause a new structure to form into a new stable pattern (or form in a stable causal manner).

This is about set-containment, metric-space properties, eg both stable shapes and (physical) properties which exist within a new context for containment, eg position in space and stable patterns conservation laws "continuity in time."

This is more like a real-world adventure into philosophy (which is a part of everyone's life) (or a real abstraction, just as D Juan tried to lead C Castaneda into a real abstraction), than a quest into the material world whose forms and realizations are determined by extreme violence, but whose descriptions are both irrelevant and descriptions of illusions and of fleeting, unstable patterns.

By whom are these new ideas (new ways of using language) opposed?

Answer: The lying thieves, who effectively own society, always oppose individual creativity which is not directed at increasing their own power. That is the social construct based on violence, wherein the so called experts express "truth" within the media, but they really express "the truth that the owners of society require that they believe" if they want to be (well-paid) experts, where the owners of society want truth expressed in the form of materialism, randomness, and non-linearity, since this suits their interests just-fine. This is an arbitrary authority, which is built into an unequal society, and it is upheld by extreme violence.

For example, the large salaries given to the policeman (and the institutions of justice) who torments and follows a person who stole some bread for humanitarian reasons.

The experts are placed into a social context of being intellectual bullies, and protected by the owners of society, so that no aspect of the revered media dares challenge that authority, eg virtually all the well meaning radical (or progressive-liberal, or conservative anarchist) voices which are allowed onto any of the media (including most of Indy media, which is supposed to support free speech) seem to protect science and math without realizing (1) science and math have failed and should be challenged (2) the basis for all social creativity, ie creativity related to investment, and the basis for a wage-slave social structure (where the rights of the owners of loaves of bread are sacred), has its origins in the failed authorities of the current beliefs of the (far too narrow) "professional (wage-slave)" science and math communities.

This is really a choice between staying technologically within a inventive context of 19th century science, a context of ever increasing violence, or expanding intellectually into new contexts, where intent and creativity is directed to mankind's true heritage (of an adventure into the abstract, a context associated [accessible] to everyone).

The propaganda system is best suited for identifying itself as the sole voice of an authoritative truth (but protected by law), ie it is the voice of the society's true religion (we must worship our paymaster, by law) where one of these religions is the authority of the professional math-science communities a group of people who are required (by their careers) to be faithful to their narrow dogmas [expressed as peer-review].

The propaganda system, though it is the seat of modern religion, expresses the two religions the two authoritative dogmas of society:

(1) the authoritarian voice of hierarchical religions, whose authority is subjective [similar to economics where there are solitary players who can control the whole game, ie the quantitative models of economics mean virtually nothing], and

(2) the measurably verified voice of science and math; where randomness is associated to rates of nuclear reactions, and non-linearity deal with the improperly defined context of feedback systems (guidance systems); but other than coupling quantum properties to classical systems the descriptions of science and math are irrelevant to the development of new technologies, or irrelevant to identifying new contexts for creativity. Peer-reviewed physical science is (now, 2013) only about illusionary worlds, and the literary contexts associated to "Physical Review" (and other professional journals of its ilk) and to Star-Trek.

Extreme violence, shrouded by high-minded institutions, upholds a delusional world of ineffective but very authoritative science (community) and social inequality, and this is done by means of propaganda, where experts are allowed to use their partial truths so as to develop technologies whose only possibility (whose only possible fate) is destruction of the planet, eg nuclear energy yet the nucleus has no valid model; manipulating DNA yet there are no valid models of an entire living-system.

Challenge to authority from a deeper truth

Not only is this a direct challenge to the authority of science and math, but it also provides these professional communities with solutions to their most difficult problems, yet the world is lost in its own concept of bullying, which is the only behavior which the current Empire allows within its social institutions, ie bullying to protect and serve the elite (selfish) interests.

Human-value is about knowledge and creativity, and humans seem to really be, in relation to the trap of materialism, a part of an abstract creative context.

This is a positive statement, it is constructive criticism (the solutions to math and science's most difficult problems are provided), it is not anger from a failed spirit, though the bullies (in their realm of selfish conceit) will interpret that way, but this is like saying Copernicus was a failure, which in its harsh reality of extreme social violence, one could say it was [where the US revolution, basing law on equality, was a weak attempt at opposing the extremely violent traditions of western thought, and trying to push the west into a civilized and reasoned state, allowing widely diverse creative expressions].

11.

Propaganda

Propaganda and language-domains

Propaganda, narrowly defined math-science, rigid uses for language leading to the decay of both language and society:

Artificially fixed symbolic-domains of iconic-value are used in a business-centered world associated to the US society's propagandistic language (suggestive language), which is held together by the "letter of the law," and a complicit world of people administering and making law (politicians and judges and propagandists), where these symbolic-domains of iconic-social-value are maintained and upheld by extreme violence perpetrated against the public (eg the policeman, in Le Miserable, hounding the person who stole bread in order to help starving people), where this has led to a nation of bullies (each bully presiding over limited domains of social-value, and each such domain is endowed with the power of social coercion) where each bully fits into a domain which coerces the public (ie the other people, whom comprise the public), and defines the bullying-person's social power (their wage-slave social-value), and this process of partitioning symbolic power helps the ruling minority to keep society relatively fixed (this is the same model of society as the model of the Roman empire.

Indeed this is the reason that the letter-of-the-law is such a great corruption of an equal and literate society, a society of equal people who are willing to challenge "fixed authority," and be interested in a wide range of different practically creative contexts. Thus, being related to a truly free-market, which is to be built on a set of equal creators.

Introduction

Artificially fixed symbolic-domains of iconic-value are used in a business-centered world associated to the US society's propagandistic language (suggestive language), which is held together by the "letter of the law," and a complicit world of people administering and making law (politicians and judges and propagandists), where these symbolic-domains of iconic-social-value are maintained and upheld by extreme violence perpetrated against the public (the public is not allowed to escape the symbolic domains provided to them by the propaganda system), where socially this . . . . (social organization, [where product is associated domains of influence, and where it is all-about a command and controlled market-place, ie it is central planning, associated to the social power of the few in the ruling class, shrouded by a propaganda system calling it a "free-market"]) . . . . , where this has led to a nation of bullies (each bully presiding over limited domains of social-value, and each such domain is endowed with the power of social coercion) where each bully fits into a domain which coerces the public (ie the other people, whom comprise the public), and defines the bullying-person's social power (their wage-slave social-value), and this process of partitioning symbolic-power helps the ruling minority to keep society relatively fixed (this is the same model of society as the model of the Roman empire, where engineering is associated to limited domains of social value, it takes away the power of the public but the trade-off is comfort and convenience in living in a limited and constrained context). Indeed this is the reason that the letter-of-the-law is such a great corruption of an equal and literate society, a society of equal people who are willing to challenge "fixed authority."

Indeed this narrowness associated to business and social control was mocked by S Jobs in his youth, where his famous statement was, essentially, 'these people running big business are not particularly smart,' where Job's business career was characterized by his development of several different products rather than building a symbolic domain of market-dominance around a single product.

Indeed, building a symbolic domain of market-dominance around a single product, is more about gaming the system and following the path of the Roman-emperors whose power is based on extreme violence directed against society, so that there are only very narrow paths, which people can follow to social-power, and they are all paths which uphold the system of arbitrary-power upheld by violence.

It must be noted that S Jobs was able to reassert himself in business, because of his capacity to rule as an arbitrary emperor.

Engineers are narrowly defined "pawns in the game," though their language, based in the measurable, is related to an ability to create, but both Godel's incompleteness theorem, as well as the example of Copernicus using a precise language in different ways from the authorities of his day, express the idea that precise (measurable) language also has great limitations. That is, the engineers are (also) caught in their own narrow context of language, where they perform as the well-paid servants who adjust the complicated instruments, which help the ruling-class maintain their power.

Engineers are also divided into symbolic domains of influence (or relevance) so that the rulers can have available, to themselves, an assembly-line of inter-changeable parts (separate engineering domains) for relatively complicated instruments. But such a use of language, in narrow limited contexts, though useful for particular instruments (or particular processes), in fact, identifies a precise language which is narrow and limited in its relevance. Thus, it (also) defines a precise language which is ultimately destined to fail (to have limited relevance), especially if it is attached to fixed ways of organizing society (organizing care for the instruments of society), and organizing knowledge in fixed ways. It also needs to be noted that the invention of these instruments is related to the culture as a whole, but developing these instruments in a fixed social context which supports primarily the few rulers of society, is a process which shows how the letter-of-the-law can be used to steal knowledge from the culture, to be used in opposition to the social forces which allowed the invention of these instruments, and then using these instruments in a narrowly defined context of social use.

The use of the computer and the inter-net, should not be about spying and social control, but rather it should be "all about" a communication system through which descriptive knowledge can be developed, so that descriptive knowledge can become a big set wherein the many different languages can be based on the ways in which precise (measurable) languages can be created and/or organized, in regard to a precise language's relation to any of a wide-range of practically creative contexts. Note, however, the usefulness of any precise language will depend on a descriptive language's relation to stable patterns, ie to stable geometric patterns, and not to (improperly defined) randomness.

Propaganda, and fixed ways of using language

The main criticism that the public should have concerning the math-science experts . . . (especially in regard to the correct interpretation of the Godel's incompleteness theorem, where the best interpretation of Godel's incompleteness theorem is, that math requires many languages if math wants to be relevant in regard to the development of precisely described patterns, ie or in regard to the development of knowledge) . . . is that the experts stay in a fixed dogma, which is centered around the language of a specific set of assumptions and contexts. and interpretations (of data, and/or interpretations of math patterns themselves), etc; wherein the focus on quantities, geometry, and randomness is (has come to be) about either the continuum and convergences, within a containment set, modeled as a continuum, which is measurable, allowing for too many types of opposing ideas to be contained in the same measurable set, eg allowing both random and geometric descriptive structures to seem to be compatible with one another, because both sets of ideas can have convergences defined on the same continuum, or about quantitative discreteness (or quantities defined on a continuum), and solvability in an unrealistic math context which is not globally separable (globally parallel and orthogonal [at each point]), thus, there are no stable patterns which exist in such a context of "solvability," or "the far too general" context of the continuous deformations of geometries, whose shapes are difficult to conceive, and which are unstable shapes, or an excessive focus on improperly defined randomness, and its associated function-space structures, which are without: spatial, temporal, or geometric properties by which to confine the properties of these function-spaces (whereas function-space techniques can work in the context of spectral-systems which possess physical properties (or are defined within a physical system), since physical-systems possess natural system-constraints).

The simple goals of the experts are defined by the big-business interests [so as to form a monopoly of knowledge for big business, in regard to both knowledge and creativity within society, a monopoly of knowledge which, in fact, they have no business managing, in regard to the public welfare of knowledge, and the sets of creative contexts to be used by the people, the market place is far too controlled (and thus distorted and manipulated, it is a market which is only slightly different from the market defined by the Roman Emperors, ie it now has gadgets invented with the knowledge of classical physics, including physical chemistry)].

That is, the narrow definition of knowledge (defined by big businesses and its "investments in established technologies") allows for only competitions within a narrow dogma (and within complicit educational institutions), where this dogma is identified by the limited creative interests of monopolistic businesses, businesses which sell one product, or are based on one-resource, or only apply to one social context, eg the military associated to violence and communication systems, and violence is all about maintaining inequality.

The business community has its small set of highly competitive (and thus fully indoctrinated communities of people, who are fully indoctrinated in a narrow dogma) these are "the experts" whose "job" is to adjust complicated instruments for the rich and powerful, where these are the instruments which help the powerful stay in power.

These people are represented as those experts who the investment community can trust.

Thus, expert descriptions in the science and math fields are about:

0. Classical physics, and coupling physical (quantum) properties to classical systems,
1. materialism and a continuum [and not stable patterns which are continuous in time, even though the property of "continuity in time" (one way in which to define a stable pattern) is a fundamental definition within physics (ie the definition of conserved energy)]

2. overly general shapes (associated to non-linearity, and instability, and randomness [usually improperly defined]),
3. complicated codes (of which native languages of the US first peoples are the most complicated), and
4. random and non-linear processes associated to unstable fleeting patterns

related to material systems (whose purpose can be defined in the context of a feedback system), which is: both

I. transitioning between two sets of relatively stable states, eg reactions, but there do not exist any valid descriptions of stable states . . . , where there is a vague thermal description of conservation of energy (yet quantum-system stability, as well as solar-system stability, are both patterns conserved in time, but whose stable structure is not describable when the descriptions are based on applying physical law (of quantum or particle-physics descriptions, nor within general relativity, or even within Newton's physics for a many-body system) within the accepted paradigm of physics (2013)), and

II. the focus on unstable fleeting patterns is related to adjusting macroscopic components to conform to metric-space distinctions within a feedback-system {even though such distinctions are not a part of the metric-space's natural set of stable properties}, where (in a feedback system) stability is defined in regard to the system's defining partial differential equation, where the "stable range" of such a differential equation is difficult (usually bordering on impossible) to determine . . . , thus the malfunctioning of such feedback systems, with unknown causes, will be a common property of such systems.

This is because these descriptions are non-linear, and thus quantitatively inconsistent, leading to (amongst other things to) a chaotic system which is being represented as a "distinguishable relatively stable system structure" within the measuring context of a metric-space, but all parts of the pattern are unstable, so the failures of such systems can enter in many different ways.

There is an alternative which is outside of corporate and investment interests

Namely, a many dimensional hyperbolic metric-space, partitioned so as to be bounded (in the partition) by stable shapes of various dimensions and related to various subspaces, in turn, defining a finite stable spectral set for all the metric-space components contained within the containing space, and where component interactions are "dimensional-level-centered" processes, so that this context is, in fact, a very complicated and very interesting space within which to study math patterns which are stable and complicated and physically relevant.

The public and the experts cannot talk about alternative ideas (such uniformity of thought is maintained by the propaganda system)

This is because wage-slaves are easy to deceive, since the strategy of the propaganda system is to associate those who help corporate, and investment, interests as those people who possess high-social-value, they are those few who "do know," while the public, or a non-funded intellect, [according to the propaganda system] cannot possibly be discussing ideas that have any social value (where social value are the big business interests, where the high-value of these narrow interests are endlessly repeated over the propaganda system so as to become the society's social doctrine).

When an editor who bases their science-values on the interests of big business, or equivalently (for example) on the interests of the wage-slave MIT experts, and not on the actual nature of the many possibilities associated to descriptive structures of math and science, then one has an incompetent editor, who is judging the important descriptive ranges of math and science as being defined by a (the) narrow dogma (expressed in the media, or propaganda system), narrow dogmas which are (really) being identified by the very narrow interests of big business.

That is, one sees the circular thought patterns of society, whose social values are being defined by propaganda communication systems, systems which are owned and controlled by investors.

The media (or business interests) defines the focus for the MIT experts, and the focus of these experts defines the beliefs of a highly-valued editor (a competitive fellow aspiring within the dogmatic rules which are associated with educational competitions in a context of wage-slavery [or equivalently, a meritocracy]), but the editor is, in fact, the incompetent-link which defines the circular structure.

Thus, the ignorance of a Boston-indy editor . . . (who basically, is not understanding the school-boy example of "Copernicus vs. the church's authority," which is really equivalent to, "new ways of using language vs. accepting the authoritative way of using language as defined by the experts," in regard to the relation that new language has to developing new descriptive knowledge) . . . is the invaluable ingredient (which supports big business interests), associated to a narrow authoritarian agreement, between wage-slaves and investors, and which holds the propaganda system together.

But the wage-slaves did not really agree to this, the social state of being a wage-slave is a fiction invented by corrupt judges within a justice system, a justice system which does not uphold the "equal and common welfare" basis for US law. That is, wage-slavery is a result of the exploitation, by the big business community, of the weak-character of appointed (or elected) judges.

That is, a math-science context which is associated to stable patterns in a many-dimensional containment space has many more possibilities . . . . , for both measurably verifiable patterns as well as in regard to modeling very complicated systems, systems which can be practically useful, since the description is based on geometry (geometric stability) . . . . , than does today's (2013) mostly irrelevant descriptive context of math and physics, which is simply data-fitting (since that is the essence of a description based on randomness), where a random description of a stable system, built from relatively few components, is a useless descriptive context.

Thus, for an editor to make a judgment about knowledge and truth is again an example of weak and exploitative personalities using an (invalid, but) established social hierarchy for their own personal advantage, in the upward climb along the meritocracy, a social hierarchy based on the fallacy that authority and/or social-position are equivalent to truth.

Note: There is much confusion about the role of the descriptions of science to technical development.

Though there are quantum properties which can be coupled-to classical systems, the science upon which technical development depends within our current society is based on classical physics.

This is because the control of systems whose described properties are based on randomness only exists in a classical-thermal context (where measured values are averages over large reservoirs of many components), and systems with relatively few components whose described properties are based on randomness do not allow for any form of control.

Building transistors and micro-chips is almost entirely about using classical thermodynamics and chemical properties of materials, as well as optics, so as to form P and N types semi-conductors by thermal processes, then etch circuits on the semi-conductors, so as to possess and maintain semi-conductor properties within circuits (where the circuits are chemically etched into the P and N material properties of the semi-conductors used in these circuits). It is a classical thermal-chemical processes (with the help of classical optics) which create both the properties of N and P semi-conductor materials, as well as the circuit structures etched onto these semi-conductor materials, so that the thermally developed quantum properties of the different semi-conductor-types can be used in a classical context of electric circuits.

Micro-chips are, predominantly, about manipulating chemical properties of materials in a thermal context, it has very little to do with quantum physics other than semi-conductors possess quantum states which can be used in a classical context when the materials of the semi-conductors and the materials adhering to the semi-conductors are carefully thermally (and optically and chemically) manipulated to form classical circuits.

Furthermore, if today's irrelevant math-science context can be so complicated that the experts themselves have no understanding of it, ie they cannot apply their quantum and/or particle-physics descriptions to a practically useful context (particle-physics only applies to the rates of reactions in nuclear weapons), then the knowledge of the experts is irrelevant.

That is, if the identified quantum properties can only be both coupled to and used within a classical context, then the quantum context is irrelevant, especially when the quantum properties can be coupled-to in a classical context.

That is, the irrelevance of quantum description, [eg solve the radial equation for an atom with five or more charged components {ie nucleus and other charges}, this is not now being done, general solutions to these types of radial equations do not exist], means that its authority should be challenged.

This really means that there can be a very large set of languages associated to many math-science communities, which can be very large communities, with well separated sets of experts associated to the different languages, which are built on different sets of assumptions and contexts, ie whose interpretations and context are quite different in their natures, where the goal of these descriptive structures is to determine valid descriptions of observed properties and to find a new context within which to create new things based on the new (observed) stable microscopic (or now called quantum) context.

If E Witten or S Hawking want to talk to monopolistic business interests about a bunch of irrelevant baloney, OK, but do not let this pointless talk interfere with the exploration of new ideas.

That is, it needs to be noted that these "icons" of intellectual-value have not been able to solve the more difficult problems related to stable many-but-few-bodied material systems, but this is a problem which is immediately solved in the new context.

But "the E Witten, or S Hawking, talk" is really designed by the propaganda system to influence the impressionable editors (eg Boston-indy editors, aspiring to higher social positions within a corporate society, aspiring within the ranks of the elite intellectual-left) to believe in both an authoritative model for truth, and to believe in the monopolistic high-value of the (selfish) big-business interests.

This is a communication system (propaganda system) which is not guided by reason and thought (careful considerations about truth) but rather it is led by icons-of-value, which are used to suggest beliefs within the public, and to create agreement, a type of cheer-leading agreement, for corporate interests (and corporate investments), and the society which the investors control.

This gate-keeper-of-truth status of "funded editors" (anyone funded so as to have a voice on the media) of thoughtless impressionable editors (or news personalities), is more like a "bad joke," since the media is opposed to the truth.

Thus, for someone to think that a voice, or gate-keeper, within the media is exercising their "competence to judge truth" by keeping other people from reading new ideas on the inter-net about science (new ideas which have not yet been associated to investment interests, and thus, it may be safely assumed, by the aspiring editors, to be wrong [since these new ideas are not, now, creating profits for the monopolistic economic institutions]), is quite a far-reach for a "rather loosely identified" gate-keeper of truth.

However, the main function of the main-stream media is to lie and mis-represent information, so why be subjected to the, so called, "high-standards" of an impressionable, thoughtless person, positioned as an editor (who is really about acting and making judgments to protect investor interests, since our model for truth is provided to us by the propaganda system, and the judgments of these editors is masked in a context of either distinguishing competence or distinguishing truth) one might see this state of affairs as a bit funny (perhaps funny to G Beck, or R Limbaugh, one might find them somewhat humorous), because the funny thing is "the sad-state of how information is controlled by propaganda" in a big-business centered society.

That is, if a math conference of professional mathematicians are curious to hear these new ideas, yet these ideas are to be excluded by some authoritative editor at Boston-indy, a person who is clearly more interested in conforming to the dominant culture than being a person who is attempting to express the truth for the public, ie an editor at a web-site claiming to promote free-speech, but who is a defender of a "fixed state for society" which will aid the dominant interests (the owners could not be made any happier), and this is because science and math are (have been) captured (by monopolistic business interests) in the same way in which the government and the justice system are captured, so as to support the interests of the ruling class.

One should also note that the language of the first amendment is now interpreted, within the propaganda system, in terms of the "letter of the law," so that free-speech is really about the right of a wage-slave "press" to espouse propaganda.
Whereas, free-speech, (eg equal free-inquiry) is about the "determination of truth," where truth needs to be related to an equal creator's "inventive interests," or developing a useful (geometric) model of existence, which conforms to the observed properties of the world, but a description which can be practically useful.

In a culture, whose religion is personality-cult, and where authoritative personalities are the icons-of-truth, one sees that the "difficult attempt to determine" (or discern) truth is an activity which has no meaning, so that one can "get away with" identifying the beliefs of the ruling class with the social attributes of both authority and (the vague idea of) personal virtue, so that the opinions of the high-valued authorities determine both the basis for intellectual authority, and the basis for a law (both knowledge and law which serves only the selfish interests of the ruling class) for all of society, and compliant editors can be found to protect this (fraudulent) truth.

Furthermore, this is done in the name of competence, since competence is defined by the media to mean that "one is 'competent' if one serves the interests of the ruling class."

When the core of a society's communication system is a propaganda system, used to support the interests of the oil, military, banking business monopolies, then social value and the value of knowledge (within society), which is expressed in the US educational institutions, has a purpose, and that purpose is to cause the public to vocationally-fit into the interests of the big-business monopolies, then when these narrow interests have reached their exploitative and destructive limits (of both the earth and society) then the knowledge base for the society fails, but editors (people who are judged to possess high-social-value) remain loyal to the fixed and established social-values (if they want to keep their jobs, or if they want to be thought of as protectors of [a corporate] truth, a truth as discernable through the eyes of those whose only focus is on propaganda, eg all-of-us), and the society fails in an even deeper manner.

Consider AARP, it used to be an effective lobby for old people's interests, then a retired CEO ran for its presidency, and, apparently, the old-people believed that the retired CEO had high social value, and they voted him in as president, but this CEO destroyed the ability of the old-people to lobby for their interests by means of the AARP organization, and the AARP started lobbying for corporate interests, as directed by the deceptive (and already rich) retired CEO, who the public accepted as a person with high-social value.

This shows the danger of following the value "defined for the public" by the propaganda system.

If one looks at indy-media, it seems to have very little discussion about alternative ideas, its articles mostly are devoted to re-hashing what the main-stream media "talks about." There is some helpful reporting concerning the deceptive practices of the US in regard to foreign relations practices, and much (pointless) talk about the legal details of legal cases, and economics, and violence, in the established context of these narrow discussions, ie the context established by the propaganda system.

But there is no substantial discussion about "the role of law in society," and "how law should be structured." That is, law is to remain in the domain of the experts, and the discussion has no substance, but is only to be related to the marginal attributes of bullying, to which the public energetically react in the context of bullying and violence, since the society has been divided into domains of domination, in language and in violent acts.

However, the US already has a legal legacy, and US law is really about equality, and the US government is about the public welfare in regard to its citizens being equal.

Note: There cannot be a free-market unless each citizen is seen by the law as an "equal creator," whereas, the narrowly defined idea about "creativity and knowledge" which is used by big business is not enough to maintain a society.

Much of the social violence has been manipulated and controlled by the US secret security operatives, those hirelings who secretly, coming out of the defense and justice systems.

But this context of spying to protect corporate interests, such as paid reactionary agents, for example, either interfering with the Florida vote-count, or creating an issue about immigration where the focus was on the people crossing the border, whereas the problem of this type of immigration was caused by businesses wanting to pay low-wages, is seldom discussed. These are not simply unemployed free-spirits solving the jobs-crisis, by violently stopping people crossing the border, these are paid undercover agents working for the interests of big business.

Rather the media talks about "ordering people around," for example it is claimed that, people cannot use violent expressions in speech (things which the secret operatives as well as political operatives were (are) doing, eg cross-hairs) but the enforcement arms of the justice system only apply these rules to liberals (the non-violent public), while the violent, undercover, agents go un-affected.

Instead of manipulating the public by terrorism, ie by these secret operatives as well as by political rhetoric, eg reactionary militias, whose loyalties are to the interests of big-business (ie whose loyalties are to the few elite in the ruling class). The rich are allowed to terrorize the public, eg kill the striking-public and destroy their protests, but the US will pre-emptive-ly attack someone who might be thinking about weakening the interests of the corporate-state.

So what is the terrorist organization, the corporate-state or the public seeking, eg in the occupy wall-street movement, equality and the freedom to know and create.

That is, if violent expression is used to change public-policy, then it is only allowed by the state and their secret operatives, eg tea-party and righteous-right religions Florida vote-disrupters etc.

Furthermore, the expression of new ideas (or how to apply established knowledge) is only allowed if the ideas are controlled by big business (the global warming scientists who interpret global-warming to mean "stop using carbon fuels" are marginalized, and this is done with the help of the professional peer-reviewed publications of science, publications controlled by the big investors).

Note: It is the propaganda system (and its solitary and, absolutely, authoritative voice) which has most contributed to the US-corporatism state to become a dominant empire (quite similar to ancient Rome), as opposed to a society about equality (knowledge and creativity) and a government which is supposed to realize this equality for the public, based on promoting the common welfare, where instead, "common" is now (in all likely-hood) been re-defined to mean "the common interests of big corporations" by the corrupt judicial system, and the letter-of-the-law is then implemented and applied in its re-defined form.

Furthermore, the narrative of the propaganda system (which plays the role of the bully) is that the propaganda system is the judge of correctness of others (any arbitrary authoritative statement can be changed as the authoritative propaganda system if the propaganda system sees fit to change the nature of authority, so as to arbitrarily change the nature of authority)

However, in the US:
Law means equality (the Declaration of Independence)
Government means to "promote the general welfare," administrating a "real equality" for a population of equal creators, for a public who are free to believe and free to express those beliefs and thus to develop a new precise language and use that language in new creative contexts, where people create in a selfless manner, so that then the "market" (trade of ideas and things) can, truly, be a free market.
Where free expression is about the many valid ways in which different measurable languages can be related to practical creativity.

Yet, instead the judges and the politicians "choose winners," they help monopolistic businesses to come into being, eg W Disney turned-in the, so called, "commie artists" to the FBI, ie his competitors, and this helped W Disney's business grow, and the government officials serve these few business-people to help them dominate, as well as frighten, society. [Note: R Reagan was also an FBI informant, and he also rose up the ranks of the "ruthless" and, so called, "morally" dominant right.]
Thus, the free-market is (has become) a controlled market, which is essentially based on the commodities and associated living styles established by the Roman empire, but also by the electronic and thermal technology which emerged from 19th century science, as well as the exploitation of oil and coal in energy agriculture and textiles, but the vision of electronic development, by corporate America, is a vision related to a complete capability to "spy on" an entire population.
Judgment of such information (about such an all inclusive invasion into people's lives) can only be used in a very hierarchical society, a society where the ruling class identifies truth, and it is also above the law.
The conservatives are for keeping things fixed, a fixed society ruled by emperors, who are the personality-cult-centers which, essentially, make them "the gods" within society.

The liberals should be about equality and (practical) creativity.

On the other hand, most human-beings are for developing both knowledge and new contexts within which to create, which should mean that most humans are for equality, but as wage-slaves this is difficult to consider.

This is because wage-slavery is based on violence, and a need to terrorize and coerce the public, so that they will not want equality, and they will faithfully serve the authority which big-business defines for society.

Within such a violent society, it is the job of editors [and educational competitions, competitions defined in a narrow authoritative context], to keep the public on-track. The public is not allowed to stand for knowledge or creativity, since these are the domains of corporate interests, yet, masked as the beliefs of the superior intellects of society.

Aspiring editors, promoted by the media as "the best and the brightest," are those personality-types who acquiesce to authority, can collude with the interests of the propaganda system, and rise through the system, and thus they can either get to bully others or they can get a job and they can get published (and thus be a part of the bullying process), since a frightened, wage-slave must get a good job, and to have security (within society) they must achieve a social-level of being able to bully others.

One sees a consistent model of the ruling class opposing expressions of equality and reasonable-ness where the groups expressing an interest in equality are infiltrated, and, if possible, led astray, or parts of the group are set-up to be betrayed and charged, or the group is not a group which adheres to equality (or the leadership of the group does not adhere to equality) . . . but rather to some form of elitism, eg they support science [but the rulers of society know that the science is too narrowly defined, and too authoritative to lead the society in any valid direction, since its main purpose of science is to support the military, oil and banking business interests] then the elite leaders (of the elitist group) are isolated and manipulated, thus leading the group to ruin, and to social gain for the leaders.

However, if the group is based on violence then it becomes a "recruiting grounds" for the enforcement and infiltration and terrorist-arm of the justice and security structures of society (though the justification for all this intervention is a violent group which is related to an elitist group which exists outside the US society, but if a spy group with unlimited resources cannot stop this, then how incompetent do these spy agencies have to be?).

The relation between "the point of the violence" (of the violent groups, which are groups which are mostly expressing self centered elitism, which is vaguely allied with "socially acceptable" "higher-principles") and social norms are carefully manipulated by the spy, recruiting, and propaganda agencies, eg: theocracy, racism, straight-out militarism, and absolute authority {this is essentially the math-science communities viewpoint, eg working to serve the ruling class, extreme violence in military viewpoints, etc}, opposition to other elite groups, etc.

Note, that theocracy and science hold an analogous authoritative viewpoint, but this is because science has been manipulated, and the manipulation has been done through the narrow expressions about truth provided to society by the propaganda system and through the capture (by the military, banking, and oil) of the management of the educational institutions, to become a narrow educational contest based on the narrow dogmas of certain authoritative viewpoints, so as to fit into the vocational needs of these businesses.

The schools are failing since the management has failed, by design, or perhaps since the plan is far too narrow and it was destined to fail.

Consider that now "peak-oil" means "$10 or $20 dollars a gallon" for gas, and the coercive social forces, eg spying and enforcement by the justice system, are in place to ensure that this becomes true.

The advantage a US citizen has is that the documents in favor of: equality, freedom of belief, and freedom of speech, freedom to create, and a government directed to serve the common welfare in a context of a law based on equality; are the central ideas of the documents which define the US nation-state.

Assert our heritage and throw the . . . lying, thieving, owners of society (or ruling class) . . . out. Security supporting equality can now be done with the spying apparatus now in place. Get rid of the corrupt "republic-empire" in Washington, and re-institute the Continental Congress, and real democracy, and get a free-market where each person is seen as an equal creator, where the welfare of each of these equal creators is protected by the governing body, only then can there be a truly free-market.

Only funded voices can speak within the US propaganda system. But now the phrase "freedom of the press," the statement about the US first amendment right to free speech, is now being narrowly defined as the right of the "press," owned and controlled by the owners of society, to provide "objective" descriptions of observed events, where these events can only be interpreted by the owners of society and the "presses" which they own ie interpreted in the context of the economic and social theories which the owners of society want the society to know. That is, "objective" information is to be interpreted, or filtered, within the contexts of arbitrary dogmas which support the elitist viewpoints, dogmas which support the idea of a violent destructive empire.

Information is filtered by arbitrary authority (whatever best fit's the owners interests), or by experts of the objective material truths of science, but material-based science has failed.

The model of society's relation to an authoritative truth is represented by science . . . . (unless science interferes with the interests of big business as the example of global warming shows where peer-review is open to questionable science which helps big business, but peer-review is closed to new ideas, ie closed to the likes of a Copernicus who presented new ways of using and organizing a descriptive language, nonetheless the propaganda system is capable of glossing over the problems with having a public which is mis-informed) . . . , whose truth is based on the ideas of materialism, randomness (improperly defined) and non-linearity, ie chaos where at best feedback models of system-control are provided. But, what about nuclei, general atoms, molecules, crystals life and the solar system which are all relatively stable systems which all go without valid descriptions based on the, supposed, laws of physics? This scientific truth is essentially absolute and authoritative in the extreme, just as the church's truth was absolute and authoritative in the days of Copernicus, but now the ideas of the likes of Copernicus, ie solutions to difficult problems in science, can be ignored since the scientific truth has already weighed-in on (or expressed their ideas about) nuclei etc. thus in our age technology is almost exclusively dependent on controlling systems by means of the ideas of classical physics, ie 19th century science, but this fundamental issue of science not developing, and thus being irrelevant, is hidden by the propaganda system. This authority of science is narrow and fixed and thus it can be used to measure "intellectual-value" on the absolute scale of truth, as identified by today's overly authoritative, but nonetheless, practically irrelevant science and math, so that only a few people measure-up (in the educational competitions, competitions based on absolute truths) as being intellectually capable people. That is, it is a propaganda system model for the necessity of having an elite group of people defined in society.

The claim one might assume that an editor working for the interests of "competence," (which really means working for the narrow authoritative interests of big business, as defined by the propaganda system which is owned and controlled by big businesses), is the Boston-indy-editor's opposition to the claim that the problem of identifying the correct context through which the stability of a many-but-few-body models of physical systems has been found, and the problem solved, and the solution is found within a new context of "many-dimensions partitioned by a finite set of very stable discrete hyperbolic shapes."

It is a given that an editor for corporate interests would have to claim that the "best and the brightest" scientists have not acknowledged this solution, and thus as an editor for Boston-indy whose job is to "protect the truth" (as seen by corporate interests) . . . . , [even though finding any truth, within the media, other than the media being a process for placing data (objective news events) into the most convenient set of (arbitrary) frames . . . . (or descriptive contexts, where each context is presided over by a set of "competent" experts) . . . , for interpreting "objective" events in order to support the interests of big business] . . . . , so the idea of "protecting truth" is a form of absurdity, and is a part of the bullying process (where the organization of the rational for the bullying is provided by the media), which ultimately upholds the great inequality which is used in organizing a hierarchical society, a society ultimately based on arbitrary violence, as was the Holy-Roman-Empire], so that apparently the Boston-indy editor is claiming that the set of scientists favored by big business interests, even though these scientists which are so very smart, "do not themselves have a solution to the many-body problem," this set of scientists, who are identified by big-business interests {through a highly controlled and highly managed education system (controlled by military interests)}, are [arbitrarily judged to be] better intellects than a person who has presented the new math context within which the problem is solved.

The only way in which this editor can have any validity is if the editor has their own solution to the problem, so please "editor of Boston-indy" provide the world with such a solution, or get your favored set of scientists and mathematicians . . . , the set of people whom you so fervently uphold as being "superior people," . . . . to provide a solution to this fundamental problem.

If not, then please explain how your very limited knowledge allows, such a person as you, to make decisions about people's value . . . [and decisions about truth, which apparently you are not capable of making, since if you did possess a capacity to determine truth, where if you knew truth then perhaps you could give a solution to the many-body problem] . . . . ., value-decisions which are no different than the arbitrary value-decisions made in the early Puritan history, where the Puritans arbitrarily judged that the "savage native peoples" could be exterminated, so as to take their lands, by means of violence, to be used by the "good and civilized Puritans" for capitalist profits ("a better use of land, to satisfy their selfish interests" where they learned their arbitrary ideas, about the "value of land," from their European culture, a culture based on arbitrariness and violence).

Apparently, editors of propaganda systems have made these judgments, about "the value people possess," based on "what the media has led these editors to believe" about "people having different intrinsic value," but what the media really expresses, is that "the value of a person is to be measured in regard to a person helping big business gain ever more power" in a wage-slave based society.

Where do editors learn about the "value, and correct context, for a well articulated truth?" from the media? Yes. It is a circular flow of thought which only supports the views of the owners of the press.

The superior race "of high-IQ people" who adhere to an authoritative dogma are to be protected in the organization of "how the society is to be bullied" by the media, and their flunky editors, who are similar in character to the character of the politicians and judges, whose personal characteristics might be best described as the worms which feed on the decay and corruption of society and its prejudices, whereas M Twain described them as thieving murderers.

Apparently the "founding fathers" thought that both the idea of promoting the common welfare and freedom of both "belief" and "speech" would be a sufficient antidote to a slave-holding property-possessing set of minority rulers; the US social order which was established, partly by the constitution and mostly by the judges of the supreme court, but "free speech" was never upheld, and high-valued wage-slave editors can be used to stop free speech . . . where "in a sign of having a bad sense of humor" . . . these editors are considered to be "the gate-keepers of truth," eg scientific truth, and objective reporting of events, which must be interpreted through the "correct" context (namely, the context which supports the interests of big business), and the protectors of certain people's high-social-value, certain people who are both favored and trained by big business, eg protecting the indefinable idea of possessing "high-IQ," where this is a term used for social manipulation (but has no meaning), since scientific-truth (the ideal of the "truth" espoused by the propaganda system) is both measurable and of interest to the ruling-class.

Whereas a better definition of scientific-truth is both measurable and, practically, very useful, and if it is not useful (in a practical sense) then it is a "truth" which needs to be challenged.

"Being smart" is best defined within the media as a person who aspires to help the rich and powerful, and since this is repeated over and over on the media, this is also how the public has come to identify the vague idea of being-smart.

Though the character of politicians, judges, and propaganda people have been correctly identified, by M Twain, as lying, murderous, thieves, and furthermore, the "directors of movies" (icons of high propaganda) have praised the violent henchmen of the: politicians, judges, and propaganda people, for their ability to instigate the extreme-violence needed for "a movie's story" to be moved-along, as well as, it is so-claimed, for the society being able to function, but what may be missing in this descriptive context is the real social-context of corruption and social-rot which identifies the true context of the western society's stagnation, and its need for violence, wherein the characters of the: politicians, judges, and propaganda people might best be identified as being like the worms which feed on the rot and corruption (which is caused by their violent henchmen destroying the people and their culture, but it is a society with a social hierarchy which upholds the corruption and destruction upon which the worms can feed and thrive, and within which the henchmen operate). This is essentially the same model of society as was the social model of the Holy-Roman-Empire.

Consider the strange nature of the European mind, eg B Franklin expressed his thanks to Puritanism rather than the egalitarian Quaker society (within which, whose communities, he thrived), whereas the Puritans fled oppression within Europe so as to set-up an oppressive society of their own, a society of corruption upheld by a propaganda system expressing high ideals. Perhaps B Franklin thrived in the egalitarian society of Philadelphia since he was a "marauding Puritan" and thus willing to exploit and destroy an egalitarian society for his own selfish gain.

12.

Science

What exists?

Is it material?

or

Does existence possess a greater than simply the "material world" ?

or

Is existence about material, which when reduced leads to fundamental randomness?

But then how can the observed stable patterns of material existence come into being? (or how can stable properties be described?)

After about 100 years the descriptive structures based on randomness and associated non-linear interactions has not been capable of describing the observed stable patterns of some of the most fundamental physical systems.

so

Is their primarily stable patterns so that reduction is ultimately about these stable patterns which exist in a many-dimensional context?

What is life?

How did life get to be so complex, if its description is, supposedly, based on a improperly defined randomness (eg evolution; mutation and natural selection)?

Why is life so stable and so highly controllable?

Is life (or existence) about materialism, reduction, and randomness?

or

Is life about stable patterns which exist in a many-dimensional context, where different dimensional levels are bounded (or confined) within a context of stable-shapes of different dimensions and different sizes, where the stable shapes can be related to stable (material or metric-space) components?

Is knowledge about a fixed authority?
Or
Is knowledge (ie a measurable description's truth) best determined by its (helpful) relation to practical creativity?

Is knowledge about a descriptive knowledge? If, yes, then should the descriptive language be precise and measurable, and should the "patterns described" be stable patterns, or should precise patterns be based on randomness?

Random patterns whose measurable properties are based on average values determined over large numbers of components, which all together determine a physical system, can be placed in a practically useful context since such a description . . . still relates (is useful only if it does relate) . . . . to geometry in a causal manner

Are the global random patterns which are related to local spectral-particle (random) events better associated to a different set of math patterns, ie instead being related to distinguished points of non-local geometric-spectral patterns, where light is modeled as a charge-cloud (not necessarily an electron-cloud), eg semi-finite and semi-infinite as are "electrons-and-neutrinos forming an electron-cloud" ?

Can angular-momentum actually couple-together what one might think of as widely different "toral components" of a complicated discrete hyperbolic shape or a shape with various (different) subspace relations?

Should (one try to relate . . .) global spatial distributions . . . be related . . . . to local, random, spectral-particle events, eg the global descends to the local . . . . , (however, in general, the descent cannot be determined, the spectra associated to sets of (too few) operators acting on function spaces is not consistent with the spectra which is observed, (. . . . does not allow the spectra to be determined (determinable)), ... , whereas the more useful information is about locally measured geometric properties (of material systems) being related (by partial differential equations, and their solution functions) to global spatial distributions, eg the local ascends to the global (leading to new information about the global aspects of the system [or pattern]).

The model of society's relation to an authoritative truth is represented by science . . . . (unless science interferes with the interests of big business as the example of global warming shows where peer-review is open to questionable science which helps big business, but peer-review is closed to new ideas, ie closed to the likes of a Copernicus who presented new ways of using and organizing a descriptive language, nonetheless the propaganda system is capable of glossing over the problems with having a public which is mis-informed) . . . , whose truth is based on the ideas of materialism, randomness (improperly defined) and non-linearity, ie chaos where at best feedback models of system-control are provided. But, what about nuclei, general atoms, molecules, crystals life and the solar system which are all relatively stable systems which all go without valid descriptions based on the, supposed, laws of physics? This scientific truth is essentially absolute and authoritative in the extreme, just as the church's truth was absolute and authoritative in the days of Copernicus, but now the ideas of the likes of Copernicus, ie solutions to difficult problems in science, can be ignored since the scientific truth has already weighed-in on nuclei etc. thus in our age technology is almost exclusively dependent on controlling systems by means of the ideas of classical physics, ie 19th century science, but this fundamental issue of "science not developing," and thus being irrelevant, is hidden by the propaganda system.

However, when the authority of science is narrow and fixed then it can be used to measure "intellectual-value" on the absolute scale of truth, as identified by today's overly authoritative, but nonetheless, practically irrelevant science and math, so that only a few people measure-up (in the educational competitions, competitions based on absolute truths) as being intellectually capable people. That is, it is a propaganda system model for the necessity of having an elite group of people defined in society.

In the context of many-dimensions, a (system) containing coordinate-set of many (independent) measuring types, if the operators, used to identify a locally measurable property, (or the system containment context in regard to the system's [locally linear] measured properties) is not commutative then one cannot have a valid description of a stable system, or one cannot be describing a stable (well defined) pattern. Which means that if one has a measurable description based on a set of many measurable properties, then these measuring sets need to be independent of one another at each point, in order for the description to remain (to be) quantitatively consistent, and, thus, for the described pattern to be stable, eg if a measure of length is contained in a set of coordinates then there is a set of (local) coordinates where by the length-function depends only on one-term for each variable of the coordinate set in regard to the length-function's representation.

13.

Power of Banks-oil

Comments about the;
Mc Murtry paper at, website: globalresearch.ca 2-19-13, Moral decoding of 9-11:

J Mc Murtry has written a paper in which he makes (what, one might believe, is a correct claim) that all of the sanctioned actions of government and the justice system are about making sure that the oil-banking (and military attaché's) big businesses continue to grow and make ever more money, whereas all the governmental and justice system power has been channeled to help the profits of these industries (the basic power structure of society is used to enhance the pursuits of the biggest banking-oil businesses),

This is the statement that "the world revolves around money"

McMurtry points out that the National Security Act of 1947 has been the blue-print for assassination and exploitation by the state for corporate advantage as well as the "national-security defense," for corporate or state interests being above the law (where the state has come to represent the interests of corporations not the interests of the public), and this was (according to McMurtry) based on the structure of the Nazi-SS's mandate to attain world domination (this seems to be something which could be quite true), but McMurtry claims that this power given to corporate interests of D Rockefeller, came from D Rockefeller funding a failed academic Leo Strauss, where Leo Strauss is, apparently, a moral philosopher (as is Mc Murtry) and Rockefeller and Strauss did this in cahoots with the Fascist business-network of the "age of WW II," where (according to Mc Murtry) Strauss provides the academic cover needed to support the idea that society is better-off, if the "banking-intellectual-investment social-class which possesses superior intelligence" has direct control over society. One might look at history and say that that mandate for European culture was established in the Roman-Empire, and it has never been changed.

Strauss is indeed a great villain
However apparently it was not Leo Strauss but rather Lewis Strauss, who conspired with E Teller, to turn physics departments into bomb engineering departments

Mc Murty is essentially claiming that Leo Strauss is a failed academic upheld by funding from D Rockefeller and thus the cooperation of government with banking-oil interests is quite flawed apparently because Leo Strauss was an inferior intellect.

Mc Murtry seems to be unaware that the academic departments at universities have long-ago (the 19th century, eg famous physicists claiming that man will never be able to fly, claims which were made just before the Wright brothers first flight) been captured by the investment community, and this capture has led to a particularly narrow way in which to use language in university departments so that the (narrow, but technically precise) language is set-up to turn the university academics into the intellects who are to serve the vocational interests of the big businesses (the investment social-class)

[Faradays and Tesla's were badgered and exploited by these academics, as well as by the business community, whom stole their ideas through the patent process so as to make a lot of money]

It should be clear that the reason for the US society to be in the position so as to back-up the banking-oil big businesses was that the Supreme Court unlawfully declared that US law was to be based on: property rights, minority rule, and that business contracts favoring the rich were to be upheld, this was done by J Marshall's supreme court around 1800.

Jefferson described the absurdity, in regard to the meaning of law, of these acts by the court, but nonetheless the "letter of the law" and "rule by the richest" has been the way in which power within society has been organized ever since, but it really simply re-establishes the basis of law of the Roman-Empire.

and
This needs to be corrected within the US.

Mc Murtry is a typical person one encounters in the propaganda system, where a failing of the governing system is exposed, and its cause essentially identified, namely, that D Rockefeller was controlling information in the propaganda systems he influenced (as well as controlling the governing institutions), but it is really an attempt to claim that the academics are not responsible for "the failing of knowledge to be related to practical creativity," apparently the academics believe that academics and poetry and lyrical knowledge, or science-fiction knowledge (or plot developers), or the world of illusion; is the basic realm of the academic.

In fact, this is what the academic community has become, a bunch of highly technical dogmatists, whose descriptive structures eventually become better suited to describe illusions than practical structures which are observed to exist.

For example, the main problem of physics is describing the stable properties of many-but-few-body systems at all size scales, but this is not trumpeted as the main problem facing physics, instead there is the big-desire to unify physics . . . , (this is an example of how the propaganda system used Einstein as an iconic-intellectual-illuminati so as to distort physics to provide physics with abstract-complicatedness so as to make science more ineffective as a source of practically useful knowledge, so as to stabilize the investments of banks in complicated instruments), . . . , based on a descriptive structure (of indefinable randomness and the quantitatively inconsistent non-linear patterns) which has no relation to any stable-system except this context having a relation to chaotic transitioning systems transitioning between two relatively stable states eg the states of stability which go without valid descriptions but, according the McMurtry's of the academic world, the world would be better if only "there would be more stringent control of valid expression by means of peer-review," where the high-value of an academic can only be measured if intellectualism is based on very narrow dogmas, ie the academics take the old-role of the church, which, due to its dogma opposed the new ideas of Copernicus, ie the old-dogmatic church opposed the new ways of using language as were presented to them by Copernicus.

Thus one must say to the "great academics whose academic value can be measured within the authoritative dogmas of the academics of the world" "please solve the problem of describing the stable properties of the fundamental physical systems characterized by the property of being systems which have 'many-but-few-components but which nonetheless form into very stable physical systems'"

This problem has been solved, and it was solved outside of the dogmas of peer-review, and outside the sanctioning of peer-review, outside of academic measures, yet they (or rather the propaganda system) refuses to pay attention to these new results.

That is, it is the peer-reviewed academics who are the ones who pave the way for the owners of society to continue to dominate the creative efforts of the "educated" society.

Furthermore, Tesla would not publish for peer-review, and Tesla, by following Faraday, brought to us the electronic age, ie two curious people who are kept at the margins of the academic structure which serve the corporate interests so effectively.

Furthermore Mc Murtry does not provide any anti-dote for the social condition, which he describes, other than his implicit statement of "believe-in and follow the peer-reviewed academics, since they are the superior intellects," and "the intellectual-bankers such as D Rockefellers should not be followed."

But how is Mc Murtry different from D Rockefeller?

The ideas of Leo Strauss which trouble McMurtry

Ruling moral absolute supranational sovereignty of an intellectual elite and bankers . . . limitless capital accumulation . . . . the highest right and moral duty (ie money makes the world go round)

But people want to know and create, and they will do so without reward if the common welfare of everyone is taken-care-of in an equal manner . . . , ultimately this leads to a truly free-market . . . . , whereas investment is about fixed social relations, and the best-bet for investment, eg stealing the inventions from the creative sources for selfish gain (when inventions work, then steal them, so as to make-money from them, this is patent law) and then limiting other creativity, creativity controlled by investment in a limited viewpoint about knowledge and what can be created, and how it can be created

Capitalism's success is about providing a productive force for development . . . , K Marx

Mc Murtry fails to identify the state of academics, and fails to provide an alternative to the investment bankers, other than one should follow the academics who have proved themselves to be of merit (by means of a limited view of knowledge upheld by peer-review), but The trouble with claiming that there is a recognizable value associated to intellectuals is that then one must then determine "who is the intellectual superior McMurtry or Rockefeller?"

Distinguishing such intellectual value by means of the value which is assigned to all aspects of society, [as well as in the context of survival of the fittest] then of course Rockefeller is the superior intellect, this is what the propaganda system "will tell us."

Furthermore, since academia is a bunch of departments which support the context of the Rockefellers ruling society eg economics and sociology and psychology as well as physics being weapons engineering departments

How is this different from indy-media editors or N Chomsky, are these social commentators any different from D Rockefeller so that these intellectual heavy-weights (at least in their own eyes) are simply quibbling over "who is the superior intellect?"

Here the essence of a quote from a young S Jobs is quite telling, "the giants of big business are of quite ordinary intelligence"

On the other hand there are the where here Hollywood moguls who have sided with the business titans by stating that it is those who know how to implement extreme violence are the ones who do the most good for society

Propaganda is like an authoritative parent where the introduction to M Ruiz's "The Four Agreements" describes a person who self-regulates themselves . . . , based on what their authoritative parents have taught them about high-social-value . . . . , so as to conform to a level of high-social-value which no one is capable of achieving. This is what: Rockefeller, Chomsky, McMurtry, and various editors of both main-stream and alternative media outlets profess to have attained, namely, a high-social-value which these people certainly do not achieve, yet the media grants to them this high-social-value

Some truths which Mc Murtry seems to be expressing (his sentence structure is never direct and always vague).

Private capital to rule the world with the US military as its instrument of force and terror (or better yet, with the US military and the US justice system as its instrument of force and terror)

Only corporate and private money have any rights, and all others are obligated to serve these interests

Reasons applies only to self-maximization, wall street led reason associated to private money is defined to be the ultimate good globalization

How is this failed viewpoint

The honesty and truly in favor of life over capital accumulation is to be commended but what alternative does he give: "is the world of men truly a reason vs. capital accumulation game?" or is life about knowledge and creativity and about the great limits of reason defined within a language whereas there need to be many languages and many viewpoints about practical creativity not adhering to a fixed social condition, where today's social condition (eg life styles) are essentially the same today as in the days of the Roman empire where Machiavelli wrote the handbook for the power of bankers around 1450 the same time in which the Constantinople empire collapsed

The western tradition . . . , up until the US revolutionary war (wherein equality was proclaimed as the basis for law) . . . , has always been the tradition of "might makes right" where in America the colonies were allowed to govern themselves until they possessed "economic value" and through Hamilton and human vanity, European thought . . . , and the value defined by banking investment . . . , rather easily won-out over law based on equality

On dandelion salad
The movie, "Obey," Based on C Hedges "Death of the Liberal Class" The choice given by the movie is rebel or Obey.

But the educated class is steeped in illusion, and they Obey the illusions, and oppose a valid basis for a new world viewpoint, a basis for rebellion which is both true and in direct opposition to the wishes of the corporate masters who dictate truth to the public through the media, including to the educated people who seek a valid rebellion.

Is it the D Rockefellers? or Is it the academics? who "know the truth," but "everyone else does not know the truth."

. . . , The police are loyal to a "proven (absolute) truth" the truth of money.

Thus, the editors and those who obey the corporate command by adhering to the university expressions of ideas who's dogma is identified by the propaganda system as being correct

Who decides the "truth" for the propaganda system?
Is it
The academic's or professors
Or
The beliefs of D Rockefeller?

Whereas Rockefeller's ideas are what get expressed in the culture, and the professor's ideas are only slightly different from Rockefeller's ideas

But the best answer is that both of these authoritative figures is quite wrong, where one is selfish and the other is obedient. Furthermore, the experts picked to judge ideas are picked due to their social badges (won within narrowly defined academic competitions) which are all essentially provided to them by the Rockefellers, and both types of thought (academic and business-based) are expressions of illusions. That is, the theoretical physicists is not providing any practical help to society, and the CEO's are looting society and destroying the earth

Those who judge value within society are themselves expressing the idea of elitism

The choice is equality
Or
Elitism
Or
Alter-elitism, where the illusion of choice within society is that people choose between either the elite or the alter-elite, ie those who are for science, knowledge, and higher-education, yet they oppose those who express ideas different from the experts, and they do not accept equality and they do not partake of the act of determining truth for themselves.

14.

The reach of JD Rockefeller

The curriculum at university departments is determined by the investments of big businesses in equipment and instruments, which are used in their production systems, as well as the advertising propaganda about the life styles and the lobbying which best suits the marketing strategies of the very big businesses for their products, and this advertising-propaganda affects the history and philosophy and sociology departments etc at universities.

Moral philosophy is an expression of arbitrary value
For example the Puritans and the Mormons would not accept each other's morality, or expressions of arbitrary moral values, but both pious societies saw that it was in their best interests (when the Mormons were led by B Young) to both steal from and exterminate the native populations, so that they could "prosper" ie fit into the trade practices related to banking investments of the greater European society.

Historians often claim that these life-opposing Puritans and Mormons are great examples of disciplined people "doing well" in a "free" social environment.

Yet, it is (really) another example of a community adapting their lives and beliefs, so as to walk in lock-step with the leaders of violence, and provides an example of the violent results which occur because of a narrow viewpoint which a community decides to follow (concerning life and violence), where the personal stability of a few will best be upheld by violence perpetrated by a particular group on other people, and this is called leadership, but it seems to be conforming to the dictates of those who one considers to be the bigger bullies.

The Quakers had some courage, while the other protestants (and similar types) sided with empire and the banking investments of the age.

When philosophers debate about the property which allows a material-yard-stick to measure the same value of a marked-off space from one day to the next, and then the philosophers wonder "how one 'knows' the same results will be obtained tomorrow?" but where it is assumed that thought itself cannot cause this, then "what does cause such a consistent set of measuring properties?"

This is used to distinguish induction from deduction.

But this is also about word usage.

One could assume metric-invariance (from day-to-day), but usually it is assumed that it is the properties of the material from which the ruler is composed, which is the basis for "that which allows" such (inductive) measuring agreements (from day-to-day).

Chemistry, in the 19th century, was finding that chemical reactions were affected by probabilities of component-collisions, then radioactivity was found, and the idea of probabilities of random-collisions became a model which was related to "rates of nuclear reactions," so that quantum descriptions, which "reduced material systems to elementary particles," has remained based on random-collisions of the reduction of material systems to elementary-particles, since this form of thought is used by the interests of the ruling class (in weapons production).

The technical experts, who work with narrow precise language, always compete to be "correct" within their narrow interpretive structures, where this obsession with "being correct" is an illusion which has been artificially created within society so that the competing experts (competing within the context of narrow authoritative dogmas) are picked (by the ruling-class) to work-on, and maintain, the complicated instruments in which the ruling class has invested, eg building nuclear weapons.

There is no "correct" within the attempt to precisely describe the observed patterns of existence.

This is because "for precise descriptions of measurable patterns" there are always existing patterns which cannot be described by the given language (this is Godel's incompleteness theorem), where this means that precise language needs to always change at the level of assumption, context, and interpretation, etc. ie there is no "correct" within the "descriptive forms of 'knowledge'" eg descriptive truths are verifiable and practically useful (measurable descriptive) truths (but big business fixes these descriptions so as to form dogmas, eg peer-review, which are used to cause science to serve the interests of big businesses).

One can consider the context within which induction can operate (or be relevant) the assumption of a space which is metric-invariant would allow for consistent length measures on a region in space.

What property allows induction? Is it mental? or Is it an assumed property of a descriptive context?

The current assumptions (or descriptive contexts) are materialism which is reducible to elementary particles whose relation to space is a set of random spectral-particle-events in space (where spectral could be interpreted to mean mass of a particle) where the particles are in a many-dimensional internal particle-state space, which must be partitioned from space-time to allow for the idea of materialism and the higher-dimensions (according to string-theory) are small and un-seeable (undetectable), where in this model, both particle-interactions and the large scale properties of space, are both determined by non-linear patterns, but non-linear patterns are quantitatively inconsistent patterns, and they lead to chaos, and thus, there do not exist any stable properties, except the one stable property of random particle-collisions, but the particle-collision is a geometric model placed into a random quantitative structure, ie it makes no sense.

Or

One can consider metric-invariance in a many-dimensional context where both interactions are dimensionally dependent, as well as the definition of material-components, but reduction stops at the level of the natural stable shapes for metric-invariant spaces, where these natural shapes are linear, stable, and identify a controllable (or stable) context. The apparent randomness is derivable, and the shapes have distinguished points, about which an apparent particle property would emerge. Furthermore, in 3-spatial dimensions the material-component interactions are spherically symmetric.

But

This second context is much more complicated and diverse than is the idea of materialism. Furthermore, materialism is used in society to divide people into a "science (belief in materialism) vs. religion (a pious but false claim to be above materialism)" set of opposing groups.

Where this has been divided further into communist socialism (of Marx) vs. democratic capitalism of "free-markets" (of A Smith) both hierarchical (and/or oligarchical) societies based on an arbitrary structure to propaganda.

Namely, these oligarchies are both essentially the propaganda structures of moralistic religious communities, where, for capitalism, Puritanism has supplanted Quakerism, ie the free-markets are highly controlled. That is, materialism is central to the propaganda structure within society, for oligarchies.

The left which is given voice (allowed onto the media) are the one's who most protect the established intellectual order, since they have competed, to get their very limited voice (as journalists or as experts), so they believe themselves to be superior within the intellectual contexts in which they believe they have been competing.

The left "who have voice" are those who try to fit into a (wage-slave) value structure defined by corporate interests. Yet it is a set of interests which are cloaked in the noble icons of:

Science,
Law,

Economics,
Sociology,
Psychology,

which have been adjusted to fit into vocational structure, of the corporate interests, for the public.

Nonetheless, the science and math community claim that they are seeking truth in a free and unbiased way, but because they are competitive, and they are advertised as those few "top intellects" which can measure-up to the standards of society. However, those high-standards of society are determined by corporate interests, and they are carefully presented over the propaganda system, ie the competition of a peer-reviewed authority is a narrowly defined dogma, which allows measuring of value within the narrow confines of the dogma which best serves corporate interests, so the "truth" to which these intellectual adhere is a very biased "truth."

That is, the condition of an authority "always being 'correct,'" is about being within a narrow dogma whose main purpose is to maintain and adjust the (often complicated) instruments in which the ruling class has invested, ie the best-bet for investment in a very conservative mind-set of an investor whose investments are better served in a very stable social and technical context.

The so called great advances in technology are all based on electromagnetism and thermal physics of the 19th century. For example, the cell-phone is about the optical and acoustical relation of digitally coded information processed by a very organized electronic switching devices, which was already established (as computers) by 1935, except for the micro-chip of the Texas instruments at about 1955, but nonetheless, all based on 19th century: thermal, voltage-waves, and electronic technology, while quantum-physics, particle-physics, and general-relativity has all been applied to weapons technology of the explosion, yet fundamental mysteries, such as the stable properties of most of the fundamental material systems, which allow our realm of experience, all go without any valid descriptions.

Nonetheless, the loyal opposition of the left . . . . , ie those with high-value social positions within society, eg professors . . . . , are oblivious to the way that the corporations are allowed to manipulate (for the selfish interests of the corporations) the so called "search for truth," which our society's, so called, "top-intellects" pursue at universities. And the intellectual left is the "great protector" of the ideals of science.

In regard to global warming, the attack from the right deals with the most weak aspect of modern science, namely, its grounding in randomness, especially, where randomness is very difficult to define. In fact, attacks on all the different ways in which data can be interpreted is a valid exercise within science, especially a science based on indefinable randomness . . . ,

(is earth axis-wobble and the relation of CO_2 to temperature increase [due to the green-house effect] enough to counter all that the right criticizes about the statistics? The real issue is that business is supposed to adapt to society, not that society should adapt to a dominant business ruling class), . . . , but in fact best tactic the left could take, is to attack the indefinably random basis for university department physics, since this paradigm is only relatable to weapons-engineering, while the main problem of physics, ie the cause for great stability precisely modeled in a measurable descriptive context, go unanswered.

Indeed the criticisms of the religious right concerning the possibility of the development of complex highly ordered systems by means of a description based on probability, as is Darwin's evolution, actually developing.

This is actually a central question of science, but the commentary of the expert authorities concerning . . . , "science and materialism" vs. religion (where religion is really also about materialism, since they do not embrace math models which transcend materialism) . . . , has been carefully selected by the business owners of society within the propaganda system, so as to enter into the education system, where the ideas about the interpretation of history (or the interpretation of science) is central to controlling the public.

That is, commentaries by so called authoritative experts within the media and within educational institutions has created a false viewpoint concerning "academic truth," so that editors and dogmatic-experts feel that they can exclude solutions to some of science's most difficult problems based on their social authority (a social authority granted to them and protected for them by their relation to helping corporate interests).

This is the same use of social authority which is used by big business to control (through propaganda and investment) what is called science in the university "science and math" departments, ie it is arbitrary use of prejudice, essentially put into place by the propaganda system, and is the descendant of the Holy-Roman-Empire, ie a marriage of violence and an authoritative propaganda system.

The social solution to this dilemma is the one offered by the US revolution

Namely, we are all equal, we are equal creators, we are equal free-inquirers (the nature of precise language, to be used for creative production, requires this viewpoint), and the concern of the government is the common welfare, which needs to sorted-out in regard to each of us being equal creators, so the idea of "the common welfare" is based on equality.

It is either this (solution), or the destruction of the world based on selfish concerns backed-up by extreme violence.

15.

State failure

One has within the propaganda system complete baloney which represents the owners of society and the things they produce, with 19th century science, but otherwise it is a society whose institutions are almost all (except for the propaganda system) complete failures

The media is presented in terms of left and right

Within the media (the propaganda system):

The right is for the rich hierarchical elitists: authoritarian, of course very judgmental, and dishing-out "arbitrary moralistic crap," all their positions are based on extreme violence. They are a bunch of bullies, who are protected by hiding behind the skirts of the rich, while the poor associates, who are also on the right, are basically prone to violence and coercion, where the rich-bullies control their poor right-associates with propaganda. The right-media can discuss ideas based on vague principles and they talk about generalities and express cliché's, they are allowed to say anything which supports the status quo, and they are allowed (or required) to have the more commanding and authoritative voice, when righties talk to lefties (within a media channel)

And

The left are the intellectual-bullies, who are disciplined, aggressive, competitive, obsessive, and stand-up for the personality-cult which represents a scientific truth, where the ideas expressed by these science-personalities are the ideas which are used within the principle production businesses of the owners of society, eg weapons and communication systems. However, the left always yields to the right in discussions, and the left must always use long-winded arguments, which do not fit into the media's way of talking quickly and glibly (since the entire context of the propaganda system is set-up to support the right).

Those on the left are less likely to use violence, but they believe in inequality, and subsequently they also believe-in personality-cult, and they believe in "the system" (but "the system" is all about the illusions of the propaganda system), and they uphold and praise the law (which, in fact, is corrupt), and they express a need for protection (of the elites from the public).

Because they are more disciplined and more moral, in regard to violence, and they defer to reason (but they believe in a dogmatic form of thought), all of their liberal ideas are erroneously interpreted by the right to be ideas in favor of an oligarchy based on the government ie some form of communism.

Thus, the media harps-on the icon of "big government being bad," ie "big government" is portrayed by the right-media as being about communism. But as communist-China shows, communism is, essentially, the same as the oligarchy which is called capitalism, in the west, but capitalism, in the west, is really central-planning based on the selfish-interests of the owners of society, where the US government serves the needs of the big property-owning oligarchs of capitalism.

In fact it is the Declaration of Independence which defines the basis for US law, and it claims that all people are equal, and that the government is dependent on the public, and (in the Constitution) the government needs to serve the common welfare (of a society of equal creators), ie free and equal people who can express beliefs and relate descriptive knowledge to practical creativity (the first amendment of the Bill of Rights), and it is only then that a truly free market can be established.

Whereas efficiency is only about a very narrow way of relating products to a narrow vision of life-style, a life style which depends on certain products, but then this is not capitalism. This narrow structure of a stable society allows the investors to usurp the powers of the government, and life-styles come to be about investment practices of the central-planning committee of investors, subsequently supported by the state.

It seems that efficiency is the only attribute which big-investors can provide to the world. So why are they so important? Because of perversions of the US law, by their being big property owners, means that they are of ultimate importance.

That is, illicitly, US law is based on property rights, and minority rule, and rule by violence and coercion. That is, the US revolution was a change from a European outlook (established by the Roman-Empire) based on property and violence, and the major change was that law was to be based on equality, and it was the egalitarian Quaker colonies which best represent the idea about the nature of the US nation at the time of the revolution, where the Quakers made agreements with the native peoples, agreements which the Quakers honored.

This section is for the ridiculous editors of all the managed and infiltrated media outlets such as indy-media (except for one outlet which is now jammed), and other more controlled publishing outlets, editors (and peer-reviewers) who say arbitrary foolish things (which adhere to dogmas) and make idiotic foolish judgments about "truth," so as to serve the propaganda system . . . , news is any arbitrary story (which is allowed on the media and labeled as news) . . . , the left must represent such "news" as a documented truth (accepted by arbitrary authorities of academic institutions, but these are authoritative institutions which have failed in terms of being institutions which, actually, can determine the truth), since, supposedly, the left is about people who mislead the public (provide them with free things {which the earth provides for mankind in a free manner} . . . ,

[while the right steals the property and steals the culture with patent law and then sells material products based on the creativity of the culture based on fixed prices within a controlled monopolistic market structure, which it claims the right {the property owners} has the right to do, because the property owners can produce things more efficiently.

Unfortunately, the assembly line and mass production can be done by anyone, and what such a system really causes (because of its narrow obsessive focus) is the destruction of certain resources within the earth due to monopolistic and domineering social powers, so that this destruction is allowed for big-money by a (big) government which has come to serve the owners of society (where now the "big government" is "big on coercion" which serves the interests of the owners of society)]),

. . . , while the right can say anything on the propaganda system, and the right mostly states things which mis-represent "the truth," . . . . , eg science, math, economics, politics, the law, and mostly arbitrary moral judgments resulting from a partnership of the-owners-of-society's with the hierarchical and arbitrarily authoritative religions (eg the Puritans) [where the partnership is based on a funding process] etc, ... , to the public.

But (implicit in the assumptions of the media, since the media is owned by the right) is that the right is providing the public with the correct and moral scenario concerning life.

Let one count the ways that the major institutions of society have failed:

0. Education has failed since it teaches the absolute authority espoused by the propaganda system this is a result of controlling the communication system especially one centralized by means of electronic communication equipment thus education does not follow Socrates and equal free-inquiry in regard to inquiry's relation to creativity nor does it follow the equal democratic ideals of science as born from Copernicus and descriptions relation to freedom of belief and truth being both measurably verified, as were the ideas of Ptolemy, but also practically useful in regard to practical creativity, nor does it follow the obvious need to always change language based on the great limitation which a precise measurable language possesses as demonstrated by Godel's incompleteness theorem

1. Science and math; no new inventions due to the information obtained from descriptions of physical systems based on physical law (except for the laser) as expressed for: quantum physics, particle physics, general relativity, and all other speculative theories derived from these three descriptive contexts, which are claimed to be the basis for physical law,

    Failure of indefinable randomness to identify valid expressions for risk despite its, supposed, rigorous setting,

    Failure of math to identify the idea of "a stable pattern" as being a fundamental math construct, and instead focus on convergences within continuums, continuums, extending number fields, indefinable randomness, function spaces related to non-linear partial differential equations, etc.

    Only further developments of the technologies provided by 19th century classical physics eg from electromagnetism, Newton's gravity, thermodynamics, statistical physics, where quantum theory only provides a vague basis for quantization of phase-space in statistical physics, where electronics is about speed of switching (on-off states), and nuclear weapons are about probabilities of particle-collisions, ie the 19th century chemical model from statistical physics which is used to model the rate of chemical-reactions, especially in an explosion,

    Note: The transistor and micro-chip are all produced using the properties of classical physics, mostly thermal physics, and physical chemistry, and optics.

1b. Failure of biology, since its "model of development" is based on indefinable randomness, so it cannot account in a useful way for the development of complex systems, so it is not a description which leads to useful results.

2. Failure of economics as a valid description based on measured values (even though money is an obvious quantitative structure). This is clear since there are a few people who have enough money to change the nature of economic structures in a matter of hours, by changing their investing, so "what is one to believe about the quantitative structures which can be so suddenly changed?" ie the quantitative models mean nothing.

    Furthermore, business is based on (business) risk (which is indefinably random) and there are no valid quantitative models of indefinable randomness.

    What there is, is the stable social structures which have remained stable since the Roman-Empire.

    It might be best to define a modern market-system as a society which has a fixed structure of providing for: food, shelter, water, and transportation held in a fixed relationship due to extreme violence and a controlled market-system has been able to exploit these relatively stable conditions (thus, a controlled market) of necessary social interactions within a society. This is what the Romans did, form a stable set of social interactions.

Wherever the Romans were, the Roman solders held society fixed by violence, but they also built: roads, water systems, and sewers, so that food and shelter fit into a fixed social order, so this fixed order could be related to a "market system."

The value of what we call "capitalism" is all about the violence which holds the functioning structure of a society fixed. The value of an economy is based on (or defined within) this stable functioning process.

The Roman structure for society was held fairly stable in the European region until about the 1400's wherein the bankers started to take over the market traditions of organizing society and its fixed way of social organization around food and shelter and transportation (and water) and trade, etc.

Economics sociology and psychology are not sciences which have any valid relation to measurable properties. All the relations in these areas of study are based on statistics and lame correlations associated to indefinably random statistics. But these institutions are associated to characteristic behaviors of their practitioners, and these behaviors are about manipulating and deceiving people, basically to serve the interests of the owners of society. Social forces act on personality types in particular ways, where these behaviors can be observed.

3. Failed propaganda system, since it has created a society which is far too narrowly focused, and too focused (or more realistically exclusively focused) on the protection of the few super-rich and their property.

The main point of the propaganda system has shown itself to be wrong. Namely, inequality is the main message of the propaganda system, and inequality leads to destruction of society and the earth as well as leading to the destruction of valid knowledge.

Furthermore, inequality needs to be upheld by extreme violence, so all the ways in which any form of coercive violence can be incorporated into society is welcomed, eg gun rights vs. gun control (is about using violence to terrorize the public), etc.

Furthermore, the propaganda system is based on academic and journalistic expressions of the ideas which aid with helping the main-stream of corporate interests, so the expressions of the, so called, "opposition" (the enlightened expert propagandists on the side of the left) mirror the ideologies of: science, math, biology, economics, and sociology etc whose measured constructs are arbitrary and meaningless, but nonetheless serve the interests of very big investors, and the relation of knowledge to creativity within a society which serves the interests of the very rich.

Thus, the discussion (by the left) remains well within the language of the propaganda system as set-up by the right, so the opposition is actually helping to keep the propaganda system within its correct framework, wherein it wants to stay.

4. The failure of spying (failure of the ironically named, Intelligence),
    they did not see the collapse of the soviet union,
    they did not see the economic collapse,
    they did not see 9-11 (or at least, that is their story), and
    the unnecessary confinement of ideas due to their infiltration into the left organizations, and subsequent destruction of ideas (or movements) which oppose the ideas and social structures which support the super-rich, and

    Thus, there are no new ideas and this leads to the collapse of the economic system (the engineering of the modern "new-Rome" is based on the ideas about controlling and manipulating electromagnetism of Faraday and Tesla, as apparently the engineering of "old-Rome" was based on the arch and cement)

5. Failure of law and government.

    The basis for social class-warfare is waged by the courts and politicians against the citizens, this is clear, and these actions are interpreted to be perfectly legal, and now there are high-level personal, ie managers, who serve the super-rich; managers who have come to be above the law, eg high level managers of the super-rich, ie the super-rich have always been seen to be above the law since murder instigated by the owners of society, eg in strikes by unions, has always been interpreted to be legal (apparently it is legal under the law of the Roman Empire).

While . . . . , politics is all about being a part of the propaganda system (the propaganda system owned and controlled by the right, and based on owning and controlling electronic instruments and controlling the regulations about message distribution, which ensure that particular communication systems [owned by the very-rich] are the dominant communication systems).

The propaganda system is represented to the public as a single, solitary voice of absolute authority, which espouses nonsense and arbitrariness, which the public is required to accept as absolute truths, because the public is terrorized by the legal and political system, and their wage-slave status (enforced by the justice system), and they are coerced by both the economy and the academic system, where all of the positions in these institutions are about bullying (intellectual or by violence) which is based on an arbitrariness, but these institutions (politics, law, and the universities) consistently fail to do the actions which are required for "the intents of their institutional name-sakes to succeed."

For example, one cannot create anything of practical value based on modern physical science and math, except maybe calculations of business risk, where these calculations are made by MIT

PhD mathematicians and physicists, whose main characteristic is that they fail to properly identify the risks, ie indefinable randomness and non-linearity are a failed math constructs.

Unfortunately, due to the mis-representations of truth by the propaganda system, the public is led to form a religion of personality-cult which follows the scientists, and their so called scientific truths.

This is done in order to ensure that there will be no competition for the super-rich few in regard to creativity coming from their expert math and science authorities, while chemically based biology studies are a bewildering mess of statistics (concerning molecular makers, ie identifiable molecules) and confusion, where successes are measured if 10% of the patients can be affected by a molecularly based plan for health cures.

The public is led to believe that they, the public, must be wage-slaves to support investment, but investment is narrow and adheres to the fixed social structures, essentially, following the fixed way in which civilization was organized by the Roman-Empire.

This narrowness leads to social failure, even though the academic structure is set-up to gradually develop complicated viewpoints (contexts) associated with complicated instruments (into which there have already existed investments) which serve (have served) the interests of the investors, and then any new developments (in regard to improving existing instruments) are stolen from public institutions, so as to result in the (perturbed, or adjusted) ideas having a relation to the investment structure of society, but the science and math ideas upon which the instruments are based are not valid science laws (in regard to addressing the descriptions of the observed (stable) properties of stable micro- and macro-systems) due to their relation to such a narrow viewpoint of serving investment interests (eg their relation to 19th century physical science).

6. The failure to protect the earth, which is a result of a dysfunctional propaganda system, which is only used to support the power of the few very-big investors, ie the message is "people do not count," ie "violence is the only way in which for a person to actually stand-up within the world," but the violent-person must support the power of the very few

Globalization is all about "the very rich few" gaining ever more power at the expense of society, so that society must conform to the investments of the very rich, and this is allowed by greater military prowess, both controlling the violence of other institutions, eg other nations and/or groups etc, and controlling the actions of the public within society ie greater surveillance so as to stop unwanted ideas, and surveillance is only marginally about stopping violent opposing groups, since these groups (usually on the right) are easily manipulated through the media.

The society is moving away from national (economic) flows of value-of-goods, to global flows-of-goods because of the failure to develop new contexts for practical creativity, ie the failure of physical science and math.

The very rich are supported by both

(A) a set of pathological people who support the rich for personal advantage, these people are manipulated by violence and the lure of social domination, and they manipulate others in the same way, and

(B) by the intellectual opposition, who are disciplined, and not prone to violence, they are the mirror-images of the very rich, but they are competitive and they argue against the policies of the rich, not the rich themselves, ie this group of people wants to be in the positions of the rich, since they too (also) believe in social inequality, since that is what is taught in biology, eg evolution (survival of the fittest), where this was originally taught at the behest of the ruling mercantile class in the 19th century.

That is, those in the opposition, who are "given voice (allowed on the media)" are those who accept the absolute basis for the authority of the very rich, an idea which was provided to them in the education system, eg accept this dogma and compete or become marginal (they are, henceforth, judged to be both not accurate and not competent). [Copernicus would not have been burned-at-the-stake, instead, today (2013), he would be marginalized and ignored by the intellectual class, the intellectual-class who serve the owners of society and who guard the truth for the owners of society]

That is, the opposition supports the intellectual basis for the current social structure, they have been indoctrinated and rewarded to believe all this baloney.

Apparently this spawned the idea (on the right) that some of the best propaganda, is the propaganda which expresses ideas opposite to what the rich want:

1. Against big government

    The rich are dependent on big government, eg regulations and the military and the justice system are extremely expensive and require big government, and the propaganda system tricks the public into big government by means of social issues, eg against abortion or against big government, but this is done by means of espousing being against big government

2. Want accountability in education

    The education system has been managed by the military personnel since the end of WW II, and the education system has provided the work-force perfectly tailored to corporate interests, so accountability means they do not need education any more, science is mired in its own stupidity, and is irrelevant to either further development, or to causing unwanted competition (thus necessitating having control over science), and/or they (the right) want the money which exists in educational districts, ie the money in property taxes, so they want to destroy education.

Furthermore, the calculation of business risks was a big component of the economic failure, ie big science and math failed big-time, so they are to be punished.

3. Fiscal responsibility

How can this be taken seriously, they want big military and a very expensive coercive justice system attacking both the world and the public, respectively, the rich are very paranoid about "who is, in fact, is allowed to run everything, namely, themselves" This social structure is to be held in place by extreme violence.

4. They are concerned about protecting the rights of the people

ie they are opposed to gun control, where gun control is really about the manipulative side of violent coercion: the A Bremmer's, the J Laughner's, the Lee Harvey Oswald's, the Sirha-Sirha's, the J E Ray's etc ([almost] all attacking the left, or an intellectual institution, or a intellectual-leftist institution), all being manipulated by their secret-agent handlers, in a world of wall-to-wall spying.

Furthermore, the only human-right they (the very rich) are really concerned about is "the right" to own and dominate property, from which they derive their social power. They support property rights over life, and this is not explicitly in the law (the constitution, where the constitution includes the Bill of Rights, however, the constitution is about minority rule, ie the Constitution is opposed to the Declaration of Independence)

The strongest statements about US law are in the Declaration of Independence and the Bill of Rights and none of these laws are (or ever have been) enforced.

The US justice system has no valid legal standing.

People have the right to religion, as long as the religion stays in a fixed structure based on hierarchical authority and domination. [This is how the first amendment is now interpreted]

Thus, the rich get the public to support an authoritarian religion over religious freedom but religious freedom is really about the freedom to believe.

Free speech is now defined within the letter-of-the-law to mean the right of "the owners of media outlets" to express what-ever propaganda they want to express, so as to be expressed by the paid and funded personnel, who it is claimed, must only report objective truths, matters of fact, to be interpreted in the contexts of an intellectual structure which has been taken over by the interests of the ruling class and its propaganda outlets, ie to be interpreted in any arbitrary way which supports the interests of the investment class,

5. They are pro-life

    But they are all about violence and inequality and domination, ie they are at war with life, where life takes the form of both "the public" as well as "the earth,"

6. They are against the left, namely the "left which has a voice in the media," but the "left which has voice" supports all the intellectual framework which, in turn, supports the social and creative framework upon which the power of the very rich depends, where it depends on (through) the market (of these very rich investors) which supports these very rich few

    The "left with voice" never express the belief-in equality. In fact, they almost always support the idea of inequality, and that there is a need to develop the merit (of people) which is needed to run both the state and the learned-institutions.

    Unfortunately, it is now seen that these institution, especially the intellectually dependent ones, are all failing in a major way.

    In fact, it is the job of the media to show that the intellectual institutions are all developing all sorts of new instruments, but unfortunately they are all instruments which are based on 19th century science and math (though 19th century math is more robust than modern math).

Nonetheless, the left expresses the ideals that most Americans have, namely, the ideas expressed in the propaganda system, so the police departments and spy agencies all need to destroy any organizations which express the ideas of those on the "left" (the left which has voice), since this is the group (the left with voice) which is best qualified to supplant the very rich, and still remain in the same social structure,

However, it is a good prediction (based on the current attitudes of the "left with voice") that this "left with voice" (if they were in power) would turn around and support all the same ideas and structures which the super-rich now support,

7. They (the very rich) are against taxes which they put into an economic scenario which is upheld by the propaganda system so the public accepts this, obvious baloney, about how it is best not to tax the very rich, but in fact the very rich support big taxes applied to the relatively poor (relatively poor when compared to the very rich), eg big property taxes, eg using taxes to build private sports arenas

    They are all for economic competition, but, in fact, they get big-government to legislate away the competition

Summary

Thus, we live in a society which is based on property rights minority rule and inequality all held together by means of extreme violence, within a communication system which espouses an absolute authoritative truth, upon which the power of the minority (of a few very big property owners) depends, and this dogmatic authority is expressed in an education system which does not serve equality and knowledge (and the relation of knowledge to (practical) creativity), but rather (the education system) supports narrow authoritative dogmas related to maintaining complicated instruments whose functioning is controlled, not by the public, but controlled by the owners of society.

Such a society is only possible if it is maintained by extreme violence and coercion, which is most effectively instituted by the "left with voice" those discipline ones who do "buy into the authoritative narrowness" and who do believe there is "no other possibility," (other than quantum theory, particle-physics, general relativity, and indefinable randomness) ie the unimaginative left, and supported by a right whose arbitrary authority is most closely associated to an authoritative-religion whose main idea is a narrow hierarchical domination.

This is an example of the clever use of personality types, where the "left with voice" are obsessive types who are obedient, with little imagination, and no vision tied in their viewpoints (except the religion of personality-cult), where their viewpoints uphold a narrow dogma, and where the right is violently dominant.

If the dogma of the rich is challenged, then both the left and right would both exclude such challenges, ie this is a result of the "societal magic" of controlling a centralized communication system.

Thus, one sees a very clever mix of the violent and the obsessive both agreeing on an unquestionable authority, which effectively excludes any differences in practical creativity in regard to markets and social domination, either by the violent of the right or by the obsessive adherence to dogma on the left, so that society (life styles) and institutions (which are supposed to be associated to new knowledge and new creative contexts) stay fixed and consistent with the structure of both markets and the societal structure which support the very rich
(We are still living in the Roman-Empire, or the Holy-Roman-Empire of the Roman-Emperor Constantine)

Thus, there is a powerless left, and a right both employed to protect the super-rich, where the super rich are quite similar to the left which the right hates.

Everyone is "played for a chump" in our, so called, "equal and free nation," where the main player is the controlled communication systems (communication channels), ie propaganda, and the context (of words) provided by the propaganda system is: unequal domination of the many by the few, either intellectual or by violence.

In Roman time (after 350 AD) intellectual domination meant intellectual domination by the Church, but now it is the failed dogmas of science, eg indefinable randomness, materialism, and non-linearity, and an apparent failure to be able to contain either the system itself, or the properties of the system, which is supposed to be contained within the construct of materialism. ***

There is a new way in which to use a precise measurable descriptive language which transcends the material world, and where the material world is but a subset of the new descriptive context. Namely, it is many-dimensional precise descriptive structure, and it is a description which is not based on materialism, so it should be of interest to the religious people [the false conflict between ideal religion vs. material science; is ended], it is geometric and linear so it can be thought about (it is possible to have ideas about geometric-shapes), and it can also be placed in a controllable context (a linear context), ie for practical creativity.

It is consistent with classical physics, and it identifies the cause for quantum randomness, as well as the point-like character of quantum properties. It is many-dimensional and unitary so it fits into the patterns of particle-physics. It gives a linear example for general relativity, wherein general relativity, ie the shape of space affects material motion, actually functions {but in a limited but fundamental, linear context}. Note: The non-linear models of general relativity are useless for any form of practical creativity (worm-holes are a joke).

The new descriptive context allows for the explicit descriptions of "stable" physical systems.

The new descriptive context requires that math consider the idea of a stable math pattern, to be a fundamental math property.

These stable math patterns (stable shapes) have been identified by the Thurston-Perelman geometrization, namely, the stable geometric patterns are predominantly the discrete hyperbolic shapes, which, in turn, have been described by D Coxeter.

These new ideas provide an example of how to use precise language in new and different ways, in a similar way as did Copernicus, back around 1550.

The reason these new ideas are excluded and marginalized from the media, both "the radical media" and main-stream and the, so called, "alternative" media, is because of the existence of the gate-keepers of truth within the propaganda system.

The communication system (owned by the big investors) is a communication system which uses language to best serve the interests of these big investing bankers and investing monopolistic businesses.

The gate-keepers of truth are the editors, as well as those in the peer-review process (of an entire professional community, why a professional community needs so much protection is beyond me, especially, since the scientists which the business interests want peer-reviewed are peer-reviewed, no matter how bad the science and math they present to the professional journals, and the fact that math and science have failed their descriptions are irrelevant to developing any new practical creative systems except in regard to systems dependent on averages or the indefinably random feedback systems).

These "gate-keepers of truth" play the same role in protecting a dogmatic and authoritative truth as did the pope in the age of Copernicus, or as the Puritans who judged the native people's to be a part of an inferior culture so that both the pope or as did the Puritans [or other gate-keepers of an absolute truth for the highly controlled media (or propaganda system)] and this gate-keeping-for-truth process is used to justify the right (of the Puritans {and the pope, in regard to the likes of Copernicus}) to arbitrarily exterminate the native peoples, so as to get their land (property), and the gate-keepers of an absolute-truth are used to exterminate the ideas (in particular ideas which are in fact superior to the current ideas) of people who are not members of the personality-cult of science and math.

This arbitrary way in which ideas are excluded and terminated from the media is the true mode of operation today within the propaganda system.

This intellectual barbarianism still happens (as the pope would have done to Copernicus) even though there are great limitations to precise languages, and this has been known from the time of Godel's incompleteness theorem (around 1935).

However, when experts talk about the incompleteness theorem, the discussion is, not about the tenuousness of any truth based on axioms, but rather the discussion is about the axioms and definitions of numbers, and "processes of creating" numbers. This is a red-herring. Usually when the experts, or the media schills, represent Godel's incompleteness theorem to the public, it is most often stated abstractly, "that a math system cannot prove its own consistency," but this really means that axioms may possess within themselves either opposing or incompatible relations in regard to the other axioms of the system (or as compared with the other axioms of the math system).

Furthermore, creativity is best related to the elementary aspects of a system of a precise-language's set of relationships, where in the new example (many-dimensions partitioned by stable shapes), the idea of materialism is replaced with the idea of many-dimensions partitioned by a fixed set of very stable shapes, so the stable shapes become the material, but the new material properties depend on both the dimension of the containing metric-space and the dimension of the stable shapes, where the many-dimensions can, in fact, be placed in either a real number (isometry) or a complex number (unitary) context.

The professionals math and scientists are only able to discuss ideas which are irrelevant to (or even opposed to) practical creativity.

That is, their Platonic truths (based on word-agreements in a particular axiomatic context) lead to a world of illusions not to the real world of experience (eg the failure to calculate business risk based on a supposedly highly rigorous math structure of indefinable randomness).

Why should there be such sheltered people, sheltered by peer-review, when their own math structures, ie Godel's incompleteness theorem, insist on considering many different ways of organizing the language of mathematics

That is, both the fixed dogmatic descriptive language, as well as the violence from which the owners of society derive their social and economic power, are limited structures. This is the nature of language, so to give only certain people an immense (great) amount of social power, forming an unequal society, results in destruction and failure.

One cannot allow either JD Rockefeller, or JP Morgan, or E Witten, or S Hawking, to amass so very much societal power, as the propaganda and legal and political system grants to these people.

This imbalance of power leads to a system in which people are willing to do both absurdities and atrocities for a system which opposes both life and creativity, and elevates a few.

Yet

The public is reluctant to believe a system of thought which represents life and creativity at its most inventive, creative, and exploratory level [the new example of a new way in which to organize the description of existence, in many-dimensions based on very stable shapes] so that

The propaganda system, which is wedded to materialism and an absolute but arbitrary high-value (whose failures have shown these "dogmatic truths" to not be (even) closely associated to a truth which can be related to practical creativity) which promotes the sacrifice of life in both action and in knowledge, where this sacrifice is used for selfish purposes to uphold a lofty few.

This is clear evidence of mind control

So why are these ideas . . . , which are substantial enough for the joint math conference on mathematics, in San Diego 2013, to be addressed by the author of these ideas . . . . , not expressed in the media, and why do editors exclude these ideas on the, so called, free-speech outlets on the internet?

Because (1) these people (right-wing editors and authorities on the left) are chumps, and (2) these ideas require a revolution of thinking which will be destructive to the current social structure, where in the current social structure "authority is considered to be the same as truth."

But the propaganda system sounds "too much" like the authoritative church in the days of Copernicus?

Indeed, it is exactly the same, and the lesson of Copernicus is not about "science and materialism vs. religion" rather the lesson is "authority" vs. "new ways of using a precise language (even though both authoritative ideas were being measurably verified eg Ptolemy's epicycles were consistent with data (after they had been corrected) and they were very precise in regards to predicting the motions of planets in the heavens)."

For example, today it is common practice (within particle-physics) for new particles to be proclaimed to exist when a new phenomenon (such as dark matter) is seen.

This "adding of new particles to the list of elementary-particles" (elementary-particles which had not ever been predicted) is very much an epicycle structure, exactly like the epi-cycle structures of Ptolemy.

That is, the intellectual basis for the left is destroyed by these new ideas, but the bulwark of arbitrary religious (or moral) authority of the right is also laid-waste.

There is a creative way in which to consider the world of the ideal (ie the religious context) and that "ideal world" is very real, and that "ideal world" is the proper context in which to consider religion.

The super-rich have the (poor) public trapped and tricked in every direction that the public turns, (this is with the help of the propaganda system and the justice system (guided by a political system which has been corrupted), so as to support the super-rich, and a major way of doing this "tricking," is to have the people be dogmatic, rather than creative, in regard to the activity of science and math.

The rich want particular types of complicated instruments adjusted by the intellectual elites, but the very-rich do not want the intellectuals (or their intellectuals) thinking too much.

16.

Comment on General Relativity

The idea of general relativity does not work since it is non-linear, and thus it is not quantitatively consistent, so this leads to chaotic behavior in actuality, ie one loses cause and affect, the very thing Einstein said he was concerned about.

The stability of nuclei implies a controlled system, ie a system governed by cause and affect.

The origin of the problem is entirely simple: When one measures then one can consistently change to a new quantitative set if one multiplies by a constant, ie y=cx where c is a constant, but this is a linear function a non-linear system means that such a (linear) quantitatively consistent pattern does not exist (but this is defined locally by derivatives).

We think of numbers as stable patterns otherwise their meaning is limited, yet, most (or much) of modern math places the descriptive context of measurable patterns into a context in which numbers lose their stable relationships and this leads to chaos and indefinable randomness.

Numbers keep their relationships stable in a fairly limited context of linear, metric-invariant (length stays fixed for allowable transformations of a set of system containing coordinates), and separable (or local linear measuring relations must be commutative everywhere). This is ignored by the authorities who look for abstract generalities of great complication so as to prove their mental prowess, but why and to who? They do this apparently as wage-slaves to impress the owners of society.

Then others, who are interested in math, are provided with an education wherein complicated patterns are memorized and the student looks-up to the "math masters" (eg the professors), and the student is afraid to posit questions at an elementary level. Thus, inequality destroys learning and it destroys knowledge itself and it thus destroys people.

There is not thought, there is only memorization of an authoritative claim which is upheld (primarily) by authority and the propaganda system, whose main message is inequality.

The elite can think, but a person cannot, since the person is not superior. This has always been a basis for extermination of whole thoughts and of whole peoples . . . . it seems that the public is next . . .

If we do not have confidence in ourselves, because society excludes us, then we become terrorized, but our terror is more physical when it comes to the justice system, eg Occupy was violently bulldozed off the streets, the message of equality and no leaders was too strong for the propaganda system to deal with

The best agenda for Occupy might be: discard the law based on property rights and minority rule, instead "base law on equality, allow freedom of belief, and freedom to create, and freedom to express beliefs related to creativity (or about social organization) but make sure the expressions are done at the elementary level of: assumption, context, interpretation, and (if possible) the creative intent, so that the messages can be categorized, thus new ways of organizing language would be apparent. Furthermore, those ideas which are continually repeated by the media, and the experts, can be sorted into there limited categories of ideas so that if one wants to hear such useless noise one can go to the appropriate category. Make sure the governing body (or governing principle) is to promote the common welfare, so as to take into account the requirement of equality. It is only if each citizen is treated as an equal creator that there can be a truly free market. Organize law to punish selfishness and allow or protect selflessness, eg selfless creativity.
There can be a 1 year transition period, wherein many of the most corrupt agents of "the mess we are in" are to be jailed, or their voices excluded.

Furthermore, the monopolistic business-centered system of creativity used by our society only needs a limited range of knowledge to manage its narrow interests, eg reduction of all material to random particle-collisions of elementary-particles, where the probability of collision can be determined, is exactly the knowledge needed for determining the size of the blasts of nuclear reactions, ie it is all that they need, so it is better to not have a possibility of new creative contexts for their business models, so that is all that people are allowed to study.
Within these "accepted" physical mathematical models the explanations (of say stability) are ludicrous, note that there is better discussion between S Hawking and the science fiction writers of star-trek, than between the so called scientists who simply put-forth unrealizable models in an illusionary world, but they (the demigods of science) are willing to be demigods, and they seem to have well intended, but deluded, attempts at pushing forward into their never-never-land, while cross-sections of elementary particles are continuing to be determined (cross-sections are related to probabilities of collision, and thus to the rate of reaction) and used in bomb engineering in the military businesses, ie the complication is a ruse.

It is exactly a world of illusion for everyone, and it is my belief (and apparently the belief of Socrates too) that only equal free-inquiry (best put in relation to practically creative ideas) which can uncover the illusions, eg occupy was bulldozed since it was causing issues with the propaganda system.

Indeed people are sacrificed, the human-sacrifice of an Empire based on the selfish interests of a few, ie the owners of society.

Math literature is fairly rich but these ideas were close to the ideas of Kepler, his music of the spheres was based on regular polygons (or polyhedrons in 3-dim's). However, Kepler had the wrong context. When (some of) these polyhedrons were related to circle spaces then interest was revived, eg F Klein, in the 19th century, but they are simple patterns and mathematicians seem to want to consider complicated contexts of abstraction and great generality but these contexts are non-linear.

The correct context is that these shapes are (when considered within themselves) metric-spaces with open-closed topologies, but while these shapes (or metric-spaces) are contained in a higher-dimensional metric-space they identify a boundary in the higher-dimensional metric-space.

When some of the regular polygons are placed into hyperbolic space, and in that space one is given a line and a point, there are many parallel lines through the point, (in contrast to only one parallel line through a point [given a line and a point] as in Euclidean space)

The idea was about a line segment identify perpendicular line segments at the vertices of the given line segment to get a square (putting-in the opposite side) then extending perpendicular line segments from the vertices of the square to get a cube, etc (I have trouble with this idea, but it could be a good thing to try, ie to think of the new direction perpendicular to the vertices of a cube) so I look at other types of patterns associated to the shapes and metric-spaces

One can also consider a line segment and bend the segment around so that the vertices touch so as to get a closed curve, but it is simpler to consider the shape as a circle, one can do this same circle construct with all the line segments of these "cubical shapes" for all the different dimensions.

Consider the simplest things.

Only lines and circles can be related to measuring gradations which can be made linearly consistent with one another.

The streets are full of people who do not compromise, and want to have their own lives, but unfortunately the propaganda system and society require hierarchy and domination for social interactions, and this is usually expressed as violence.

The mind is quite capable of doing many things.

Consider the idea that the mind is actually associated to a "physical structure" which is higher-dimensional (and today that is not considered physical, yet the physicality can be isolated to each dimensional level, ie it is difficult to interact with the higher-dimensional levels by material components [or metric-space shapes] of a lower dimension), however, our will can operate in such a context (because we are actually higher-dimensional), it seems that a will can only exist in such a context. The idea of the religious statement that one should try to [perceive the world as it really is] is a claim that we are actually composed on the higher-dimensional (relatively stable) shapes.

For a slightly over simplified model of the mind;

There are two spaces the metric-space (the external) and the metric-space's associated metric-invariant Lie group the group of local coordinate transformations which leave the metric-function invariant ie length stays the same after the transformation. Lie groups can be associated to each dimensional level, so one can be associated to our being (thus one can think of the Lie group as being an internal space), where living systems are based on these same types of shapes, but life is sure to be of higher-dimension than 3-dimensions. That is, life is organized in dimensions which are above the dimension of "what we call" a material existence. The over-all high-dimension metric-space is based on the very stable circle-space shapes which have a fixed set of spectra, so that all the circle-shapes contained in the high-dimension space (or at any particular dimensional level) resonate with some subset of this finite spectral-set, so spectra and the shapes and "sizes of these shapes" make-up what we see and perceive.

For example, in our particular 3-space subspace material is the size of atoms etc.

In the Lie groups there are toral shapes (doughnut shapes) which are simpler and not as stable shapes, but which can easily carry spectra by resonance, this is the mind or memory, as well as being necessary for perception, but there seems to be an interpretation (or glue or an identifiable inter-relationship) which goes with the sets of spectra (which go-together with one another), so the mind sort-of resonates with the stable spectra of the external shapes and our other sensory data allow us to make-out a geometric image.

This is quite similar to the cosmology which C Castaneda described in his book "The Fire From Within," but it is entirely a math model, and the shapes and math is relatively simple, but the context is extremely complicated.

It is entirely a math model (ie it is a part of what is now considered to be the western culture, apparently this is due to the violent domination of others by the west, and how the west coddles and protects its math knowledge, ie it controls knowledge)

17.

Empire of oil

Consider that both: (the elite-intellectuals and the elite investor-class)

(1) S Hawking (or E Witten, or A Einstein, etc.) represent today's (2013, or the recent past's) leaders of the "aristocracy of intellect," where a hierarchy of intellect (an "aristocracy of intellect") is identified and measured in regard to narrow dogma, a dogma which supports the interests of the "owners of society," and which identifies an "absolute truth" associated to the idea of materialism, where measurable descriptions are based on the idea of materialism . . . , [and the containment set-structures of measurable descriptions which this implies (as well as the reduction of all material to elementary-particles)] . . . , is called (physical) science, which is backed-up by narrowly defined "rigorous" set of properties deduced from a fixed set of both assumptions and contexts for a rendition (or an expression) of a, so called, absolute truth concerning the material-world and the . . . . math . . . , {the study of quantity (including probability) and shape (where quantity and shape are supposed to both be stable patterns, but the usual math contexts of both indefinable randomness and non-linearity defined on quantitative sets which are too-big, eg the continuum, excludes a description of any stable properties which are either observed in the world or expressed in such a (randomly based) mathematical context)} . . . , math . . . , rigor which only depends on agreement concerning the meanings of words and acceptance of contexts and interpretations, but such a "language dependent" vision of truth can result in a limited descriptive context which essentially only describes general abstract contexts are only a part of an illusionary world (which such a limited and precise language describes)

{unfortunately there is a relatively small group of (well-funded) intellects, which composes the leadership for the "aristocracy of intellect," whose descriptive contexts . . . , only have a very small relation to practical creativity, where any relation to practical creativity is in regard to classical physics {based on geometry and measuring (and average measures)}, . . . , while their modern ideas about physical science, based on indefinable randomness, non-linearity and convergences within a quantitative continuum (a set which is too-big, thus, its logical consistency is to be deeply questioned), is basically irrelevant in regard to practical creativity, furthermore, these math techniques cannot be used to describe the properties of general quantum systems based on the so called laws of quantum physics . . . , (and thus the, so called, verification of modern physical science is not (effectively) any different from the verification process used by Ptolemy to verify his model of the planetary motions)},

and

(2) J D Rockefeller (or J P Morgan, or W Buffet, or King of Saudi Arabia (who knows, exactly, who the very-very-rich actually are?)) etc.) are examples of a very small-number (perhaps as few as five) of extremely-very-rich, "owners of society," which compose the ruling class (for the entire Empire).

Where these two groups . . . , the second one being infinitely more powerful in society . . . , together identify the creative and coercive activity and the knowledge upon which creative activity (and the intellectual coercion) upon which the power within society depends.

While the right, the social arm by which society is maintained and the society's most powerful people are protected . . . , by means of both the arbitrary "moral or social" authority and the extreme violence . . . , where these social values can be controlled by the very rich because of the wage-slave status of the public, a social status which is (also) maintained by means of extreme violence against the public.

Why should society adopt such a narrow view of (intellectual) dogma, and monopolistic ownership to run and organize society?

Could this narrow context for society . . . . , which is essentially the narrow fixed traditional context for society which the Romans constructed for Europe beginning around 100 BC . . . , actually be the reason that the equivalent banking-monopolistic Empire is failing so consistently in a continuous manner whose affects are felt in a periodic manner (ie periodic economic collapse)?

Wherein it needs to be noted that the recoveries from such periodic social failures, simply results in the re-shuffling of the managing-personnel and, perhaps, new expansion-conceptions for Empire to be based upon the same-old dogmas and traditions and technologies.

The Romans seemed to make roads, build water systems, and built a few (fancy) public buildings (with many sculptures and mosaics), and create instruments and techniques used for extreme violence . . . , where this became the basis for a stable society essentially upheld by extreme violence . . . , while today there are gasoline-engines (the oil-monopoly), and electrical communication instruments, ie now we are stuck with 19th century engineering capability, eg note that the nuclear-reaction and the 19th century model of chemical reactions are the same model.

In fact, such a narrow expression of an "authoritative" truth, are the types of truths which the Declaration of Independence was addressing, in regard to both:

(1) science, in regard to belief, inquiry, and expression within ever new interpretive contexts and assumptions which are to be the basis for new precise languages,
(2) as well as addressing the rights of evangelicals to read the Bible and express their beliefs about how to interpret the Bible, so that there was

   (1) the Quakers who tried to follow the main teachings of Christ (Yet, in regard to the native people, when W Penn was gone, the Quakers were as opposed to the native people as the rest of European colonists, though they always claimed tolerance for the Native people's religions) and
   (2) the Puritans who believed in a hierarchical society based on harshly enforcing arbitrary (subjective) moralistic laws . . . , unfortunately, a natural interpretation of the Bible, since Constantine wrote the Bible as a propaganda tool to support the Roman-Empire.

Why should society adopt such a narrow view of (intellectual) dogma, and monopolistic ownership to run and organize society?

Answer: This is because this way of organizing society, and its intellectual activity (based on narrow authority [which can only follow tradition]), follows the social (and propagandistic) organization of the Holy-Roman-Empire which was established by Constantine around 350 AD, . . . , and then (this Empire) transitioning to a banking-empire eg C Medici, around 1450 AD, . . . , then the oil-electronics monopolies from 19th century science, ... , and then around 1950 (under Truman in US) to an oil-banking-military Empire (where the military and oil interests [where oil dominates agricultural interests] dominate the thinking in science)

Note: The US revolutionary war . . . , based on the Quaker model of an egalitarian society . . . , tried to re-write (European) law, so as to be based on equality, and freedom of belief, and freedom to inquire in an equal way, and freedom to express one's beliefs, where beliefs are best defined in regard to using knowledge (based on belief) for practical creativity, but arrogant factions representing absolute-authority (based on authoritarian churches and banking interests) as well as an absolute Holy-Roman-Empire arbitrary viewpoint enforced by extreme violence, associated to the European culture around 1775, identified the viewpoints of the Revolutions leadership after the war was won.

This arbitrary, authoritative social structure . . . , which was effectively controlled by means of extreme violence . . . , identified how those from the European culture (ie the immigrants who violently stole the land from the native peoples) viewed "governing institutions," and subsequently, ignored the true basis for US law.

Where the right for a nation to determine its own law was won in the context of European tradition that the victors had their own-right of rule (or to determine their own set of laws) [this right belonged to the victors of wars (apparently, a tradition of Roman-Empire)].

In a society based on equality, a truly free-market is based on a population of equal-creators (placing their creations (built in a context of selflessness, often for social improvement) into the society {or the market}) governed by a consideration for a community-system which identifies and promotes the common-welfare for everyone in an equal manner, since it is the public which empowers the government.

The documents defining the US society still exist, and they can be the basis for changing the society:

from Empire
to
an equal society.

But this seems to need to be implemented when there is a disaster or a collapse, but then (after the collapse) the new ideas must be forcefully asserted, where the law opposes both inequality and selfish behavior.

However,

The empire, which now exists (2013), requires a hierarchical social structure which, in turn, requires narrow dogmas and prejudices to exist within the population, which, in turn, can be used to identify military-commands expressed by the Empire, in order for the Empire to run smoothly, and so that the narrow dogmas and hierarchical social structures (upon which the monopolistic-Empire depends) can be protected.

The two types (intellectuals and owners-of-society) are both associated to dogmatic or monopolistically driven social forces and language forces (within society) which are used to identify, for the public, an absolute social-value, attributed to a personality-cult . . . , (the [de facto] true religion of the age of expansive-Empire, where the bankers of the 1500's identified intellectual-value by means of their paid-for social-projects, in what was already a wage-slave type market-place), . . . , where this narrowly defined language is formed with (by) the propaganda-education system (ie a highly controlled media system) begun with Emperor-Constantine, where the church was the propaganda-education system at that time).

Both of these groups of the personality-cults (intellectuals [including propagandists] and owners-of-society) are veiled . . . , (in the media, and administratively within: government, legal, and educational institutions) . . . , by complexity, and nonsense, and arrogance . . . , (ie maintained by means of social inequality, where the idea of inequality is the main message of the media-system and of an Empire), . . . , which are character traits associated to those "in charge" of (those individuals who agree to) a limited viewpoint, upheld (within the public) by the power of extreme violence (the same model as the Roman-Empire).

Does it matter if there is either:

equality-peace-knowledge-creativity
or
narrow-selfish-violence

(identified in society by a social-value defined within (or by) a monopolistic economic-governing model associated with only one authoritative viewpoint concerning assumption-interpretation-context for (precise) languages (ie the language of science and math which are supposed to be associated to practical creativity within society)?

Answer:
It depends on one's viewpoint concerning "the nature of life" and how life naturally fits into both nature and creativity.

Is (human) life based on, either Violence and narrow selfishness associated to both the ruling-class and their "aristocrats of intellect," an "aristocracy of intellect" which is defined by narrow dogma, and which the rulers identify as the narrow context of precise language which best serves the selfish interests of themselves (the rulers of society)

Or

Is (human) life about a deep (truthful and controlled) relation between nature and creativity? Can this human creativity transcend the idea of materialism.

Is the idea of materialism central to the idea of an expanding Empire?
Answer: It is!

That is, religion has given a none-effective voice to the idea that idealism is valid (as opposed to materialism, ie material properties determine existence), but their religious (spiritual) ideas do not lead into a practical context of creativity which transcends the idea of the material world.

That is, religion has failed in regard to its knowledge being relevant to the human condition of being practically creative, and people being selfless in this natural internal interest in practical creativity.

And now, after over 100 years of both life-science and physical-science being based on materialism, and (mostly indefinable) randomness, and non-linearity, and function-spaces acted-on by sets of operators, and convergences defined within a continuum (a set which is too-big to be logically consistent) the ideas of these modern sciences, based on materialism, have not led to any valid new contexts for creativity, as well as no valid descriptions of the observed stable properties of the most fundamental material systems, where they put-forth the authoritative claim that these "fundamental stable systems are simply too complicated to describe" since they arrogantly claim that they do possess the "absolute truth" concerning the nature of physical law; and that law is based on: the reduction of all material to elementary particles, randomness, non-linearity, and that material does define existence.

Thus, there is both the ineffective ideal based religious-knowledge and an ineffective material based reductive science.

This leads to stagnation, and domination (by the owners of society), in regard to "what is thought and said by the public," which is controlled by a very narrow propaganda-education system (the social position of the church in the Roman-Empire) and by a wage-slave social-status of the public, realized by an extremely violent justice-system which is at-war with the public so as to protect the property of the very-rich, from whom the justice system takes its orders.

It is Marx's material-based view of both economics and sociology (or history) which has created a pair of Hegelian-opposites used in the propaganda system "the creative elites" vs. "the dependent worker."

In the propaganda system (media-education system) dialog is based on "the right" and "the left:"

The right supports arbitrary religious authority and extreme violence, while the left supports material-based science, {a descriptive context which is both practically useless and has no valid descriptions of the very stable fundamental material systems of: nuclei, many-charged atoms, molecules, crystal properties, the solar system, the motions of stars in galaxies (dark matter) etc.}

Furthermore, the left believes all the nonsense about economics, sociology, psychology, linguistics, being related to a measurable verifiable description, as well as the relation that, the so called, evolutionary biology, plays in these subjects, eg the-survival-of-the-fittest which is interpreted as supporting hierarchy etc

Furthermore, both the left and the right, seem to think that science is capable of understanding the inventive contexts of Tesla, even though in the hierarchical social structure, it was Faraday and Tesla who ushered in the electronic-age, where both of these men died relatively poor, while the bankers were able to monopolize (a rather complete knowledge of electromagnetic phenomenon) by using their "stables" of obedient professional scientist, and patent law structures (based on deceptions about the nature of newness, used to steal the knowledge of culture for their selfish interests in investments), and methods of control by funding, and to subsequently, determine how these technologies were (and are) used in society.

However, S Jobs was an interesting personality in the instrumental set of structures of: TV's, fast switches, and thermal control of P- and N-doping properties within some semi-conductor materials.

S Jobs was an example of a more creative personality (rather than narrowly defined monopolies) guiding technical use of science and engineering knowledge, though one would expect that 25% of the public would be able to do as good, if not better, job at innovation, than S Jobs was able to do.

In the propaganda system (media-education system) dialog is based on "the right" and "the left:"

The right

Is for violence (called protection or security [which is all about security for the ruling class and their selfish monopolistic interests throughout the globe])

And they are for social inequality, while

The left

Is for "more nuanced forms of (intellectual) inequality." They (the left) consider themselves non-violent (but their mid-level high-social rank means that the Empire protects them (and their arrogance is upheld) because of their beliefs, which they gained (or accepted) by competing in the narrow frames of social-value defined by the propaganda-education system) but these beliefs, which are defined by narrow dogmas, require protection and security based on extreme violence. These are the beliefs upon which the "aristocracy of intellect" (including material-based science) is defined.

Consider the absurd fight "to save the planet from destruction," where the destruction is caused by the poisoning (or polluting) of the planet by narrow exploitative monopolistic social-economic concerns; the "conflict" is framed in a context of absolute truths

The absolute truth of material based science (the left)
vs.
the absolute truth of economics, and the relation of economics to social conditions (the right)

Thus, the argument regarding poisoning the planet due to both pollution and the destructive use of "partial knowledge," . . . , eg using DNA as an enzyme to make chemicals which chemical-science cannot make (due to the ignorance of chemical-science concerning chemical processes, in turn, due to an invalid model of atoms and molecules, etc) as well as the poisons of nuclear wastes, etc, . . . , is narrowed to the context of:

Global warming being caused by increased concentrations of CO_2 (and methane, etc) in the atmosphere (but the atmosphere and the earth are complex systems, so the argument is easy to mire in details, and where the propaganda system, itself, questions the absolute authority of these ideas . . . . , ie the narrowly defined authoritative context, related to statistical models, which, in turn, are based in a context of indefinable randomness . . . . , where the propaganda system, itself, associates such authoritative presentations to science, leaving the personality-cult of the "worshippers of science," ie the public, quite confused).

That is, though the correlation of vitamin-C and the disease of scurvy is quite strong, it would still have to be an example of indefinable randomness, since the cause, required by such a strong correlation, does not exist. This is because there are no valid chemical models of the chemistry of "chemical reactions induced by catalysts," ie as in the case of DNA (there is now an actual expression of the epi-genomic context in regard to the functioning of DNA and its true relation to molecular structures in living systems), similarly the catalysts may also have a greater context beyond the existence of enzymes, so the cause is not explicit.

The nature of markets, and trade, and the definition of arbitrary social value [which is always tied to a fixed property of the products which society uses (fixed) over time] cause the idea of slavery (unpaid help in a market process) to be very valuable.
There has been:

Slaves—owned property
Serfs—people attached to the land and associated to production
And now
"Wage-slaves"—people attached to a propaganda-education system,

A propaganda-education system which guides the public into having a narrow vision about an arbitrary social-value associated to a market, and an arbitrary social-value associated to science, etc.

A few (of the rich "leaders") took over the 1776 American revolution, so as to make sure that the European model of society be instituted in the US society, where that European social model was that of a banking-mercantile empire, which was to be maintained within the US society, despite the revolution being about basing US law on equality.

Many, if not most of these (rich 1776) American leaders were "free masons," a society (or club) which understood the relation that violence can have, in regard to controlling capital markets.

18.

The fight

Politics, economics, and the law, need to all be placed in a new language, and new interpretations of observed properties (new interpretations of history) need to be provided (here is a clear example) similarly

Science needs to be placed in a new language and new interpretations of observed properties need to be given [see Succinct]

To fight the rich (ie the owners of society who dictate how society is run):

If law is based on property rights and minority rule (as the US supreme court has interpreted and guided US law to so be) then the rich will always "win," the rich will always have their contracts upheld (in law), while the contracts held by the poor are not as valid.

Thus, for the other 99.999% (ie not the owners and their immediate set of business managers) to have a life of their own, which the earth can nourish in a harmonious manner, ie for the public to prosper, the law must be based on: equality, and the freedom to believe, and express one's belief, where belief is primarily about an expressed language's relation to practical creativity, furthermore, government needs to depend, for its power, on the public, and the government is to serve the common welfare in a context of equality.

There cannot exist a truly free market unless each person is treated as an equal creator, who is motivated by their own selfless need to create in a practical manner. [and the value of creation cannot be judged]

History:

Western culture has been based on the model of the Roman-Empire, and the Emperor Constantine turned a military based Empire into a propaganda-military based Empire, when Constantine made Christianity the Roman religion, and then Constantine wrote the Christian

Bible which was a document of propaganda for the Roman-Empire. Thus began the Holy-Roman-Empire, and its various representations.

This Roman-Empire expanded militarily to take all the wealth of the conquered peoples and then the Empire built roads and public building and public art and water-works and sewers, and they occupied their conquered territories so these conquered regions remained stable, where the stability was maintained by Roman violence.

Apparently, the people appreciated the Roman engineering achievements.

This condition was essentially upheld from 100 BC to about 1400 AD (Constantinople, the continual center of the Roman-Empire (since around 400 AD, fell around 1450 AD) and during this time of stability, the stable living traditions associated to: water, and food, and shelter, and art, and engineering (where engineering served the interests of the Emperor), and trade, subsequently these (stable) traditional social activities came to be organized around money and banking, and the Holy-Roman-Empire was transformed into a Holy-Banking-Empire around 1400 AD.

That is, the markets of finance depend on very stable and narrowly defined social institutions: farming, iron-works, art, engineering (controlled by the few), travel-trade, and a military based security system, and a propaganda system (the authoritative church) controlling social-thought, and thus forming (in the public's minds) the allowed social context in a hierarchical (and unequal) society.

A social hierarchy built upon propaganda and the extreme violence of a ruling-class, where stable markets become the context of social interactions, which were not based on violence, bust which depended on violence for their stability.

Art becomes narrowly defined, art is defined within a stable-market based life-style, and practical knowledge (or science) was associated to either productivity or the military, which fit into the same narrowly defined stable social-market context, associated to particular (narrow) types of (risk-free) investing.

Thus, European society is about inequality and violence, within a narrowly defined social, life-style context upheld by an authoritative propaganda system, in turn, upheld by violence.

The printing press allowed the Bible to be printed, where the Bible is the central document of the propaganda system, upon which social authority is defined. Thus, if the Bible is read and interpreted by the individual reader, religion was questioned and re-interpreted.

Furthermore, in regard to other questions about existence (and the nature of man); Copernicus provided the answer upon which material based science would be developed.

That is, the authority of language was being challenged, and the relation of language to belief and to practical knowledge, in turn, related to practical creativity was being explored.

Thus, for the American colonies, which were self-governed (since the lost-souls of Britain were sent to form colonies) and based on "religious groups interpreting the Bible as they wished," where one such religious group followed the golden rule, and based their self government on equality, namely, the Quakers.

Where the Quaker communities was the most prosperous and energetic of all the US colonies.

In this context of science and freedom of religion, the US revolution was about changing the idea of western society so that society was to be based on equality, self governing, and where science and its relation to practical creativity was to be based on equal free-inquiry, and not based on dogmatic authority, and violence, and social hierarchy.

But the constitution was made by the men of property, but it had a preamble supporting the public and a Bill of Rights defining individual freedom; freedom of belief and the freedom to express one's beliefs, and thus individual equality.

But, the Bill of Rights was never upheld, and those people who were for equality, apparently (since those voices for equality were also from the upper class, eg T Jefferson), did not have the needed resolve to maintain it (equality) as a law, and the elitists could wait and scheme, since hierarchy was implicit in the European viewpoint which also existed in the population of the European public, which was the pre-dominant population in the 13-colonies, since most of the colonies were at war with the native populations, only the Quakers negotiated with the native people and honored their agreements with the native people.

With the western hierarchy re-established in the US, then expansion could continue within the narrow traditions of the European-bankers vision of social growth, the US became rich, but the rich became undisputed emperors of investment and wealth, and they came to rule the US.

Thus, there was a self governing set of colonies associated to each of the different colony's beliefs, and a new land in which to expand (or to continue) the exploitation of resources for a European economic-related life-style of expansion (the model of Empire), most of the Europeans possessed a very narrow vision of life-styles associated to money-related social hierarchies, most often built around violence, prejudice, and intolerance.

This was due to their emersion in the European culture, which was based on a social hierarchy built and maintained by violence and narrowly defined propaganda concerning life.

The predominant idea about life, was that people were horrible sinners and generally not worthy of life.

However, the religious colony of W Penn, the Quakers, tried to live by the golden rule, and they formed an egalitarian society, which was the most prosperous, and the most energetic of the colonies in the US.

Furthermore, R Williams (of Rhode Island) provided an example of the difficulties in regard to believing in religious freedom, though R Williams himself believed in narrow dogmas, but he also believed people should freely decide for themselves about their own beliefs.

But most of the colonies where extremely narrow, exploitative, and intolerant (as were European societies of that same age). The culture was based on intolerance and inequality (ie deep immorality in regard to the golden rule, but nonetheless justified by its propaganda system)

The revolution of 1776 was truly about equality as the basis for US law, and limited government, but the government was to serve the interests of the public, whose interests were to be expressed democratically.

Equality is (was thought of as being) about freedom of belief, and freedom to express belief, and freedom to create in a practical context, and in a selfless manner. This is best for science, and religion, and governing, but regions of jurisdiction needed to be made explicit, and laws of the golden rule and tolerance needed to be generally applied. The people tended to not be religious (religious-ness was formed by the propaganda system) but rather practical, but with narrow vision concerning social organization.

Free markets only exist if everyone in society is an equal creator.

It is not clear what type of life style actually supports populations better, living in harmony with the land, as the Native people's did, or exploiting resources to be consistent with a narrow viewpoint about life and social-class and the relation which money has to social-class, a hierarchical society wherein people are (the public is) considered to be bad (or people are suitable prey for, and to be exploited by, and to be turned into wage-slaves by, the upper social-classes), and people are considered to be not capable, which is consistent with the European viewpoint, which came from the Roman-Empire, whose leadership was taken over by (ie the new Emperors were) the bankers of a narrowly defined life-style, and banking and markets were associated violent military regional domains, where extreme violence ensures the stability of the narrow life-style (upon which power was based).

Then after the successful revolution, the rich took-over, where A Hamilton stood for the old narrow-violent Roman-banking social-order, wherein the public was to be attacked by the elites, and T Jefferson stood for equality and self-reliant, free, people living close to the land.

Because the issue was about Europeans, and European viewpoints about life, due to the violence with which most colonies attacked the native peoples for the exploitative gain in resources for the Europeans in their narrow viewpoint about life.

This played out, in a swan-song for equality, in terms of A Jackson attacking A Hamilton's bank, but at the same time A Jackson was exterminating the native peoples, ie the Roman way of life won-out, and the banking investment class came to rule the US. (under Jackson, equality was expressed as mob-rule)

But

What can be concluded from the US history; ie the take-over of an egalitarian society by a social hierarchy based on violence?

Answer: Those very few people in the ruling-class are the only ones who one can be sure are: sinners, and bad corrupt people, who are indifferent to justice, and who are motivated to selfishly exploit a narrow vision upheld by both "propaganda based on delusions," and "extreme violence" . . . , the very few people in the ruling-class have shown themselves to be completely incompetent, and most of their, so called high-valued-knowledge has been shown to be a total failure, eg science, economics etc, yet the propaganda system claims unending scientific and economic successes. The very few people in the ruling-class are the ruling banker-oil-military social-class are clearly at fault since they are in control of society, and thus they are completely responsible for the corruption, and life-style failure, which is poisoning the planet, due their creative actions being based on a failed science and failed thought.

For example, these failed rulers have used partial truths (or clearly incomplete knowledge) in contexts which are resulting in ever greater destruction, eg nuclear energy (and weapons) and trying to manipulate genes, and run-away exploitation of resources (oil, coal, metals etc), and the propaganda-education system is the main source which has caused the failure of science, the failure (by the education system) to allow the expression of ideas in an equal way.

Economics and psychology are not valid models of existence (since the basis of measurable properties, upon which these types of [floating] descriptions claim to be based, are not valid quantitative structures) but rather operate in the narrow context of thought created by the propaganda system.

The: moralistic vs. materialistic pair of opposites, upon which the language of the propaganda system is based, are both words which have no content (they have no useful meaning within the contexts in which they are being used), and (subsequently) the creative context which they are supposed to define is a failed context.

The conclusion: Go back in US history to the egalitarian vision of T Jefferson, and more particularly of W Penn, and the golden rule (perhaps better re-stated as, "all people are equal,") and a government which works for the people, and it sorts out equality by promoting the common welfare, guaranteeing equal freedoms to inquire and create (and placing a limit on possessions [or on wealth]), so as to foster selfless creative actions in the public.

The media represents discussion as that which concerns a set of Hegelian-dialectic-opposites, eg right and left; science vs. religion, etc

The right is for inequality-arbitrary values (including the arbitrary value of dysfunctional physical and biological science)-and extreme violence, so that the state and society are designed to support the arbitrary value which is defined by an unequal society. The right is about "might makes right"

The left is for reason-fairness-valid competitions (or what appear to be valid competitions)-and the high-value of current traditional intellectual endeavors (even though all intellectual activity is narrowly defined to support the interests of the owners of society, ie intellectual efforts are designed to support inequality). The left is more for "the golden rule. (maybe)"

Here is how the propaganda system works:

The right (the owners of society) say a bunch of baloney over the media (expressing how society is going to change) while the left does not say "that is a bunch of baloney," but rather discusses the ideas at the level of reason . . . , so society changes as if the changes are reasonable.

The crux of the matter is that the A Hamilton's, the J Jay's etc (the propertied people, the right, those who want to maintain their dominant social positions by means of violence) never upheld the law of equality, as proclaimed by the Declaration of Independence, the society was corrupted from the time of the Constitution, since the Bill of Rights was never enforced. Now society is mired in corruption.

Subsequently, science and intellectual endeavors are also failing, since they are too narrowly defined . . . . , (for example, please describe, based on the accepted laws of physics, the observed stable system properties of nuclei, general-atoms, molecules, crystals, the stable solar system, and the motions of stars in galaxies. If this cannot be done (and it cannot be done) then physical science . . . , and all the other sciences which depend on the, so called, laws of physics . . . , are not valid, and their logical foundations are to be questioned in a deep way) . . . , and the public, and all of culture, is controlled by the single solitary voice of absolute authority, an authority which cannot be questioned, and that voice is the media, ie the voice of the owners of society.

The voice of the owners of society is the narrow irrelevant authority about the nature of existence (society, morality, success, and material based-truth) where the society, built on such arbitrariness, is mired in corruption and its existence is conditional on extreme violence, which is needed to maintain the social order.

This propaganda system "of the owners of society" is the societal "jail of language" imposed by "the owners of society," on the public, to motivate the public to be wage-slaves, who are to serve the narrow interests of "the owners of society."

This absolute language is perpetuated by its circular re-enforcement due to the one-voice of the propaganda system which continually repeats its own authoritative dogmas.

In order to exit, within . . . , the moralistic and materialistic . . . , night-mare provided to the public by the owners of society

Consider (truly valid) scientific-mathematical ideas which transcend materialism

That is, the 11-dimensional hyperbolic metric-space partitioned by a finite bounding (or existence-limiting) set of existing stable discrete hyperbolic shapes, of different dimensions and different subspaces along with a (metric-space-size) size-containment-set construct (defined on 11-dimensional hyperbolic metric-space), where together these different dimensional levels and different subspaces of containment . . . , where higher-dimensional levels are difficult to perceive when one is within a given dimensional level and if one has lost the realization concerning the size of one's own system (this is related to the origins, so called, myth of the giant Orion), . . . , define the stable space of existence, and it is upon such a model of existence which a creative extension or re-arrangement (of this same structure) is possible [is the command upon life made by existence] due to living, intending, energy-generating, existing-systems (such as humans).

This context of adventure and creative possibility defines human existence, not the material world, though the material world is a proper subset of the new descriptive context.

This is the true abstract basis for existence, and upon which the creativity of humans is best suited, and about (or to) which the owners of society have no relation, and their actions oppose such a viewpoint.

Instead, humans have surprising "dreaming and doing" capabilities, but the context is an abstract but truthful context for existence, not limited to the material context, though the abstract context contains the material context as a subset.

The new context is already proven at a deeper level than the current, so called, "accepted laws of physics" are "proven" since the proof of the truth of the currently accepted set of ideas (namely materialism) is exactly the same type of (data-fitting) proof as the (epicycle) ideas of Ptolemy "had for their proof" in the age of Copernicus.

It is a data-fitting exercise which exists between language and observed patterns, the language is adjusted to fit data, ie new particles are found and then new forces are proclaimed, and the new particles are fit into the language, but there are no predictions as to what the new forces would be, eg dark matter was not predicted within particle-physics, but now there is claimed to be a particle associated to dark matter, after the affects of dark matter were observed, ie the motions of stars in galaxies.

The proof that the new descriptive context is true, is that this new descriptive structure identifies the structures of . . . : nuclei, general atoms molecules and the solar system . . . , as all being stable systems, where it has been observed that all these systems are observed to be stable, whereas the currently accepted "laws of physics" (as given to society by the owners of society) simply state that these (just mentioned) systems are "too complicated to describe using the, so called, laws of physics."

Furthermore, the new descriptive structure provides a likely model for dark matter, whereas dark-energy seems to be a result of a galaxies (an 11-dimensional hyperbolic metric-space's each of which define an SU(11,1) Lie group) relation between Lie groups and the conjugation transformations between these Lie groups

Simply put, the stability of the solar system is proof that the new ideas are "correct!" while the currently held beliefs about the material properties of existence are essentially irrelevant as a viable intellectual context for consideration (what the owner of society believe deserves consideration [in the current set of narrowly defined {dogmatic} scientific beliefs] is that the particle-cross-sections of are related to the rates of nuclear reactions which are models of weapons, ie the current set of scientific beliefs are serving business interests (and that is why they are expressed in the propaganda system).

19.

The only voice

When one identifies a new context for measurable description, which is both accurate and practically useful in regard to creativity, then one is faced with the social problem of "being heard" in a society wherein the propaganda system expresses the absolute truth (about which the public has been trained to be in lock-step agreement), so that within the language-context construct of the propaganda system, both social organization, and (funded) practical (or productive) creativity are defined for all of society (by means of the investments by the owners of society (ie the very few: big bankers, monopolistic corporations, wealthy individuals, and the national security state, which serves these very few people, within society).

This limited absolute truth (of the propaganda system) is, essentially, the authoritative voice of the owners of society, ie the very few who determine what is funded in society, for a society which is organized around both/either wage-slavery and/or the investments of the few owners of society ie those few who control the monopolies, banks, and the military (all very narrowly defined) within society.

The absolute truths of the propaganda system are of sufficient, but narrowly defined range, to be used to practically create the products which support the social power of the owners of society, in regard to the investments of these owners of society.

That is, new ideas would interfere with the interests of the owners of society, and thus they are not allowed to be heard, or if stated somewhere, then they are ignored.

The propaganda system is the solitary voice of absolute authority for society. This is one of the main causes for the destruction of society . . . , because this voice is far too narrow, and despite its obvious failings, due to its narrowness, the social power-structure manages to hide these dismal failures from the public, ie the public is confused and is willing to follow an authoritative voice.

152

But clearly the leadership, the owners of society, or that handful of people controlling important social institutions and controlling social power within society, is even more confused, and they simply maintain the narrow social structure from which they derive their social power.

Thus, leading to social failure.

However, there is an inertia of a relatively stable life-style within which people continue to exist, ie this narrowly defined very stable social structure (maintained by extreme violence) has the same effect that the Roman-Empire, and the narrow interests of its emperors, had on the regions which they controlled. These are geographic regions whose communities the Romans made stable, and these communities were narrowly focused and compartmentally organized, so that when such a social structure is maintained, it can be turned into markets and cycles of investments, without much risk.

The form (template) which the ruling class follows in designing its propaganda-strategy is the Hegelian dialectic opposites; of science and religion (which developed as the printing-press, banking, science, and religious reform also developed), ie small diversity in a stable social structure can be created and exploited, where strategies for social exploitation revolve around the extreme violence and associated inequality which had been modeled (or formed) within society by following the social structure of the Roman-Empire, which brought it into being.

However, Now (2013), science has failed, it is not capable of describing the stable fundamental physical systems which exist at all size scales, eg nuclei to solar systems etc. and, that these systems are stable, in turn, implies that these systems are quantitatively consistent and controllable, ie these physical systems are to be modeled by the simplest of the math patterns. Nonetheless, physical science and biology are both based on the quantitative (math) models of indefinably random descriptive contexts, along with the quantitatively inconsistent non-linear constructs.

These ideas are presented as vague ideas within complicated contexts, so as to pull the wool over everyone's eyes.

Nonetheless physical science is serving the current needs of business (or military) production.

Biology is based on the story about (narrow) competition and the survival of the strongest, or most virtuous (ie those who are superior, ie it is used as a basis for social inequality), where this idea is explored on a molecular descriptive context.

This model is so abstract and complicated, that one must ask, "How can one take it seriously?" Especially, since living systems have an almost absolute control over themselves, which again implies a very simple quantitatively consistent and controllable descriptive context (at some level of system organization).

This same scenario of baffling complexity, is also used in the political media, where there are vague ideas about inequality, eg the need for competitions, ie competition fits into narrow rigid dogmatic contexts, along with vague ideas about "the market" and its related economic quantitative properties.

However, the basis for these economic quantities are arbitrary-value, which are created by a propaganda system, ie the quantitative structure of the market (or the economy) is arbitrary and unstable.

But associated with the political media is religion, with its basis in arbitrary morality. For example, in the catholic church the spirit is dependent on a person's relation to the church, while in (Calvinistic) Protestantism, a guiding principle (in regard to "the virtue" of a person's soul) is success in business or success in the military (ie can a community get what it wants by violence or investment?), though this relation between spirit and social success within a narrowly defined society, would be implicit in the catholic church too.

The Roman emperor Constantine made the Christian Bible a handbook which supports Empire. For example, the golden rule is not the very simple statement, that "people are equal," but rather, that "one should treat others as one wants to be treated" an almost baffling statement.

The Roman-Empire also provided the violence needed to create and maintain stable social conditions, and separation of a narrow set of skills, which together got organized around markets and investments.

Essentially, around the 1400's the bankers took-over the emperor's-seat from the Roman-Empire, eg Constantinople (then the center of the Roman-Empire) fell around 1450. The take-over by the bankers was related to being able to adjust to new ideas which could be used to perturb a stable social structure in ways that the ruling class could exploit.

One might say that the US dominance in capital has been related to being able to change to a new technology . . . , which classical physics and the machining of metal shapes, was able to generate during the 19th century . . . , whereas Europe had more social inertia concerning its organization of engineering technology and the relation of this technology to expanding investment models, in order to stay strong an Empire (or an economy which is too narrowly defined) must expand.

From these institutions of science and religion, both of which have come to be defined by absolute dogmas, are now both failed institutions, where religion is all about vague abstract promises, but its descriptions have no practical value, they only have social-value in regard to social manipulations within the propaganda systems. Within these contexts of absolutes, but arbitrary, authority (in regard to both science and religion) the propaganda system weaves the absolute truth which best serves the interests of the chief investors in these narrow social structures of business and its relation to community, interpreted as a narrowly defined market (so that investment risks are small), and the propaganda system endows the communities (by controlling the context which language can create within a community) with the faithful attempt of its people, to strive for (a narrow range of) societal success, ie serving the interests of the owners of society, ie since the private bankers own the media instruments.

Inequality leads to failure of society, including the failure of the cultural knowledge, where this cultural knowledge has been stolen from the culture by means of copy-right laws, whereas society is stolen from the public, by having law based on property rights, and by an associated form of organized violence, ie the usual traditions of the Roman-Empire, where there are many hidden groups which can be activated to administer focused terrorism onto the public. *The US gun culture is about manipulating these hidden groups to terrorize the public.

Note: Today's (2013) ruling business class developed the relation that the basis of their engineering had to "new" products was developed in the 1800's, when the US big businesses were developing. Thus, the business ruling class today has an engineering relation with the classical physics of the 1800's. Subsequently, little new technical innovation has occurred since the 1800's, and though micro-chip development was slightly significant, S Jobs and (less significantly) B Gates entered the business scene, but it was still based on the science of the 1800's, but it has been organized as a spying device for military and propaganda control of the public. Note: The social structure only allows individuals to rise to power, this allows for more control from above and it selects domineering personality types to become socially prominent.

In an unequal society, controlled by propaganda and violence, one cannot express a new descriptive truth which is contrary to accepted dogma, since social relations are managed and controlled by both language and violence, eg intellectual domination, since such a new truth would undermine:

The value structure of society
The ability to control and manage language within society and
Thus, it would undermine
The social inequality of the society, unless the provider of the new truth [produced a new product, and] sought personal aggrandizement, and used the social structure to become a dominant player within the unequal society.

The power of the ruling class in an unequal society, in a society ruled by a few, rests on dogmatic absolute truths expressed by the propaganda system, and these truths can be organized around the dual truths of science and religion, or materialism and (arbitrary) morality.

The experts in this game are numerous, they are aspiring, ambitious, competitive types, most often with social advantages from the start, who seek recognition of being of high-social-value, and they are people who may think that they are seeking truth, or the more powerful are those authoritarian domineering types, those whose personalities are suited for an unequal society. That is, psychopathy seems to be a learned behavior within an unequal society, which can become well rewarded in an unequal society, where authoritarian domineering types seem to be the personality types who end-up on the mythical "council of planners." For example, E Teller in conjunction with the military, together they turned university physics departments into bomb engineering departments. As well as JD Rockefeller and the Bankers etc.

As S Jobs (remember S Jobs was a manager for micro-chip development) once said, despite the hype, "these people are not all that impressive," rather they are manipulated by propaganda and social power, in a social structure which suits their domineering personalities, so as to become the personal face of the same narrow dogmas which were forced upon them, and by their intent to dominate, and yet they must obey the Empire, so that they can acquire the reins of social control, where the complete failure of their narrow endeavors are obvious to see (the destruction of the world and the people in society), yet they (within the propaganda system) falsely claim "to help so many people, by efficiency and inventiveness" in their superior positions, where they manage other superior people, rather there is:

Destruction of people and resources,
failure to invent
failure to "market" well intended products,
. . . , and creating a society based on mistrust, jealousy, and violence, and
a society which is intellectually fitted into a straight-jacket.

Where development and control of the food system has been their strategic focus (both markets and militarily), and it is still organized similar to how the Roman-Empire organized the food system.

That is, the bankers are the Augusta Caesars in essentially the same social structure which is ultimately based on violence and the control of the solitary voice of absolute authority (ie the propaganda system).

That is, it is a system designed to serve and maintain the power of the few, and their world design is too narrow and exploitative, ie it is based on absolute truths which are not true, so it consistently fails, while it continues to float upon the, essential, social structure of the Roman-Empire: food, transportation, water, sewers, a nod to (civic) art and cultural-entertainment, and old engineering traditions which can be fit into the power structure of the Empire.

The narrow structure of this social structure demands expansion, and thus it also demands continual war with those with no (or very little) social power, ie war on those who do not directly serve the Empire's interests.

Essentially, a society cannot realize a measurable described truth which is practically useful unless the society is an equal society.

This is the essential conclusion of Godel's incompleteness theorem, which states that precise languages have limits to what patterns which they are capable of describing, so practically useful precise language, related to measurable, useful, informative, and couple-able descriptions, must be continually renewed, based on new: assumptions, contexts, and interpretations, etc, ie the language needs to be built from an elementary level, and everyone is well equipped to join-in on the game.

20.

. . . the follower, is the one who "must learn to fill in the blanks"

To see the judgmental words of those who follow, or is this a high-ranking meritorious person's beliefs . . . oh they are the same types of people . . .

to the mind swindlers

Is this to be addressed to either E Witten or the "protection-racket detail" of E Witten, you arrogant guards of the truth (or equivalently pillars of nonsense, or more to the point, "protectors of social class") please provide a solution to the stability of the general nucleus, or a solution to the more than 4-body general atoms, namely provide your solution to the radial equation for this context, or provide the equations which show that the solar system is stable.

These new ideas concerning geometrization applied to a many-dimensional context (which allows both macroscopic and microscopic higher-dimensional math-physical constructs) provide such solutions.

Why you let others do your thinking for you, has always been a mystery to me, since you are protecting the beliefs expressed by the propaganda system (which ignores the interests of the public), and since we, supposedly, live in a free country, where there is freedom to believe and freedom to express helpful thoughts, helpful in regard to accuracy of a detailed description, and its relation to practical creativity.

If you (thugs of intellectual-nonsense) have nothing positive to add to the discussion, ie solve the problem (or provide a solution to the problem), then be quiet and let freedom ring-out for all to join the discussion about the range of description concerning measurable and practically useable descriptions.

Main part

The trouble is there is no expressions of substance anywhere, but a new foundation for physics is a good try, expressions which possess meaning are ideas which are related to creativity

This is not academic, rather it is communication based on assumption, context, interpretation, and an implied relation to creativity, or an implied relation to life

creativity is first about:

"what is existence?" then
"what are some of existence's properties?" and then
"how can these properties be used to create?"
but most important
"what is the context of creativity for humans?"
It is not the material world.

The issue in academic science has been about measuring in relation to the properties of material, and it now has degenerated to randomness and quantitative-inconsistency, so as to form the, so called, measurable basis for material (and life) properties, but this context only describes complicated nonsense.

furthermore, the academics ignore the requirement of the stability of patterns which is needed for this measurable set-up

The context of stability (in math) is (happens to be) the context of stable shapes (they are very simple shapes) . . . (this is expression at the level of assumption, but to the academics, the idea of geometrization now implies this idea), . . . , and these stable shapes are the primary basis (or focus) for a description of existence's properties (and the basis for actual properties) (in many-dimensions), and they are patterns which also allow both stable measuring, and stable systems, which together define the context of (actual or attainable) "relations" or "changes."

so

what is creative?

Perhaps (creating) a new containment context and simple shapes with new properties (ie new relations) or new simple shapes (ie changes).

The many-dimensional, but simple, shapes include within themselves (within the containment-set defined by existence) both material and life systems.

life uses knowledge (communicates) to create in this bigger context
(this is about the meaning and substance of life).

there is no way to make new ideas concise in the propaganda system, unless one is an expert and communicating in the categories which are allowed to be expressed, in which case being concise means writing in code, "in the aurora of the expert"

creativity is about geometry and measuring, and understanding these concepts at the level of assumption and interpretation, and conciseness means going along with the current assumptions and interpretations.

all people take the propaganda system as the absolute voice of authority, and they babble mindlessly, and often in very careful detail, about categories in which intellectual endeavor is defined by that absolute institution of the propaganda system. that is the owners of society are defining the boundaries of people's communicative inquiry through the propaganda system

consider left and right:
the right is for the owners of society and traditions, while
the left is for reason and supporting the experts (who serve the owners of society)

if there is a valid division in political thought it would be:
equality and freedom
vs
inequality, violence, and absolute authority (the Roman traditions)

no one sees equality as the correct basis for freedoms (though this was close to the true mind-set of the public at the time of the American revolution), rather people have been trained {by the media and education system} to consider . . . absolutes, which are really backed-up by violence, ie the violence of the restriction of communication and the violence which allows one set of thieves to prosper and, in turn, attack the public if they try to "steal," . . . [this is what is seen] as the basis of freedom for both "left and right" (note: In China the leaders which would be chosen by Marx rule, but there are no differences)

Who speaks English?

In regard to social issues I do not think anyone speaks English, everyone babbles about baloney, and the owners of society do not want any expert to make any sense, since the owners want the status quo.

The social issue is:

Is law to be based on equality, as the Declaration of Independence states?

or

Is law to be based on property rights and minority rule, enforced by using extreme violence, (as the federalists wanted, and as the Roman-Empire was, so, based)?

If the latter, then the 99% always loses, and the babble remains ineffective for the 99%.

21.

Propaganda and the destruction of the mind

New ideas are not allowed by the owners of society in the criminal national security state, which serves the few owner's interests. This is the crux of why society is failing. This hierarchical and criminalized way of organizing society is accomplished, mostly, by the propaganda system, where the public (and public institutions, such as the press and the education system), as well as many secret organizations (which guard against new ideas, as a part of the national security state) support this social structure, but this "support" is mostly a result of the fact that "the language of the society has no meaning (due to the propaganda system),"

(another new attempt at being concise, but it might not be comprehensible, because it is a description which uses too few words)

and this (non-meaning within "the language people use") is seen in the context of modern science being dysfunctional (physical science of quantum systems is unrelated to practical creative development), almost all technology of the big business interests is based on 19th century science (including micro-chips). Nonetheless, the science authorities are mistakenly worshiped by the public, in terms of personality-cult, as this is suggested to them by the propaganda system (based on the charlatan-magician's deception of the implication of the experts possessing (secret) knowledge of an unknowable mystery).

Just as the Romans, today's owners (emperors) of society provide the public with engineering niceties, eg the Romans made water systems, and sewers etc, the technology of bricks.

Consider that equality is the main principle taught by most (if not all) religions, and yet in the US society the propaganda system has developed a set of people . . . , perhaps they are mostly paid agents, yet they are prevalent, and thus they most likely (also) represent the twisted mind of the public (twisted by the propaganda system) . . . , who deride equality as a social-evil.

But then "the Emperors religion" does not serve religious interests, but rather it serves the emperors interests.

This opposition to equality is also related to the political agenda of the propaganda system, where inequality, and narrowly defined competition, is the basis for the main message of the propaganda system. The propaganda system is the sole voice of absolute authority within society, which is used to guide everything which is done inside public and private institutions of the society.

Consider the main sets of pairs of opposites upon which the propaganda system is built:

Science vs. religion
Or
Materialism vs. idealism
Or (more to the point)
materialism vs. arbitrary moralizing

Where
The owners of society, the banker-oil-military interests, readjust the idea of knowledge (by means of the propaganda system) so that the study of material properties is related to the production associated to big corporations, ie knowledge of the material world is organized (in all educational institutions, including universities) to serve the interests of the owners of society [apparently, unknown to the experts themselves].
The main way of doing this is to:

1. Base most (all) engineering on 19th century science
2. support the material based view that arbitrary randomness is needed for the descriptions of both
2(a) physical systems (so that the focus of physical science is on the rates of reactions in unstable systems, which are in transition between stable states) and
2(b) in order for the material-based living systems to "evolve" into complicated systems they must change in a context of competition for limited resources, the survival of the fittest (so as to provide, an invalid, scientific proof that the world is based on inequality).

This, in turn, is used by the propaganda system to prove that inequality is a scientific fact.

That is, materialism, and a material based description of biology, are ideas used to support the idea of inequality having a scientific basis, where physical science is based on indefinable randomness and non-linearity and/or quantitative inconsistency.

Furthermore, indefinable randomness is related to probabilities of collision, and this is the main idea used in the military industries.

Thus the pairs of Hegelian dialectic opposites upon which, the way language is used in, the propaganda system is built, might best be identified as:

Materialism vs. equality
(since equality is the main virtue taught by religion)

But
In order to organize society to serve the narrow limited interests of the few, "arbitrary moralizing" is substituted for equality, and this "arbitrary moralizing," and an invalid claim that inequality has a scientific basis, is used to manipulate the public by means of the propaganda system.

Materialism is the basis for classical physics, and for some unknown reason it remains the basic assumption of physical science today, but now physical science wears a cloak of complicated mathematics which is not capable of describing the observed stable properties of fundamental systems, eg nuclei, the stable solar system etc.

Math begins by being based on stable patterns; the stable patterns of quantities; and the stable shapes of geometry.

One can change measuring scales in a consistent way if one multiplies by constants, or if one uses linear inter-relationships between quantitative sets, which are used to measure and identify physical properties.

Then measuring came to be about (due to Newton) the values of functions which represent measurable properties of material systems, which are contained in coordinate domain spaces (for the functions), and measuring was given a local linear model, the derivative . . . ,
(and an equation, where for equivalent but different measuring processes [used for defining the same property] are set equal to one another, so that the (partial) differential equations of material systems is defined, and where material is defined [or identified] as a scalar factor associated with some of the terms in the system's (partial) differential equations), . . . , which in the limit process, which defines the local (linear) structure in regard to measuring local function values, transforms functions, in such a way so the function transformations (of the derivative operator) can be inverted (by the integral operator), and thus (partial) differential equations can be solved, but only

for "system shapes" where the measuring model is continuously independent and linear (diagonal everywhere) in the coordinates.

When material was observed to be reduced to smaller stable charged components, which form into stable systems, but if these systems are broken apart, or observed by a probe, or if energy is emitted by these stable systems (in the form of light), the components (associated to these possibilities) seem to possess random properties, ie as random spectral-particle events in space. But bounded charged particles must be bound together by forces, and forces applied to charges cause the charges to emit energy in the form of radiation, and thus such a system would decay, for material defined by (partial) differential equations associated to a system's shape.

So it was assumed, by Bohr, that the components of these systems formed into stable orbits defined by discrete values, in turn, defined by (or related to) integers (discreteness) and the Planck constant. The values of energy found for such orbits when perturbed into elliptic orbits, provided agreement with the system's spectra, which is of about the same precision, if not in better agreement, with the observed values, as the wave-equation models of the same system can do, but the wave-equation is now preferred (where the wave-equation is based on function spaces and their operators, eg the derivative acts [or operates] on functions), because (it is believed that) the wave-equation methods identify a good model for the indefinable randomness of point-particle-spectral events. However, the wave-equation cannot be solved for general quantum systems, but rather it is solvable only for the simplest, separable systems, eg the H-atom. That is, indefinable randomness does not work as a valid description, which can be applied to general quantum systems, and subsequently solved, ie the general quantum systems cannot be solved.

It is true that randomness can be determined for some types of random events in a valid manner, such as the toss of a coin, where the random system depends on some finite stable state of material, which defines the random event, and which determines the set of random-event outcomes.

Though general quantum systems are stable, the math methods used to identify these stable states cannot find these stable states by means of calculation.

That is, these quantum systems are being described using the math context of indefinable randomness, and for which these stable systems cannot be solved.

This means that these quantum systems are having their descriptions based on indefinable randomness, so that in this context "the meanings of words" start being changed in order to describe the observed properties of systems (where it is assumed that these systems can be reduced to unstable components), eg renormalization (where space no longer means coordinate containment in regard to a physical quantum system). This is quite similar to the epicycle structures by which Ptolemy "predicted" the motions of planets.

However, some of the components (or elementary-particles) are stable, eg the proton and electron which, apparently, do belong to the material world, thus the unstable components (or the unstable elementary-particles) do not belong to the dimensional level which is defined by the material world (this is perhaps the best interpretation of these results, concerning the data from particle-collision experiments).

To summarize the mathematics: both indefinable randomness and non-linearity are properties associated to quantities and shapes, but this association is done in a quantitatively inconsistent manner, and this tends to move the description outside of the realm of valid math, where math is built upon stable patterns, and math must try to describe stable patterns (the observed stable properties of fundamental physical [quantum and classical] systems). The descriptions based on indefinable randomness and non-linearity become both unsolvable and of no practical creative value.

On the other hand, assuming that components fit into the stable shapes (which are metric-spaces, and) that these are the stable shapes that the "metric-space containing-spaces of physical systems" naturally determine, is a much more natural viewpoint (and a better way to reformulate Bohr's idea about the atom, and at higher-dimensions apply it to the stable solar system), and that such a structure of metric-space shapes can be put into a high-dimension containing space, so that unstable elementary-particle events can be described within their proper containing space, where such a metric-space model would exclude radiation from being emitted. This is a context which can be associated with the idea of idealism (many-dimensions) which in turn, contains the material world as a subset to itself. That is, it is a very absorbing, and relatively simple (due to the simple stable shapes of the metric-spaces) alternative to materialism and indefinable randomness. Furthermore, the shapes of this context immediately provide the stable orbital geometries needed for modeling the stable physical systems of: nuclei, general atoms, molecules, (crystals), as well as the large-scale stable solar system. The many-dimensions do not make this a simple model to consider, it is much more complicated and more varied context than is the idea of materialism, yet it gives understanding concerning the structure of stable physical systems which exist at all size scales.

That is, if the subspace-size-containment set structure of this model of existence is sorted-out in the 11-dimensional hyperbolic over-all containment metric-space structure of existence, then the partial differential equation (or operator acting-on function-space) model of determining measurable-forms, which exist physically, becomes unnecessary, since, though these shapes are solutions to partial differential equations, they are also simple stable shapes of a particular type (circle-spaces), and thus they are simple to identify, if the finite spectral-size set of the over-all high-dimension containing space is identified, and thus stable systems can be related to existence (within the particular 11-dimensional hyperbolic metric-space), through resonances, which exist between the spectra of circle-space (metric-spaces). The partial differential equation (or operator acting-on function-space) model of existence ends-up only revealing the main structure within metric-space subspaces, and that measurable structure is mostly indefinably randomness and/or the non-linear structure of material interactions, because a description confined to only one subspace (of existence) is a descriptive-context which does not possess a bounding stable set of structures (identified on both subspaces and dimensional levels) upon which stable shapes can be determined, ie or exist and identify a stable quantitative construct within which measuring is also stable.

The math and science related to these ideas, ie ideas which were presented at the joint meeting of math in 2013.

In regard to the language of the propaganda system, Marxism, capitalism, and biology based on indefinable randomness (and thus is interpreted to support inequality) are all on the side of materialism, while equality is the most significant virtue taught by religion, but, in fact, equality is also related to a process of determining a precisely described truth, where a described truth "is a truth which is limited (by the language itself) in regard to the set of patterns to which it can be related," which, in turn, means that descriptive truth must be always changing its relation to: assumption, context, and interpretation, etc, and any person is equally qualified in doing this ie no hierarchy can be defined on the unknown.

Thus, one sees that when one supports equality and one is derided, for such a belief in equality, then this is a criticism by a person who implicitly supports lying, stealing, and violence, ie it is belief structure which is in favor of criminality, a criminality which is mitigated by the, so called, "scientific law" of "survival of the fittest," which is defined within a context of indefinable randomness (interpreted to support inequality, but indefinable randomness cannot be used to describe a stable pattern). In turn, indefinable randomness implies an invalid math (or invalid measurable) descriptive language, where such a dysfunctional language should be changed.

The uppity types who self-righteously vent their prejudices exist at all levels of the discussion from the: secret society of exterminator Puritans, to the rational liberals who blindly worship scientific authority (which they believe is beyond the reach of ordinary people), to the A Goodman's and R Limbaugh's of the media, all (funded) pawns in the few owners' grand design, a design which can be thwarted by applying to society the core of religious teaching, namely the idea of equality (we are all equal). Furthermore, if one wants to find truth, which is useful in regard to practical creativity, then one also wants equality of belief and equality for expression, and equal free-inquiry, associated to either practical creativity, or associated to the design of society, so as to determine which human attributes that the people want their society to develop within the public (within their society of equal people).

No comment (to the mind-phantoms which emerge from the propaganda system)

22.

The illusions of propaganda

Answering the deceptions of the propaganda system.

In this, so called, age of communication, one finds that words do not have dictionary meanings rather they are icons and stimulate deep belief and emotions, so a matter-of-fact communication based upon word-meanings and ideas, which words are supposed to identify, (ie categories of thought) cannot be realized.

Rather the public has a mind which is more like chards of pottery than like vases, where one is to thinks of a vase as representing a whole thought about which many ideas can be related. For example, consider the words capitalism vs. Marxism, these are both words about how society chooses to organize itself about material, where material is thought of as the only medium of practical creativity, for both of these ideologies.

The propaganda system has infused the word "capitalism" with the idea of individualism and efficiency, and "Marxism" is a word which implies government controlling the social use of material, where this use is determined by a small committee, then in such a scant representation of ideas one is leaving out the fact that capitalism is about monopolistic domination of society, so that all other businesses of any type need to have a (deep) relation to the monopolistic banking-oil-and-the-military interests, ie all capital depends on a few dominant individuals.

Note that, this organization of society by capitalism is essentially the same organization of society as a controlled-economy planned by a few, the (implied) idea of Marx. That is, due to the domination by monopolies of all society, especially dominating the market-place, a capitalist society is centrally planned society based on selfish interests of the owners of society (the bankers), and again it is not valid to represent such an idea about capitalism as being an economy governed by a quantitative model (eg resources, added-value, and the subsequent relation that this process has to the relation of supply and demand associated to prices in a free-market), because there are individual players (or individual stake-holders) who possess enough monetary dominance that all aspects of the market can be changed by these players changing their investments.

Thus, in regard to the intended (or proclaimed) virtues of capitalism, Marxism provides (might provide) a better way of both planning, economic flow within society, and it is closer to a quantitative structure which can be adjusted by economic principles of a free-market, instead of a corrupt state which builds on the illusions about capitalism so as to support a highly controlled society which is advertised to be capitalism.

The proof of these claims is that Marxist China is now winning the game of, so called, capitalism.

One could say that "the key" to a valid model of capitalism, is to have a truly free-market, which allows a wide range of creative efforts, where some such efforts could be based on creativity related to material, and some creative efforts which should not have a material under-pinning, as an important part of the social organization of these activities in society, such as art, where art in the capitalistic society of today, is based on the instruments of communication, ie communication channels and the wiring of these instruments, which the propaganda system, or the owners of society, it is claimed, must control, so as to have a "free-market."

But a truly free-market can only be realized if everyone is considered an equal creator, and to have equality of material resources and to have the value-added capability of anyone else. In this case creativity will be directly related to a valid form of knowledge, as opposed to the currently accepted model of science.

The currently accepted model of science and math is narrowly dogmatic, so that the dogma is related to the production of nuclear weapons, and in turn, that dogma is based on indefinable randomness and non-linearity; which are two mathematical ideas which have not been shown to have any relation to their being able to describe the observed stable properties of material systems which are observed at all size scales. Thus, today's science and math has almost no relation to practical creativity, because they are too narrowly defined, where university is organized around finding obsessive obedient types who can be used to service the fairly complicated instruments which serve the interests of the owners of society.

That is, this narrowness is done to please the narrow selfish interests of the owners of society. It must be pointed-out that the structure of society today (2013) has essentially the same narrow structure as did the Roman-Empire, where Rome was based on both property rights and military coercion, so as to be organized around a narrow way of organizing society, which included the engineering of brick-laying, and then violently requiring such a narrow way in which to organize society, so that this is all done to support the Roman-Emperor. Now the engineering provided to the public by the owners of society (the bankers) is based on: 19ᵗʰ century science, oil, and nuclear weapons, where this narrow viewpoint is violently maintained to support the owners of society.

That is, the individualism of the so called capitalist societies is about narrow selfish interests which require expansion of use of resources and markets expanding to steal the resources and expanding to sell product all within a fixed narrow model of society, a model which serves the owners of society.

Language has become filled with fragments of ideas, words with value-laden meanings, which are used to express loyalty to the high-values of society so that these high-values are defined by the propaganda system or by the owners of society. For example the high-value of intelligence but intelligence is defined in regard to the process of acquiring the knowledge needed to adjust the instruments used by the owners of society used so as to serve their selfish interests. That is, intelligence is not defined as the capability to discern truth. Instead fragments of ideas which are locally related to instrumentation are memorized in a very disconnected way so that competitions concerning either funded research interests, ie grades, or "discovery" within the context of the language, eg discovering elementary-particles, is dominant, and not issues about describing fundamental properties which the descriptive language is supposed to describe, but which it cannot describe.

If a society's science is failing, especially in regard to accurate descriptions, then the more pressing problems of finding a cheaper and cleaner energy source based on knowledge, ie finding a cheap and non-polluting way to provide energy cannot be solved. Or of equivalent importance requiring that society provides an energy-source based on a less-destructive process. That is, since science and math are failing, there needs to be challenges to science and math, and these challenges need to be made at the elementary level of: assumption, context, interpretation and etc. otherwise science and math are playing a role of helping to create the fragmentation of our communication structure and knowledge structure, and the main issue for the institutions of science and math degenerates into selfishly seeking to acquire social-intellectual domination for the expert authorities whose main purpose is to influence the propaganda system so as to have experts who are incompetent but who the public are told to worship.

So experts, please provide the descriptive language in which the stable properties of fundamental physical systems (which exist at all size scales) can be described in a way in which the description is related to useful practical creativity, especially in regard to (if possible) providing cheap clean energy production.

However, it should be noted that there does exist a new descriptive language which accurately describes these stable physical properties of physical systems at all size scales, and they are very simple geometric ideas.

Thus, these new ideas should be a central part of the expert discussion, or a better descriptive language needs to be put-forth

The idea of intelligence is now used to identify superior people, ie people with superior DNA, so that this has now taken the place of racism. It is the same social mechanism of identifying superior people and then this can be used as a justification to exterminate those who do not serve the owners of society.

Expansion of narrowly defined set of products, in turn, needs consumers, but wage-slavery and its associated narrowness leads to failure for such a model for society, which can now be seen so clearly, ie 16-trillion used to cover-up the fraud and stealing of the bankers.

That is, the idea that society is a meritocracy and that people get what they deserve, is an illusion perpetrated by the propaganda system, and this illusion has a number of fatal flaws:

the knowledge base (science and math) is actually failing, and
thus, it does not really define merit (or value) [it is a society based on fraud],
it is too narrow, individuals base their activity on their own personal narrow vision, and
it requires expansion, which the world cannot sustain and still sustain life, and
it requires endless violence, which is motivated by a fraudulent social-value structure,

where all of these properties together lead to destruction of both the earth and the destruction of society.

23.

Reply to Abstract

Reply to an Abstract submitted to professional physics journal

Reply to an Abstract concerning a non-linear model concerning physical descriptions, a subject of central importance to the professional math and physics communities, as the author was asked to consider, by some physics research journal.

This abstract is absurd

It makes claims about the pattern of interest being described as a non-linear pattern, ie it is an unstable pattern, so that only the conditions in which the partial differential equation is defined can be related to some sort of transitory, and usually difficult to define (as being valid), range for a feedback system. That is, the descriptive effort is for a context of possible feedback control which is unstable, ie the effort is essentially useless.

To re-iterate, this abstract is about a non-linear, quantitatively inconsistent, non-commutative math context which describes a transitory unstable state, no explicit pattern is being described, only operators which are being associated to transitory unstable properties.

On-the-other-hand stability of spectral-orbital systems at all size scales is very prevalent, but it goes without valid descriptions.

The stable patterns for geometric shapes in space are the discrete hyperbolic shapes where this is the conclusion of the Thurston-Perelman geometrization, so the properties of the fundamental stable systems of existence must be related to these stable geometric patterns. (This has been done, but it is a new math construct, so it is not allowed in the peer-review literature.)

Science is about stable patterns which have practical uses, eg the control of circuit properties modeled as linear (solvable) systems. Such knowledge about the conditions of stable patterns can go a long-way!

Though it is interesting that transitory patterns can be related to properties of differential operators these types of math structures will not (be able to) describe the observed stable physical systems, which are so fundamental to the properties of existence.

So much effort for unstable curiosities which have such limited relation to useful practical creativity.

24.

Commentary (comments concerning N Chomsky, and the academic-propaganda viewpoint)

The main division in regard to the structure of language in the propaganda system is the "science vs. religion" dichotomy, which is a false expression of opposites, since a many-dimensional macro-micro model of containment of observable properties of existence is better suited to being able to describe the observed patterns of the world than is the very restrictive idea of materialism, in regard to "physical" science, and its subsequent relation to molecular biological science.

The division of personality types in regard to the "allowed types of voices of the propaganda system" are:

(1) the rational types (the liberals) who express obedience to the authority of the Emperor's science, and a belief in an objective quantitative analysis of both issues and problems (of politics) being faced by (of) an unequal society,
   and
(2) the loyal to-social-class types (the conservatives) who defend the emperor's position in regard to the use of violence and the required consequences of social inequality (a scientifically lawful belief, in the sense of the, so called, "natural law" of biology [the survival of the fittest]).

There are the more consequential intellectual stance of the liberals on the left . . . vs. those intellectuals who are endowed with the weight of the propaganda system on the right, those (on the right) who misdirect, misrepresent, and delude, the public by their use of language within the propaganda system (which the right controls), and by the issues upon which they focus, . . . , yet it is the left which most strongly deludes the public, since it is they who most represent and defend the, so called, objectivity of a material world, and the, so called, scientific authority which rests upon the idea of materialism (and its relation to reduction indefinable randomness and non-linearity, upon which the work of the intellectual illuminati depend).

The left demands that public policy be based on rational implementation of the scientific truths concerning the world, as they interpret scientific truth.

For example, scientific truth demands that global warming be stopped by the method of "stop burning so much carbon."

The left loses track of the fact that, "what has trapped civilization" is the narrowness of the Imperial-material social context, where social dominance (by the owners of society) and its narrow intellectual-practical basis demands expansion of this (the) narrow context.

That is, "science based on materialism" is far too narrow a viewpoint upon which to base the creative efforts of a society, unless one accepts the fact that such a society must expand and, subsequently, exploit both the earth and society, so as to allow this expansion (within n such a narrow social context).

By obeying narrowly defined scientific dogmas they are also supporting the narrow way in which society uses resources. For example, the hunter-gatherers used a wide range of plant and animal knowledge so as to nutritionally support themselves, Thus, the question is: Does the earth produce more in its natural, balanced (but tended) state, or in a narrowly defined cultivated state?

There are the dedicated and serious progressives in the media, on the left, who express moral indignation at the brutality and lack of rational concerns of the ruling class, which must be matched on the right by a more appealingly sincere opposition to sometimes similar and sometimes made-up outrages, where R Paul's opposition to war and support for individual freedom . . . , (but by "individual freedom" he seems to really mean the freedom for the ruling class to dominate society, where the ruling class supposedly derive their power from a free-market) . . . , shows more back-bone than most of the left's complaints (though the left resists, but the issues about which they complain would be better motivated if they focused on those properties of both government and law which makes government and law so compliant with the will of big-business interests, ie property rights vs. social equality).

However, R Paul seems to be used as a way in which to stir-up a deeper rational-emotional basis for violent support for a, non-existent, market-freedom and its relation to the individual freedom of the ruling class.

The manipulation of a rational-emotional context of thought seems to be a central technique of propaganda.

Thus one sees "what might better be represented" as a true set of opposites:

materialism
vs.
equality

being misrepresented as:
materialism
vs.
inequality,
(ie inequality vs. inequality)
associated to language (of left and right, being based on)
either
rationally (assuming materialism is true)
or
based on loyalty (to the natural unequal order of the material world) and the need for violence associated to inequality.

The violence of the right is best modeled as the relation that (the violence associated to) the, so called, secret societies, and the contrived laws and their selective enforcement, have to the justice system of the government.

By an organization being secret, it can be centrally controlled, but be made to appear random and, apparently, based on an individual-perpetrator (in regard to its causes).

The problem with trying to promote the idea that, "what in philosophy" is represented as the opposing language structures of:

materialism vs. idealism,
essentially equivalent to
"material based science" vs. religion,
. . . , and this is essentially a red-herring . . . ,

where the most subtle (or spiritually deeper) forms of religion would be representing religion as an attempt to see "the world as it really is" which "in the language of Judeo-Christian-Islam" might be stated as "knowing God directly."

Whereas in the "religious" cosmology presented by C Castaneda in the book, "the fire from within" this would be about "experiencing the relation of a living being's energy to a spectral structure of existence" . . . ,

. . . , that is, idealism is consistently represented so as to appear to not have a measurable basis upon which its ideas (or world models) can be expressed and measured, and subsequently used.

Thus, the opposites, within which everything (within the propaganda system) is contained, are represented as:

measurability verifiable vs. subjective use of language.

The problem with this set of, apparently, all-encompassing pair of opposites (ie opposites in a Hegelian dialect of opposites) is that Ptolemy's model of the planet's motions was measurably verified, yet it would have never led to the physical sciences being based on "solution-functions to differential equation representations of material systems," as Newton so positioned the "new" science, with which Newton tried to explain the planetary model of Copernicus.

That is, measurable verification also needs to be related to descriptions which allow couple-able and practically (and creatively) useful information, in regard to the just mentioned solution-functions and their context of containment, for such measurable descriptions to really be considered, "to be true."

To become a valid critic of the overly narrow "material based science," so as to break-free of language designed for an Imperial system of propaganda, one needs to criticize both material based science and the way in which the patterns of math are used and interpreted.

Science cannot describe the stable patterns of general fundamental spectral-orbital material systems, whereas classical physics is very useful because it is so general, and classical physics allows system-control and inter-system coupling, and thus it is practically useful, within the limited confines of solvable systems.

That is, modern science is neither accurate nor practically useful, other than useful in regard to manipulating arbitrary contexts of language (just as the emperor's religion has been socially useful).

In the propaganda system science represents the high-value of intellect within society, where the claim that such a high-valued intellect exists and is detectable, is assumed to provide further evidence in favor of the claim, that inequality of the high-valued few (vs. the incompetent public) is a natural context for the world.

It should be noted that 19th century science is central to the engineering products of the monopolistic products of today's (2013) ruling class, just as masonry was central to the engineering "gifts" Rome gave to its empire (deluded the public into believing an obligation [or gratitude] toward the Empire).

Furthermore, math based on non-linearity and randomness needs to also be criticized, since the prevalent non-linear pattern (associated to material geometry) is a quantitatively inconsistent pattern, which, essentially, cannot be used to describe stable patterns (this is the correct conclusion of Thurston-Perelman geometrization). and randomness is only valid in a very limited context. Furthermore, if (general) randomness is being described in the context of "function spaces and operators," wherein the operators are non-commutative, then again, there cannot be a valid description of stable patterns within such a quantitatively based descriptive context.

That is, math patterns which have a valid relation to observed stable properties are to be narrowly defined (similar to the solvable context of classical physics), but the descriptive (or containment) context needs to be widened.

The structure of non-positive constant curvature metric-spaces whose metric-functions have constant coefficients (and the containment spaces are metric-invariant) can be exploited, since they possess shapes which are linear and separable, and the discrete hyperbolic shapes, which have been dimensionally characterized by D Coxeter, are very stable patterns, they are the most prevalent shapes within the geometrization classification of the stable shapes.

The idea of equal free-inquiry is central to an equal free society; as it deals with knowledge, creativity, and the equal consideration by the public (about) as to what social forces one wants guiding a society, where these social forces are, in turn, guided by the laws of the land.

Furthermore, the public needs to know where to begin to challenge authority, the authority of a described truth, a truth determined from words has meaning in so far as the descriptions, the precise measurable, couple-able descriptions, of tangible systems relates to both actual states, or measurable properties, of "physical" systems, and subsequently to practical creativity.

That beginning point is at the level of context, interpretation, and assumptions; concerning the words (and their precise meanings) which are to be a part of a description.

One questions the "claimed mastery (by the experts)" of fundamental ideas; such as: measuring, stability, sets of assumptions, what one can conclude from assumptions, sizes of sets, ideas about containment, the nature of the (partial) differential equations, the nature and models of infinity, etc.

An opposition to challenging the accepted authority, associated to the creativity of the society, comes from those few (owners of society) who want the social structure to support their own (selfish) interests, which deal with investments (and risk). This social structure of investment determining creativity within society, and based on currency, seems to only work if this structure is backed-up by great violence (supported by the community). Investment with less risk is done within a stable, narrow, and controlled social context.

The American solution to social oppression . . . . , such as intellectual oppression by authoritative dogmas . . . . , where (as is now the case for physics) the dogmas neither describe properties nor are they related to a wide range of practical creativity . . . . , or the oppression of needing to live based on providing money to shop keepers (eg grocery stores) . . . . , shop keepers who are consistent with the interests of the owners of society . . . , the same money upon which the oppressive owners of society depend for their power, to the owners of society, in both of these fundamental types of social oppression (creativity and needs for survival) . . . . , the American solution to this type of oppression is to base the society's law on equality.

This viewpoint is essentially a copy of the very egalitarian and thriving communities of Quaker society in the Pennsylvania region of the American colonies. A society which both made and honored agreements which it made with the native people.

The propaganda system determines both truth and a, so called, social debate between the narrow range of the opposing sides of those wage-slaves who are valued enough, whose beliefs are known, to be allowed to express ideas on the media, [the media is the message and that message is that the owners of society own and control the media]. This is presented in the context of right and left, or religion vs. science described at the beginning of this paper.

There are certain voices which hold the attention of the left, within the public at large, with unreasonable attentiveness. Namely, the voices of highly-touted academics, where the perfect example is (N Chomsky) an MIT linguist who is a "scientist," studying language based on the quantitative and philosophical trends of the time (allowed at MIT), eg indefinable randomness, and an evolved trait of an animal (use of language), but little concern about the meaning of language, other than meaning being about "providing for islands of order in a sea of chaos," rather than "meaning" as being principally associated to fundamental practical creativity.

That is, it is of value to conjecture that, language resulted from both order within the world and order's relation to practical creativity, but this hypothesis might be outside the trend to describe things as being indefinably random, and thus, essentially, indescribable.

That is Chomsky's intellectual efforts are best characterized as being obedient to authority.

The rule of the propaganda system is (seems to be) that Foreign policy can be criticized, since it is not affecting the US public (other than improving their standard of living (so the myth goes)), but domestic criticism, other than expressing "the state of social power," is not allowed, unless one is willing to lose one's privileged job (at MIT) within the part of society which is endowed with social influence.

Though Chomsky is morally indignant over foreign policy (a very good thing) and he is honest about the structure of social power in the US, ie the owners of society control virtually all of society, however, he seems to believe in the same, basic inequality in regard to the condition for life, as did W Lippmann {or, in all likely-hood, as did both K Marx, and A Smith}, and that Darwin's material-based idea of . . . , (mutation and) survival-of-the-fittest basis for . . . , evolution proves the unequal condition of individual life-forms, and he supports the intellectual web (of illusions) upon which society is built: inequality, (invalid) economics, and violence, where he claims that the oppression of society to support the owners of society is, in fact, a result of carefully considered law (by the founding fathers) and that this condition was not achieved through violence. Whereas the best model for the existence of "owners of society" is the emperors of the Holy-Roman-Empire, where the emperors were transformed into bankers within the narrow, violent society (with few risks for investments) which Rome created.

Though the social changes, in the transitional period, also allowed for the printing press, literacy, social discourse, and the relation that the public had to knowledge, and in regard to public creativity.

In turn, the bankers controlled public creativity by means of inventing the professional scientists, and the relation of science to "tend to" the complicated instruments into which the bankers invested.

However, the revolution of classical physics during the 19th century changed the relation of public creativity to investment, but the banks, by using copyright laws (legal theft of the knowledge developed within a culture, based on using the letter of the law), quickly controlled the revolution, especially in the US, and put an end to "creativity based on science," by having science focus on indefinable randomness and non-linearity, which was also the model for reaction-rates.

This is what Chomsky supports in his position of media privilege, and apparently his mindless support for MIT.

Chomsky seems to be a representative of dogmatic authority and the belief that "there are no alternatives" the totalitarian viewpoint expressed by M Thatcher. Careful reading of Chomsky one finds an acquiescence to the idea that the bewildered heard should be controlled by means of propaganda (eg p127 of his 1987 Managua Lectures book). The viewpoint of the MIT-Chomsky concerning creativity is in lock-step with the corporate over-seers of MIT, eg the opinions about creativity of people like B Gates.

It needs to be noted that the 19th century science-engineering products which modern emperors provide to their colonies, similar to the high-tech brick-laying the Roman Emperors provided their colonies, emanated from a handful of scientists (the few is related to the hierarchical social structure [of the illuminati] imposed on professional science) thermal physics and electromagnetism, where Faraday was a main figure in developing the language of electromagnetic fields, as well as Tesla putting circuits together, but none of these creative types are the recipients of great riches, rather by means of the letter of the law associated with copyright laws, their knowledge was stolen by the bankers and oil-men for their more centralized and narrow profit structures, and then with nuclear weapons 19th century science was focused into military applications. That is, investment wants monopolistic domination resulting in a narrowness of product and a product-associated creative need, ie the creativity of the banker's society is narrowed to minimize their investment risks, yet the emperor also needs to provide engineering conveniences to the market (public).

This is really (as much) about the narrow set of sustenance-products needed for a stable empire along with a few engineering niceties (the prescription for stability for the Roman Empire), which remain tied to a fixed (narrow) representation of knowledge, so as to curtail creativity, which could be "competitive," so as to destabilize the monopolistic dominance of the investor-class . . . , as opposed to the ways of hunter-gathers, (or equivalently equal free-inquirer creators), which utilize a much broader range of (knowledge) products or viewpoints (descriptive contexts) to survive (or create).

Monopolies demand narrowness of both product and knowledge, and a domineering structure in regard to "the" public language "as defined by propaganda."

Though it is a good-deed that Chomsky provides dissenting views, they are from a mentally "very weak perspective," ie he follows authority in his linguistic research (one cannot get a position at MIT without demonstrating obedience to narrow dogma; Note: Tesla and the Wright-brothers had disdain for the academics), since these criticisms of his (Chomsky's) viewpoints are essentially based upon the ideas of Socrates and/or from simple conclusions to be drawn from Godel's incompleteness theorem, ie issues concerning language and its relation to both survival and practical-creative use of "described knowledge."

The unfortunate fact is, that the owners of society are as bewildered as anyone else, but they see their selfish advantage clearly.

Thus it is all about:
Materialism (and the strategy to dominate based on controlling material and instruments)
vs.
Equality.

There seems to be a dis-connect between the relation of precisely described truths to practical creativity which are the main ideas of Socrates' equal free-inquiry, the example of Copernicus expressing a new descriptive context for planetary motions, and the conclusions of Godel's incompleteness theorem that descriptive truths must always be questioned at the level of: assumption, context, interpretation, logical structures, containment constructs, etc, in order to have a robust relation between precisely described truths (or measurable properties) and practical creativity: measuring, coupling, geometry and timing in regard to inter-relating systems in a practically creative manner, ie his analysis of education has no grounding in the traditions to which he, apparently, wants to appear to be so loyal.

Though Chomsky plays the role of a courageous honest intellectual, yet his intellectual identity depends on his being obedient to an academic dogma (of linguistics and its fundamental relation to indefinable randomness), ie he ignores the above examples which relate knowledge to creativity, so as to attain his social privilege, and furthermore, his analysis of the US society completely ignores the institutional violence which is extreme; it is a policing system which will not only hound someone for stealing bread for a lifetime, but it is a justice system which is involved in organizing violence to terrorize the public, this is the legacy of southern justice, and is the basis for the relation between gun control and the part of the population from which most of the justice system personal come.

Social organization

How the organization of society opposes new ideas concerning both practical creativity and social re-organization, unless the (social) re-organized to support the existing power structure of society.

How is social power within society organized?

It has the same structure of social hierarchy as did Rome during the Roman-Empire.

It is narrowly focused, it is very violent, in order to maintain its narrowness, and it provides for the building of a material based infra-structure (roads, buildings, sewers, art, etc), which it is believed by the public, to facilitate an easy life-style.

Its law is based on minority-rule, property rights and upholding the formal agreements made by the upper-

social-classes (contract law), so that these laws are enforced with great violence.

The superior people dominate the inferior people, where superior means being in an upper social-class, though by being a competent helper in the maintenance of this system one can rise in social stature.

The knowledge needed for this society is narrow, technical, and arbitrary, and it remains stable, and it is controlled by the values and the wealth of the ruling-class, and this process of maintaining the narrow dogmas of the intellectuals within society is done by the academics whom have positions at public universities.

[There is now the belief that markets are the best means to determine how the society should be organized materially, eg if the public wants roads then they must pay for roads. However, this is now more about deception and propaganda, and the relation of this deception and propaganda to violence, than it is about material product, but this violence is done in a hidden manner by the justice system, and its arbitrary relation to the enforcement of laws.]

The narrowness of focus of such a society results in relative stability, but it is maintained by violence, and thus a vague market based on a few products can form. Within this narrow context, of products needed to maintain a community, technical skills are teamed-with both deception and violence so as to form into monopolies, so as to keep the knowledge base of society narrow, and (in a sense) the products are to be inter-changeable through-out the empire if one uses the road system (or transportation system) for trade.

In regard to knowledge, and its relation to creativity, knowledge is placed on a very high social value-structure and this knowledge is narrow and dogmatic.

Thus, there is the dual attributes of high-valued cultural knowledge . . . , (ie not knowledge used for practical creativity for everyone),

. . . . , (1) one must follow the dogma (the rules of the academic contest) or not be in the game (of using knowledge for creativity), and

. . . . , (2) the knowledge used by the culture is, considered to be (or represented in the media to be), so advanced and complicated that only a few superior intellects, whose mental capacities, IQ's, can be discerned by modern psychology, are capable of achieving a mastery of this highly difficult and technical knowledge so as to relate this knowledge in a useful manner to society by developing new products for the corporations owned by the owners of society. [where the (social) proof of this is the very complicated instruments used by our civilization, and how they are used].

That is, the high-social-value of knowledge is being controlled by the ruling class, since they can define the context of social value, and they can judge the usefulness to which knowledge can be related (effectively controlled by investment).

If one is allowed to be on the media, or, equivalently, in the upper social levels of the academic community, one must accept the assumptions which will define the context of the ideas which are allowed to be expressed.

This narrowness is more about convenience (by the management-class) for the ruling class, and the stupidity and selfishness of the ruling-class, than it is about the capacity of the public to possess knowledge, ie the public is capable of developing knowledge in a better and wider-range than is the ruling class.

The (Roman) empire could expand and remain relatively stable if it provided technical niceties, such as sewers and roads, to the regions where it expanded, but the niceties and the empire required narrowness of knowledge, and narrowness of product development, based around such a stable social context, formed in patterns of monopolies in regard to craftsmen-guilds, so that these monopolies worked with the ruling class, but eventually banking controlled this narrow relatively stable social structure, taking-over the ruling positions, the banking-class whose money had value if it was kept invested in a narrow social context, and its value was violently maintained, but still in a narrowly defined social context, in regard to products and resource use. The bankers provided art, while the merchants and trades-people provided products if there was enough investment.

Today (2013) the knowledge of craftsmen is defined around 19th century physical science (motors, TV, phones, computers, micro-chips, refrigerators, etc), but whereas biological-chemistry (of enzymes and proteins) is now defined around DNA, but it is a bunch of vague correlations defined around a few chemical markers (a few chemical for which a chemical-detection process exists) and no understanding of how such a complicated, yet highly controlled, system can operate.

Religion is about the institutional development of a civilization's language, the (religious) institution continually repeats the language of an empire. Religion deals with origins, cultural interests, and life's relation to death. Eg Paul created an institution about which a language was centered and then continually repeated, giving the language stability and a religious community stability. The church was socially about helping the poor, but the church realized that this required that there always be a sizable poor population.

Calvinism (or Protestantism) essentially re-sanctified the narrow life-style which was associated to the virtuous ruling-class, but it also associated life with the more common virtues of: obedience and faith in hard-work. This is essentially what the protestant reform movement provided to the European culture in the 1500's.

The printing press, literacy, and the more demanding spiritually inspired people, re-interpreted the Christian religion (the bible) in terms of the golden-rule . . . , or more succinctly, people are equal . . . , such as the Quakers . . . , who were allowed to express their ideas in America, under (the protection of) a very egalitarian, W Penn.

Knowledge is destroyed by judgments of value based on arbitrariness, ie the value of meaningless language (which is nonetheless continually repeated).

This can be contrasted with the development of language related to practical creativity, ie a language which precisely describes the (stable) patterns which are observed and the description is based on a few laws which identify how to control the observed stable patterns.

Then the question: "What is the real value of a new product built from a knowledge which allows one to identify and acquire new materials, and then control the properties of materials so as to build a new product?"

What is a "product's value" in relation to the nature of life, ie within life's place of existence?

What is life's place of existence?

or

Does material define existence?

or

Does life (or does something else) define a context in (of) which "the material world" is only a proper subset?

To describe the state of the world and subsequently to try to realize our creative capacities within the world measurable descriptions of (stable) properties is needed so that a new systems parts can be put-together and controlled. This requires the stable properties of numbers which are used to describe (stable) patterns, where the stable patterns are associated to shapes, in fact, to a relatively limited number of very stable shapes which are shapes which remain consistent with the (stable) number system. There are also issues about set-size so that the context of containment remains logically consistent.

Part IV
(more emphasis on technical, measurable, description of existence)

25.

Succinct

Science needs to be placed in a new language and new interpretations of observed properties given (here is a clear example)

similarly

Politics, economics, and the law, need to all be placed in a new language, and new interpretations of observed properties (new interpretations of history) provided (see The fight)

A physical system's measurable properties.

For a "material" system's containment in the coordinates of space and time (or space-time) [these restrictions on the dimension of a material-system's containment set is due to the idea of materialism] a system's measurable properties are functions defined on the coordinate system-containment-set eg space and time and material . . . , {or as in classical thermal physics: volume, pressure, temperature, and number-of-particles, (together define the containment set [or domain space for energy and entropy functions] of the measured coordinates of a thermal system) in an organizing context of the scalar functions of energy and entropy; for thermal systems whose measured values are averages over large numbers of components which are contained in a closed system (in a containing shape (or volume))} . . . , these measurable properties would tend to be geometric measures within the physical-system containment-space measures, and temporal measures, as well as other measured values one can either identify, eg mass, temperature, etc, or "invent measurable properties" by means of formulas, or with the operators of calculus (the operators, some of which can be interpreted to identify a rate of change, and which define the constants of {or associated to the} local linear transformations which act on of the local directions of the (local) coordinates so as to represent the function's-value at that local-coordinate point [at the domain point of the local measuring process], by means of derivatives, or measured values can be formulas whose domain is the coordinate values of the system-containing coordinate-set, eg thermal functions (or algebraic formulas, as found by R Boyle, in about 1650), eg pV=kNT.

An equation is about two different ways in which to represent the same measurable property (this can be thought of as the relation (or property) being in two different containment sets), where equality is conditional on the values of the variables in the two representations, so as to find either the set of numbers, or the function, which make the equation(s) true.

If the equation is algebraic then the solution is a local relation between numbers.

One can consider either numbers to be measured values of system properties, or functions can represent the measured values of a system's measurable properties, where function-values depend on the points of the domain space upon which the value of the function's formula depends, but the coordinate contain all the points which compose the system (or measurable pattern) one is describing, where the description is based on the measured values of the coordinates within which the system is contained.

For functions representing measured properties then measuring can be modeled as a local linear relation between the system containing domain space (the directions of the local coordinate properties, which are not necessarily rectangular coordinates) and the function's values. These local linear measures defined on functions can be of relations between different derivative orders eg 1st-derivative, 2nd-derivative, etc, of the derivatives (or, equivalently, local linear measuring relations).

When derivatives of various orders are placed in an equation, and if there are invertible operators (or invertible processes) so as to be able to solve the (linear) equation, then the solution to a differential equation can be found, and it (the solution function) would be a global function.

(in order to remain quantitatively consistent, the differential equations, which represent locally measured values of a system's properties, need to be linear, and if not quantitatively consistent then the numbers of the system do not (cannot) be consistently related to one another, and the system's properties (or their changes) are chaotic, and its measured properties are indefinably random, and the pattern is without any stable form.)

[This is an elementary idea.]

Furthermore, in a coordinate system, one must also require metric-invariance, so that the gradations of a ruler remain consistent, for all orientations and/or translations of the ruler in the coordinate system.

That is, one wants geometric coordinates to be consistent with the geometric measures defined on the coordinate space.

Furthermore, the local linear and metric-invariant measures of a (relatively stable) system's properties need to be continuously independent (or orthogonal) at all points in space, in order to either solve the differential equation by separable coordinates or so that the system's equations are always consistent with the metric-function (ie the ruler's gradations) (general relativity is non-linear, and thus it has no informational value which is valid, ie it is irrelevant).

That is, one wants the geometric information of a solution function for a system with geometric properties in space and time to be consistently measurable, couple-able, and controllable. The measures of physical systems must be consistent with geometry in the system's containment space.

Without such a simple math set-up then one cannot describe in a valid manner . . . , (ie quantitatively consistent, and in consistently uniformly measurable context within the pattern-(or system)-containing-set (of measurable properties) (or when the system is contained in a coordinate space)) . . . , the stable patterns which are both observed in physical systems and which are fundamental to mathematical descriptions, too, where math tries to describe stable patterns which are measurable couple-able and practically useful.

So

What is the proper context of existence and of practical creative use?

Is the context materialism?
or
Does existence transcend materialism eg into high-dimensions so that materialism is a subset?
Answer: The latter is true.

Consider:
One wants to be able to use the information of a physical system's solution function, to the physical system's differential equation, in both an accurate and a practical way so the system needs to possess stable controllable properties.

Note: The notion of material and dimension can be changed if material systems are changed into material represented as different "metric-space shapes" of given dimensions, which, in turn, are contained in metric-spaces of one-dimension greater than the dimension of the "material" component (or metric-space shape).

History:
Newton using the ideas of Galileo (first two laws) and the answers provided by Copernicus and Kepler and identifying three laws: inertia, force, reaction; as well as inertial-system containment-frames so as to identify force as the definition of mass in terms of a local linear set of material-property measuring structures (derivatives), which were related to global geometric properties (spherically-symmetric force-fields [emanating from material point-sources] where the geometry was restricted to a plane) represented as local vector properties (associated to force-field formulas), ie a linear 3-dimensional differential equation, modeling F=ma, eg of a two-body model

of gravitational interaction placed in a center-of-mass coordinate system (reducing the two-body problem to a one-body problem).

Then, within this context, Newton's math achievement was the definition: that a function's local linear definition of its slope (defined at a particular point in the function's domain space) identifies a (local linear) transformation: from an original function to a new function (the derivative), defined in a limit process defining convergence to the local domain point of the local linear measuring process (within the derivative), so that the new function, in turn, has an almost inverse-function defined by a local linear summation process defined in a continuous manner over the region (upon which the new function is defined), where the inverse function of the new function is the original function, but only if (or provided that) either boundary conditions or initial conditions are given.

This is about continuously defined processes of either the local slope of a function's local linear approximation, or (the inverse process) regional summation of partitioned areas "under" a function's graph.

This summation inverse operator transforms regional original function properties to a new function's (where the derivative of the new function [after the summation process] is the original function) boundary properties. This is about continuity ie the summation formula working consistently (for the area of the function's graph being summed) in the small and this local process is extended over the entire region.

This math structure takes local information (which might also be related to global shapes of force-field interaction) and finds a global solution, and, thus, provides more information, gained from the local linear measures defined at a point of the solution function's domain space (by derivatives).

These methods can be applied for material systems associated to continuous translations or rotations and for the local mechanics of stationary material structures (or rigid material shapes), but these equations for physical systems, eg F=ma, most often lead to non-linear equations.

In a 3-or 4-dimensional system-containment set, each dimension needs to be continuously separate and linear for the global shape (or region) of the system, in order to find a solution to a system's differential equation.

But

Is a differential equation (or a partial differential equation) about local linear measuring?

or

Is it about operators acting on function-spaces?

Answer: Neither of these models is quite correct, but rather the derivative is a discrete operator in regard to discrete changes of the descriptive context in regard to the various contexts of the (many-dimensional and many-metric-space-states) containment set.

Faraday defined both the electric and magnetic fields, and their inter-related patterns in the context of linear partial differential equations, but in a new containing metric-space, eg space-time, for charge and current related to an invariant wave-equation defined on space-time.

The Lorentz-force related the electromagnetic fields to inertia (where mass is best defined in Euclidean space).

However, again only simple separable (continuously independent and linear relations to) geometries of charge and current systems (as well as to conducting circuits whose linear differential equations are defined by constant coefficients) stay linear (separable) and solvable and controllable, and thus only such simple linear, separable physical contexts are descriptions of practical creative value.

There are also charge distributions related to simple lattice-forms in space, or multi-pole distributions of charge, dipole, quadri-pole etc, in space, where the summation of these different-power approximations of charge distributions represents a system's static-charge distribution in space can also be useful for such rigid multi-pole arrays of charge sources, which also seem to also apply to wave equations, in general, and are used in acoustics, with sound sources also having a multi-pole array expansions (or multi-pole summations).

The relation of force to energy and using KE and PE terms in representation of energy-related expressions, such as action (ie px or Et etc) and applying variation techniques, defined on action, and associated to a minimum (or an extrema) allow some additional system equations, which represent physical systems in general coordinates, ie more easily placed into a general geometric coordinate systems, but only linear separable coordinates are solvable and "controllable by controlling a system's set-up, eg boundary conditions or initial conditions."

Then there is general relativity, an attempt to place physical description in its most general geometric context, (in part, general relativity was motivated by the variation techniques applied to an action-property) but general relativity is non-linear, and thus it is only relatable (or applicable) to a one-body problem which possesses spherical symmetry, ie it is a description which has no practical value.

However, there are stable systems at all size scales: nuclei, general-atoms with more than four-free-charges, molecules, crystals, and the solar system, and the motions of stars in galaxies, and this stability implies that these systems are linear, metric-invariant, separable, and controllable.

Quantum systems (math which does not work)

Nonetheless, for small systems, ie quantum-systems, by reducing micro-systems to their point-particle components, and associating the description to the probability of random (materially reduced) particle-spectral-events in space and time. In this context, the quantum-system becomes a function space, ie one function for each observed spectral-value, and linear operators are applied to function spaces, where the quantum-system probability representations are wave-functions which have complex-values, so that the operators are supposed to be related to physical properties, in relation to a quantum-system's spectral-values (eigenvalues), and the subsequent ("diagonal") spectral-representation of the quantum system's function-space.

This is an energy invariant construct if it is unitary-invariant.

But this math technique, of modeling a quantum system as a function-space (which have complex-values), has not been applicable in a successful way to general quantum systems.

Furthermore, this method goes from global (ie the functions of the quantum system's function-space) to local random spectral-particle events, so as to lose information.

The limited capabilities of mathematicians (and their authoritative dogmas)

Though mathematicians delve into the abstract, eg in regard to function-spaces, but their techniques do not work (this is the point of the above discussion of quantum physics).

Thus, it is clear the, expert, mathematicians do not understand what they are doing.

So, what is a (partial) differential equation?

and

What is a solution to a (partial) differential equation?

Answer: Solutions to (partial) differential equations (can, and often do) determine the context of engineers' descriptions of systems, which they are adjusting, and adapting, and controlling, and using, but quite often the knowledge of the system exists before the solution function is found. For example Faraday was building electric and electro-mechanical devices as he developed the "correct" language for the electromagnetic field. Thus, the point is that (partial) differential equations are sometimes quite relevant, if they are solvable and controllable, but many (most) systems are not of such directly controllable types. Furthermore, it can be difficult to interpret the context of a solution function so that its information (and subsequent control [if linear] by boundary conditions) is useful.

But the very many non-solvable systems are associated to math patterns, which the mathematicians do not understand, (or perhaps non-solvable systems are properly outside of the stable-quantitative range of valid mathematics, eg measurable descriptions of system properties would not be mathematically valid if these patterns were both not quantitatively consistent, and not stable [shapes]).

Thus, a valid measurable description of fundamental physical systems may not be as directly related to solutions to (partial) differential equations as about the shapes and stable bounding patterns of the containment-sets of the measurable patterns of existence, and differential equations are not being properly defined on the contexts of either (1) linear continuity, or (2) spectral vagueness, but rather (3) derivatives are "discrete operators" defined over

1. small time intervals, and
2. defined between discontinuous dimensional levels of existence, and
3. between discrete folds (on [the lattices of] discrete hyperbolic shapes), and
   . . . , their solutions are directly related to the substantial restrictions imposed by the containing space, which defines the bounds of stability and quantitative consistency (quantitative sets are based [built upon] on a finite set of quantities).

That is, the context of differential equations, is about the measurable properties of systems contained in a measurable context.

but

This is really the context concerning the stable measurable patterns (both math and physical patterns) which bound existence (or mathematically set-containment of the pattern whose measured properties are being described) allowing for (stable) quantities and shapes (to exist) as well as physical law (allowing it to exist), where physical law in the new context guides both interaction structure and (dimensional) perception, where interactions deal with the organization of set-containment and the local geometry and structures of ("material-interaction") dynamics, including dimensional relationships and/or fiber Lie group relationships, but only slightly related to "stable dynamic structure," eg elliptic orbits of condensed "material" in a metric-space. However, orbital structure and orbital possibilities, which exist either during interaction-resonances or in regard to a (high-dimensional) system's high-dimension geometric structures, may have complicated and far-reaching inter-connections, in regard to properties associated to a containment context of stable shapes, such as angular momentum "orbiting" between various dimensional levels of a high-dimension stable orbital shape (ie a discrete hyperbolic shape of a high-dimensional systems, including its various "lattice-folds"), or between various molecular possibilities in chemical reactions involving catalysts.

Continuing with the invalid descriptions of quantum physics

The quantum interaction model (or particle-physics) further reduces, a spectral function-space representation of a quantum system, to a space of internal particle-states attached to the quantum-system's spectral-functions (Note: Spectral functions which cannot be determined for general quantum systems) and a non-linear operator acting on these vector-states, causing the wave-function to be perturbed and summed over the internal symmetry-states of the colliding particles of the interaction (Bosons and Fermions colliding), but the summation diverges, and (a required finite summation-value for the perturbations) requires re-normalization (or the arbitrary re-scaling of physical processes for small sizes, due to the quantum break-down of space, eg this is equivalent to saying an interacting quantum-system cannot be contained in a consistent math context) and subsequently, adding and subtracting infinities.

(So this is not really physical law but rather an elaborate epicycle, data-fitting process)

This is a unitary and many-dimensional, but nonetheless remains based on materialism, where the internal-symmetries are called math-fictions (represented virtual events, or whatever) identifying the random particle-spectral (or random particle-type dependent processes) events in space (or space-time).

String-theory tries to identify a geometric and many-dimensional context for the particle-unitary-symmetries, but it, somewhat arbitrarily and in an ad hoc manner, maintains the idea of materialism, and it is based on quantum physics which cannot describe general systems, so string-theory is doomed to failure.

The relatively stable properties of nuclei are well beyond anything approaching a valid description based on the math techniques of the particle-physics context (or based on the, so called, laws of particle-physics).

Comment

This description cannot be taken seriously by anyone who thinks for themselves.

Nonetheless it is required dogma for the "intellectual aristocrats."

Furthermore, these "aristocrats of intellect," who obey dogmas, clearly cannot think for themselves.

Their motivation must be either obsessions within a memorized abstract language in a world of illusion, or the will to be intellectually dominant (ie psychopaths).

However, it needs to be noted that the cross-sections (or probabilities of collision) for elementary particles can be measured and related to bomb engineering (rates of reactions), and thus it is so heavily funded, and subsequently, its authoritative dogmas define high "intellectual-value" within society.

Consider the media; when a probability based description, which is provided by the (weather) experts, is in opposition to the interests of the owners of society, then the probability associated to these models (which oppose the interests of the owners of society) are questioned by the media, and subsequently, questioned by the education institutions, very vehemently and strongly, even though the correlation of higher CO_2 concentrations in the atmosphere and the earth's increased absorption of sunlight (ie global warming) is a strong correlation (perhaps as strong as the vitamin-C and scurvy correlation).

However, when a probability based description cannot be used to describe the stable properties of some of the most fundamental physical (quantum) systems, and thus clearly probability is a very poor model for one to try to use in order to try to describe the properties of the very fundamental and very stable quantum-systems, but the probability and particle-collision model (of quantum material-interactions) is useful in regard to "probabilities of particle-collisions being used to determine rates of nuclear reactions," and, thus, probability in the context of quantum physics and particle-physics remains an "absolute truth," which must be accepted in order to become an expert, and to be published in professional (peer-reviewed) journals.

Believing oneself (and not believing the dogmas of the propaganda system)

Thus, one needs to ask (if one wants to believe in one's own ability to discern truth); According to, the so called, physical law, "How do stable many-(but few)-body orbital-spectral systems exist?"

Where it is observed that such stable physical systems exist: from nuclei, to general atoms, to molecules, to the (apparently) stable solar system.

Furthermore, the stability of these fundamental systems implies that these systems are formed in a controlled context, which is associated to linear, metric-invariant (or unitary-invariant), separable shapes, and related controllable processes.

In regard to classical physics (ie the useful range of currently accepted physical description):

In the useful range of physical descriptions and its associated math constructs,
Either
There are Galilean frames for mass and inertia: F=ma, conservation of momentum, and interacting material-pairs experience equal and opposite forces (the action-reaction duality); And if the system's differential equation is: linear, metric-invariant, and continuously independent (locally linear) coordinates (or separable), in regard to the [partial] differential equations used to define a physical system's measurable inertial properties, then one has a controlled set of geometric-solution functions for these limited types of classical inertial systems.

Or

In a space-time frame (or, equivalently, in a hyperbolic metric-space) there are the set of linear equations: d*F=j and dF=0, so that if these partial differential equations are: linear, metric-invariant, and continuously independent (locally linear) coordinates, which are associated to these (separable) [partial] differential equations, which are used to define a physical system's measurable electromagnetic properties, then one has a set of controlled geometric-solution functions for such a classical for electromagnetic fields. For these simple types of (partial) differential equations along with the linear, constant coefficient, differential equations related to electric circuits, one has a lot of control and a lot of useful information related to these types of classical physical systems.

Is there a similar succinct descriptive context, and direct assessment of its validity, in regard to its accurate and practically useful information which such a new descriptive context provides, which is (also) a much more inclusive new descriptive set which is based on a new set of: assumptions, contexts, and interpretations?

Answer: Yes.

For the full answer, one needs the list of both the finite spectral set and the subspace containment organization (or determining how the size of the metric-space dimensional-partitions are organized in regard to set-containment), of the over-all 11-dimensional hyperbolic metric-space containment set of "physical" description, in order to provide answers about the new description's accuracy and practical usefulness, but the new description provides a new context in which to re-consider many elaborate, already known, and/or easily adjustable, math constructs for some very complicated, and very perplexing, but apparently controlled, physical properties and physical processes.

For example, the stability of the solar system is proof that the new descriptive context is correct, since the current descriptive context cannot describe the solar-system as being stable, ie planetary resonances disprove KAM (Kolmogorov-Arnold-Moser). However, the new context re-defines ways to think about envelopes of orbital-stability for the planets.

The new descriptive context depends on classical categories of both metric-spaces . . . ,

[either
Euclidean (or Galilean) frames, and associated local coordinate transformation properties of inertial and position-indicating frames, ie frames of spatial displacements (translational or rotational).

or

hyperbolic metric-spaces ie generalized space-time metric-spaces, and their associated (physical) properties associated to charge-current systems, and energy-time systems, as well as the very much needed (extra) property (of the existence of and) containment of "stable patterns"]

. . . . , and types of partial differential equations associated to the descriptions of physical systems, either their field properties or their dynamic properties, eg the three-types of 2nd-order differential equations [elliptic (orbital), parabolic (free or semi-free, angular momentum), and hyperbolic (eg collisions or mechanical-waves)].

It (the new descriptive context) identifies the sources of quantum randomness, as well as an apparent material reduction to point-particles, ie the vertex of a discrete hyperbolic shape, about which all interactions for the shape are centered.

It shows where 3-dimensional spherical-symmetry comes from, for (inertial) interactions in 3-space.

It re-defines general relativity for metric-invariant metric-spaces, which are linear, and which possess the geometry of discrete hyperbolic shapes.

It is a many-dimensional descriptive context, which can be defined in either metric-invariant (SO) or Hermitian-invariant (SU) contexts for the system's, Lie, fiber-groups.

The new descriptive context is both stable and geometric, and the interactions are most similar to classical descriptive contexts, but the new calculus-operators are discrete, and based on both small discrete time intervals, as well as being related to discrete dimensional changes, which are a natural part of the containment set for physical systems (or for existing systems). The new descriptive structure of the containing metric-spaces have physical (or mathematical) properties associated to itself, so that these metric-spaces now possess a pair of opposite metric-space states, upon which a fundamentally discrete interaction structure depends. This leads to system-state (or metric-space) containment in complex-coordinates, and, subsequently, related to unitary Lie fiber-groups.

The (solution) shapes for the radial equation have been found, and they are discrete hyperbolic shapes, which are consistent with the fundamental finitely-based quantitative structure for spectral-geometric systems, for systems contained within an 11-dimensional hyperbolic metric-space, which contains the spectral-geometric basis for the "existence of physical systems" . . . , (where systems exist due to the system's resonance with a subset of the finite spectral set of the over-all high-dimension containing space [which is partitioned into a finite set of discrete hyperbolic shapes]) . . . , in each of the different dimensional levels, as well as the different subspaces of both the same and of different dimensions.

The solution-shapes (ie discrete hyperbolic shapes whose spectra are consistent (or in resonance) with the given finite spectra; set for the over-all high-dimension containing space for physical (or existing) systems) depend on the folding of the solution discrete hyperbolic shapes by Weyl transformations (contained within the metric-space's fiber Lie group), and this, in turn, identifies a geometric solution shape (or region of stability) which is similar to Bohr's planetary atomic model. Such a shape allows for geometric orbital adjustments to be defined within the stable discrete hyperbolic shape (solution), which identifies the envelope of orbital stability for the stable system.

Functions contained in function-spaces will need to be both discrete hyperbolic shapes, as well as discrete Euclidean shapes.

New description (re-iterated)

Only lines and circles are quantitatively compatible in a linear context for measuring. Thus the stable shapes are "cubes" and circle-spaces, which also happen to be associated to "cubical" simplexes, which form the fundamental domains of the lattices of discrete isometry and unitary subgroups which determine circle-spaces. These circle-spaces are stable shapes and also models for metric-spaces.

The stable controllable patterns of math which are quantitatively consistent are translationally invariant lattices and may include discrete rotational invariance's, and these patterns are the [associated (or related)] circle-spaces (defined in a moding-out process (or a topological equivalence relation) on the lattice "cubical" simplex structures).

These are the metric-spaces which are metric-invariant and of non-positive constant-curvature, ie flat or hyperbolic metric-spaces $SO(s,0)$ and $SO(s,1)$, as well as $SO(s,t)$, where s is the dimensional of the spatial subspace and t is the dimension of the temporal subspace. That is, metric-functions with only constant coefficients.

Properties of spatial position and the existence of stable patterns (math or physical) {where stable patterns are the basis for ideas about quantity and shape} [including the quantitative probabilities, but stable probabilities need to be related to stable system properties, such as the well-defined faces of a dice], eg patterns which possess the property of continuity in time (or conservation properties). Such stable patterns are the basis for physical laws, which apply to a wide range of existing systems, and whose subsequent descriptions (or solution functions) are practically useful, but the different types of metric-spaces also allows for the identification of either inertia or charge, and this, in turn, allows for the definition of pairs of opposite metric-space states. This leads to both system containment in complex coordinates, and to a unitary fiber group (or a unitary context for containment).

By partitioning an 11-dimensional hyperbolic metric-space by discrete hyperbolic shapes for each dimensional level and each subspace, and so that there is a conformal (constant) factor defined between subspaces (and dimensional levels, and between subspaces of the same dimension). Then such a partition defines a finite set of quantities upon which measured properties of geometric-spectral shapes depend (for their existence by resonance in the containing metric-space) and along with using the rational-numbers all measurable (or quantitative) descriptions can be identified for the properties of all existing systems (which must resonate with the finite spectral set) within such a high-dimension containment set.

Interactions are discrete operators defined for short discrete time intervals which allow for Newton's ideas to be adjusted and used this also allows for the descriptions of system formation into stable forms (or shapes) by the mechanism of resonance, and it allows for the solution of the many-(but-few)-body problem in the form of a stable, linear geometric pattern.

The actual calculations of resonances and energy-ranges . . . , needed for a stable discrete hyperbolic shape to form during interactions, which allow stable physical systems, either interactions by collisions or orbital existence, both of which depend (for their emergence) on a form of resonance in regard to (of) either charges in stable spectral-flows (of a stable discrete hyperbolic shape which defines the system) or condensed matter whose dynamic path resonates with an orbital-path within an existing discrete hyperbolic shape which represents the containing metric-space for the condensed material (in regard to condensed matter, this is a linear model for general relativity), so that this all . . . , depends on finding the finite spectral set for all the possible existing discrete hyperbolic shapes which are contained in (or they define) the various dimensions and various subspaces, as well as depending on finding the subspace set-containment array for all the systems (represented as the very stable discrete hyperbolic shapes) that exist within the full containment context of the 11-dimensional hyperbolic metric-space.

Note: Only hyperbolic shapes can be related to a finite spectral (or quantitative) set, while Euclidean lattices can have many spectral values.

Conjecture: the process of crystallization is a movement (or a change) of the components (within the crystal is formed into a Euclidean lattice) toward an "allowed stable context" of a "discrete hyperbolic shape" for (or associated to) the crystal system.

Note: The new interaction structure is essentially a non-linear connection . . . , but interactions are affected by both (1) metric-space states, and (2) resonances (of stable systems) with the finite set of spectral quantities associated to the over-all containment set, ie an 11-dimensional hyperbolic metric-space and its partition by dimension and subspaces into discrete hyperbolic shapes, . . . , so, except for the shapes and energy requirements which fit with allowed stable discrete hyperbolic shapes, most interactions are non-linear.

The new context of many-dimensions, partitioned into a finite set of discrete hyperbolic shapes for each dimensional level, and for each subspace, is very much more complicated than is the context of materialism.

The, apparent, great complexity of the currently accepted, material-based, descriptive context is a result of its decent into indefinable randomness and non-linearity, where this is done in an attempt to both describe everything (every detail), and to describe the properties of material systems, which a descriptive basis in materialism is not capable of describing.

Subsequently, the currently accepted context of the expert's, so called, "physical law," has no practical value, and it is entirely devoted to ridiculous nonsense eg point-particle-collisions described by elaborate schemes of convergences in a fundamentally (indefinably) random descriptive contexts.

Expanding on the nonsense of the authoritative descriptive structures:
The convergences needed for point-particle collisions in quantum physics, apparently, are defined within the Poincare Lie (fiber) groups (apparently, the Lie groups of the domain space of the functions of the function-space, which is the containment set of the quantum system), where each elementary-particle is associated with a different representation of the Poincare group (of different dimension (?) whose internal-vector-spaces (of the elementary-particle's internal states) are attached to the functions of the function-space, and where these internal symmetry-vector-states are transformed by the interaction connection term, which can also possess various dimensions). The perturbations on the wave-function, of the given quantum-system, are due to the particle-collision interactions, and associated non-linear connection transformations of particle-states, which are defined on the internal (individual) particle-space-states, which are the "internal vector states (of the colliding-particles)" which are added-on to (or placed onto) the quantum-system's wave-function [where this adding-on of vector-states to the wave-function is done in an ad hoc manner, but is likened to a particle's spin-state].

One sees nothing but failure when the authorities try to identify either (as in physics) stable patterns of physical systems, eg relatively stable nuclei, or as in biological evolution, the development of highly ordered and very complicated living systems by the evolutionary process, where both descriptions are defined within a descriptive context of indefinable randomness and/or non-linearity, and this descriptive context, whether used in physics or in biology, has (for over 100 years) shown itself to be an unsuccessful (or a failed) effort in regard to describing the observed stable properties of physical and living systems.

That is, it has no relationship to a valid descriptions of any significant observed physical properties which general fundamental systems possess. That is, it is clearly a bunch of nonsense.

Measuring, quantities, and "quantities modeled as stable patterns," eg numbers are assumed to be stable

Quantities can be modeled as counting of stable properties or stable forms.
or
Quantities can be modeled as geometric measures associated to a linear scale, a uniform scale of (uniform) graduations defined on a ruler.
Do these graduations remain the same as one changes one's position in space, ie Does the coordinate-space possess the property of metric-invariance?

This is about local linear, metric-invariant, and continuously independent (or continuously orthogonal) local coordinate properties, which are associated to the shape of a solvable physical system, where the physical system is contained in the coordinates, and where one wishes to describe the stable measurable properties of such a (an assumed) stable system, in a context of (physical) measuring which stays stable and consistent.

However, there are the issues of identifying a pattern which may not be a stable pattern (eg an unstable non-linear geometric pattern (which decays), or a set of probabilities for what one might believe is a distinguishable, but random, pattern).
If one is dealing with a geometric patterns then the stable geometric patterns are the circle-spaces.
If one is considering a probability based description, then the random events need to be related to a stable context so that counting of events can be assured to be have a valid relation to stable (and consistent) quantitative sets which contain the set of random events one wants to describe (which allow a consistent identification of a stable, measurable property, so that such a property can be counted), eg the faces of a dice for determining probabilities associated to rolling a dice, etc.

In regard to probability patterns, identifying correlations between pairs of relatively identifiable: (1) system states and (2) system components, which are contained in a complicated containment (or descriptive, or system) context, eg the chemical enzyme vitamin-C (system component) (where enzymes are catalysts, or facilitate certain types of chemical reactions) and the disease-state of scurvy [or some other identifiable molecule type and what is believed to be an associated (or correlated) disease, etc].

Correlations are statistical attempts to identify a "cause and effect" relationship in complicated system interactions.

That is, a correlation needs to be associated to a definitive (system component) cause in regard to the associated state (or disease, or effect) for the correlation to have any validity as a truly identifiable pattern of cause and effect, ie vitamin-C must be identified with a (or several) chemical processes (chemical reactions) within the system which relate vitamin-C to the bodily symptoms of scurvy.

However, present chemical knowledge, and its relation to quantum physics, does not allow for a valid model of chemical reactions in general, and for catalyst-induced chemical-processes this is a very distant wish, for such detailed chemical information to be modeled within a precise measurable description.

However, in the new descriptive context it is easy to consider the existence of a state of partial resonance within which there can exist controllable (orderly) chemical-state paths (linked-together by [possible] angular momentum properties, which are to be defined upon the orbital-spectral properties of the system's discrete hyperbolic shape) so as to transition between different chemical products, in a set of well defined pathways for chemical processes (linked by angular momentum), as a model of an enzyme chemical reaction.

Need the list of both the finite spectral set and the subspace containment organization, of the over-all 11-dimensional hyperbolic metric-space containment set of "physical" description, in order to provide answers about the new description's accuracy and practical usefulness, but the new description provides a new context in which to re-consider many elaborate, already known, and/or easily adjustable, math constructs for some very complicated, and very perplexing, but apparently controlled, physical properties and physical processes.

26.

To the point

Know upon what one needs to focus, then be direct in communicating

New ideas provide new ways in which to solve fundamental problems of physics and math. These new ideas can be used to solve the "many-(but-few)-body system which is stable" problem for both classical physics (or general relativity, ie a linear model of general relativity wherein there is non-localness) and quantum systems, and finding a finite basis for the set of quantities used to describe a system whose containment set appears to be a continuum and redefining what a mathematical model of a stable physical system might be, ie a discrete hyperbolic shape which resonates with the high-dimension quantitative containment set of the system, as opposed to using differential equations to model such systems, since differential equations become superfluous (though that method can still be used)

The professionals do not want to listen to someone who is not in their social class of the "intellectual aristocracy," though they do not have a solution to the "stable many-(but-few)-body system" problem, they just claim they possess the absolute truth concerning the laws of physics and using these laws such a "stable many-(but-few)-body system" their claim is that these systems are too complicated to describe, yet these systems are stable, and thus linear and controllable, and Thurston-Perelman geometrization suggests that the experts are quite wrong about their own methods, and/or their own formulations concerning physical law. The experts do not really understand (partial) differential equations, and the context (or meaning) of measuring.

So why the extreme expression of social-class concerning scientific inquiry?

The universities are not about truth, rather they are contrived contests defined by authoritative dogmas. which are related to the business needs of the owners of society. The education system got that way because it is destroyed by the propaganda system, which has the social position of being the sole voice in society which always expresses the truth to the people, and this voice, of course, is owned by the owners of society, and the owners of society want the graduates of university to compete to help further the interests of their (the owners) big businesses.

What is the issue, here?

The issue is law.

If law is based on property-rights and minority-rule then one gets inequality, and the subsequent extreme violence, either physical or intellectual, needed to maintain this inequality.

Note:

The owners of society (big-bank CEO's) today are in the same social position as was Augustus Caesar in the Roman-Empire, or more realistically as was Constantine who absorbed the church so it became the propaganda system for the Emperor, where the Bible of Constantine was propaganda for (the) Empire.

Society is based on violence and narrow ways of considering how to live, leading to stable markets, based on narrow vision about life, and based on some forms of engineering, eg building transportation systems and providing water etc, now engineering is associated to complicated instruments which are used to serve the interests of the owners of society. The technical departments of university are about providing the help needed for the narrow vision of using complicated instruments, though the micro-chip and S Jobs did cause the main-way to be perturbed, though communication-channels and their associated instruments are all about military control in the counterinsurgency methods of the justice system, where the insurgents are the public, just as the justice system would also be an arm of the Emperor's military in the Roman-Empire.

But the main insurgents, which "the owners of society" are concerned about are (would be) those as smart as (or smarter than) "the owners of society," ie 95% of the population.

Law needs to be based on:

equality,

People are to be self-governing, and

the government serves the people, and

equality is to be sorted-out in relation to the government promoting the common welfare.

Note: Only if each person is considered an equal-creator can there truly be a free-market, where people create in a selfless manner to help others.

This is the intent of the Declaration of Independence and the Bill of Rights, it is about freedom of belief and freedom to express belief in the context of being creative, not creative to make a buck for the owners of society but to be creative so as to selflessly provide the world with gifts of gratitude for life neither of these legal documents has been upheld by those in the government formed under the Constitution, and thus the current law is invalid, and there is needed a 3rd Continental Congress to re-instate the correct US law

The energy of a society is about imagination, creativity, and ingenuity but not in a social context of theft and selfish gain channeled to the few due to a set of dysfunctional laws, ie laws which destroy society, instead of providing the society with an order which allows each individual to express their creative interests

27.

Bird's eye-view

If one tries to express original ideas about physics and math then one finds that one will be excluded and oppressed by tradition and authority.

First one needs to realize that modern science has no valid relation to: both accurate descriptions of the prevalent observed stable physical (and living) systems at all size scales, and it has no (identifiable) relation to practical creativity. This has been noted by the press, but the reporter was almost immediately sent to the margins, and this was done by J Horgan in his 1995 book, "The End of Science," wherein Horgan noted that math and science had become irrelevant to practical development, and science has, essentially, come to be a "society of literary criticism" (ie essentially a science describing an illusionary world, ie science-fiction).

However, when one has actual new ideas about science, then the criticisms become stronger and more direct, and one sees that science has out-right failed in regard to both its descriptions and its relation to practical creativity.

Why should it be true that a dis-functional authority, in regard to science and math, exclude new ideas? Apparently, this is because of both science and math's relation to traditional authority and to the authority of the propaganda system?

Because there is a dis-connect within the US society between the relation of precisely described truths to practical creativity our society is failing.

The relation (between valid description and practical creativity) motivates the main ideas of Socrates' concerning equal free-inquiry, and is central to the example of Copernicus, wherein Copernicus expresses a new descriptive context for planetary motions, and the relation of precisely described truths to practical creativity is the focus, in regard to the conclusions about Godel's incompleteness theorem (ie that precise language has great limits as to the patterns which it is capable of describing), wherein the conclusions would be; "that descriptive truths must always be questioned at the level of: assumption, context, interpretation, logical structures, containment constructs, etc, in order to have a robust relation between precisely described truths (or measurable properties) and practical creativity: measuring, coupling, geometry and timing in regard to inter-relating systems in a practically creative manner."

Why would such a clear dis-connect, between the relation of precisely described truths to practical creativity, exist?

Because the owners (or the ruling class) of society do not want any interference by any significant efforts, or expression of significant ideas, which interfere with their narrow interests.

Note: This is the same model used by the Roman-Empire with the church providing the arbitrary moralizations and authoritative pronouncements by which the propaganda system supported the actions of the emperor, a Holy-Roman-Empire, but now it is an absolute authoritative but dis-functional physical science upon which the authority of propaganda rests, ie modern engineering is still based on 19th century science principles (including the nuclear bomb and the smart phone).

That is, modern science has no valid relation to:

> both accurate descriptions of the prevalent observed stable physical (and living) systems at all size scales, and it has (almost) no relation to practical creativity.

According to Chomsky, (and it seems to be correct) everything within (every aspect of) the US society is organized to serve the interests of the owners of society. (but Chomsky's analysis runs thin after this revelation, science is not thriving, yet Chomsky believes it is (he is an authoritative practitioner of linguistics), the inequality of the US society rests on the extreme violence of its institutions, an issue which Chomsky cannot see, etc).

Furthermore, and though the owners are as bewildered as anyone else about making sense about existence, nonetheless they have clear viewpoints about their own social power and their interests, which they enhance and defend. Whereas everyone else must first serve the emperors (their clear model for purpose in life), and then try to make sense of the world.

There is the belief that this organization of society, around the interests of the owners of society, has become a tradition with deep authority, and thus, "there is no alternative."

[Apparently, all people who believe that they are "articulate types of people," eg possess institutional positions, believe that thought-control must be applied to the bewildered herd, but it is clear that those people instituting (or controlling) the thought-control are as bewildered as anyone else, if not even more bewildered, as the destruction caused by the institutions of this current society destroy everything; both the earth and society, yet they can only see their own selfish interests]

Thus, the idea that there is no alternative, and that people's stupidity (the bewildered herd) is proof of social inequality, is quite a lie.

The so called stone-age (native) cultures of the Northern American continent were highly individualistic societies, and basically they were equal societies, many of them organized around a dominant principle of the female (so all the results of sexual activity could thrive in the community), and the wide knowledge of biology, and ideas about natural harmony, and the fellowship of life allowed for an easy capability for sustainability and survival leading to a very "deep" religious-scientific viewpoint about life, and such a viewpoint's true reaches.

It was the highly centralized hierarchical cultures of South America, which the narrow exploitative European belief structures found easiest to destroy and take-over.

In fact, the core issue of the relation of the US to European society during the US revolutionary war, was about the idea that law be based on equality, a government built both "by and for the people," where government is to be used to promote the common welfare, and to promote freedom to know and create, was the only way (or a very good way) in which to dull the socially destructive edge of both inequality and arbitrary authority. The American revolution was about the relation that knowledge and practical creativity had to an equal society. It was about both exploring the ideas of both religion and science.

The model of equality represented by the Quaker community was the most equal and energetic of all the early US colonies. While the Puritan communities fought within itself about the expressions of new ideas about religion.

The solution of the US colonies was to base law on equality, and learn and create (selflessly), ie the Declaration of Independence and the Bill of Rights.

Is civilization about being centrally-run empire?
or
Is it about knowledge and (practical) creativity?

Is the market to be narrow and run by investors who do not allow risks so as to have a narrow society which must expand?
or
Is a market about equality and creativity so as to be related to a wide range of knowledge and new (practically) creative possibilities, as well as a community's relation to food?

Is knowledge and creativity to be narrow so as to only serve the rulers?

or

Is knowledge and creativity expressed in an equal society to be about that which an equal free-market should be, in a society where everyone is treated as an equal creator?

This boils-down to making decisions about
equality and inequality,
or
Science vs religion
or
materialism vs. equality

The US revolutionary war is still of central importance to the American society, not the Constitution, since the Bill of Rights was never enforced.

[but, since the main teaching of religion is "equality between people," but in regard to "the religion of the Emperor," such a religion serves the Emperor's interests, while the idea of evolution, or of biological materialism, implies inequality (survival of the fittest, ie the superior organism wins-out)]

Thus, in regard to what words mean within society: science vs. religion, is really about:

materialism vs arbitrary moralizations

But placing these words in a more grandly philosophical context, it would be
materialism vs. idealism

But this is not the whole story, since there are new math and physics descriptive structures which makes the material world a proper subset of a higher-dimensional containment set, wherein the issue of stability is made concrete (the core idea [ie stability of patterns] about which both math and physical science are not properly dealing), where this relation to stability is made clear (within the patterns of math) by the Thurston-Perelman geometrization, so that the higher-dimensional containment set, whose different dimensional-level contexts are both macroscopic and microscopic, is related to the stable and very prevalent set of geometries of the discrete hyperbolic shapes.

Thus, in the new math-physics model (equivalent to the model of Copernicus and Kepler which gave the answers to the reasons for the motions of the planets) both the material world and an ideal world (a world which transcends the material world) and this is done in higher-dimensions in a stable context which may be either macroscopic or microscopic in its constructs.

The current claim by modern science that their models are measurably verified, would essentially be the same claim that Ptolemy could have made in regard to its defense against the ideas of Copernicus, where the ideas of Copernicus won-out due to the fact that the ideas of Copernicus were considered to be simpler. (The claims in favor of modern science today (2013) are: that modern science is measurably verified, and other ideas are too simple; [but more fundamentally {it is about the dominance of language} the ideas of modern science are backed-up as their being considered to be authoritative due to their relation with the propaganda system])

The value which a society really possesses is its ability to both

(1) organize itself to be in harmony with the earth, and
(2) to possess knowledge which allows for the expansion of "practical" creativity within the society, where "practical" creativity does not necessarily mean only in regard to material-creativity.

It needs to be noted that "the new vision of math and physics" will upset the narrow and highly controlled investment constructs of the ruling class. Thus, there is the simple investment rule: Keep society narrow to both lessen risk and to stabilize existing investments.

The new ideas about science and math will support the idea that each person is an equal creator, and the practical context of creativity might not be in the material world, ie the creativity of others is not judge-able (a meritocracy is an illusion).

But then the media-voice of Chomsky (an elitist, whose intellect is overly protected) can be challenged too.
Chomsky's claim that (or something similar to the following)
"MIT-science is considering all possibilities, which science can consider, so that science is developing technology, and then it is given to private interests to develop for profit" is not true, rather MIT is developing the very narrow set of "science ideas" which the corporations want developed by MIT.

(However, if one challenges the authority of a media icon, such as Chomsky, then others in the media ({and adherents of the media . . . , but the public is given only one choice, which is both in the media and in the education system, ie training for corporate jobs} will scorn you, ie this is an example of the power of personality-cult [which is a main theme presented by the media to the public].

However, if one challenges the authority of science and math then . . . . such a statement cannot be heard . . . it is outside of the myth, ie of social mind control, especially, since math and science are failing at such a horrendous scale [and the evidence is clear {see above, both inaccurate and practically useless}])

This, in a nut-shell, are the social forces which oppresses new ideas, both personality-cult and authority determined by the investment-class, and new ideas which are not within tradition and within established authority, will not be allowed to be expressed in a context of authority, where the context of absolute authority is a central component of the "sole authoritative voice" of the empire, ie the propaganda system, or the media, controlling language and acceptable thought, where the media is a set of instruments controlled by the owners of society, and used to control thought and language within society.

Where all the institutional authorities: academics, journalists, engineers, lawyers, etc are the enemies of equality and freedom, ie they effectively oppose the likeness of a modern Copernicus, those who change language at the elementary level of context, and thus challenge authority.
ie those with proven-merit (the articulate few) still serve the Empires of Emperors (since they are shown to have merit within the narrow scope of the narrow interests of the emperor), and subsequently, they see the public as an emperor sees the public, as a bewildered herd, where the life of these high-valued experts has the purpose of striving (in a narrow and exclusive context) to help the empire, as the emperors strive to uphold their own high-social-value.

But the emptiness of these purposes is demonstrated by the fact that engineering-creativity is stuck in the 19th century, held in place by a requirement of the highly-touted risk takers (the owners of society) not wanting any risks, and "not wanting any creative competitions, ie not wanting equality and not wanting equal free-inquiry."

One sees what happens when the ruling-class opens-up to the idea of risk . . . , and the owners of society believe that math is providing them with valid measurable-models of risk within the economy (but any indefinably random math model is invalid {stable patterns do not exist within such a math structure}) . . . , one sees devastating losses, and then the expression of, so called, legalized-fraud, and then legalized-theft, within the empire (so as to make-up for these losses).

There can be a very well-defined (and very different) sets of ideas concerning the main issues of: food, shelter, travel, (relations between permanence, individual dwelling, and change), communication, and practical creativity, which a modern society can devise . . . , so as to both allow for a wide range of different individual choices and still be in harmony with the earth and life . . . , and these are ways of social organization which society could adopt.

It should be pointed-out that the web-sites where the notion of "free speech" was to be the guiding principle, where these web-sites grew out of the 1999 Seattle anti-corporate demonstrations, ie Indymedia, have changed into sites of controlled-speech, wherein expressions of both value and truth have become very limited and narrow. Apparently, the now-captured editors of these web-sites, are deferring to the authority of the propaganda system, so as to only value (and allow expressed) the narrow ideas about science, education, and the progressiveness of the, so called, left (where those allowed to speak on the left are those few who have demonstrated [to the owners of society] that they possess the "accepted authority" to be allowed to express [the allowed] ideas on the media).

That is, the, so called, left is now as much about personality-cult (and its associated authority), as is anything else which is allowed to be expressed in the media.

Context of descriptions

What provides the best model for practically useful descriptions of observable and measurable patterns of existence?

Do . . . ,

Differential equations associated to physical law, do this,

or

Does . . . ,

simply identifying the correct: shape, size, dimension, and containment context for the discrete hyperbolic shapes in a many-dimensional over-all containment set

. . . , provide the best model for practically useful descriptions of an observable and measurable existence?

What is the process of finding the observed characteristic (measured) properties (or qualities) of a stable system within a math model of containment and measuring, ie a context of quantitative sets associated to measured properties of systems.

It is assumed that:

The measurable models of physical systems can either be:

formulas within the context of containment (also associated to local measuring of formulistic properties),

or

laws based on local measuring (vectors and "physical" properties defined at a point in the containing coordinate space), and represented as (partial) differential equations,

ie (classical physics)

or

laws based on random patterns being contained in a function space,

or

Random patterns imposed on an assumed set of internal particle-states, upon which non-linear changes of the particle-states are defined, in turn, these internal particle-states are imposed on the individual-functions of the function-space, (where, apparently, these internal particle-states . . . , which are associated to unitary particle-state symmetries [or particle-family symmetries] . . . , model the data from particle-collisions very well), ie (quantum).

However, these quantum descriptions or solution functions are not capable of identifying the stable properties general fundamental quantum systems, such as: Nuclei, atoms with more than 4-particle-components, molecules, crystals, (quantum systems) . . . as well as the stable orbits of the planetary system (a classical system).

The stability of these systems indicates that these systems are quantitatively consistent and controllable, ie linear, metric-invariant, and separable [(partial) differential equations, which can be used to model these systems], ie solvable (partial) differential equations, but "What is the better context of description (laws of measuring represented as (partial) differential equations, or shapes, sizes, and a many-dimensional containment context)?" Answer: The context which provides information about these systems "most" (or more) directly, and the information is more practically useful. Namely, the geometric structures which identifies the stable context within which the reliable patterns of existence are identified.

It should be clear that (partial) differential equations and their relation to function spaces or to non-linear geometry are not understood math constructs (ie they cannot be used to identify the observed stable patterns of material existence). Furthermore, the idea of materialism has failed.

If

The nature of math's descriptions of measurable (physical) systems (patterns) all of whose (the systems) measurable properties can be placed into formulas of variables associated to a finite dimensional containing (coordinate) space, so that such descriptions of (stable) properties are quantitatively consistent and the measurable context describes the stable (actually existent) measurable patterns (or quantitative properties) of physical systems,

Then

The math structure must be very simple:

(locally) linear,
metric-invariant,

with the coordinates, which are associated to the system's shape (within the containing coordinate space) having a continuously independent set of relationships between (1) the [formulistic] system properties [represented by a formula, or function, in containment coordinate variables], and (2) the containing spatial coordinates values (or coordinate-variables) associated to the system's shape, ie the shape of the material, in space, where the material is that of which the physical system is (assumed to be) composed. That is, the measurable properties of the system should emerge directly out of (or determined by simple relations to) the system's shape.

When the shapes (or systems) are bounded, then these simple math properties are related to simple shapes which are "cubical" simplexes, and subsequently these "cubical" shapes are associated (by a moding-out process) to circle-space shapes, which are contained within the spaces of constant curvature (whose metric-functions have constant coefficients), eg of Euclidean metric-space, ie tori, and of discrete hyperbolic shapes [a shape composed on toral components] of the hyperbolic metric-spaces (but not spheres whose metric-functions do not have constant coefficients and thus the spheres are non-linear shapes).

There are also the unbounded discrete hyperbolic shapes.

These various types of discrete hyperbolic shapes may be bent (or folded) by discrete sets of angles defined by the Weyl transformations, in turn, defined on the maximal tori of Lie groups, ie the Lie fiber groups of the containing metric-spaces.
These folds may be related to angular momentum properties.

The infinite extent discrete hyperbolic shapes identify a place (or subspace) of existence for stable patterns, while bounded discrete hyperbolic shapes identify the positions of stable patterns.

What these shapes describe . . . , are the stable patterns (or stable shapes), so that when these shapes are bounded-shapes, then "in Euclidean space" the positions of these stable bounded shapes can be identified, and changes, in regard to spatial displacements, can be identified in Euclidean space, where the spatial displacements of these stable shapes would be due to material interactions (or due to inertia).
Position related material interactions refers to a global relationship which is associated to action-at-a-distance (or non-local) interactions, which are defined for both "discrete time intervals" and in relation to other identifiable (implied opposite) patterns, within a context of "pairs of metric-space states," associated to discrete spatial changes for each "discrete time interval," where the "discrete time intervals" are defined by the period of the spin-rotation between the pairs of opposite metric-space states.

That is, for every spatial displacement of a bounded stable shape, there is an opposite displacement (Euclidean space), and for every stable pattern in time, there is a stable pattern in opposite-time (hyperbolic space, or equivalently, generalized space-time).

Thus, the metric-spaces are associated to pairs of opposite states (or opposite properties), where "the state and its opposite-state" can be fit into the real and pure imaginary subsets of the complex coordinates, so as to be spin-rotated (or to define a spin-rotation) between these pairs of opposite metric-space states (this is part of the discrete new model of system-interactions, which is a model which is quite similar to classical interactions). Now, within complex-coordinates the new description has an invariant Hermitian-form in the complex coordinates, and this complex-coordinate structure of (real) opposite metric-space states, is related to unitary fiber groups (and unitary invariance, or energy invariance, because the discrete hyperbolic shapes are very stable conserved patterns).

The over-all high-dimension (real) containment space is an 11-dimensional hyperbolic metric-space, or an associated containment space with a unitary fiber group.

(Note: An 11-dimensional hyperbolic metric-space is equivalent to 12-dimensional space-time space).

There is a finite partition of the different dimensional subspaces of this 11-dimensional hyperbolic metric-space by discrete hyperbolic shapes of different sizes, where the size of these stable shapes is controlled by the multiplication of a constant defined both between dimensional levels and between subspaces of the same dimension.

There are containment sequences of subspaces depending on the sizes of the discrete hyperbolic shapes which form the partition, where the dimensional-containment sequences can be identified in relation to the sequences of increasing-dimension of the subspaces; note that:

charges fit into nuclei (2-dimension),

nuclei fit into both atoms and molecules (2- and 3-dimensions),

atoms and molecules condense into crystals and planetary shapes (3-dimensions), and

planets fit into orbital envelopes of material-containing metric-spaces (4-dimensions),

these solar-system sized metric-spaces fit into galaxies (5-dimensions),

galaxies into the local universe (6-dimensions) etc

[but the last bounded discrete hyperbolic shapes is a 5-dimensional discrete hyperbolic shape (this property was identified by Coxeter)].

Finding the stable measurable pattern, eg a solution function to a system's (partial) differential equation is about finding a simple quantitatively consistent stable shape, and its:

1. Size,
2. Spectra, and
3. Angular-momentum inter-relationships,

where angular-momentum are relations of either flows, or condensed material, that can exist between angled-toral components of discrete hyperbolic shapes.

Another consideration about stable geometric shapes (which are allowed within a measurable descriptive structure) deals with the size-measures (of the containment space) which can change between adjacent dimensions, which, in turn, allows either (1) orbital-spectral-flows or (2) condensed material within the stable orbital envelopes of a material-containing metric-space.

Furthermore:

Do angular-momentum relations exist between:
Toral components
Different subspaces
Different dimensional levels
Different over-all containment sets ? Answer: Apparently, Yes.

This type of containment construct is very much more complicated than the construct of materialism.

However, materialism has been placed into descriptive structure of very complicated nonsense.

The very complicated math patterns associated to indefinable randomness and non-linearity are quantitatively inconsistent math patterns, and when material is reduced to elementary particles, then the elementary particle-properties can not be re-attached to material description so as to describe the observed stable properties of fundamental material systems, such as nuclei etc.

Particle-physics is best interpreted to only demonstrate that the observed properties of unitary-ness and high-dimensions exist.

In particle-physics the relation of space and material geometry to the properties of force-fields lose their meaning, eg renormalization of the particle-collisions, which are assumed to be a part of (quantum) material interactions, is about the break-down of the structure of space, caused by the assumed local (energetically uncertain) structure of particle-field models of material interactions, related to the supposed properties of both randomness (the uncertainty principle) and local unitary particle-state symmetries.

To remotely, try to, make particle-collisions a viable model of (quantum) material interactions, the particle symmetry properties are added . . . , as perturbations to the particle-states, in turn, added to the wave-function structures, where internal particle-states are associated to the absurd model of particle-collisions, in a random based description . . . , and then subtracted, based on spectral bounds imposed due to uncertainty of the random properties of size and momentum (or energy).

This is absurd, and means that containment cannot accommodate the idea of materialism and its reduction to particles.

Rather the high-dimensions and the unitary patterns of what appear to be events emanating from a collision-point, are unstable properties, and:

(A) thus, they are properties which properly belong to a higher-dimensional containing-space, which has metric-space states associated to its (real) metric-space containment structure, ie it has a unitary construct associated to such a descriptive structure.

(B) The collision point is a distinguished point of higher-dimensional shapes, which do not attach themselves to the dimensional level, upon which the idea of materialism is defined.

The (partial) differential equation is fairly unimportant since the stable patterns are the discrete hyperbolic shapes, ie those discrete hyperbolic shapes which are resonate with the finite set of discrete hyperbolic shapes, and an associated finite set of multiplicative constants (defined between dimensional levels and between subspaces) which partition the 11-dimensional hyperbolic metric-space, as well as the set-containment structures of the dimensionally-sequenced, set-containment context of the new descriptive structure.

It is the very simple discrete hyperbolic shapes (in resonance with the finite spectra of the containing space) which would be the solution-function shapes for the stable-systems, which one is trying to model by:

either

(partial) differential equations {mostly defined for condensed material}*

or

(sets of local linear operators applied to) function spaces (used to model the physical properties of the material system).

Thus, for solution geometries to be knowable then one needs to know the finite spectral set and the "subspace-dimension-size containment" construct of the over-all 11-dimensional hyperbolic metric-space containment set.

The construct of interactions when applied to small (molecular) systems result in the type of randomness associated, by interpretation, to the fundamental quantum randomness of random point-particle-spectral events in space and time, so as to imply the uncertainty principle (in a limited context), and non-locality, where the uncertainty principle is an approximation (or only appears to be true) within a context of many small-body interactions, but non-locality is fundamental to the new description. The apparent point-like structures of small systems are a result of the distinguished point of the discrete shapes (associated to the vertices of the moded-out "cubical" simplex fundamental domains of the discrete shape's coordinate space lattices).

The interaction is based on the context of the derivative which is to be modeled in a context of local measuring (as is done in classical physics), but the derivative is a discrete operator (in regard to time-intervals, dimension, subspace, and discrete angles defined between a discrete shape's toral components), the derivative within the model of material interactions has a structure similar to a connection, ie local displacements (on coordinates) are determined by the fiber group. It is this interaction structure, ie it is this connection, which in 3-spatial-dimensions requires that the interaction be spherically symmetric.

If the derivative operator is placed in a function-space context, then the functions would need to be sets of the discrete shapes of non-positive constant curvature metric-spaces, placed in a containment context which has many natural constraints of: size, dimension, subspace containment structures, and the allowed sets of stable: spectra and shapes.

These new ideas transcend the idea of materialism so that materialism is a proper subset of the new containment set, they are based on higher-dimensions, which are actual (or real) parts of our existence, but they are difficult to perceive when only "the context of material is considered,."

This new descriptive construct is an expression about an "ideal" world, ie wherein the material world is a proper subset, and where "the ideal world" is the context wherein the idea of "a spirit" can be mathematically modeled, and these models will be (are) shapes within containment sets whose properties are measurable. Thus, the notion of idealism can be placed into a measurable descriptive language, whose properties can effect the subset (or subspace) within which material is defined.

The above math construct of existence shows that the structure of materialism is a proper subset of a math model of existence, which is consistent with an interpretation of a particular notion of idealism, ie a many-dimensional construct which identifies both spectral-geometric limits of (eg physical) existence, as well as ways in which size-subspace-dimension-containment structures may be organized in regard to these same spectral-geometric restraints of the (useful) geometric and spectral properties of the over-all high-dimension containment set.

Thus, now, physical law is about the finite spectral sets upon which stable measurable patterns can be defined, within the various ways of organizing a dimension-size-containment set of coordinates, all of which are contained within a dimensional-subspace partition of an over-all 11-dimensional hyperbolic metric-space, along with an associated complex-coordinate containment of (opposite) metric-space states, for the various metric-spaces, which have properties associated to themselves, and which are part of the containment-existence construct (thus the description relates to both real isometric transformations and complex unitary transformations of the local coordinates).

Interactions between component-shapes are defined in both (1) various (adjacent) dimensions, and (2) between discrete time-intervals, which are contained in the set (and, thus, these components are in resonance with the finite spectral-set of the containment set, and they are spin-rotating {in time} between pairs of opposite metric-space states), and these interactions are now modeled as discrete operators, which are very much like the (classical) models of local measuring of geometric-material properties, which are defined in classical physics.

It is assumed that:

The measurable models of physical systems can either be: formulas within the context of containment (also associated to local measuring of formulistic properties), or laws based on local measuring (vectors and properties defined at a point in the containing coordinate space), or (classical) laws based on random patterns being contained in a function space, or based on internal particle-states imposed on the functions of the function space, upon which non-linear changes of the particle-states are defined (quantum).

28.

To sum-up

New ideas vs the propaganda-education-expert system

Existence is about partitioning an 11-dimensional hyperbolic metric-space (as well as its unitary structure, which is associated to the pairs of opposite metric-space states, defined on each metric-space [or shape]) into different dimensional subspaces, where the partition is done by a finite set of discrete hyperbolic shapes of specific sizes, where size is controlled by constant factors defined between both subspaces and dimensional levels, so as to form a construct of set-containment based on dimensional-size sequences of the partitioning shapes, so that the containment-tree-of-subspaces determines the type of geometric properties . . . : charges, nuclei, atoms, molecules, crystals (condensed material-components), (life), planetary bodies, (life), solar systems and stars, (life), galaxies, etc . . . , which can (might) exist within any of the containment-trees.

Interactions between material-components is defined by a set of discrete connection-operators, ie locally defined derivatives related to local coordinate transformations (or fiber-groups of metric-spaces), and by the usual 2nd-order dynamical and geometric (partial) differential equations of classical physics, where the stable physical structures, which result from interactions, are the discrete hyperbolic shapes which are both (1) in resonance with the geometric-spectral properties of the finite set of discrete hyperbolic shapes partitioning the over-all 11-dimensional hyperbolic (containing) metric-space, and (2) fit into the containing subspace (due to relative sizes).

That is, the currently accepted (2013) context of (partial) differential equations defined in either a geometric context, or in a context of sets of operators acting on function-spaces, which are used (today, 2013) for the descriptions of physically observed properties, is really being defined in either the indefinably random or the non-linear context, and this is because these operators are too narrowly confined, and thus they ignore the full set of existence, ie it is a context which ignores the bounding (or confining, or restricting) very stable context of containment, and the relation that a "finite spectral-geometric set" can have to allowing (or requiring) a stable quantitative basis for a measurable description of the (measurably stable) observed properties of the physical world, as well as the world containing stable physical systems.

That is, a discrete operator context is needed for the physical descriptions of the underlying, fundamentally stable set of physical systems, which are defined on each of the different particular discrete contexts of , . . : dimension, subspace, toral-component of discrete hyperbolic shapes, and discrete time-intervals . . . , where each of these discrete contexts needs to be considered in order to fully integrate the bounding stable discrete hyperbolic shapes . . . , within which the containment of existence is defined . . . , so that a measurable description can be correctly related to physical descriptions of the observed stable systems.

The new descriptive construct transcends materialism and enters into a measurable description of the ideal (or spiritual) context of existence, ie it unifies measurable description with a greater vision of existence (ie science and religion are unified!).

However, this would disrupt the propaganda system of the owners of society, where the propaganda system is the sole voice of absolute authority within society, and it expresses the language structure for the public, which best suit's the owners of society, and their petty interest in their own dominating (psychopathic) social power, which they possess over society.

Where the propaganda system is based on a Hegelian dialectic of opposites: of science and religion (or, equivalently, materialism vs. arbitrary moralizations). {or left [equal freedoms] vs right [obedience to the owners of society]}

It should be noted that many of the descriptive categories of university departments: science, economics, psychology, sociology, politics, etc are placed in an descriptive structure which appears to have a basis in a quantitative description, an apparent measurable context, but this is really not-true (for any of these categories, in which these technical experts [with their overly technical and incomprehensible languages], are defined), since the quantities are defined in an "unstable math context" for each of these descriptive constructs. Thus, the quantitative claims of these experts are mostly arbitrary claims, ie counting is not defined in these contexts so the quantitative measures have no meaning (no true basis in numerical value) including what are now considered to be the laws of physics.

   . . . .

[however, correlations, which might be identified on such a floating quantitative structure, may be observed to be true, eg vitamin-C is strongly correlated with scurvy, one should note that indefinable randomness is assailed, in regard to global warming (though atmospheric concentration of CO_2 is very strongly correlated with global warming), but indefinable randomness is not strongly assailed, in regard to quantum physics, even though the mythological object of a very highly-valued figure within the cult-of-personality, ie Einstein, was opposed physical description (ie quantum physics) being based on indefinable randomness. However, it needs to also be noted that, Einstein supported indefinable randomness imposed by non-linearity in general relativity. That is, Einstein was interested in ideas, but as usual, people are often wrong when they actually think for themselves, but the authority of the experts who serve the owners of society, are never allowed to be considered to be wrong, yet the owners, themselves, are (clearly) always blundering (they destroy both the earth, and the independent creativity of the people), but the dogmas of the experts which serve the interests of "the owners of society" are absolute truths.]

. . . .

That is, the classical laws of physics have limitations, mostly leading to non-linear systems, while the quantum laws are all (essentially) useless, since they are based on indefinable randomness, and such a description is irrelevant in regard to understanding the stable, controllable properties which are, in fact, a part of the observed properties of the most fundamental existing physical systems (which are very stable) [both large and small systems].

It is not valid to claim that "all such stable systems are too complicated to describe" based on the currently accepted laws of physics. Especially since there now exists a math model within which the stability of these systems can be described.

However, to express new ideas about the culture's knowledge about a measurable existence, the ideas must be filtered through, what are essentially the very few owners of society, who are essentially in exactly the same social position as the Emperors of the Roman-Empire, but now they are bankers and extremely wealthy private investors, who preside over a "far too narrow way" in which to organize a society, in regard to both life-styles and practical-creativity (a narrowness which is causing great destruction of both the earth and society), since all of these allowed endeavors now done within our society, which are associated to knowledge and creativity, are designed, and organized within society, to serve the social-power of the owners of society, and these owners of society are both ignorant, concerning knowledge (or act as if they are ignorant), and selfish in their outlook about how society is to serve them, ie they are immoral in a scale of [arbitrary] moral-value which is different from their scale of arbitrary value (though their scale of arbitrary value is the scale of value which is used within society, which is run by both investment-money and the violence needed to uphold such arbitrariness and associated inequality).

That is, new ideas are not allowed if they are not going to serve the interests of the current owners of society, or if the new knowledge will destroy the propaganda system and expose the ignorance and selfish motivations of the owners of society, in a society whose "real" religion is the personality-cult, or the worship, of a set of (failed) experts, and the relation of these experts have to the pronouncements by the propaganda system of what has (arbitrary) high-social-value (such as the high-value of the absolute dogmas which define the experts), and which the public is not allowed to question.

Unfortunately the public has been trained (by the society's solitary authoritative voice of the propaganda system [and educational system], and associated wage-slave status of the public) to "nod their heads in agreement with," or "in respect for," the experts which serve the Emperor(s).

29.

Most affect actions

Consider the properties of propaganda within society which most affect actions and attention

One sees all the time, a set of institutional bullies (of the national security state, ie the employees of the justice, military, propaganda (ie politics), and banking-oil-corporate rulers) doing bad things to people, and then others (usually, those in these very institutions who observe these actions by the bullies within their institutions) report these bad things, and [if allowed onto the media, then the media] (rightfully) express moral indignation at what the bullies are doing.

But seldom do the bullies stop it.

The propaganda system most often discounts these claims (or misrepresents these claims, and marginalizes the messengers), and there is never any consequences, in regard to the immoral actions of the bullies, ie the rulers henchmen, where this inaction within the political-justice system results from the structure of law within society.

Yet all societies always claim to be the most moral society ever to exist.

So go to the source:

the laws and its enforcement (administering law) in society, which allow these bullies to do these things.

The crux of the matter within a society, which is characterized by non-stop bullying, is always about basing law on

(1) property rights . . . , ie stealing which is allowed if one is a member of an upper social-class . . . . , and ,

(2) minority rule . . . . , eg a Roman republic . . . . , where these laws serve the interests of the owners of society, ie the bankers-oilmen-military business people, and subsequently

(3) the imposition of wage-slavery on the public, imposed by a combination of both a (corrupt) justice system and the propaganda system which proclaims the sole voice of absolute authority, which the public must obey (or not get a paying job).

This single voice of authority is the social force most the cause for stopping the expression of new ideas within society, and it also stops a set of wide-ranging possibilities for individual creativity. Knowledge and creativity serve the interests of the owners of society.

Since, within society, money now equals property, and the standing-military takes its orders from those "with the most monetarily-derived social-power," then it is good idea to ask (when did) is the US society (become) ruled by the richest few people in the world, who may or may not be Americans?

However, the principle (or the main idea) of the American revolution was that;

Instead of law being based on the control of property by the ruling class, law is to (should) be based on equality, so that there does not exist a social class of oppressors.

This is central to the logic of the US Revolutionary War, in which the Declaration of Independence declared that US law was to be based on equality, so that the systematic oppression of the public, by the upper social classes, could be stopped. However the upper-classes wanted a republic, but the Republic they sought, ie the US Constitution, was not allowed to be law, unless there was the Bill of Rights attached.

However, the Bill of Rights was never upheld, since a social-class of oppressors was allowed to manage the Republic. The US Republic was supposed to provide for the common welfare in the context of law being based on: equality (everyone is an equal creator) and its citizens possessing freedoms to believe and to express one's beliefs, which are principles about freedom which are to ultimately be related to selfless practical creativity based on new knowledge.

That is, in the age of enlightenment . . . . , which was also at the beginning of the scientific revolution, where the scientific revolution reached its zenith when electromagnetism and thermodynamics as well as statistical physics were developed in the 19th century (where statistical mechanics was based on closed, bounded systems with a fixed number of, N, neutral particles, which possess only mechanical properties, eg mechanical collisions, settling into an equilibrium associated to a constant thermal energy of an isolated system) . . . . , the founders were well aware of the relation of scientific revolutions to both freedom of belief and freedom of expression, ie new ideas must be considered in order for science to develop.

Equality and free expression allows for the development of truth, where a described (measurable, or verifiable) truth is about both accuracy as well as the information within the description being practically useful in regard to creativity, ie using observed measurable couple-able properties to build new (material) systems.

Equal free-inquiry (in turn, the information is to be related to practical creativity) was the idea which Socrates supported, and that society should be committed to such a simple principle and to such a humane principle.

The current law should be suspended, since

(1) it has not been administered in a manner consistent with the law of equality which the US stated, "was to be the basis of US law," and a 3rd-continental congress (or whatever number it should be) needs to be re-instated, so as to restructure law and government, so as to have equal institutions, and to promote the common welfare based on equality, where a free-market is not possible unless everyone in society is accepted as an equal creator, and

(2) the gambling casino has collapsed, some of the "owners of society" lost their shirts, and criminality (since the ruling class are all thieves) has become the basis for these big-players to retrieve their fortunes, where the big-players are needed in the game of empire being based (waged) on (baloney, or fraudulent) economic principles.

An empire is most concerned with military constructs through which the empire can expand, so as to feed its narrow mental construct of world-properties by means of a highly controlled set of mechanisms for exchange, and which are the basis for the power of the emperors, ie the emperors are those few who direct the expansion and a mercantile society's narrow context (upon which are defined the need for stealing and its associated violent support) upon which earth's destruction and society's destruction are both guaranteed.

That is (in reference to (1)), one wants a society where law supports the idea of human beings being fundamentally about acquiring knowledge, and using knowledge in relation to the development of practical creativity.

That is, humans are not basically about property and the use of violence to steal property for one's selfish gain.

The main division concerning communication within the propaganda system (the language mechanism through which the context of a human life of wage-slavery is expressed) is that of: Materialism vs. idealism

(where idealism is a belief that the properties of existence transcend the idea of materialism)

In the math-physical description given below, materialism is a proper subset of idealism, where the containment-set of an "ideal" existence is an 11-dimensional hyperbolic metric-space, wherein stable shapes of the various dimensions (and the various subspaces) exist

Usually the division is
Science vs. religion

But the main teachings of religion is, simply-put, the idea of equality, whereas science has insisted that existence is determined by material properties,
Thus the division should be
Materialism vs. equality

Where the material model of biology implies survival of the fittest, and thus, Darwin's evolution implies inequality,
Furthermore, the emperors religion has always served the emperor, and it has never served the purpose of religion, ie the Emperor's religion has never served the purpose of equality.

One could say for a "described truth," where truth is a description's relation to both accurate information (concerning observed properties), but more importantly the description's relation to practical, useful, creative efforts, and in trying to get-at a valid "described truth" the social condition of equality is the principle (or social-state) which Socrates supported (equal free-inquiry), while Godel's incompleteness theorem also implies that truth depends on an equal expression of ideas.

So in the way things are used in the propaganda system, both "the emperors religion" and "material based science" are viewpoints which support the idea of both materialism and inequality, where science has been manipulated within the constructs of social-class, so as to become authoritative and dogmatic, and in turn, this is used to define a competition, within which, supposedly, a class of intellectually superior beings can be determined. Unfortunately, the main idea expressed by the, so called, top intellectuals is that they are obedient to dogmatic authority, namely, the same authority upon which their social positions depend.
But this is not science, science requires equal free expression and equal free-inquiry.
However, based on the dogmatic science competitions, people are found to fill the positions of authority within society, who are aggressive (seek dominance), inclined to like to use language, and are very obedient types of people.
(Those few who win the narrowly defined "science competition," are not a class of intellectually superior people by any means, since the representation of physical science (or physical law) upon which the authority of the competition is defined, cannot describe any fundamental

physical system (it is not capable of describing the stable: proton, general nucleus, general atom, molecules, stable solar systems etc) and these, so called, intellectually superior people are oblivious to this pathetic state of affairs within which science and math are to be found, rather they are people who mostly seek fame and fortune).

Essentially, there is no person who supports equality (in no uncertain terms, as did the principles of (equal) free-inquiry of Socrates) who is allowed to have a voice within the propaganda system, or such a person's statements would be carefully edited, and such a person, who believes that equality is the correct social model, and which best express the intent of the American revolution, and there is no person who (also) presents new ideas based on the elementary (language) level of assumption context and interpretation, (the same level of precise language within which Faraday developed electromagnetism), and if a [person tries to do this they are attacked, based on an arbitrary belief in high-value of superior truth, in regard to how words are used by the propaganda system.

Such a belief in arbitrary high-value was the basis for the Puritans to exterminate the native people of the Americas, and also for Columbus to enslave and murder the native people of the Americas (ie where America is a European derived name) in the name of a very narrowly defined European market model of exploitation, whereas the native people's were doing quite well living in harmony with the earth (can we do equally well living in harmony with the earth, but the context has been changed).

This extermination of native peoples is now (2013) always presented in historic accounts as evil actions which was not a result of personal knowledge, but rather as a result of what the society considers to be a superior cultural knowledge, it is always formulated by an expression of an idea (supposedly consistent with the society), yet it is a viewpoint which is blind to the nature of (precise) descriptive truths, where Socrates' principle of equal free-inquiry allows many ideas to exist, so as to not allow one (murderous) idea to become dominant.

The intellectually superior experts . . . , (where other countries try to replicate such an intellectual capability, so as to also indicate their "superior status as a society") . . . , represent to society a vastly superior culture, and this superiority is implicitly used as a justification for any actions which the society might take.

Yet, it is an intellectual capacity which exhibits only failure, otherwise please describe the stable:

proton, general nucleus, general atom, molecules, stable solar systems etc.

New ideas

There are several books and papers, including the speech delivered at the math conference in San Diego (Jan 2013).

There are new ideas, which are about measurable descriptions based on the principles of mathematics, where a new emphasis on the interpretations of certain ideas in stricter ways, stricter in regard to the importance of the properties of both stability of patterns, and stability of the containment context (wherein measuring is reliable), so that these newly emphasized math principles of stability are applied to the observed properties of physical systems, and in doing this, the most fundamental problems of physics are (can be) finally solved.

Namely, describing the stable properties of orbital-spectral systems at all size scales, wherein it is assumed that these systems are composed of many-(but-few)-bodies, and which are observed to possess stable orbital-spectral properties, (the solution is) a feat which modern science is not capable of doing (or of solving).

Stable properties imply a quantitatively consistent and controllable context . . . , (linear, metric-invariant (for metric-functions with constant coefficients), and the local directions of the shape of the system are independent in a continuous manner everywhere, ie (geometrically) separable and algebraically diagonal), . . . , in regard to a system and/or its creation through an interaction process.

In math, (where math is [should be] about precisely describing the measurable properties of the world) the mathematician is always encouraged to seek the most general context (for a pattern), and the most complicated aspects of "a general, and abstract context," and then to list simple patterns of algebra, and relate the complicated contexts (complicated patterns) to sets of simple algebraic patterns, ie the ideas are framed in a very abstract manner, in turn, the simple algebraic properties of the complicated context are often associated to simple geometric symmetries. But, this close expression of math ideas within a carefully described context of algebra ignores (or veils) the need, in math, for both the properties to represent stable patterns (within a descriptive context) and the need to emphasize the math properties which make a containment-context both stable and measurably reliable.

But the idea of stability, of both descriptive context (reliable measuring), and of the stable patterns (protons, general nuclei, general atoms etc) being described is never considered, except in an overly general and overly abstract context, eg assuring a set containment property so that the algebraic structure is maintained in the containing set. However, very large quantitative sets, defined within continuums, may not be properly relatable to a stable uniform unit of measuring, ie is stability considered only in a vague and abstract context which ignores the stronger needs for a property to be stable. (linear, metric-invariant, independent local coordinate properties continuously related to the containment set).

The math properties which allow for stability of patterns (in the context of world experience) are very restrictive patterns which are about geometry, since measuring and quantity are geometric (or worldly) set constructs (stable properties of the world).

A brief history concerning the math methods of physics (or for science in general)

Measurable properties of systems (or patterns) are represented by function-values (or related to function spaces) whose formulas (for each function) are defined by the variables of a system-containing coordinate space (or domain space).

System containment of material-geometry based systems (ie systems composed of material) is the assumption of materialism, and its associated restricted relation to the idea of dimension, or the realizable independent directions for material's containment coordinate space.

So how does one get at the function's formulas?

(1) by local measuring, and
(2) by equivalent ways of expressing the same concept by different quantitative relationships (ie forming a differential equation), or by
(3) identifying measurable quantities and discovering (by experiment) the different relations that these quantities have to one another, eg $pV = kNT$, where with the aid of thermal functions and associated thermal laws these relations can also be found by local measuring relations,

where local measuring is either

(1) about linear local relationships between functions and their coordinates where the coordinates depict the system's shape (or the laws of thermal physics), or
(2) about sets of operators acting on function-spaces, where the functions in the function-space are related to the system's spectral values, and where the system is contained in the domain space of the set of functions which compose the function space. It is not clear if this is local measuring, or (alternatively) measuring the spectra of a containment set.

Is this the correct context for determining the measurable properties of physical descriptions? Is materialism the correct context for physical description?

The above principles of physical description are supplemented by an assumption of reduction to (ever) smaller components, where such reduction depends on the interpretations of high-energy small-material-component collisions.

This assumption of reduction of physical systems to smaller components also leads to an assumption of an (apparent) random set of properties in regard to observing the components, to which collisions (of the collision experiments) reduce material systems, in turn, this leads to physical description, in regard to (2), being based on indefinable randomness (built upon function spaces upon which (local linear) operators are defined), while the idea of differential equations, ie (1), leads to the most prevalent pattern observed in regard to local measuring, the pattern of non-linearity.

However, indefinable randomness and non-linearity cannot be used to describe the stable patterns which are observed for the most fundamental of physical systems, wherein stable spectral-orbital properties are observed.

So what is the mathematical nature of stability of patterns?

Answer: Stability and quantitative consistency are related to the patterns of simple shapes, namely, cubes and circle-spaces

The two shapes which are quantitatively consistent with numbers (or quantities) are lines and circles, where local linear independent directions can maintain a continuous linear and independent (direction) relation which can exist between the lines and the tangent directions of the circle(s), when these (types of) shapes are the coordinate shapes which compose the quantitative sets (or the system-containing coordinate-space).

Quantitative sets are characterized by sets of algebraic properties in regard to operations of: order of operations, existence of identities and the existence of inverses, and the distributive property where the distributive property gives number systems the structure of polynomials defined on quantitative variables (the bases, eg 2, 10, etc, upon which a number system is defined).

Thus, there is the real numbers of the line and the complex-numbers of the plane (or measuring based on a circle and a ray) which are quantitative sets which can be made to have quantitatively consistent relations to each other, ie the lines and circles are the quantitatively consistent and can form into stable math patterns.

This leads to a simple rule, the stable and quantitatively consistent shapes are the circle-spaces defined on metric-invariant metric-space whose metric-functions have constant coefficients, eg tori on Euclidean space, and strings of attached toral components defined on a hyperbolic metric-space [or on R(s,t), see below], ie the discrete hyperbolic shapes.

What happens when one tries to turn higher-dimension local rectangular coordinates into consistent quantitative sets (such as the complex-numbers)?

Consider the quaternions, a number system based on 4-independent directions, have all the same properties, in regard to number operations, as real and complex numbers possess, but there are zero divisors in the quaternions (ie a and b are both not zero, yet ab=0, so a and b are zero divisors), so equations cannot be solved.

Geometric measures are necessary in regard to containment of physical systems within geometric models of coordinate sets, which represent "physical" systems, these are the local measures of length and they are also local alternating forms, which are consistent with length measures, but instead of measuring length the alternating-forms measure: area, volume etc.

If there is to be quantitative consistency within a description of a material-geometric system, which is contained in a metric-space coordinate system, then the local transformations of local coordinate values need to be metric-invariant, whereas this also implies the invariance of the other geometric measures.

(one expects that) The metric-invariance of metric-functions, where the metric-functions are symmetric matrices, implies there are always local coordinates in which the matrix representation (or the local linear representation) of the metric-function is diagonal, ie the metric-function is independent in each local direction of the coordinates.

Thus linear functions which are also diagonal in a continuous pattern (of local diagonal linear properties) which exists on the global shape of the coordinates (or the global shape of the system being described) are the local functions which allow quantitative consistency in regard to a function's values being consistent with the geometric measures defined on the system containing coordinate set (or coordinate space) [where a function's values represent the measurable properties of a stable system being described].

Note: Local linear maps, which are not continuously commutative everywhere (ie not diagonal) determine quantitatively inconsistent (or non-linear) measurable relations, ie they identify patterns which are not stable.

The stable solutions of dynamic equations for physical systems composed of material components are related only to the set of discrete hyperbolic shapes whose spectral-geometric properties are consistent with (or resonate with) the finite spectral-geometric set of values defined for the entire over-all 11-dimensional hyperbolic metric-space containment set (of what is essentially, for these material components, a model of all existence).

Thus, for the three types of 2nd-order metric-invariant partial differential equations . . . ,

(1) elliptic (bounded and/or orbital systems),
(2) parabolic (confined but free systems, eg projectiles [rockets], as well as angular momentum),
(3) hyperbolic (component collisions, essentially free and unbound systems) . . . , the new description provides solutions which are related to bounded orbital-type component-like systems so as to possess a new context within which angular momentum can inter-relate either

(1) to the toral components of a stable shape or
(2) higher-dimensional stable geometric-spectral components (or shapes, or metric-spaces).

There are two categories within which these shapes can be achieved (or solutions determined),

(1) where the lower-dimension components have spectral measures which fit into the spectral measures (structures) of the (new, if interacting) higher-dimensional components,
(2) where the lower-dimension components are too small to fit into the orbits of the adjacent higher-dimensional component (which is a part of a dynamic interaction), in which case there is condensed material, composed of smaller components, which may or may-not fit into their containing metric-space-component's geodesic (or orbital) structure, so as to identify stable orbits for the condensed material that the metric-space contains.

(1) is essentially determined by resonances, while (2) is defined either in the parabolic or hyperbolic context, wherein the pairs of opposite time-states identify a local involutive transformation (ie the positive-time position and the negative-time position are interchanged in order to determine the spatial displacements of the two opposite dynamic states, [where involutive means the transformation is its own inverse]) between the positive and negative time states of a "discrete dynamic local transformation," but because the interaction for such free-states is (in general) non-linear (determined through a connection, which, in turn, is defined within the isometry fiber group) the orientation and spatial position of the next time interval are changed, yet locally they remain involutive.

It seems that this would define a dynamic structure which is a symmetric (manifold) structure, and it is such a symmetric structure whose local opposite displacements might define a path which is in resonance with, ie tangent to, the stable and restricted orbital structure of the material containing metric-space (where, to re-iterate, the metric-space would possess a stable shape, and thus they also possess stable "orbital" properties, into which condensed material components, ie existing components which are smaller than the size of the (containing) metric-space shape, would [eventually] enter into an orbital structure).

That is, the entire context of the (partial) differential equation is qualitative, wherein the context of their (current) definition these operators are simply steering the underlying stable components (which the containing space does contain) into a spectral (adjacent spectral sizes fit together) or orbital (small components condense) relation with an underlying finite set of stable shapes, which define both a partition of an 11-dimensional (hyperbolic) containing space, as well as a finite spectra for that same containing space.

Furthermore, the differential operators are better defined as discrete operators, which are defined between the discrete separations of:

time intervals,
subspaces,
dimensional-levels,
toral components (discrete angular folds), and
in relation to other sets of 11-dimensional containment sets.

However, there is (in the new descriptive structure) a new context for angular momentum.

That is, angular momentum is defined on the various toral components of the stable shapes, which are allowed by the containment set (which defines a finite set of stable spectra-geometric measures, to which the existing stable shapes must be in resonance), and on possible links, defined by angular momentum.

There are unbounded stable discrete hyperbolic shapes, which exist on all dimensional levels, and these unbounded shapes are associated to stable material components, ie stable discrete hyperbolic shapes defined by their being resonant with the finite spectra of the various subspaces of the containing space [which is partitioned by (into) a finite set of stable discrete hyperbolic shapes of all the dimensional levels of the over-all containment set].

On the other hand the 2-, 3-, and 4-dimensions are relevant to the descriptions of "material" components contained in hyperbolic 3-space, where these stable shapes are also related to both bounded and unbounded, or semi-unbounded, discrete hyperbolic shapes, where an example of a semi-unbounded shape would be the neutrino-electron structure of an atom's (2-dimensional) charged components (which is also called an electron-cloud of an atom, eg for an atom the nuclei are bounded shapes while the electron-clouds are semi-unbounded), so that all "material" systems are linked to an infinite boundary of the over-all high-dimension containing space. Thus, one can think of angular momentum as a controlled (or controllable) link between the many different

11-dimensional hyperbolic containing metric-spaces, by means of such unbounded and associated bounded (angular momentum) links (between 11-dimensional hyperbolic metric-spaces).

Thus, one can consider a possible consciousness for people would be to examine the different creative structures of these different universes, where the individual 11-dimensional containment sets for the different universes (or perhaps different galaxies) might be perceived as intricate bubbles of different types of perceptions, through which we can control our journey since we are in touch with the infinite reaches of these types of separate existences. (see below for a high-dimensional model of life-forms, eg models which allow all life-forms to possess a mind)

Is this the true context within which the human life-force is to develop knowledge and intend creative expansion of such a context?

To elaborate (more)

The stable shapes are dependent on the metric-functions of the coordinate containment spaces having constant coefficients, and also having non-positive constant curvature, ie the zero-constant-curvature and (-1)-constant-curvature metric-spaces, or on either the R(n,0) or R(n,1) coordinate spaces, or more generally on R(s,t) coordinate spaces, where the metric-functions have constant coefficients, and also have non-positive constant curvature, and their associated isometry fiber groups.

The stable shapes are the circle-shapes composed of tori (doughnut shapes) and shapes with linearly attached toral components, or the toral components can be folded upon themselves.
The shapes which are strings of toral components are called discrete hyperbolic shapes, and they are discrete in their allowed size-shape toral components, and they possess very stable spectral-geometric-(size-of-toral-component) properties.
The tori are continuous in their sizes, but the discrete hyperbolic shapes are rigid and discrete in their geometric properties.

Both the discrete Euclidean shapes and the discrete hyperbolic shapes are the natural stable shapes which are part of the metric-invariant structure of a metric-space. Namely, they are the shapes which are related to the discrete isometry subgroups of a metric-invariant space's associated isometry fiber group.

D Coxeter studied these discrete hyperbolic shapes using reflection groups and classified them by dimension.

The only resulting shapes, upon which the formulas are defined for the results of the measurable properties of material system interactions, where such interactions result in (stable material systems with) stable shapes, are the discrete hyperbolic shapes.

Thus the discussion about solutions for a physical system's defining differential equations can be restricted to the discrete hyperbolic shapes, so as to be stable systems which are both linear and solvable, and thus controllable and useful.

That is, the useful information about existence, ie shape, and spectra, which math can provide, in regard to a measurable description of the world, deals with the very stable patterns associated to discrete hyperbolic shapes, and their containment within (and their partition of) an 11-dimensional hyperbolic metric-space. It should be noted that an 11-dimensional hyperbolic space is chosen, since the highest dimension discrete hyperbolic shapes, identified by Coxeter, are the 10-dimensional discrete hyperbolic shapes, which are contained in an 11-dimensional hyperbolic metric-space.

The stability of a shape, which is measurable, and quantitatively consistent, and stable . . . , and thus couple-able, and useful, in a descriptive context wherein measuring is reliable . . . . , is determined by a complex of a subspace-size-containment set structure within an 11-dimensional hyperbolic metric-space, which has been partitioned by a finite set of discrete hyperbolic shapes, whose sizes are adjusted by multiplying by a constant . . . . , where multiplication by a constant is defined between either

(1) subspaces of the same dimension, or
(2) dimensional levels of different dimensions, as well as
(3) the sizes of the toral components changing between the discrete folds which exist between the toral components.

. . . ,

That is, size adjustments, in regard to the subspace set structure, are done by multiplying by constant factors which exist between these natural discrete divisions of the over-all containment space.

That is, the quantitative structure of a measurable description of stable patterns is generated by a finite set of spectral-geometric properties defined by partitions and other discrete actions (or discrete divisions) within the descriptive containment context.

That is, depending-on the containment-size-subspace-dimensional structure of an 11-dimensional set, there can exist, in a particular 3-dimensional (hyperbolic) subspace, condensed material, composed of small components, ie the atomic hypothesis, which can be molded or manipulated to form shapes.

Metric-space states

A circle on a plane can identify a pattern (the circle) which has a position on the plane, wherein the two states of translation and rotation are distinguishable, and rotation of metric-space states is necessary.

The two time-states of charge, in regard to position or spatial displacement, as well as the two types of charges, require that one-dimensional models of charge be defined on a stable 2-dimensional discrete hyperbolic shape so that there is a geometric rotation between opposite metric-space states.

Thus charge-size and the size of 2-dimensional discrete hyperbolic shapes within a size-containment-dimension context are linked together.

Thus, isolating and cooling charges (as cool as the cooling-process allows), as did Dehmelt, might be significant in regard to measuring the range of possible 1-dimensional spectra of charge, though holding the system to a plane might make the context irrelevant.

There are natural stable shapes, those of odd-dimension (3,5,7,9) and with an odd-genus (where genus is the number of holes in the shape, eg the torus has one-hole, or a genus of one, ie the genus is the number of toral components of a discrete hyperbolic shape) which when fully occupied by its orbital charged flows are charge imbalanced and thus would begin to oscillate, and thus generate their own energy. This would be a simple model of life.

Thus such a shape which possesses a higher-dimension could cause the lower dimensional components to, in turn, possess an order which can be controlled by the higher-dimensional shape, through angular momentum states (properties).

Down in 3-dimensions this control by a higher-dimensional structure could be the complicated microscopic-and-macroscopic structure of life, which appears to be run by complicated molecular transformations.

This is simply about assuming that stable shapes determine the underlying order and stability which is observed, and the fact that these stable shapes (mathematically) have a dimensional structure associated to themselves.

However, according to the currently accepted laws of physics both the stable properties of quantum systems and the stable control which is possessed by life are unexplained (or unexplainable).

The patterns of stable physical systems are unexplainable within the current dogmas about the material world, since the current dogma is based on the dimensionally-confining idea of materialism, and within such a confinement, descriptions seem to be based on indefinable randomness and non-linear systems (or patterns) defined on a (quantitative or coordinate) set which assumed to be a continuum.

Such patterns are fleeting and unstable, though the decay times can, sometimes, be of relatively long duration.

Suppose human life is associated to a 9-dimensional shape of an odd-genus, then such a shape is an unbounded shape (noted by Coxeter), and thus it could well be relatable to many such 11-dimensional hyperbolic metric-spaces (why should an unbounded 9-dimensional stable shape, generating its own energy, be confined to any particular unbounded 11-dimensional containing space?) wherein the living system's lower dimensional (material) structure may be quite different (in the new containment structure), and thus the living system's perceptions and interactions could also be quite different within other 11-dimensional containing spaces.

This is a possibility which is not based on DNA, rather it is based on stable shapes being related to stable contexts of existence, which exist in a quantitatively measurable (and stable shape dependent) context of containment. Indeed, it is the molecule actions and living system actions which are conforming to the ordered context of a (high-dimensional) living system.

To re-iterate

Existence is about partitioning an 11-dimensional hyperbolic metric-space . . . , (as well as its unitary structure, which is associated to the pairs of opposite metric-space states, defined on each metric-space [or shape]) . . . , into different dimensional subspaces, where the partition is done by a finite set of discrete hyperbolic shapes of specific sizes, where size is controlled by constant factors defined between both subspaces and dimensional levels, so as to form a construct of set-containment based on dimensional-size sequences of the partitioning shapes, so that the containment-tree-of-subspaces determines the type of geometric properties . . . : charges, nuclei, atoms, molecules, crystals (condensed material-components), (life), planetary bodies, (life), solar systems and stars, (life), galaxies, etc . . . , which can (might) exist within any of the containment-trees.

Interactions between material-components is defined by a set of discrete connection-operators, ie locally defined derivatives related to local coordinate transformations (or fiber-groups of metric-spaces), and by the usual 2nd-order dynamical and geometric (partial) differential equations of classical physics, where the stable physical structures, which result from interactions, are the discrete hyperbolic shapes which are both

(1) in resonance with the geometric-spectral properties of the finite set of discrete hyperbolic shapes partitioning the over-all 11-dimensional hyperbolic (containing) metric-space, and

(2) fit into the containing subspace (due to relative sizes).

That is, the currently accepted (2013) context of (partial) differential equations . . . , defined in either a geometric context, or in a context of sets of operators acting on function-spaces . . . , which are used (today, 2013) for the descriptions of physically observed properties, is really being defined in either the indefinably random or the non-linear context, and this is because these operators are too narrowly confined, and thus they ignore the full set of existence, ie it is a context which ignores the bounding (or confining, or restricting) very stable context of containment, and the relation that a "finite spectral geometric set" can have, in regard to, allowing (or requiring) a stable quantitative basis for a measurable description of the (measurably stable) observed properties of the physical world, as well as the world containing stable physical systems.

That is, a discrete operator context is needed for the physical descriptions of the underlying, fundamentally stable set of physical systems, which are defined on each of the different particular discrete contexts of . . . : dimension, subspace, toral-component of discrete hyperbolic shapes, and discrete time-intervals . . . , where each of these discrete contexts needs to be considered in order to fully integrate the bounding stable discrete hyperbolic shapes . . . , within which the containment of existence is defined . . . , so that a measurable description can be correctly related to physical descriptions of the observed stable systems.

Note: No comment to the mind-phantoms of the propaganda system, where the point of the commentators is (essentially) always about issues of social-class, or equivalently, the commentator is claiming to possess a socially dominant position in regard to the language of the propaganda system (the current rational interpretation of the world, as proclaimed by the propaganda system).

Apparently many of these comments are fundamentally based on the commentators' inability to read (or does not read) and comprehend.

Perhaps these "careful" commentators should read Physical Review (which is supposed to be about physics, but which is now (2013) about speculative math, which, in turn, is modeling an illusionary world) and clamor to the editor-authorities about the poor abilities of the Physical Review authors to express ideas in a comprehensible manner. Indeed, physics is failing since the ideas its authorities express are not comprehend-able by anyone (including the authorities themselves), as they are certainly unrelated to practical creativity. Perhaps these critics (who apparently believe in the absolute truth of the propaganda system) should smirk about an indefinable probability structure for a descriptive language which can never get at the correct answer concerning: nuclei, atoms, and molecules, as the oil companies whine about CO_2 concentrations in the atmosphere only being a statistical correlation with global warming, and thus not an absolute truth . . . with the implied message being, that everything presented on the propaganda system is an absolute objective truth.

That is, the probability based ideas of particle-physics and string-theory are more closely related to wild-speculations than is the correlation between CO_2 concentrations in the atmosphere and the heating of earth.

The authorities claimed clean cheap fusion-energy would be available by the end of the 1950's. It is still only two-years away.

But ideas are simply ideas. However, ideas can be related to (practical) creativity, and it is the propaganda system which places such emphasis on the upper social class (the owners of society) possessing a "proven" absolute authority, but if authority requires that the "authority of the propaganda system" be associated to a limited number of ideas, then, in turn, this leads to a limited the range of practical creativity, but if it is true that a described truth can only be related to a certain fixed set of assumptions, contexts, and interpretations . . . , (ie if the upper social classes actually did possess an absolute truth, which is implied within their propaganda system, then please solve the obvious set of fundamental problems which go unsolved, and because there is an obvious set of fundamental problems which go unsolved) . . . , then this has resulted-in science and math being so ineffective, especially, in regard to science's relation to practical creative development.

But a "precisely described truth" does not have to only be related to a certain fixed set of assumptions, contexts, and interpretations.

Current science and math is trapped in a narrowly defined set of dogmas, which result in ineffective math models, which are placed in a . . . . quantitatively inconsistent, unstable, and chaotic . . . context, and thus precise descriptions are being based on an ineffective descriptive context in regard to both accurate (and sufficiently precise) descriptions and the descriptions being related to practical creativity.

Current science and math provide limited descriptions which have neither descriptive-value nor meaning (they are incomprehensible [in regard to valid interpretations of observed properties], and they are of no practical value).

This is, in fact, a much more positive statement than anything which the propaganda and its army of experts, and army of nay-sayers, are capable of expressing, since this expression gives both a solution to the failures of sciences, and it provides a viewpoint about society consistent with the human nature of people wanting to seek knowledge and wanting to be creative, and it is a math model which transcends the idea of a material world (allowing for a model of sustainable creativity, where expansion does not depend on destroying vast reserves of material resources).

30.

Physical descriptions

The propaganda-education system teaches the public to obey authority, but whatever authority in which a person might believe, the new ideas, based on many-dimension being partitioned by a finite set of discrete hyperbolic shapes (and expressed in this paper), describe the stable structures of the very prevalent many-(but-few)-body systems, and this is something which modern physics, based on (indefinable) randomness cannot do.

It is only the stable "geometrically-separable" shapes defined in a linear, metric-invariant context . . . , [where the metric-space must also be a metric-space of non-positive constant curvature . . . , where the metric-function only has constant coefficients, and the metric-function, ie symmetric 2-form, is a symmetric matrix] . . . . , in which

1. stable (math, or physical) patterns exist,
2. measuring is reliable, and
3. there is quantitative consistency,

and where linear metric-invariance implies a need for a separable shape (or a geometrically-separable-shape) ie continuously locally diagonal coordinate transformation relations or locally orthogonal local coordinate relationships.

The stable shapes of the set of discrete hyperbolic shapes can be used to determine subspaces of dimensional levels (where this can be done by partitioning an 11-dimensional hyperbolic metric-space with a finite set of discrete hyperbolic shapes) within which:

1. measuring is reliable,
2. stable systems exist,

3. these stable systems can change between different types of stable shapes, or
4. the containing metric-space identifies stable orbits for condensed material,

where condensed material is material components which are too small to be interactive (material) shapes within the containing metric-space, but they are components of a particular dimension which resonate with some aspect of the finite spectra (of the proper dimension), which is defined for the over-all 11-dimensional hyperbolic metric-space which is the containing space, That is, some stable material can exist as condensed material in stable planetary orbits.

In a containment context, where measuring is reliable, local linear models of measuring, ie (partial) differential equations, make sense, and stable solution functions are related to resonance of either low-dimension system-shape models, or to the orbits of the containing metric-space's (orbiting) condensed material (which can often be charge neutral), but the resonant metric-space, and its orbits are actually contained in a higher-dimensional subspace.

That (partial) differential equations make sense in a context within which measuring is reliable can be interpreted to mean that because feedback for non-linear systems . . . , based on the critical points of non-linear (partial) differential equations (and an associated limit-cycle convergence [or divergence] structure) . . . , works—then this seems to indicate that the context within which we exist is, in fact, a (narrowly defined, or highly constrained) context within which measuring is reliable.

Though this math context is built from the simplest of math patterns,

1. They are stable patterns (shapes)
2. So that measuring is reliable, and that
3. The stable solutions . . . . to either orbiting dynamics or to stable system component shape, ie the patterns fundamental to existence (or which determine the properties of existence) . . . . are related to lists of spectral-geometric stable properties (or quantitative sets), which are both microscopic and macroscopic properties, which are defined by resonances with the finite spectra of the over-all high-dimension containing space.
   But
4. The over-all high-dimension context . . . . , in regard to:

   a. metric-space-shape size for the dimensional levels and
   b. subspace structure, and/or
   c. the (dimensional-size) tree-structures of set-containment,
      (in regard to metric-space shapes which partition the 11-dimensional hyperbolic metric-space),

. . . . . can be very complicated, and

5. The relation of a component (or shape) to the containing metric-space . . . ,
    . . . (where the metric-space is also a shape of some size) [eg a metric-space containing subspaces of particular dimensions (ie contained components within [bounded] metric-spaces)] . . . , in regard to:

    a. infinite extent discrete hyperbolic shapes, eg the infinite-extent neutrino of the electron-cloud, contained within a bounded metric-space, as well as,
    b. the property of action-at-a-distance, as well as
    c1. the determination as to whether an infinite extent shape of low dimension is either bounded by a higher-dimension containing metric-space, or
    c2. if it extends out to an infinite subspace,
    d. and thus such a shape (defining an infinite-extent subspace) is also relatable to other infinite-extent stable shapes (other 11-dimensional hyperbolic metric-spaces, within which finite spectral sets define other existences) . . . , can be difficult to actually determine.

Though charge is likely not a 1-dimensional construct, but rather a set of charged 1-flows which fit into (at least) a 2-dimensional discrete hyperbolic shape, so as to allow spin-rotations of opposite pairs of time-states.

On-the-other-hand mass (or inertia) can be 1-dimensional, a circle, since a circle's center is a distinguished point, in regard to position in space, for translations or rotations, but any point on the circle could be a distinguished point for rotations, or a pair of opposite points, a diameter, or two pairs of opposite points so that each diameter is orthogonal to the other diameter, and furthermore, the orthogonal pair identify the circle's center. Thus, such an orthogonal pair represent both rotation frames (rotating stars) and translation frames (fixed stars).

So that, the circle and its center can be mapped into one another, so as to represent the map between translational and rotational frames of the circle on the plane.

A new context

Though the new descriptive context agrees with particle-physics that the description is unitary, due to metric-space containing opposite metric-space states, these opposite states are related to spin properties of material components, and that the containment space is an 11-dimensional hyperbolic metric-space, but that such an 11-dimensional hyperbolic metric-space can be related to other such 11-dimensional hyperbolic metric-spaces, and that the stable properties of "material," which are contained in each such a space, must be in resonance (and in the correct dimension) with the finite spectral set defined by the metric-space subspace-partition of each of the over-all containing 11-dimensional hyperbolic metric-spaces.

In the new descriptive structure there is a new context for angular momentum.

That is, angular momentum is defined on the various toral components of the stable shapes, which are allowed by the containment set (where the high-dimension containment set defines a finite set of stable spectra-geometric measures, to which the existing stable shapes must be in resonance), and on possible links, defined by angular momentum, in turn, defined on the various toral-components of the system's shape.

Links

There are unbounded stable discrete hyperbolic shapes, which exist on all dimensional levels, and these unbounded shapes are associated to stable material components, ie stable discrete hyperbolic shapes defined by their being part of the partition of the various subspaces of the containing space [which is partitioned by (into) a finite set of stable discrete hyperbolic shapes of all the dimensional levels of the over-all containment set].

On-the-other-hand the 2-, 3-, and 4-dimensions are relevant to the descriptions of "material" components contained in hyperbolic 3-space, where these stable shapes are also related to both bounded and unbounded, or semi-unbounded, discrete hyperbolic shapes, where an example of a semi-unbounded shape would be the neutrino-electron structure of an atom's (2-dimensional) charged components (which is also called an electron-cloud of an atom, eg for an atom the nuclei are bounded shapes while the electron-clouds are semi-unbounded), so that all "material" systems are linked to an infinite-boundary of the over-all high-dimension containing space. Thus, one can think of angular momentum as a controlled (or controllable) link which can exist between the many different 11-dimensional hyperbolic containing metric-spaces, due to the existence of such unbounded and associated bounded (angular momentum) links (between 11-dimensional hyperbolic metric-spaces).

Thus, one can consider a "possible consciousness" for people (or their realm of creative intent) would be related to their ability to examine (perceive) the different creative structures of these different universes, where the individual 11-dimensional containment sets, for the different universes (or perhaps different galaxies), might be perceived as intricate bubbles of different types of perceptions, into which our awareness can enter, and within which we can control our journey, since we are in touch with the infinite reaches of these various types of separate existences. (see below for a high-dimensional model of life-forms, eg models which allow all life-forms to possess a mind)

Is this the true context within which the human life-force is to develop knowledge, and to intend a creative expansion of such a context?

To re-iterate

Though charge is likely not a 1-dimensional construct, but rather a set of charged 1-flows which fit into a 2-dimensional discrete hyperbolic shape, so as to allow spin-rotations of opposite pairs of time-states.

On-the-other-hand mass (or inertia) can be 1-dimensional, a circle, since a circle's center is a distinguished point, in regard to position in space, for translations or rotations, but any point on the circle could be a distinguished point for rotations, or a pair of opposite points, a diameter, or two pairs of opposite points so that each diameter is orthogonal to the other diameter, and furthermore, the orthogonal pair identify the circle's center. Thus, such an orthogonal pair represent both rotation frames (rotating stars) and translation frames (fixed stars).

So that, the circle and its center can be mapped into one another so as to represent the map between translational and rotational frames of the circle on the plane.

That is, the different 11-dimensional "bubbles of hyperbolic metric-spaces," . . . . , between which human life might be able to enter (or exist) so as to travel between [or link between] these different 11-dimensional bubbles of different perception-types, so as to do this with an intended purpose, that is, if one's higher-dimensional structure is understood and/or perceived, ... , seem to depend on sets of 2-planes which can carry the essential "inertial orbital-structure" for the various bounded regions of an 11-space, wherein (on these 2-planes) the pairs of opposite states on inertia (matter and anti-matter) which can be defined [for each of these particular regions sliced by 2-planes which determine the organization of inertial properties of the region (or for these particular bounded regions)].

These sets of 2-dimensional regions are bounded since inertia is defined in relation to only the bounded shapes of discrete Euclidean shapes, where Euclidean space is the space of position and spatial displacement, ie Euclidean space is the space in which inertial properties are contained.

That is, these sets of 2-dimensional regions could be used to map the different 11-dimensional "bubbles of hyperbolic metric-spaces."

Life

There are natural stable shapes, those of odd-dimension (3,5,7,9) and with an odd-genus (where genus is the number of holes in the shape, eg the torus has one-hole, or a genus of one, ie the genus is the number of toral components of a discrete hyperbolic shape) which when fully occupied by its orbital charged flows are charge imbalanced and thus would begin to oscillate, and thus generate their own energy. This would be a simple model of life.

Thus such a shape which possesses a higher-dimension could cause the lower dimensional components to, in turn, possess an order which can be controlled by a higher-dimensional shape, through angular momentum states (properties).

Down in 3-dimensions this control by a higher-dimensional structure could be the complicated microscopic-and-macroscopic structure of life, which appears to be run by complicated molecular transformations, eg relations between the structure of the living-system and enzymes, proteins, and DNA.

This is simply about considering the results in regard to assuming that stable shapes determine the underlying order and stability which is observed, and the fact that these stable shapes (mathematically) have a dimensional structure associated to themselves.

However, according to the currently accepted laws of physics both the stable properties of quantum systems, eg nuclei atoms molecules etc, and the stable control which is possessed by life, are unexplained (or unexplainable within the currently accepted descriptive constructs).

The patterns of stable physical systems are unexplainable within the current dogmas about the material world, since the current dogma is based on the dimensionally-confining idea of materialism, and within such a confinement, descriptions seem to be based on indefinable randomness and non-linear systems (or non-linear patterns which are quantitatively inconsistent) defined on a (quantitative or coordinate) set which is assumed to be a continuum.

Such indefinably random and non-linear patterns are fleeting and unstable, though their decay times can, sometimes be of relatively long duration.

Suppose human life is associated to a 9-dimensional shape of an odd-genus, then such a shape is an unbounded shape (noted by D Coxeter), and thus it could well be relatable to many such 11-dimensional hyperbolic metric-spaces (bubbles within which perception and action might take place) (ie why should an unbounded 9-dimensional stable shape, generating its own energy, be confined to any particular unbounded 11-dimensional containing space?) wherein (each different bubble of perception) the living system's lower dimensional (material) structure may be quite different (in the new containment structure), and thus the living system's perceptions and interactions could also be quite different within the other (different) 11-dimensional containing spaces.

There can be many of these 11-dimensional hyperbolic metric-space containing spaces in regard to a stable reliably measurable set of experiences.

These 11-dimensional sets could either be in the same space, so as to be organized around different sets of finite spectral-geometric sets, or they could be related to a set of fiber-group conjugations, which could be defined between these different 11-dimensional (hyperbolic metric-space) sets (or spaces).

This second construct would be similar to a model of conjugations between (say) galaxies so that the galaxies drift apart due to the structure of the group-conjugation.

This would be an account of (a new interpretative context for) the, so called, expanding universe. This structure, which focuses on galaxies actually being 11-dimensional spaces, would be organized around a primary 2-plane for the galaxy's inertial properties (but this primary structure would only be) in regard to our planet's (or our galaxy's) inertial structure.

Elementary considerations

Consider the observed patterns of the physical world which have been associated to sufficiently precise descriptions so that other systems can be built by using reliable measuring processes associated to a system's properties, and the system placed within a describable context, so as to be related to practical useful creations.

Or

Such precise descriptions fitting into some type of descriptive context, thus forming an informative descriptive context.

Classical physics

Thermal descriptions associated to closed systems composed of a particular number of components and associated to thermal properties, eg temperature, pressure, volume, component-number, energy, entropy, etc, of a closed system, where differential-forms can be used to define thermal equations.

Newtonian

Material components of mass or charge contained in space and/or time with positions in space or measurable properties of the system-component which can be associated to measurable displacements in space, in turn, related to a causal material-geometric relationship (or changes of state) or pattern sufficiently precisely related to spatial displacements (or changes in state) of position in space and time. This is the ma part of Newton's F=ma definition of either force or mass.

And then there is the relation of material geometry and material motions surrounding a material component and its relation to force, ie the F part of F=ma equation.

These ideas are developed in both the forces (or force-fields) of gravity and electromagnetism.

In inertial systems there are the conserved systems of a spherically symmetric force-fields constrained to a plane associated to planetary orbits for a two-body system transformed into a center-of-mass coordinates so as to identify elliptic orbits, but for the 3-body system, the equations are non-linear and there are no stable solutions, and "why is a spherically symmetric force-field constrained to a plane?" yet the solar-system has, apparently, been stable for billions of years, why?

Faraday's model

For charged systems it is electromagnetic waves (hyperbolic wave-equation) and currents defined in linear circuits, as well as motors, where solutions to solvable equations allow for a great deal of control over system properties, and thus being able to use these properties.

There are also the relatively still charged systems which have spherically symmetric force-fields, as well as some simple, usually cylindrical geometric, geometries associated to currents, wherein the force-fields define useable relatively stable properties.

Often these systems of waves and circuits deal with oscillatory signals (or couple-able properties).

These descriptions of force-fields depend on local linear relations which exist between local geometric measures, ie alternating-forms (or differential-forms), and (of) a charged system's (or charged component and current) geometric properties.

But, accelerated charges imply the emission of electromagnetic radiation and the charged system's loss of energy, ie such a system would be unstable. This would mean that bounded interacting charges would define an unstable system unless the system was actually an internally closed metric-space which confines the charged components.

Thus consider atoms:

These are considered to be quantum systems, where a quantum system is characterized as small components composing a stable system where the small components (or particles) are related to random spectral-particle-position events in space and time (or in space-time). Thus a quantum system becomes a function space represented in a spectral construct (or system wave-function) associated to sets of linear differential operators, eg wave-operators (or energy-operators) and the point is to find the operators associated to the quantum system's spectra, or to diagonalize the function space, ie give the function space a spectral representation by applying sets of operators which commute. The spectral functions of the function space are supposed to represent the probabilities of random spectral-particle-events.

But

One cannot find such diagonal operator-function-space constructs for general quantum systems, ie there are no valid measurable descriptions of general quantum systems, eg nuclei, general atoms, molecules, crystals etc.

It is assumed that quantum systems reduce to particle-components governed by (probability-energy) waves in a context of stable spectral (wave) properties of quantum systems.

Quantum interactions are descriptions based on particle-collisions, which are high-dimensional, non-linear, and unitary-invariant.

Particle-collision experiments in particle-accelerators find many unstable particles and a few stable particles eg electron, proton, etc, where it is assumed that these particles are associated to the quantum system's reduced particle-components, which are assumed to compose a quantum system.

Is it valid to assume that unstable particles are a part of a stable systems component structure? [No! This implies higher-dimensional macroscopic metric-space structure]

Then model the wave-function with internal particle-state properties associated to a non-linear (wave) equation representing the particle-collision interactions, where the interactions perturb (or alter) the particle-states of the wave-function, where the resulting perturbation series is summed to identify the stable systems new properties dependent on the new interaction structure brought about by particle-collisions and changes in the internal states of the (assumed) quantum system's composing particles. This adjusts a wave-function, where the wave-function is now divided into particle-states if the original wave-function is close, but virtually no original wave-functions are close to the system's spectral properties which have been observed.

It is claimed to perfectly adjust one of the energy levels of a H-atom's energy structure. Such a claim, where the answer is based on subtracting infinity, is a dubious claim, especially since there are so few contexts within which the particle-theory is relevant.

The containment space of quantum physics (there are various ideas about this(?)) is a non-commutative function-space, along with a non-linear, but unitary-invariant, and higher-dimensional equation (or operators) of quantum interactions, ie particle-physics ensures that there will be non-commutativity, and thus the spectral identifications of any quantum system is impossible.

In fact, the math structure of particle-physics only identifies the properties of unitary invariance and the possibility that higher-macroscopic-dimensions exist, since unstable particles can be best interpreted to mean that higher-dimensions exist, and these higher-dimensions would have a macroscopic structure.

.

31.

Proof of . . .

Proof of an idea (science or mis-representations)

Editors (and/or journalists) are the gate-keepers of society's truth, the expression of ideas in the media, where that truth is determined by the beliefs and interests of the editors' paymaster, whereas the editors and journalists defer their gate-keeping judgments concerning "what society is supposed to identify as the truth" to the expert authorities, ie those high-paid wage-slaves who express the authoritative dogmas upon which the production of the ruling class's investment-structures depend, ie these are the allowed absolute truths in regard to how society is organized in relation to knowledge and creativity so as to serve the interests of the investor-class.

This is an example of how the "owners of the instruments" of the communication system, who are also acting as paymasters for the editors, are able to control the thought and language of their (well-paid and loyal) wage-slave editors, who report on the dogmas which the experts (whom are themselves wage-slaves), within the society, follow.

This is still the Roman example of a professional army (eg editors, journalists, and expert-authorities, as well as the military personal) being controlled by the ruling class.

In the US, the ruling class still directs the army, but now it is an army of wage-slaves. Furthermore these high-paid wage-slaves (who serve private investor interests) are now most often paid-for by the public (in the military, and in research institutions), where the individuals within these armies are duped into a mis-directed sense of self-importance (apparently, measured by relative salary-size, and a mis-guided cheer-leading, provided by the media, for the expert knowledge of our culture, where narrowly defined expert knowledge provides another example of the organization of knowledge and creativity, which, in turn, is used to oppress the public's individual creative efforts) and a subsequent desire of these (misguided) expert individuals to desire to be protected from the (other) inferior people.

Since the ruling-class controls this social set-up, and yet they are only tangentially involved, the process becomes a self perpetuating closed-circle within society, driven by: conflict, competition, the definition of high-social-value by the media, and often extreme violence (but it is the public institutions which serve the interests of the ruling-class which most express an extreme violence against the public [as well as expressing extreme (intellectual) violence against ideas which interfere with the interests of the ruling-class]).

The public takes on the values of the ruling-class, since the communication system continually repeats these values, and no other ideas are allowed (with the implication being that "the other ideas" are not correct [only the experts whom work for the ruling-class are capable of expressing truth]).

Perhaps the editors (or the experts whom the editors represent) should provide a brief "rational explanation" as to why they believe their beliefs are superior to a new set of ideas (about math and physics) presented here (latter in this paper), so as to justify their extermination of the free expression of ideas, where the extermination of these ideas is done by exclusion (or by delayed publication), of these expressions.

Is the reason for this exclusion (or hiding) simply that these "gate-keepers of truth editors" believe-in a dogma . . . . , (and part of that dogma is that the professional math and physicists determine "the truth" about math and physics ideas [though the math community were interested to listen to these new ideas by m concoyle at one of their professional meetings], yet these new ideas are excluded by editors even though these new ideas are presented at an understandable [relatively non-technical] level, concerning general categories of language, in fact, it is a very educational discussion for the public about both math and physics, and the relation of knowledge to practical creativity) . . . , and these editors have become well-paid personal, within the media, because they do uphold this dogma.

Thus, due to their high-paid social status they are arrogant, and subsequently, these editors insist that "everyone else must believe what they, "the editors," (or the beliefs of the ruling-class) believe (or everyone else must hear, or read, only the ideas that these editors believe-in),"

. . . , or do these editors believe that others must follow the beliefs of they, "the editors," and that "this is a good-enough reason to exterminate all others (or all other ideas)?" [Similar to how the puritans arbitrarily exterminated to native people, based on the puritans believing that they were superior people compared to the native people.]

That is, it is difficult to distinguish how exclusion based on arbitrary belief, eg possessing a high-salary, is any different from the Puritans exterminating the native people, in order to acquire the property of the native peoples, so that they, the Puritans, could, in turn, "prosper" within the context of European social norms (narrow arrays of products which are familiar to the markets, which are based on investment in a European culture), so as to demonstrate to other Puritans in their community that the Puritan-God identifies the intrinsic goodness of the prosperous Puritans, and thus God allows them to prosper, where prosperity is defined in terms of European social norms (this is Calvinism).

This is an example of "the use of extreme violence" for the purpose of upholding a narrow, authoritative, dogmatic language construct (business prosperity implies godly virtues, where prosperousness defined in narrow ways, the few products which define a narrow market, in turn, related to investment within a European culture, a culture dominated by the extremely violent (and narrow) social structures, which Europe inherited from the Roman-empire).

This is obsessive, psychopathological behavior, based on an arbitrary belief in the superior authority of a very narrow and arbitrary viewpoint.

Similarly, today (2013), the oil-companies are allowed to exclude the business development of alternative clean-energy sources, eg thermal-solar sources of energy (and associated batteries), and this is done (mostly) by controlling the media, which are the instruments of communication, both public media as well as the expert communication publications (constrained to be relevant to a narrow set of products [knowledge is related to a narrow definition concerning technical creativity] which define monopolistic market structures).

Furthermore, the issue, in regard to the over-use of oil, may still be about "the extermination of life" vs. "the selfish interests of the ruling-class (to squeeze oil resources for all the profits they can get)."

Perhaps the experts should re-evaluate Godel's incompleteness theorem, and see that the development and use of descriptive knowledge requires equal free-speech concerning assumptions, contexts, interpretations and set containment, etc, upon which a precise descriptive language is based.

Placing written expression onto a computing media could allow for the classification of "expressed ideas" . . . , (where one of the classifications could be,

1. arbitrary belief in dogmatic, or personal [or racial], superiority,
   another classification could be about
2. "how math relates to measurable models of existence upon which practical creativity can be based")
   . . . , and the realizing new ways of organizing language, in order to better realize accurate descriptions and their relation to practical creativity,

ie not only the creativity which supports the ruling class (and the over-seeing . . . , by a narrowly defined class of experts . . . , the instruments upon which the ruling class relies for its social power).

3. Etc,

. . . , in regard to the classification of written expression . . . . and the principle concern for a society of equal creators . . . . namely, the relation of the expression of ideas to practical creativity . . .

But the experts are not in control of how the communication instruments are used and organized, in society, rather they form an army of loyal competitors, where their competition is defined by a narrow dogma (which, nonetheless, is claimed to be about considering all types of "valid" ideas, but "'valid' means consistent with the dogma") (competitors who are loyal to the high-social-values defined by the ruling-class).

On the other hand, (in contrast to the Puritans) the Quakers (mostly) lived in harmony with the native people, and the Quaker community was an egalitarian community and it was a more prosperous, and more energetic community than was the Puritan community, in colonial America . . . (though prosperity is [was] still being narrowly defined [by the Quakers] in regard to the European social norms for prosperity, ie the result of this was "that the Quakers got integrated into the narrow vision of the European life-style"),

. . . , the Quaker colony gave more careful considerations to the wisdom which was possessed by the culture of the native people . . . ,

. . . , where the biological systems of the Americas were more abundant than were the biological systems of Europe (the American biological systems were upholding "as large a population," if not a larger population than, as was the Europe population at the same time, and this was because of the very diverse types of natural resources that the native people used for their sustenance (the native people used a wider-range of practical knowledge in order to survive), and yet, the leisure time for the native American people was probably greater than was the leisure time for most Europeans at that time, eg around 1650 {or before there was a lot of European contact with the native American people, ie before 1500}).

The European (Roman) model of society was narrow, exploitative, and destructive, (the bankers had become the equivalent of the Roman emperors), and it required growth and a subsequent ever-greater destructiveness, in order to uphold the less-risky narrow model of society based on investment (within narrow and highly controlled markets, or contexts of actions [ie practical creativity is required to be narrowly defined]).

Current dogmas concerning physics are not related to practical creativity, ie and this failure of current (2013) physics show that physics is not a valid model of scientific truth,

The failure of currently accepted dogmas associated to physical science . . . , dogmas which only relate to "nuclear reaction rate" dependent technologies, eg nuclear bombs . . . . , in turn, requires that the current authorities of the physical sciences prove that alternative ideas are logically unworkable . . . .

Quantum physics is both inaccurate and unrelated-able to practical creative development.

In classical physics general laws apply to a wide range of systems, so that the solutions to the defining differential equations of classical physical systems provide a sufficiently precise description of the system's observed properties (for these many systems) so that various other classical subsystems can be:

1. made,
2. placed in relation to one another in space, and
3. their different properties controlled by adjusting the conditions associated to the subsystems, so as to
4. realize (control) and use these properties in a practical way.

In quantum physics the spectra of "general quantum systems" cannot be found to sufficient precision, and these quantum systems are most often composed of many-(but-few)-components, which are usually small components, but these many small components are being described by the properties of probability within quantum physics, so that no control can be realized for these systems . . . , (and when placed in relation to one another their quantum properties can be changed) . . . , but nonetheless the spectral properties of these quantum systems are very stable, and this implies that these, so called, quantum systems are formed in a controllable context, ie the descriptions of these quantum systems need to be both determinable and controllable. [and, This is what the new ideas about math and physics allow.]

Proof that current scientific dogmas are wrong is that they cannot describe, in a general way, the stability of physical systems which are observed to exist . . . , where the currently accepted scientific dogmas are based on indefinable randomness and quantitatively inconsistent non-linearity, and where both of these math constructs are quantitatively contained in a continuum, where the set-size of the continuum is allowing logically-inconsistent structures (math patterns) to be defined by convergences, associated with various different contexts, into the same continuum containing-set, eg the existence of point-particle-collisions in a description based on randomness (this is an inconsistent math construct),

. . . , furthermore, the current scientific constructs are mostly proper subsets of the new context . . . . , (except for the quantitatively inconsistent ones, the logically inconsistent ones, and the unstable contexts, though unstable non-linear systems are allowed, but they are still unstable in the new context).

The proof that the new ideas are correct . . . . ,
. . . , is that there are (do exist) stable systems: from nuclei, to general atoms, to molecules, to the solar-system, etc, . . . ,
. . . , where the new context is based on: many-dimensions which are partitioned by a finite set of very stable discrete hyperbolic shapes of both macroscopic and microscopic size scales, and where the new context is capable of describing the properties of stable systems.

This failure of modern physics is an example of the limitations of a language, where the language is based on assumptions and particular contexts which, in turn, limit what patterns that a given precise language can describe, eg this is essentially the same example as: Copernican system vs. the Ptolemaic system (both are descriptive languages, but each is based on a different set of assumptions; furthermore (at that time, in fact, Ptolemy's system provided more precise predictions than did the Copernican system), both are measurably verified, but the data is interpreted in different ways [where each descriptive language depends on the model within which the data is interpreted]).

Since that fact that there is a set of fundamental physical systems, which are "stable many-(but-few)-body systems" eg:

general nuclei,
general atoms, and
molecules,
as well as the macroscopic solar-system,

. . . , where all together the stability of these systems proves that the new description defines a containment set which is capable of providing a more accurate description of the stable properties of these systems, since the current dogma (based on indefinable randomness and non-linearity) is not capable of providing valid descriptions of the very stable spectra . . . , of these various types of very stable systems . . . , to sufficient precision, wherein such a description is actually based on physical law, and a subsequent process of deduction, which is logically consistent.

(furthermore, the currently accepted dogma cannot claim that "the containing quantitative context of their own descriptive context is valid for some aspects of their own theory," and subsequently their own descriptive structure must be adjusted, eg renormalization; that is, the "particle-collision model in a context of uncertainty" causes logical (or structural math) problems (indeed it does cause problems), {where the particle-collisions are defined as converges of mixtures of positive and negative time states (or coordinate frames) for the colliding particles (in a context of electromagnetic [or Lorentz-invariant] wave-functions)} and this convergence of sums of sets of opposite time-state (of hyperbolic wave-equations) should not have been allowed in the first place. It is logically inconsistent, and this logical inconsistency leads to arbitrary descriptive processes . . .).

Another better interpretation of the particle-collision data, which is obtained from high-energy particle-accelerators

That is, The current dogma must prove that the following statement is incorrect.; . . .
. . . , U(1) x SU(2) x SU(3) is the natural structure of component (or, equivalently, particles which possess unstable shapes) decay, in regard to the unstable particle-components which are contained within 3-space after high-energy particle-collisions (in which higher-dimensional shapes were broken-apart) . . . .

If the dogmatists have no such proof, other than claiming that their context is an absolute truth (and everything should be deduced from (or interpreted within) their assumptions), . . . then please point to papers, based on the currently (2013) accepted physical law, and a valid deductive process, which identify, to sufficient precision, the entire stable spectra of general: nuclei, atoms, molecules, and/or crystals . . . .

That these (just mentioned) systems are stable, as well as the [many-(but-few)-body] solar-system being stable, proves that the new context is the more accurate context (the more correct containment context) within which to describe (and use in a practical manner) observed physical properties. Since the new descriptive context provides these systems with stable and sufficiently precise, spectral properties.

The new descriptive context can describe the stable properties, of these just mentioned systems, because the properties of these systems, as well as their new descriptive structures, result from their being stable, "discrete hyperbolic shapes," which might be folded, based on Weyl-angles (which are defined within the metric-space's fiber Lie group).

Furthermore, upon the various toral components of the "discrete hyperbolic shapes" angular momentum can be defined.

A many-dimensional context is best given a context of being related to stable shapes (so that stable systems can be fit into the descriptive context), so that these stable shapes can be of all size-scales . . . , (whereas string-theory requires that higher-dimensions be small shapes) . . . . , which are related to various "dimensional-and-size determined set-containment 'trees,'" defined within the 11-dimensional hyperbolic metric-space, which is the over-all high-dimension containment-set, for the set of stable shapes of various dimensions, sizes, and contained in the various subspaces, within which they can be contained.

These stable shapes are defined in such a containment set by resonance with a finite spectral set, which is associated to a dimensional and subspace partition of this (same) containment set, where the partition is determined by a finite set of stable "discrete hyperbolic shapes," which also form into "trees" of set-containment, wherein these allowed "discrete hyperbolic shapes" define both material-components and metric-spaces, depending on dimension and size (of the shapes) so as to allow for material containment within a metric-space.

Except for issues of resonances and subsequent containment of stable components of lower-dimension within higher-dimensional metric-spaces, these dimensional subspaces are independent of one another (due to the structure of material interactions).

Both reliable measuring, as well as stable system properties: . . . only exist in a many-dimension containment set, which, in turn, is related to a finite set of stable shapes (or equivalently a finite set of stable spectra) . . . .

But a many-dimensional containment set, wherein the dimensional levels are given shapes (for each dimensional level and for each subspace) of various size-scales, is a descriptive context which transcends the idea of materialism, yet it contains (can contain) the material-world as a proper subset.

This is an important property, in regard to the relation that life has to existence . . . .

How science works in the corporate era:

Some technologies . . . . derived from scientific models . . . . work, and they might become associated to complicated instruments, and/or complicated methods, where these (complicated) processes fit into both a descriptive context provided by science and the productive interests of some big (monopolistic) business.

. . . . So these working instruments become ever more complicated, and used in many contexts.

This complication of instrumentation and it descriptive context, in turn, is used to define the vocations associated to the technical departments at universities eg engineering and physics and math departments etc.

At university institutions, the physics and math departments claim to be tolerant of many ideas, yet they are based on narrowly defined dogmas (associated to particular instruments, eg physics is associated with the data from particle-accelerators).

Essentially, new ideas are not considered in their own context, but rather new ideas must be consistent with the assumptions upon which the narrow dogmas are based (this is what peer-review actually is), ie one cannot compete with the professionals unless one plays by their fixed [crooked] rules, [a competitive professional community becomes a dogmatic community which adjusts specialized instruments], eg it now would be required that the ideas of Copernicus must be assumed to be consistent with the assumptions of the Ptolemaic system and then the new patterns (of Copernicus) are to be proved by deduction, ie this cannot be done.

That is, the university departments are blind to their belief in the "absolute truth of their dogmas" and the: assumptions, contexts, and interpretations, which the (instrument-adjusting) professionals . . . of the academic system . . . , demand, ie they are arrogant and intolerant (they are a well-paid army of professional wage-slaves who adjust the instruments of the ruling-class), yet claim to consider all valid ideas, but their dogma determines what they consider valid.

In regard to the functional aspect of science:

Is the functional (or practically useable) part of science to be about?

Either

The solvable aspects of classical physics, which determine all technical development? Eg technical development based on the ideas of: materialism, local-measuring, eg physical law, and the useful aspects of this descriptive context are the solvable models (or if non-linear then the differential equations define a system's limit-cycles, which, in turn, identify converging or diverging dynamical behavior of the system components, which (often) can be controlled by feedback) However, the non-linear theory of general relativity has no capacity to describe a stable system, yet, the solar-system is (apparently) stable.

Or

In quantum physics:

Materialism which can be reduced to sets of random particle events in space and time, in turn, this context is modeled as function-spaces associated to local Hermitian operators (or modeled globally, to a group of unitary operators, ie function-values [or the quantum systems' wave-functions] are complex-numbers), eg the globally defined equations, eg the energy-operators of quantum systems are energy-invariant systems.

It is assumed that this math structure can be associated to a set of stable patterns of particle-decay associated to particle-collision models of quantum interactions. But this is not given a math context which allows for the description a stable system, but rather the context is of randomness and instability.

Math patterns

Quantitative patterns are considered by math, and perhaps the most important quantitative (algebraic) pattern is that of commutative property, ie ab=ba, especially, in regard to a many-dimensional containment coordinate set (for a system and its "defining" measurable properties, ie conserved properties), ie a and b are matrices, which allows for local independent measuring sets for each separate-dimension, which, in turn, allows for quantitative consistency in a context of geometric measures on the system containing coordinate space.

The continuously commutative at each point, set of linear differential equations, which model physical systems, are the linear, solvable differential equations of classical physics.

But the commutative property for operators acting on function-spaces also determine if a quantum system's spectral properties can be found.

In general, such commutative structures cannot be found for general quantum systems.

Thus the model of particle-collision interactions, used to perturb the energy properties of a quantum wave-function, is mostly an irrelevant idea.

The assumption of quantum physics is that, "whatever the observed spectra of any quantum system "might be" then that is the spectra which the probability-based quantum theory is to be consistent with," but general quantum systems are not related to the observed spectra by means of a relation between the system's function-space and its set of "spectral defining operators."

This means:

(1) that quantum description is a description based on indefinable randomness, and
(2) this assumed "agreement with observed spectra" is, apparently, the basis for the (thin) claim that "there is not any quantum system which has violated the laws of quantum physics,"
ie quantum physics assumes the spectra for its system-defining operator (based quantum description) is that same spectra which is observed in the lab (for each system), but the (operator)-(function-space) descriptive context of quantum physics cannot identify this spectral set by (valid) calculations (commutativity of operators cannot be realized on the function-space which is modeling a general quantum system).

Math that works (allows commutative quantitative patterns)

(in regard to bounded systems) It is only the toral shapes and the shapes composed on toral components, ie the discrete hyperbolic shapes, which are the linear, solvable shapes for systems [as well as the linear differential equations with constant coefficients, often associated to electrical circuits].

Thus, there is less need to determine solutions to equations, the solutions to equations of stable structures are already known (ie the allowed set of discrete hyperbolic shapes, as well as the interpretation that these stable underlying shapes of existence determine an envelope of stability in which the usual second order differential equations are defined and partly applicable until the stable shapes become the more relevant context of the system, ie the system bumps into the envelope. That is, the envelopes of stability are more obvious (and intrusive) in quantum orbital structures than in the macroscopic structure of the solar-system.), and thus the issues of description are about relative sizes of the stable systems, and trees-of-containment in regard to both dimension and subspace size, as well as finding the finite-set of spectra which is associated to an entire 11-dimensional containing hyperbolic metric-space, where there can be many of these types of over-all containing spaces.

Furthermore, there are now new ways in which angular momentum . . . , now defined in regard to tori (in Euclidean space, ie associated with inertial properties), and toral components of stable (hyperbolic) shapes, . . . , being related to infinite extent properties of stable systems eg electron-clouds of charged quantum systems (wherein the neutrino is to be modeled).

Thus, angular momentum can (may) be used to identify both high-dimensional control over lower-dimensional (contained) components (whereas component interactions are confined to adjacent dimensional levels), and there is also a new context of control in regard to an infinite-extent stable shape being connected to another, unbounded, 11-dimensional containment set.

That is, stable systems must be resonant with the finite spectral set of the 11-dimensional containing hyperbolic metric-space, while other unstable properties are still allowed by the different second-order dynamic (partial) differential equations associated to both Newton's laws and material geometry, but now there is a discrete toral geometric structure associated to material interaction (discretely defined by the property of action-at-a-distance), and its associated local equations, ie local measuring defined on both material components of the interaction and the toral (action-at-a-distance) geometry which is associated to the discrete interaction, ie a connection (or local-measuring) operator structure for material (or system-component) interactions.

Classical physics concerning material in space and time can be succinctly summarized as:

Inertia's locally (linear) measurable definition at a given (material) component's given position in space being related to the force-field at that (same) given position due to the material geometric properties surrounding the given component, where the force-fields are (can be found in regard to) local relations between local (linear) geometric-measures of the force-field being equated to surrounding properties of material geometry (and in regard to charge, including material motions).

Physics is all about how math relates to the precise descriptions of observed physical patterns.

The biggest problem with which science has had to deal, is that the differential equation was found and solved (by Newton), but it was only understood in a limited context, and that limited context is where there are sufficient restraints on a physical system that a global solution to a system's defining differential equation can be found, and this has led to confusion about:

1. local measuring,
2. function-space techniques,
   [leading to non-linear and indefinably random contexts for measurable descriptions, which are defined on quantitative containment sets which are continuums, ie but these continuums are sets which are "too big," so that logical consistency can be breached by convergences defined within such a containment context], and
3. the relation that a differential equation (or models of local [linear] measuring) has to discrete and stable descriptive contexts (ultimately related to quantitative structures based on finite sets, or [at most] on rational approximations), where the containing metric-spaces have (physical) properties associated to themselves, and thus the description also has a metric-space state structure (of pairs of opposite metric-space properties) associated to itself. That is, there are distinct metric-spaces wherein inertia is described (or contained) [Euclidean space], and there are separate (and different) spaces in which energy is contained [hyperbolic space, or equivalently, space-time]. Missing this pattern is one of the main failings of classical physics, where local linear measuring and its relation to geometry is central, while for function-space techniques, when the description is in a metric-space coordinate containment context then the functions (of the function-space) need to be related to the relevant "discrete hyperbolic shapes" as well as the toral shapes (and the toral components of the stable shapes) which are part of the description of the stable system. [Note: That, mass is equal to energy, is a result of the relation that the momentum-energy 4-vector has to the position-time 4-vector in space-time. This is about the spatial-displacement-inertial symmetries and the temporal-displacement-energy invariance symmetries associated to E Noether.] The math issues of the apparent classical and quantum non-compatibility are resolved by giving the classical notion of local linear measuring a (new) context of being given a local discrete operator structure, in relation to an underlying stable structure of discrete hyperbolic shapes, and structure of a Euclidean action-at-a-distance structure, which is allowed by the results of a Aspect's experiments, of which whose data can be interpreted to mean that there is a non-local (discrete) structure to material interactions.

Unfortunately, the confusion which has existed in mathematical constructs (concerning differential equations) is used as smoke-screen to make the "descriptive structures of math patterns" relations to "useful measurable information about physical systems" even more difficult to understand, ie it is (has been) used to define an arbitrary hierarchy of math-science-value.

This allowing an arbitrary domination of certain math ideas is about identifying authorities and allowing those particular experts to become dominant.

In fact, today, the authorities are more like balloons, they are mostly filled with hot-air, while they are encouraged to prance-around and "posture as being very important" within the settings of "important" institutional buildings. While their "knowledge" contributes nothing to practical technical development.

This is about arbitrary authority, and its primary use is to exterminate other ideas . . . ,

. . . , whereas if the people of the world are not allowed to form into living styles which are harmonious with the earth,

. . . , the narrowly defined (and arbitrary) hierarchy of both intellect and market-products, which is organized in this narrow manner so as to support only the investor-class, will accelerate the society's relationship to its own extermination.

A hierarchical social structure is intrinsically intolerant, and the hierarchical position of its ruling-class is based on the ruling-class's willingness to exterminate, but when it is a measurable description, and the nature of math and its relation to measurable descriptions of observed patterns is being carefully discussed, then the exclusion of these ideas without a rational basis and only expresses intolerance of ideas which are not a part of the accepted dogma so that the extermination of such expressions is a form of extreme violence based on arbitrary issues of social-class, in a hierarchical structure of wage-slaves protecting the interests of their paymasters (furthermore, such intellectual intolerance is in no-way different from (or is worse than) racial intolerance). (in fact, the ideas discussed in these papers are open physics and math issues about which no one has an absolute authoritative answer),

The main function of the media is to confuse the public and mis-represent information to the public, so that only the interests of the ruling-class are considered by the public, who get their information from a privately owned media

When one considers what passes for rational dialog which is expressed through the media one sees that everything is misrepresented:

1. the right to bear arms is about "there not being standing armies," but the discussion is instead about
2. the relation between guns and terrorism, where
3. the justice system has essentially adopted the FBI (of J E Hoover, or the KKK, or other hate-group militias, etc) viewpoint about controlling a violent sub-population, which, in

turn, is to be used to terrorize the rest of the population, this is about policing and its system of: informants, provocateurs, and hidden coercive actions, all used (by the justice system) to control the public, where this is a social context of "an army (within the justice system) being controlled by the owners of society."

3b. The point is, that if there is no way to have private communication through communication systems within the US, and if the FBI, etc, knew that these brothers (of the Boston-bombing, 4-15-13) were potential terrorists, then either the surveillance systems are useless, or this is another example of manipulation of people by the justice system, eg L H Oswald or Sirhan-Sirhan etc . . . hidden coercive actions . . . . etc, so as to induce either terror into the nation, or to eliminate unwanted people. Indeed, this manipulation of individuals by the authorities is the most likely way in which this total surveillance is going to be used.

In other words, public safety is of no concern, rather this information is going to be used for the purpose of individualized public terror, which is real the intent of the justice system, since, after all, it is the justice system which has imposed wage-slavery on the public.

This pattern of legalized murder by the authorities was put in place by G Washington in regard to Shay's rebellion (unpaid veterans acting against the banks who were oppressing the "broke" "veterans of the revolutionary war," or J D Rockefeller murdering the miners of Ludlow CO, etc . . . . etc . . .

4. Foreign relations is not about human rights, as it is portrayed on the media, rather it is about taking resources and exploiting labor, by means of violence, to maximize the profits of the ruling-banking-class

5. Science is always framed in terms of "material based science" vs. "spirit based religion" (but religion has no precise definition of spirit), whereas religions in general always "does teach" about the idea that "people are equal" but

6. equality is never (seldom) a representation of religion within the media, rather

7. religion is represented in the media as being about the right of an authoritative, arbitrarily-moralistic, hierarchical social-club structure, which supports the interests of the ruling class, so as to dictate what "proper" behavior should be to the, so called, bewildered public, apparently

8. so the public can cope with the oppression imposed on the public by the ruling minority, and where

9. this oppression of the public is upheld by a justice system, which oppresses and terrorizes the public,

10. eg forcing them to become wage-slaves, by laws which are based on minority rule and property rights.

11. Modern physics (2013) is about models of indefinable randomness and non-linearity, a descriptive structure which cannot possibly describe the stable properties of the fundamental systems which are observed to be so very stable, but it is a descriptive context which is related to rates of nuclear reactions.

12. Math and science are represented in the media as being about absolute dogmatic truths, which only a few are capable of attaining (or realizing), ie it is about an intellectual ruling-class (which serve the knowledge interests of the ruling class)

13. Rational thought is not about careful considerations about the limitations of (precise) language, but rather about an idea about intelligence which is related to the rate of learning in regard to the knowledge which the ruling class wants the public to possess, a step-by-step acquisition of the knowledge used by the ruling-class to create the narrow range of product upon much of their social-power rests, so as to both define an intellectual elite who represent "what would be of the greatest value" in regard to supporting the interests of the ruling-class based on a narrow definition of knowledge,

14. Science is about materialism, and materialism is most strongly identified, in the media, in regard to evolution based on mutation and survival of the fittest (where both are indefinably random constructs and essentially have no meaning) so that this biological-evolution model of materialism is interpreted to mean that people are scientifically found to be unequal, while

15. religion is supposed to be about "the creative design of the lord," that is, religion is always portrayed in regard to arbitrary-moralizations which are handed to the sinful-public by the virtuous religious authorities, where the virtuous religious authorities include the virtuous members of the ruling class (eg Calvinism)

    But, where this use of high-sounding religious language has no relation to practical creativity. But now there is a similar criticism of the language of material based science, now based on indefinable randomness and non-linearity.

16. Economics: In fact, there are players in the economic game who possess such large pockets that the economic game can be changed by their monetary actions . . . , who could these people be, eg the king of Saudi Arabia, the Rockefeller family, the Rothschild family (if they exist), etc . . . thus, quantitative measures defined for such an economic structure, eg the value of dollars, have no validity in a quantitative model, ie the economy is not relatable to a valid a measurable-system model, it is an arbitrary system governed by the "owners of society" (though the social structure of money does have a large affect on the behaviors of the public). In a society where value is arbitrary (and controlled by the owners of society by means of money and propaganda) but the actions of the people within the society are essentially all controlled by money (where the value of money is forced to have value, due to the coercive actions of the justice system against the public, eg do not steal bread from the oligarchs) then market values are either controlled, or if temporarily

not controlled, then these market-values only define an indefinably random quantitative context so that attributing patterns by math processes to such a (quantitative) context will not be valid, ie quantitatively inconsistent. [thus, the failures of models of risk provided by MIT math-physics PhD's working for the banks] That is, if certain flows are identified as crucial then the whole monetary enterprise could be pinched-off {the hundreds of trillions of dollars in hedge-funds confiscated}, only allowing the crucial monetary flows, and new structure of law-governing-market's put into place, such as a continental congress with law based on equality and a government by the people "promoting the common welfare" for the public (so as to sort-out equality, where survival, and an access to a creative context, must be a right) where the point of individual freedom within society is to be a society of equal-creators, so as to form a truly free-market, (ie advertising is simply listing a products attributes [or properties] and its social context, or use in society) as opposed to a centrally controlled, and narrowly defined system, which is irrationally directed to the destruction of the earth, which the owners of society are determining (directing), and where the government serves these few owners, since law is now based on minority rule and property rights and violent domination.

17. There are the intellectuals vs. the violent types "where the violent types are those who know how to use terror to dominate others." This was also the basis of Roman power, we are still only an extension of the Roman civilization.

    The intellectuals are supposed to stand for creativity, but in fact they stand for traditional authoritative beliefs (since there is no advantage for an intellectual to believe in particle-physics or general relativity or any of the other physical theories which are derived from these (math) constructs (eg string theory) [unless they are getting paid for their beliefs], since these theories have no relation to practical creativity, and their authority should be challenged, while 21st century technology (now, in 2013) all depends on the knowledge base of 19th century physics, even micro-chips).

    While the violent types are controlled and manipulated by others of their ilk (essentially, they exist within an army), but the violent types are supposed to stand for traditions and authority, but they are really about domination and terror and destruction.

    Thus, there is the well-paid stand-still intellectuals [who believe in arbitrary authority], and the terrorist enforcers of arbitrary authority, which together chart a course for world destruction [both sides guided by the owners of society, ie the very big economic players], ie eliminating the earth-nourisher of all of life, so as to maintain a social position of domination, which belongs to today's owners of society, ie the inheritors of the power of the Roman emperors of the western culture

18. What are women? They either try to separate themselves for self-protection and/or they seek dominant social positions for themselves, or they are, essentially, prostitutes who are

to be controlled by (violent) domination, subsequently they tend to support the, so called, winners within the narrow competitive society.

In a matrilineal society the woman more easily becomes the person who nourishes life, winners and losers, ie allowing a wider-range of viewpoints.

19. A precisely described truth, concerning observed and measurable physical patterns, must be both accurate, so as to be of sufficient precision, and it needs to be relatable to practical creativity, especially, if the systems it is describing are uniformly stable, where such a property implies both linearity and controllability, in regard to the physical system's properties. Thus, in regard to both particle-physics as well as general relativity, if such knowledge is not relatable to practical creativity . . . . then it is a descriptive language without any definitive meaning, and "at best" it is about speculation on a sea of confusion, without providing any guidance. This is true even if particle-physics might be based on the actual patterns by which unstable particle-components (made form high-energy particle-collisions) decay in 3-space.

Languages which are unrelated to practical creativity are languages which are equivalent to the language of religion, which is also without any relation to practical creativity, but rather it (both religion and modern science) is only relatable to social manipulations, in a delusional context of rationalism.

So science, which is associated to particle-physics and general relativity and all the other theories which are derived from these two theories, eg string-theory, are not a valid part of science but rather share the arbitrary moralistic authority of a religion.

Thus science . . . , by following particle-physics and general relativity and the other derived theories, it . . . , becomes an arbitrary, speculative, but nonetheless authoritative and dogmatic institution represented in the media as an institution which is espousing an absolute authoritative truth, which can only be understood by the few superior intellects of the world. (Superior intellects which modern psychology has been able to uncover). This is all a bunch of rubbish.

Thus material based science which is defined by "survival of the fittest"-based-evolution, and by its representation as being the set of superior intellects of the world, has come to represent inequality (but science emerges from an equal voice which challenges authority), the very thing for which the emperor's religion, eg the Roman-Catholic-church, has been used within the (Roman) empire, and this dys-functional-ness which is called superior intellectual capacity has very effectively been used to eliminate equal free-inquiry which is about the relation of knowledge to practical creativity. This is done by requiring that editors only allow inquiry to be voiced by the authoritative set of experts, ie the superior intellects (ie people worship the indefinable notion of intelligence, or IQ, so as to make it appear measurable)

It is in this context that the arrogant editors . . . , who apparently believe that they are protecting the public from dis-information which does not emanate from the mouths of the intellectually superior class of professional scientists . . . . , are, in effect, opposing equal free-inquiry, so as to protect the intellectually superior people of the world, but more to the point, they are excluding any practical creativity which might come from equal free-inquiry, and this means less competition for the ruling-class, which has a monopoly over knowledge, and how it is used in a creative context within society (this control is based on investment and salary, where wage-slavery is coercively imposed on the public).

Etc.

Is meaning in language essentially about the relation of language to practical creativity? {Yes} Thus, arbitrary (moralistic, or based on indefinable randomness) pronouncements have no meaning, and their emotional value (morals associated with spirit, and indefinable randomness associated with the society's elite intellectuals) is used to manipulate the public by means of the propaganda system, where the propaganda system is the sole voice of absolute authority within society, ie the voice of the ruling minority.

Social conditions, and its relation to language

1. Those who support equality, and freedom based foremost on equality, tend to believe that the main characteristic of people is that they want to be creative and productive. The value which is defined by such a culture, centered on knowledge and creativity, is the value of the culture's capability to create (creative range and its useful applicability), and social-value is not based on either: property, money, or shiny-rare-metals (which now identify a commodity market, for currencies whose values float, and thus their values are upheld by extreme violence) {the economic-military take-over of the world by the US is based on violence being used to maintain the value of the dollar}

2. Those who support tradition and authority tend to believe that people are characterized by their "badness." (eg T Hobbes) But this badness is really a circular construct of a hierarchical system, where the ruling-class instill the values of the ruling class upon the public ie thus it is assumed that the ruling-class are superior in intellect and in virtue, and thus the ruling-class has the right to steal and to uphold their theft by violence, (since the implication is that the ruling-class will do the smartest and most virtuous things for society) and then the ruling-class has the right to exploit and attack the lower social classes, so that the lower classes are, by definition, bad, in an authoritarian and hierarchical society. Thus, many in the lower-social-classes act as if they are mirror-images

of the ruling few, ie violent self-righteous thieves, and of course, these are considered to be the worst people.

3. Those who only support freedom, apparently often also believe in the biology ideas which are based on "survival of the fittest," which implies a natural hierarchy, and their definition of freedom is most often about the freedom of superior people to exploit inferior people.

    If they believe that "freedom must be based on equality" then they belong to group 1.

Note: The new ideas about math and science provide a new context for science, which is not based on materialism (though the material world is a proper subset), and thus the development of life as well as "what life is?" have a new context within which the living systems are contained, and that new context is mathematically based, ie it is a measurable description, whose descriptions can be tested by reliable measuring. But the idea of "survival of the fittest" (as well as the idea of life being solely based on DNA) is not the best description for the (existence and) development of life. However, the new descriptive context is also concerned about the molecular function of DNA within living systems, and the molecular actions of DNA, in the living system, are also of interest in the new descriptive structure, but now they can be linked to a macroscopic higher-dimensional system, which is linked by containment and by angular momentum. But, if one's attention is only on material (molecular and condensed) which is contained in 3-space, then the interaction properties of these material entities will stay in 3-space.

In fact, in regard to life's development, it might be the capacity of a living system to be creative which might be the property which is determining the development of life, yet that creativity might not be creativity in the direct context of the material world, since the material world is a proper subset of the new (mathematical) context for existence.

That is, both science and religion have new contexts in this new math construct of material and its containment space where the new math focus is on stability of patterns (so that language has meaning) and the idea of materialism is a proper subset of the new ideas, where classical physics is consistent with the new math structures but instead of reduction of material to random quantum the math operators become discrete and the underlying context is (are) the stable math patterns of existence.

Free speech is about:

1. One's beliefs about the nature of people and how this is related to how one believes society should be organized, how society's laws should be structured, so that "what one wants from society" will be consistent with "what one considers to be" the nature of people.

2. One's beliefs about the nature of "all of existence" and the relation that this has to practical creativity.

Practical creativity is about observing the patterns of nature which exist, and it is about measuring, and the containment of a measurable descriptive language to be used for describing various types of systems (or patterns) which exist, so that the (reliably measurable) properties of a system can be used in a practical manner. But, this does not necessarily mean the idea of materialism.

This is about: assumptions, contexts, interpretations, containment, definitions, quantities, variables, formulas, equations, shapes, and the stability of (math) patterns, which are used, or organized, in various ways to form a (new) measurable descriptive language, by those interested in both description and the relation of a reliably measurable description to creativity, ie using the properties which exist, and are controllable, in order to create new things . . . , which are to be selflessly given to society as gifts.

Using or developing language at this elementary level of assumption etc can be easily related to the development of practically creative designs. For example, Faraday developed electric motors, and circuits, and circuit components, at this elementary level of a descriptive language, but do not be fooled, it was the correct mathematical context. Faraday and Tesla essentially ushered-in electronic technology.

This is the social context in which we are all equal creators.

32.

Particles

Over-view of particles

Freedom of belief and freedom to express belief is being placed into a context of the letter-of-the-law in regard to the use of the wording free-press, so that this is becoming "the freedom of the owners of the press to express their ideas," whereas freedom of belief and expression is about an idea's relation to practical creativity in regard to appreciation for one's existence.

Though the probabilities of particle-collisions can be associated to rates of (nuclear) reactions when the system is transitioning in a chaotic and indefinably random manner between the stable initial and final states of a larger (transitioning) system, it is known from the start that this is about a weapon, and it is not clear to me that this is about practical creativity as a gift to the world, or about creativity yoked to a terrorist state. It is so easy to guide human thought, especially, if social value is being artificially determined by the selfish interests of the owners of society by means of their domination of "the press."

What gets continually repeated becomes internalized, we cannot change this within ourselves, so the first fight is within oneself, and in this fight we are greatly self-opposed by our perceived condition of being wage-slaves. [and social value is controlled through the media by the owners of society]

Main part of Particles

A model of particles is useful to nuclear reaction models, but not useful in regard to either describing material interactions or, subsequently, in regard to describing the stable properties of material systems, eg the spectra of general nuclei. The media and high-positioned administrators are able to destroy knowledge, and this is done for corporate interests, while the professional experts actually do the damage, since they were guided by the propaganda-education system into participating in a process of narrowing the reach of knowledge.

People who are placed in subservient social roles, ie the public, insist on the truth which comes from authority . . . , to which they believe they serve (they internalize the authority which they are forced to serve) . . . , be addressed to their satisfaction, ie the ideas expressed need to be consistent with the authority to which they (the public) are forced to follow.

Thus, one must deal with particle-physics in regard to the particles themselves (and the math theories [quantitative containment structures] within which they are interpreted).

However, the dogma of particles has to do with the relation which the probability of particle-collisions have in regard to nuclear reaction rates for systems transitioning, "within a chaotic state, which exists between two different, but relatively stable states," and the real issue is that "neither of these two relatively stable states have valid measurable descriptions," ie the math containment constructs do not really apply to the world's observed properties.

Properties of (high-energy) particle-collisions are not even remotely related to properties of stability, but rather are believed to identify a stable pattern (which is random) associated to all types of particle-collisions, upon which to place a probability based descriptive context of material interactions (based on random particle-collisions, so that these particle-collisions have a fixed set of possible outcomes).

But since the stable states of fundamental physical systems are not defined, such a description as particle-physics, which is based on indefinable randomness, cannot have any relation in regard to "using the random properties (structures) of particle-physics" to identify stable patterns, which are observed in the world. That is, the program of particle-physics has failed, eg it has failed to identify the stable spectral structures which are observed in general nuclear systems. But the propaganda system makes sure that the public is unaware of these failures.

Why does propaganda work?

Propaganda works because the instruments (the media) through which ideas are expressed in the US society are owned and controlled by "the owners of society," and thus the voice is the sole voice of authority which expresses the interests of "the owners of society" to the public, and the method of repetition of the same ideas is the central technique used within this propaganda system. However, because the US public has been forced by the extreme violence of both the justice system and the governing system to become wage-slaves, the US public is also easy to manipulate by means of behavior modification, ie incentives and hidden coercive acts.

This extreme violence is similar to the violent actions used by the Puritans against the native peoples around the Massachusetts area in the 1600's, where their extreme violence was justified based on an arbitrary belief system (which excluded other ideas), eg the Quakers lived in relative peace with the native people in Pennsylvania in the 1600's.

Orthodoxy

Particle-physics comes out of regular quantum physics, where regular quantum physics is about finding the spectra of a quantum-system, which is assumed to be composed on very small components, wherein these components possess random spatial-spectral-particle event properties (associated to measuring and wave-function (or system) collapse of a global function to a local event).

That is, quantum systems are modeled as a "function-space and set of operator" pair, where a (the) set of local Hermitian operators (Hermitian so that the energy-wave-equation is unitary invariant) is to be "found," which allow for a spectral representation of the quantum system, ie diagonalize the function-space.

Unfortunately, for real (general) systems, this can never be done, wherein the math context stays non-commutative.

Particle-physics is about quantum interactions of (what are assumed to be) quantum components, and it is based on adding-on a (new) unitary, non-linear, connection term to the energy operator, which now models the changes of internal particle-state properties of the interacting particles during a collision-interaction (or during a virtual field-particle exchange, and subsequent change in a particle's internal state, during a particle-interaction) and this requires that one add onto the (spectral) wave-function an internal particle-state vector structure, in a similar way in which the Dirac operator "adds onto a quantum wave-function" the vector of spin-states.

The big contrast

Classical physics is about the local linear properties of motion related to a mass's spatial position which are related to a local force-field caused by the geometry and motions of the material geometry (both mass and charge-and-current) surrounding the moving mass. This law applies to a wide range of systems so as to provide sufficiently accurate descriptions of system properties, and this, in turn, can be related to a wide range of practical creative uses. It is a descriptive context which begins locally and (when solvable) finds global solution functions, ie it increases one's information about the system's properties.

Particle-physics

The particle-interaction descriptive context is based within the wave-function of a quantum system, and is about the internal particle-properties of: charge, flavor, color (analogous to the property of mass in the classical description) related to one another (in regard to particular energy levels) by virtual energy transfers (by field-particles), where internal particle-states are transformed, and this is caused by the energy-transfer, so that all of the possible particle-changes and energy transfers together perturb the internal particle-state structure of the quantum system's wave-function, so as to adjust the energy-spectral properties of a quantum system's wave-function.

However, the descriptive context is, itself, deformed by the high-value of the local-energies of the interactions, so that the values upon which the description is based are distorted, and thus it is supposed that a readjustment to the quantitative structure of the containment set must be done, ie the results of the perturbation must be renormalized.

(even if one allows for renormalization, which, in fact, means the quantitative construct has no validity)

There is still the problem that there are virtually no real quantum systems . . . , (other than the H-atom) (where real quantum systems have many-(but-few)-bodies [most often charged bodies] which form into a stable system) . . . , whose energy-wave-functions are at all close to the observed properties of a general physical system.

Thus, a perturbed wave-function (given an internal-particle-state structure) is not relevant.

Thus, it is a description which does not have a wide range of sufficiently accurate descriptions, and it is very-very limited in regard to practical creative usefulness.

However it is a descriptive structure which is related to large amounts of, essentially, useless technical literature.

Furthermore, this internal-particle-state and field-particle-transfer context of particle-physics is best suited for determining cross-sectional properties of the elementary particles, so as to be able to determine collision rates (or collision probabilities) for elementary-particles in nuclear reactions.

Orthodoxy

Particle-physics is about the patterns of particles usually found in high-energy particle-collision experiments. These particle properties are, in turn, related to non-linear operators. Fundamentally, these ideas emerged from the Dirac operator, which identified (whose interpretation was used to identify) both the opposite metric-space states of matter and anti-matter, as well as the spin-rotation states of material particles.

New

The new, high-dimensional macroscopic and microscopic, ideas, concerning discrete hyperbolic shapes, have matter and anti-matter as fundamental, in regard to the physical properties associated to the metric-spaces themselves, and a set of different metric-spaces to properly contain the different physical properties, and a subsequent need to identify an opposite-state (or opposite metric-space property) within each type of metric-space, while the spin-rotation is a spin-rotation of metric-space states, which are present in stable constructs of system-components, but in such stable components these opposite states stay in a permanent orthogonal relation to one another, and spin-rotate so as to be relevant only in a locally symmetric manner, ie in regard to dynamical changes within a greater context of a orbit defining metric-space shape, which determines the orbits of (lower-dimensional) condensed material components.

Either the local symmetry of opposite displacements, or the orthogonal opposite-state relations on stable shapes, only exist in stable components associated to the spin-rotation of these opposite metric-space states ie the opposite states are mixed in discrete intervals to allow local opposite dynamic changes, but the nature of the "real" state of the metric-space is maintained within the component, or within the physical system.

Orthodoxy

Particle-physics is based on extending the idea of spin particle-states to general internal particle-states, where the internal particle-states include anti-particles, and the particles also possess spin properties.

New

In the new ideas, there are stable components, which are, in fact, stable metric-spaces, and these components do interact with one another in the characteristic types of dynamical patterns (or second-order (partial) differential equations of simple dynamical systems) of:

1. elliptic (stable orbits),
2. parabolic (free, angular momentum [and this would be the context of quantum energy-waves but this is no longer necessary]), and
3. hyperbolic (collisions),
   as well as
4. in regard to hyperbolic (and mechanical) wave-equations.

Note: For components, which are most often charge-neutral, the interaction is most often collisions, while elliptic orbits are associated with angular momentum properties, where angular momentum can couple to either a distance structure or to a containment space.

Orthodoxy

The approximate structure of particle-physics is that spin distinguishes material (Fermions) from fields (Bosons), where spin is incorporated into the Dirac operator, while the internal particle-states are incorporated into a non-linear unitary connection term of the wave-operator for quantum interactions . . . , in regard to the energy operator (or set of operators) which is (are) supposed to identify the spectral structure of the (a) quantum-system . . . , where the non-linear connection term transforms between particle-states when collisions (or energy transfers) occur.

These collision-related transformations of the connection term are the, so called, (particle) symmetries of the system of colliding particles which possess internal particle-state vector-structures. These symmetries are low-dimensional unitary transformations.

The standard model divides the particles into "distinguishable particle types" associated to U(1) x SU(2) x SU(3),

1. where supposedly SU(2)-particles interact with both the charges of U(1) and the quarks of SU(3),
2. while SU(3) is about how quarks build other particles, eg protons neutrons (and many more), plus all the anti-particles of this type,
3. while the electron is mostly related to the neutrino and their anti-particles of SU(2),
4. and, where electrons must be able to interact with the photons of U(1) etc.

This context describes a few particle interactions, which may (or may not) be related to particle or nuclei disintegrations, and the authors of this descriptive context, claim it to be related to three or four very precise measurements of stable states of electrons and the H-atom, ... , but the calculations of spectra of: general atoms, nuclei, molecules, crystals, etc, are not sufficiently accurate so as to claim that these local models about particle-physics have any relation to the observed stable properties of general material systems, eg the stable states, which characterize the beginning and ending physical states of a nuclear explosion, while the probabilities of particle-collisions are (only) related to the rates of nuclear reactions and subsequent explosion sizes.

Orthodoxy

The particle-physicists often say that the standard model accounts for all the data that is seen in the lab, by this they really mean that it accounts for a great deal of the data which comes from particle-accelerator experiments, but that data, other than being useful for determining particle cross-sections (rates of nuclear reactions), has no relation to valid descriptions of the stable properties possessed by fundamental physical systems of: nuclei, general atoms, molecules crystals, etc.

Social comment

That is, they ignore the fact that the stable states of material systems, which are so prevalent a part of material system structure, all go without valid measurable descriptions, or their descriptive contexts are not even considered.

This seems to imply that this is a professional community of a set of overly obsessive types of people who are separated from reality

This type of social construct, isolating and protecting obsessive people who command intellectual discourse about science within society, can only be related to the non-stop repetition and great authority associated to the propaganda system, and that the personal who are picked to work on these narrow problems are people who rely most on a picture of the world which is not full of "a general world reality," but rather hold in their minds a distorted world picture, which can be a reality which results from participating in "narrowing competitions" based on authoritative dogmas and associated memorized models of "reality."

The context of orthodoxy is questioned

Many of the elementary-particles observed in high-energy particle-collisions in particle accelerators are unstable, with particle-lives which are only small-fractions of seconds, while on-the-other-hand the electron and the proton are relatively stable material components, yet they also seem to need some further stable context.

What is that further context of stability? ie a stable context is defined by the stable shapes contained in stable metric-spaces. These stable shapes are the substructures to which stable charges are related. (see below)

Orthodoxy

What characterizes particle-physics theories?

They are about collisions (or close-by energy transfers between elementary-particle which are assumed to be the components of which quantum systems are composed) represented as invariance's, where the invariance is a unitary invariance (the system's over-all energy is supposed to remain invariant in a unitary-invariant transformation).

Note: Quantum representations of quantum-systems are about "operators acting on function spaces" the local operators, which are Hermitian are, in turn, related to operators which are unitary, and these unitary operators are energy-invariant operators,

ie particle-physics is a model of particle-collisions which are energy invariant,

ie the pattern is the pattern of conservation of energy which is consistent with the descriptive context wherein the function-space is made into a "Hermitian space," upon which act local Hermitian operators, in turn, related to (global) unitary operators, eg the energy equation (which has unitary invariance).

However, renormalization assumes that energy is locally changed (ie does not stay in equilibrium) so that only after the perturbation process is complete does the system (miraculously) comes back to a state of energy-conservation.

That is, the mathematical containment context becomes completely invalid until after the interaction process is complete.

So, why even have a myth of mathematical containment?

The descriptive structure as a math structure is a fraud.

Yet it is complicated, so as to filter-out people looking at these systems from a rational point-of-view, and it forms the dogma of professional physics journals.

Orthodoxy (re-iterate)

To summarize particle-physics findings:

Particle-physics is based on material particles, and field particles (causing [during a particle-collision] internal particle-state changes and energy changes), and all the associated anti-particles.

There are the particle-types of . . . :

1. charge,
2. flavor,
3. families (where, apparently, these lepton families, associated to flavors, are energy hierarchies), and
4. color (quarks)

. . . , are all particle-state types which are related to symmetries or unitary changes in these types of particle states.

(these particle-type symmetries [or local unitary matrix transformations of internal particle-states, caused by their collisions with field particles]) are associated to internal particle-states of:

(1) electromagnetism (charge, U(1)) (where the particle-state change is either energy-change or spatial-displacements) [this is an extremely limited model for describing electromagnetic properties],

(2) electron-nucleon (flavor, SU(2)) [this is supposed to model certain types of nuclear decay processes], and

(3) pure nucleon (gluons, SU(3)) [and energy-hierarchies of nucleon particles], this pattern deals only with creating nucleons, but it is claimed to also be related to the so called strong nuclear force, but it provides no valid models of nuclei and their associated stable spectral properties.) where each particle-type is associated to particular types of field-particles (where the field-particles "cause" the changes in internal-states),

The field-particles are:

Photons (related to charge),
(Weird) Field particles (W(+), W(-), Z(0)) (associated to flavors)
Gluons (associated to color),

and where the field-particles of the color-symmetries also have color themselves (and thus they have an associated set of color symmetries associated with the colored-gluons)

Particle-collisions (or almost collisions) take place wherein (virtual) field-particles are (miraculously) transferred between the material-particles (even though the model is based at a single point in space) and the internal particle-states of the material particles are transformed, so that these "energy transforms" and "changes of particle-states" affect the internal particle-state structure of the quantum system's wave-function so that the energy of the quantum-system (associated to the wave-function) is perturbed by this process, and when the math containment structure is ignored so as to allow for a math method of renormalization to take place which then allows the "perturbed and renormalized" quantum-system to be energetically consistent with the observed values of the system's spectra (sometimes to 16-significant digits). This method is claimed to be significant since it has been applied to three or four such quantum-systems which do possess a wave-function which is already close to the observed spectral values of the system. This is not widely applicable, nor is it of any practical value, other than being related to particle-collision cross-sections (ie probabilities of particle-collisions) related to nuclear reactions. Ie the weapons industry is determining the structure of physics.

Orthodoxy
There is also the, so called, scalar-particle with zero-spin, the Higg's particle, which is supposed to give the property of mass to the particles (this is caused by a degree-4 polynomial, ie a scalar-function, with obvious symmetries about zero), the solutions to the Higg's mechanism, is claimed to possess mass, but apparently, the symmetries of the particle interactions of (regular) particle-physics equations have zero mass (or some such fanciful model of: charge, flavor, and color), inventing scalar fields to artificially change the position of zero, where apparently, mass has lost its relation to "changes in motion," since the descriptive context of particle-physics is supposed to be about energy.
But the formula, mass= energy,
Apparently, does not work, thus the need for a scalar field. (does one detect more epicycle structures)

That is, there is an epicycle structure (and a deception) at every turn in quantum-physics and particle-physics.

Note: The acquisition of mass by "charged" components is related to resonances between hyperbolic space and Euclidean space, and this would instantaneously occur for each period of a spin-rotation of metric-space state. Thus, this could be modeled as a collision event, and thus could also be associated to an apparent constant, H, ie "charge" = H x mass, defined between hyperbolic and Euclidean spaces (where "charge" in case of a particular high-energy Higg's-particle might be "color," where "color" would be associated to an unstable discrete hyperbolic shape within the 3-space the color-component has entered in its decay process. That is, flavor and color are unstable 2- and 3-dimensional component-shapes, of the equivalent of charge, which are transitioning [or decaying] within the particular 3-space subspace of our material-3-space.).

Commentary

None of these quantum interaction models provide any added ability to enhance (or understand) the quantum wave-equation model, and whereas the quantum wave-equation fails to provide valid models (to sufficient precision) of the very stable properties of: nuclei, atoms, molecules, and crystals. That is, this descriptive structure is a failure.

Furthermore, a probability model of a system with many, but few, components has virtually no "informational relation" to practical usefulness of such a probability based precise descriptive model (or structure). Yet these systems are very stable, which implies that they are formed under controllable conditions.

It is a particle-state context of local symmetry within which the equations of material interactions in quantum systems are to be described, but in quantum-interactions material-interactions are not about local measures (in turn, related to a global solution function), but rather they are an artificial process which conserves energy in a non-local, discrete [but distant] change (or virtual field-Boson exchange), but the field-Bosons are basically non-linear adjustments to the (energy) spectral properties of a wave-function.

However, if one desires to have quantitative consistency in a math description then the local measures (in a math containment context) for a measurable description need to be linear. The interpretations and models of quantum interactions are such that they assure the fundamental stability of quantum systems will not have a valid description.

Comment

It is surprising that the unstable particle products of particle-collisions in particle-accelerators do so closely follow a unitary pattern of particle-compositions.

New interpretations of the particle-physics Orthodoxy

The descriptive context of particle-physics still leaves indescribable the properties of "original stable state" and "final stable state" . . . , adding no useful information to this more pressing descriptive context.

What can be taken from this unitary context of unstable particles, with internal particle-states, is that the context should be unitary, and that since most of these particles are unstable, though some are stable . . . , namely, the electron and (in all likelihood the) proton . . . , that is, there is a good possibility that the model of unstable "particles" in 3-dimensions are unstable component-shapes which are unitary, composed of opposite states which might be associated to higher-dimensional energetic properties but which are unstable after the particle-collision, ie not in resonance with the containment space's finite spectra.

Thus, these can be unstable:

3-dimensional components (shapes),
2-dimensional components (or 2-dimensional faces of unstable 3-shapes), and
1-dimensional components (shapes).

These (various dimension, 1-, 2-, or 3-dimensional) component shapes are unstable, since they are not in resonance with the containing space's finite spectral set which defines the stable patterns within the high-dimension containing hyperbolic metric-space, but these unstable (inertial component shapes) do seem to descend from higher-dimensional (stable) structures, apparently the stable shapes are decomposed during the collision.

An inertial shape (stable or unstable), which is 3-dimensions or less, will be related to the type SU(3) or SU(2) or U(1)

. . . , fiber groups, defined on complex coordinate spaces of:
C(3,0), or C(2,0), or C,

and where the real shapes would be related to
SO(3), or SO(2),

fiber groups on real spaces:
R(3,0), or R(2,0).

Thus, the unstable inertial remnants . . . , of what were stable shapes, which have been decomposed by a (high-energy) collision . . . , would be unstable shapes, within the (unstable) geometric-patterns defined by SU(3), or SU(2) . . . , or SO(3), or SO(2), etc.

That is, in 3-space these unitary (or the "real" metric-invariant) fiber groups would be related to the natural patterns of disintegration for unstable shapes which are contained within 3-space.

That is, the fixed set of particle-collision patterns, seen as data in particle-accelerators, is the natural low-dimension decay structure of unstable shapes. It is not related to an interaction process other than these unstable patterns briefly existing in an unstable context of transition characterized by the random collisions of components, either stable or unstable components.

New

The new descriptive context agrees with particle-physics that the description is unitary, due to metric-space containing opposite metric-space states, these opposite states are related to spin properties of material components, and that the containment space is an 11-dimensional hyperbolic metric-space, but that such an 11-dimensional hyperbolic metric-space can be related to other such 11-dimensional hyperbolic metric-spaces, and that the stable properties of "material" which are contained in each such a space must be in resonance (and in the correct dimension) with the finite spectral set defined by the metric-space subspace-partition of each of the over-all containing 11-dimensional hyperbolic metric-spaces.

In the new descriptive structure there is a new context for angular momentum.

That is, angular momentum is defined on the various toral components of the stable shapes, which are allowed by the containment set (where the high-dimension containment set defines a finite set of stable spectra-geometric measures, to which the existing stable shapes must be in resonance), and on possible links, defined by angular momentum.

Links

There are unbounded stable discrete hyperbolic shapes, which exist on all dimensional levels, and these unbounded shapes are associated to stable material components, ie stable discrete hyperbolic shapes defined by their being resonant with the finite spectra of the various subspaces of the containing space [which is partitioned by (into) a finite set of stable discrete hyperbolic shapes of all the dimensional levels of the over-all containment set].

On the other hand the 2-, 3-, and 4-dimensions are relevant to the descriptions of "material" components contained in hyperbolic 3-space, where these stable shapes are also related to both bounded and unbounded, or semi-unbounded, discrete hyperbolic shapes, where an example of a semi-unbounded shape would be the neutrino-electron structure of an atom's (2-dimensional) charged components (which is also called an electron-cloud of an atom, eg for an atom the nuclei are bounded shapes while the electron-clouds are semi-unbounded), so that all "material" systems are linked to an infinite-boundary of the over-all high-dimension containing space. Thus, one can think of angular momentum as a controlled (or controllable) link between the many different

11-dimensional hyperbolic containing metric-spaces, by means of such unbounded and associated bounded (angular momentum) links (between 11-dimensional hyperbolic metric-spaces).

Thus, one can consider a "possible consciousness" for people (or their realm off creative intent) would be to examine the different creative structures of these different universes, where the individual 11-dimensional containment sets for the different universes (or perhaps different galaxies) might be perceived as intricate bubbles of different types of perceptions, through which we can control our journey, since we are in touch with the infinite reaches of these types of separate existences. (see below for a high-dimensional model of life-forms, eg models which allow all life-forms to possess a mind)

Is this the true context within which the human life-force is to develop knowledge and intend creative expansion of such a context?

** Though charge is likely not a 1-dimensional construct, but rather a set of charged 1-flows which fit into a 2-dimensional discrete hyperbolic shape, so as to allow spin-rotations of opposite pairs of time-states.

On-the-other-hand mass (or inertia) can be 1-dimensional, a circle, since a circle's center is a distinguished point, in regard to position in space, for translations or rotations, but any point on the circle could be a distinguished point for rotations, or a pair of opposite points, a diameter, or two pairs of opposite points so that each diameter is orthogonal to the other diameter, and furthermore, the orthogonal pair identify the circle's center. Thus, such an orthogonal pair represent both rotation frames (rotating stars) and translation frames (fixed stars).

So that, the circle and its center can be mapped into one another so as to represent the map between translational and rotational frames of the circle on the plane.

That is, the different 11-dimensional "bubbles of hyperbolic metric-spaces," . . . .

. . . . , between which human life might be able to use (travel between [or link between] with intended purpose) if one's higher-dimensional structure is understood and/or perceived, ... , seem to depend on sets of 2-planes which can carry the essential inertial orbital-structure of various bounded regions of 11-space wherein the pairs of opposite states on inertia (matter and anti-matter) which can be defined for each of these particular regions sliced by 2-planes which determine the organization of inertial properties of the region (or for these particular bounded regions). These sets of 2-dimensional regions are bounded since inertia is defined in relation to only the bounded shapes of discrete Euclidean shapes, and Euclidean space is the space of position and spatial displacement, ie Euclidean space is the space in which inertial properties are contained.

That is, these sets of 2-dimensional regions could be used to map the different 11-dimensional "bubbles of hyperbolic metric-spaces."

There are natural stable shapes, those of odd-dimension (3,5,7,9) and with an odd-genus (where genus is the number of holes in the shape, eg the torus has one-hole, or a genus of one, ie the genus is the number of toral components of a discrete hyperbolic shape) which when fully occupied by its orbital charged flows are charge imbalanced and thus would begin to oscillate, and thus generate their own energy. This would be a simple model of life.

Thus such a shape which possesses a higher-dimension could cause the lower dimensional components to, in turn, possess an order which can be controlled by the higher-dimensional shape, through angular momentum states (properties).

Down in 3-dimensions this control by a higher-dimensional structure could be the complicated microscopic-and-macroscopic structure of life, which appears to be run by complicated molecular transformations.

This is simply about assuming that stable shapes determine the underlying order and stability which is observed, and the fact that these stable shapes (mathematically) have a dimensional structure associated to themselves.

However, according to the currently accepted laws of physics both the stable properties of quantum systems and the stable control which is possessed by life are unexplained (or unexplainable).

The patterns of stable physical systems are unexplainable within the current dogmas about the material world, since the current dogma is based on the dimensionally-confining idea of materialism, and within such a confinement, descriptions seem to be based on indefinable randomness and non-linear systems (or patterns) defined on a (quantitative or coordinate) set which assumed to be a continuum.

Such patterns are fleeting and unstable, though the decay times can, sometimes, be of relatively long duration.

Suppose human life is associated to a 9-dimensional shape of an odd-genus, then such a shape is an unbounded shape (noted by Coxeter), and thus it could well be relatable to many such 11-dimensional hyperbolic metric-spaces (why should an unbounded 9-dimensional stable shape, generating its own energy, be confined to any particular unbounded 11-dimensional containing space?) wherein the living system's lower dimensional (material) structure may be quite different (in the new containment structure), and thus the living system's perceptions and interactions could also be quite different within other 11-dimensional containing spaces.

Note: The authors of M Gell-Mann (at least his "quark and the jaguar" book) and Y Manin, are a very small handful of authors who can provide clear descriptions of the essential models of particle physics. S Weinberg is mostly confusing, but he does emphasize the properties of equations in his popular works, which Gell-Mann does not do. Yet, Gell-Mann goes through the details of particle-physics; internal-particle-states and their relation to matrices, and their relation to field-particles, in about 10-pages, which are easy to read.

33.

Meaning in math

There is a book, "Meaning in Mathematics" which is a book of essays about the philosophy of math, edited by J Polkinghorne, a physicist-theologian, and reviewed by M Heller a philosophy professor (and also a theologian), where this book review appeared in AMS Notices, May 2013, where in order to form this book, a small set of "so called" "qualified experts" J Polkinghorne, R Penrose, M Leng, M Steiner, T Gowers etc (altogether about 15 authorities), who discussed, in the book, various things about math, including "the independent realm of mathematical reality," and some doctrinal subtleties about math, as well as other questions given below, as presented in the book review.

The review begins with the question about the dimensions needed to describe measurable patterns of "reality" vs. "the idea of materialism," and the question, "Are the dimensions of materialism incapable of describing the observed physical patterns?"

This would be a good question, if it were placed in an open context. However, it appears that these experts are concerned about the math structure of particle-physics, not in regard to open questions which invite the exploration of new math ideas. For example, new ways in which to mathematically organize a physical description, so that the new ways of organizing math patterns possess more dimensions than are allowed by the doctrine of materialism and its companion particle-physics (where particle-physics requires that the random particle-events be contained in 3-space, or in space-time).

But the reviewer indicates their bias toward "the interpretations which qualify them to be a member of an expert team," where the reviewer leaves the review with the idea that . . . , particle-physics is about how the phenomenon of the physical world all depends on a mathematical-model of the internal transition patterns in regard to both the internal states as well as the decay patterns of elementary-particles (where these patterns are modeled in terms of a set of internal-dimensions associated to elementary-particles), . . . , but since there are no valid models of the spectra of a general nucleus based on the laws of particle-physics, it is difficult to see "how such a statement has any relation to truth," but nonetheless it is claimed that these patterns of transitions of internal particle-states are all defined mathematically, and imposed on the data of particle-collisions in particle-accelerators, so as to be done with great (math-pattern) consistency.

And then the final question of the reviewer is, "Does this tell us something about the nature of mathematics itself?"

[Answer from the author of this paper; that math patterns allow us to interpret our world of experience, but what is the correct interpretation of that experience, that unstable shapes in 3-space (or space-time) decay according to the patterns of U(1)(x SU(2) x SU(3), or that these patterns are involved in material interactions at the quantum level, but these interactions provide no valid descriptions of observed stable physical systems (they need to be widely applicable and provide descriptions to sufficient precision in regard to large sets of very stable systems [nuclei, general atoms, molecules, etc], but if these patterns are "to be true representations of the world" then they also need to be related to practical creativity {the stability of these physical systems implies they are solvable and they form in a causal context}.]

Though the philosophical questions put-forth by these experts, associated to the book "Meaning in Mathematics," could lead to new ideas, the experts prefer to organize them into an "envelop of fixed interpretations" which define the domain of the experts.

So this book is really about imposing the arbitrary thought of intellectual authorities, concerning the (regular) decay patterns of unstable shapes which are the products of high-energy particle-collisions, onto an educational endeavor, which is now focused on trying to understand physical phenomenon based on these regular decay patterns of elementary-particles (but physical phenomenon is unrelated to the decay patterns of these elementary particles) . . . , so that the educational endeavor cannot succeed in realizing a practically useful description of observed very stable physical properties.

Particle-physics has been around since the late 1940's and though it fills technical journals with complicated math patterns (actually logically inconsistent, and apparently in-comprehend-able, math patterns) the stable systems, which are supposed to be the focus of these descriptive structures, are not being related to sufficiently accurate descriptions nor to useful applications (other than being relatable to rates of nuclear reactions in the design of bombs).

A better philosophical discussion . . . , one not so beholden to a failed authoritative doctrine . . . , might go as follows:

Meaning in math is about how "precisely described patterns, concerning measurable-properties and shapes, can be related to practical creativity."

If math patterns cannot be related to practical creativity then such math patterns have no meaning.

Measuring deals with reliable comparisons, ie relating standards of measuring to stable systems so that the system's properties can be identified based on general law and these properties used in a carefully measured and put-together new systems which are controllable (or useful).

"The independent realm of math reality" is an idea which is about deception.

The main issue of math is about the stability of identified math patterns, and the context within which measuring is reliable (or stable).

When measuring is reliable, then a precise description of a pattern has a valid context, if the pattern is stable. If a describable (or measurable) pattern is stable then it can be used in a practically useful context.

Some of the questions posed by the contributors in the book "meaning in mathematics," and mentioned by the reviewer, in Notices, go as follows: Note: where the answers to these question given by the author of this paper are given in [ ]:

Are numbers objects?
[Only if they are used in a context where measuring is reliable.]
Can math patterns be compelling, (in that, if proven true then it is a pattern which will be found to exist objectively, ie in a measurable context)?
[This would be a true statement if the math context is stable and measuring is reliable, and if the pattern has useful practical creative value.]
Can consciousness be a factor which is applicable to math descriptions?
[Only if there is a valid, stable mathematical model of life and its consciousness.]

Does materialism define the context of existence, or does existence depend on higher-dimensions?

This is posed as an open question but it actually will be put into the traditional authoritative context of today's intellectual authorities, ie it will be related to particle-physics.

Another set of questions which could have been asked in the article, are:
What is material?
and
What about the different types of material (charge, mass)?
Are different types of material associated to different containment spaces?
For example, inertia belongs in Euclidean space, while charge and energy belong to hyperbolic space (or equivalently, space-time space).

Can a globally commutative function-space be best found by associating the individual functions of the function-space with discrete hyperbolic shapes (with certain set properties) and relating . . . . the usual commutativity (in regard to a pattern-defining set of operators acting on the function-space) of . . . . the global function-space, instead, to sets of continuous locally commutative (individual) discrete hyperbolic shapes which possess various properties of: size, dimension, and set-containment relations?

This is about the process of fitting stable shapes (ie discrete hyperbolic shapes) into various quantitative structures which are generated from a finite quantitative set.

That is, it is doubtful if both the observed properties of system stability and a reliable measuring context can exist simultaneously in a consistent math context (consistent both quantitatively and logically) unless a finitely generated set of stable shapes exist in a metric-invariant containment set.

That is, physical law, which is based on non-linearity, also requires an underlying context of stability and containment within a metric-invariant context, in order for both stable systems and reliable measuring to be observable attributes in a math containment context (or in a context of math descriptions). Furthermore, the operator constructs, in regard to the measurable properties of patterns (physical systems) within a quantitative containment-set, depend on the underlying stable shapes as a part of a discrete definition of such operators, which identify the measurable changes and/or measurable patterns, but with the set of stable shapes within the containment set are to be dominant in regard to what measurable patterns can exist in a stable context. That is, the discrete construct which becomes associated with local operators supports (and depends on) an underlying stable containment context.

These are very interesting questions about language-systems of math description,

But, unfortunately,

The point of the article (ie the book review, in Notices) is that abstract ideas about math and science can only be considered within the context of traditional math and science authority.

The article is about the dissection of math language into a fixed set of abstract categories, which are provided (to the [allowed] intellectual communities) by the traditional authorities of math and science . . . ,

. . . , and then "the identifiable set of learned (or authoritative) communities are" to study math patterns in such a narrow context.

This, dissection of the language of math into categories which have been mapped-out by the authorities, is similar to studying molecular properties of biological systems based on the narrow set of "detectable molecules" and basing descriptions on the correlations which can be made in this narrow context. Instead, one really needs a valid model of life, and since life is a highly controllable system and . . . . this property of being controllable . . . . implies that such living-systems actually exist in a solvable and causal math structure.

[Such a viewpoint (of living-systems needing to be associated to simpler math models) is never considered. However, geometrization allows new math contexts, which can be based on the idea of stability, to be considered, since such a formal authority for the idea of stability has now been formally established within the intellectual community. However, the formal structure of geometrization is not really needed to consider the idea of "reliable measuring and stable patterns" as being the central attribute of math descriptions]

In the conclusion, of this book-review in Notices, there is the statement that particle-physics is about how the phenomenon of the physical world all depends on the mathematical-model of the internal transition patterns in regard to both the internal states as well as the decay patterns of elementary-particles . . . , but these properties are not relatable in any valid way to the observed stable patterns of nuclei, atoms etc. . . . , where it is claimed that these patterns are all defined mathematically and imposed on the data of particle-collisions obtained from particle-accelerators, so that these hidden math patterns allow us to classify and identify data from particle-collisions in particle-accelerators. But are these patterns simply the decay-patterns of unstable states which are transitioning to stable states, and thus, these patterns are of no value in regard to identifying the stable spectral-patterns of nuclei, general atoms, molecules etc?

This way of allowing fixed authority to set meaning concerning observed patterns and an associated set of math patterns, and their interpretations, is really about imposing the arbitrary thought of intellectual authorities, concerning the (regular) decay patterns of unstable shapes which are the products of high-energy particle-collisions, onto an educational endeavor, which is now focused on trying to understand physical phenomenon based on these regular decay patterns of elementary-particles (but physical phenomenon is unrelated to the decay patterns of these elementary particles), so that the educational endeavor cannot succeed in realizing a practically useful description of observed very stable physical properties.

One seldom finds such a stark characterization of elementary-particle properties as primarily being related to decay paths, namely . . . , particle-physics is about how the phenomenon of the physical world all depends on a mathematical-model of the internal transition patterns in regard to both the internal states as well as the decay patterns of elementary-particles . . . , perhaps "because Polkinghorne is a theologist" the dogmas of useless models are weighing on his viewpoint . . . , if he knew that one could also considers the math structures of stability, essentially identified by the Thurston-Perleman geometrization, and he also saw that they lead to a much wider array of ideas and constructs which incorporate models of life and re-establish the context of "the ideal" (as opposed to the idea of materialism) . . . . , perhaps he would be someone who would be interested in considering the new context within which to describe our existence (based on many-dimensions but related to the math patterns of geometrization).

The art of communication is about how to acquire cheer-leaders, or about "how the owners of society abuse the culture, and manipulate people within a very controlled and narrow social context" for their own selfish advantages . . .

The academic professional is more or less a cheer-leader for the current academic dogmas, but the professionals (associated to the book "Meaning in Mathematics") whom are also theologians, one would expect that they would want an intellectual basis for the philosophical viewpoint of idealism . . . . , but academics are known for their arrogance, which they are allowed to show-forth if they are associated with the current reigning intellectual dogmas, so it would not be expected that academic theologians to be any different . . . , though an "ideal" viewpoint would be a better context for theology, but then again, theology is itself very dogmatic so an ideal viewpoint may-well not be something which a particular religious viewpoint would support.

The wage-slave and hierarchical nature of society and its organization of social-power cause havoc with both language and descriptive knowledge, since social-authority becomes such an integral a part of one's behavior in such a social context (ie a social context of hierarchy), wherein people compete and are manipulated at such a childish level.

34.

To the point 2

Descriptive knowledge

Knowledge of both the relation of mankind to state and the relation of math to practical creativity are fundamental to an: equal, free, creative society.

For the public:

One must inform them, to their dis-belief, that the social power of our society is organized around the owners of society, a supposedly economic-political power basis, where this social-power structure is succinctly put as: "the point of society is to assure that the owners of society get what they want." (this is an approximate quote from N Chomsky), where in the US this organization of social power was put into place by deceitful manipulation of the American revolution by the ruling-class (at the time of the American revolution), a ruling-class whom were well trained in the use of violence to control people . . . , but what Chomsky seems to miss (in his analysis of society's true rules) is that the power of the owners of society was originally derived from the same violence from which the Roman Empire was based. Thus, stealing and the control of social-value can be determined by the use of extreme violence. In the US this violence has been mired by the enforcement of institutional commands, mostly commands which violate the basis for US law, but which are enforced with great energy (the huge amount of energy used to protect the owners of society, ie the ruling class, as opposed to the energy invested in forming a civilization which is widely creative and knowledge-able). In this context of veiled institutional violence, the social construct of wage-slavery was instituted within the US, and thus the phrase "free speech" comes to have no meaning (social-class trumped the expression of ideas, ie the important ideas are only those ideas which support the interests of the ruling-class).

However, a logical conclusion of such a social organization is that, due to people's wage-slave social status, all knowledge and creativity within society is to be determined by the "owners of society," and it is narrow and based on selfish interest (it is knowledge which supports the social position of the owners of society).

We live in a society of central-planning based on selfish interests, where the expert learned-types live in agreement with the values and interests of the owners of society. Thus being "for science" then one is manipulated by intellectual bullies whose words have meaning limited to the support they give to the owners of society, and being "for religion" means being manipulated by arbitrary authority and violence since western religious language is based on domination [where western, is Judeo-Christian-Islam religions].

The people within society are, essentially, controlled by means of both the (1) propaganda system (which is viewed as a sole "trusted and authoritative" voice of society, but within it, only arbitrary ideas which support the social hierarchy are allowed to be expressed) and the (2) financial system, because of the wage-slave status of the public, where the propaganda system is a set of instruments owned and controlled by the investor class, and which, by regulations, the government helps the ruling-few to monopolize the propaganda system. Ultimately, this is about institutionally applied violence, as well as institutionally fomented (organized) violence.

That is, all public institutions of higher-education are made-up of departments of technical knowledge where all the knowledge taught in these public schools (universities) is to be used to adjust the instruments from which the social-economic power and the security of the owners of society depend.

For the professional class:

Thus, the main issue in regard to professional wage-slaves (associated to narrowly defined technical knowledge) is that their narrow precise languages are really about the elementary issues concerning the development of knowledge which center around: assumption, context, interpretation, containment, etc, where these fundamental issues are forming a "narrow defined bound" in regard to . . . . what a precise (or overly technical) language can be related . . . in regard to practical creativity. That is, the narrow range upon which technical knowledge is focused within society, is all about pleasing the owners of society in regard to what knowledge they want expressed, and how they want that knowledge to be used within society, but the point of knowledge is for it to be widely applicable, sufficiently precise, and practically useful, but these simple criterion are not being met, and thus the development of knowledge is the best way in which to remedy these deficiencies (within society, so that the public has a wider range of knowledge about which creativity in society can be based)

[In regard to the development of knowledge, paying attention to the elementary properties of language is the correct conclusion of Godel's incompleteness theorem . . . . , which the professionals need to re-visit.]

That is, the dogmas of the professionals . . . , dogmas which define their professionalism . . . , where these dogmas are mostly arbitrary, and their old, traditional, fixed descriptive-language has become un-relatable to the descriptions of the observed stable patterns. Yet, these traditional authoritative patterns of intellectual activity are still being used by big-business interests (eg particle-physics and its relation to nuclear-bomb engineering). And to not "arrogantly support these narrow dogmas" means that such a person will not get the job.

One should note that, if there is progress in science and math, then it is due to the work of marginal personalities who show competence, where a traditional upper-class, intellectual, academic, mind-set is so narrow and arrogant (they are always considered to be correct) that they cannot be anything but incompetent.

It is this incompetence, in regard to knowledge, which leads to the failing of "capitalism," where capitalism is a form of social organization . . . , based on the social domination of the many by the few . . . , where markets based on a limited set of products are used to support the ruling-class, and the economics of capitalism is not a valid quantitative descriptive structure. The economics of capitalism is claimed to be about a measurable descriptions of economics, but the economics of capitalism is being determined by an unfair game, mostly determined by (the control of society allowed by) violence, and thus not a valid quantitative structure.

In fact, capitalism is mostly a collection of arbitrary statements, and it is a social organization which cannot drive development and growth, which would result from new inventions, including new contexts for invention (since new knowledge is not allowed), the investor class does not want such competition (due to inventiveness), because (rampant) inventiveness makes their investments more risky. Capitalism is based on stealing: both property and the ideas (which emerge from the culture).

What are called "new inventions" are really adjustments to existing instruments, all the inventions associated with the ruling-class's control of knowledge are derived from 19th century physical science, including micro-chips, so "new products" can mostly only be adjustments to existing instruments.

On-the-other-hand practical creativity should be conjectured within the context of "any new elementary descriptive construct" (where the development of such a new language depends on considering the elementary issues of language about: assumption, context, interpretation, containment, etc).

This was true for both Faraday and Tesla, Edison, the Wright Brothers, etc ie their creativity was directly related to their own elementary language constructs.

In fact, education is exactly about the relation that (such an) elementary use of language . . . , in regard to assumption, context, interpretation, containment, etc . . . . , has to either organizing society (so as to serve the interests of the people), or to practical creativity.

For example:

The media should be filled with discussions about "the nature of people" [are people by nature, creative and desirous of knowledge, or are they "sinful" (an arbitrary attribute depending on how one frames a behavioral context) and they either want to take-the-place of the selfish-rulers of society, (whose social power depends on the arbitrary use of descriptive knowledge) or help the rulers dominate the public (so as to exclude the public from ruling positions [those who believe-in an arbitrary behavioral context, a context wherein an idea about "what has high-social-value" is a hidden assumption, or vaguely expressed])] and the discussion needs to be about the elementary-basis for a new language, within the social domain, about both "how to realize high-ideals, and in regard to creativity," and in regard to a harmonious social-organization within society, ie each person needs to be considered to be an equal creator, where focus should be about both the relation of a "population of equal-creators" to a truly free-market, but also about how knowledge about the earth can best be used so as to allow the earth's abundance to nourish its people, and the justice system should be about opposing people motivated by selfishness, and to support people whose motivation is (selfless) creativity.

That is, the language needs to be understandable to everyone, and the "spirit of the law" needs to be administered.

Understanding how quantities and measures relate to valid and reliable measuring in regard to practical creativity, is a central theme of both elementary and advanced education, in regard to using the properties of existence for creative purposes.

Understanding how the knowledge of quantities and measures relate to social organization, namely, how that knowledge is used by the ruling-class for their narrow interests, should help one deal with the fraudulent intellectual façade that is put forth by the propaganda system.

Math and the descriptions of observed physical patterns are endeavors which are trying to do the same thing, that is, these two subjects emerge from issues about:

I. identifying reliable measures (comparable sets) in a (relatively) stable geometric (geometric-material) context,

II. the relation of "a set of such measures" to "the containment-set of a description" (of a system, or of a pattern), [these sets of measures, within which a system's properties are contained, are most often sets of geometric-measures, where a system is being contained in space and time, as well as (the system) depending-on material definitions],

III. the further properties (eg function-values) which can be related to such a context of containment . . . , where

(1) functions and their domain spaces, and local measuring-methods whose local vector structure commute locally, but they commute in a continuous global manner, [on the domain space's coordinates];

(2) function-spaces and sets of commuting, linear operators, where the continuously commuting local vectors of (1) are analogous to the global diagonal functions-space representations of a linear spectral system, . . . , the continuously commutative (or globally commutative function-space) math patterns are the only math patterns which relate to a stable context in regard to a system's measurable properties,

(ie the stability of math patterns is a very limiting property, but it is necessary for quantitative consistency), and

IV. what simple rules, and methods of measuring, or of identifying a system's quantitative properties, which when placed into equations and solved . . . . , where the equations relate, either

(1) "what appear to be separate descriptive properties," (the laws of classical physics) or

(2) relate to a diagonalized function-space, which is associated to a system's defining spectra,

. . . , can lead to reliable, useable information (descriptions of [stable] patterns) by quantitative and geometric (or spectral) properties, in regard to practical issues concerning the "subsystem structure of a practical invention" ie and there are the subsystem issues concerning: coupling, control, and creativity.

Note: A diagonalization of a system's function-space cannot, in general, be associated to the system's spectral properties (to sufficient precision), ie in general, sets of commutative operators cannot be established, when a (quantum) system is modeled as a function space.

Yet, these (unsolvable quantum) systems do have stable spectral properties, which have been observed, and this implies that there is a stable geometric basis for their descriptions.

In a stable geometric basis for their descriptions, these systems do have solutions, in the new descriptive, many-dimensional, context, but this would require stable shapes contained in higher-dimensions, than "are allowed by the idea of materialism," and in these higher-dimensions, the geometric constructs can be both macroscopic and microscopic (whereas string-theory, which assumes particle-physics, only allows microscopic higher-dimensional geometries), but now (in the new context, in regard to the containment of existence, which does not assume materialism) the nature of the operators . . . , which act on such a geometric and many-dimensional context . . . , are changed, since the new operators need to be discrete, and the descriptive context is to be governed by (or to be fit into) actual geometric envelopes of stability, but these higher-dimensional properties are hidden by the discrete operator structures, which are defined in a geometric context.

Note: The geometric and inter-action properties (associated to the new discrete operators) of the material realm, traps the 4-dimensional material context within a particular open-closed topology, when considered from within that context (of materialism).

However, the stability of the solar-system is evidence that macroscopic geometric envelopes exist in higher-dimensions.

A globally commutative function-space, ie finding a set of operators which allows a function-space to realize a diagonal (spectral) pattern, might best be determined by associating individual functions of the function-space to be represented with a set of discrete hyperbolic shapes, and then relating "global function-space commutativity" with continuous local commutativity of sets of "discrete hyperbolic shapes" of various sizes, dimensions, and set-containment structures, which are related to these sets of stable "discrete hyperbolic shapes."

One of the main issues of physical description (or of mathematical description) is about:

1. the relation of practically useful description to either a geometric context or a geometric-material context.
2. the dimension of the containment set, and what restrictions are placed on the macroscopic-microscopic properties of the higher-dimensions, ie higher-dimensions than is allowed by the idea of materialism.
3. "what properties are the "better" measurable properties to identify, in regard to obtaining valid properties (or stable patterns) to be used in both a descriptive and a practically useful context. (this is related to E Noether's time and/or space displacement symmetries, as well as metric-space states)
4. Then of course one wants to consider the nature of the operators which act on representations of quantitative properties (or measurable patterns), they must be operators which depend on (or operators whose properties best suit) the properties of the containment set.

If one is describing unstable patterns, in what might be interpreted to be a context where stable measuring properties exist, then the unstable patterns are illusions, since they are not contributing to the stable context which is needed for both reliable measuring and a practically useful context, as well as such unstable patterns having the (further) property of being unrelated-able to the observed, very stable, patterns of fundamental physical systems, eg nuclei spectra, a sufficiently precise spectral description of atoms which are composed of more than four-material-components, and the properties of: molecules, crystals, living systems [which (in a relative sense) are very stable as well as being highly controllable], the stable solar system, . . . , etc.

That is, one needs to try to identify the fundamental context in which these (just mentioned) observed stable patterns can be contained and described, so that these stable patterns can be related to practical creativity.

If a math pattern is not stable and/or if the pattern is not relatable to a reliable measuring process then such a math pattern is an illusion.

Feedback systems . . . , non-linear and/or non-commutative systems

All feedback systems can fail if (since) they are based on an unstable math pattern.

So why does feedback for patterns-of-illusions work at all? Since neither the local model of measuring is valid nor is the law that "motion changes are related to a local force-field's relation to material geometric properties" valid.

Yet, why can feedback in many such cases appear to work so-well?

Answer: They almost work, so if the time-interval within which "both sensing and feedback is determined" is made smaller then the transition of the system from what are considered (or assumed) to be its stable properties (within the feedback system), namely, both physical law and that a local linear measuring process is reliable, can be identified, and then corrected by means of "feedback-related system adjustments," so as to see (by the feedback system's sensing process) if the properties of the system's differential equation still apply to the (unstable) system after the (feedback based) system-property adjustments are made (within the unstable feedback system).

That is, the solution to the (feedback) system's differential equation is the pattern which is the illusion, while the context of such a solution function might have a wider-range of (relatively stable) applicability, and it is this wider context (of the system's differential equation) upon which a feedback system (associated to unstable patterns) is based.

The question as to the validity of local linear models of measuring (ie the validity of a derivative) and the validity of an approximation to a (classical) physical law (ie related to the validity of a differential equation) are properties which are approximately valid, and this would be because the system is contained in a macroscopic containing metric-space which is the size of the solar-system.

"Is the context (within which quantitative sets are being used) stable?" so that measuring is reliable.

Apparently, stability is determined by shape, and by the shapes which are consistent with quantitative sets, ie lines and circles, eg right rectangular shapes (cubes), cylinders, and circle-spaces, ie discrete Euclidean shapes and discrete hyperbolic shapes, ie non-positive constant curvature metric-spaces whose metric-functions have (only) constant coefficients, where the local properties of being linear and independent can be continuously extended to the global shape, ie the equations are always locally commutative . . . , ie represented by diagonal matrices . . . , so that these properties are continuous on the shape's (or system's) containing coordinates.

Now, in turn, these shapes are used to partition the subspaces and different dimensional levels of a many-dimensional containing space, so that there are many "trees-of-containment" dependent on the dimensions and sizes of the stable shapes which are used to partition the subspaces of such a many-dimension containment space, which is an 11-dimensional hyperbolic metric-space. This partition defines a finite set of fundamental partitioning shapes. This, in turn, defines a finite spectral set, which by resonance to this finite spectra it can be determined as to what stable shapes can exist within such a containment set.

education

1. In math-education one wants both the elementary processes of math itself, counting, adding, grouping, multiplying, the place-structure of numbers, fractions, etc.
    measuring with rulers, making circles, measuring-angles, determining areas, identifying independent lines, ie lines which are perpendicular to one another . . . ,
    defining variables, forming equations, finding formulas, . . . , etc,
    defining function-values in regard to "coordinate places," ie coordinate positions, in the function's domain-space (where the domain-space is the set within which the function-values exist (or are defined)), function = formula {in the variables of the system-containing coordinate (domain) space}, etc,
    One needs to also consider the place-value structure of numbers, and this structure's relation to polynomials and to (converging) power-series (represented in variables) and whose terms of different degrees (different powers) represent different orders of magnitude (in regard to a number's place-value), to be used in regard to valid "precise enough" approximations of a number-type's value, in turn, to be used (in a practical way) in a descriptive pattern.
2. but one also needs consideration about big questions, like "what math is?" [it is the study of quantity and shape]

"How does quantity depend on the context of shape?" [there are discrete sets (which can be counted, and they give quantities their defining properties in regard to comparisons), and there are geometric comparisons, which depend on discrete calibrations, where (in a geometric context) units now have many fractional divisions to their calibrations, and where quantities are given the geometric model of a line, or along a circle]

"how can math descriptions be contained in a math-set structure?" [set containment within a measurable context of independent directions and coordinates, where coordinates are measuring curves]

"what is the number continuum?" [the set-structure needed to identify individual points on the number-line]

"Is the continuum a valid construct?" [or is it too-big of a set?] That is, the geometric model of "quantities modeled as a number-line" has logical problems in regard to the line's point-structure.

3. and issues about:

number-type, uniform units, and changes of number-scale, where y=mx, is the only quantitatively consistent way in which to change a number-scale, and for coordinates this linear relation must be established for each independent coordinate direction at a coordinate point, and this property must continuously extend (point-by-point) to the entire shape, in order for the shape to be quantitatively consistent and stable.

4. nd axioms:

order of operations, inverses, distributive property (related to the place-structure of numbers), and the commutative property of number (or matrix) operations, this is related to the independence of the local coordinate directions at each point of a stable shape.

5. Furthermore, one needs questions like:

What structure of stability is needed to obtain a context where measuring is reliable? [metric-invariance so as to allow (locally linear) geometric measures defined on coordinates, models of local measuring must be linear when relating function values to its domain (or coordinate) values, probabilities need to have (need to be associated to) stable well defined easily distinguishable states, which result from a random event (or toss), so that counting these events can be related to valid probabilities]

What patterns can a quantitative description describe, and must these patterns be stable? [valid patterns of description must be stable and it must be locally commutative in a continuous global manner]

What is a derivative?

Is it a model of local linear measuring, in regard to functions and their domain spaces?

or

Is it an operator which transforms functions either between function-spaces, or transforms functions within a function-space?

or

Is it a discrete operator defined in relation to a bounding set of stable shapes associated to metric-invariant metric-space containment sets of (physical) systems?

6. Mass is defined as a scalar value which opposes changes in motion, and which is attached by a center-of-mass coordinate to spatial positions (in Euclidean 3-space), yet when a material shape is rotated, its mass distribution relates the mass to independent rotational axes and to a symmetric inertial tensor, [which, in turn, is related to distinct axes of rotation] in 3-space. The 2-body center-of-mass gravitational system, though related to a pair of spherically symmetric force-fields, defines an elliptic inertially dependent orbital shape (or relationship), not circular.

On-the-other-hand charge is related to force-fields by a current-and-stationary-charge geometry so that the changing current can form electromagnetic waves, which can be in both positive time (advanced potential) and negative time (retarded potential) states, but charged shapes discretely move to new neutral states,

Is there are dimensional structure to the measure of inertia? That is, is the mass of a massive body determined by the radius of an inertial circle in a plane, which, in turn, is related by resonance to the energy of charges occupying stable spectral-flow states within a 2-dimensional discrete hyperbolic shape, so that in conservative systems, ie energy is conserved, changes in motion and transformation between energy-type (kinetic vs. potential) requires that the, (energy x time) sum, stays minimum, ie as integration of the energy vs. time graphs identify a fixed stationary value in time,

Can this be about either a galaxy (or a solar-system) having a smallest circle (radii) identifying its mass so that the hyperbolic charges . . . , to which this mass is related, by an over-all 11-dimensional hyperbolic containing space . . . , fit into a finite set of discrete hyperbolic shapes whose energy is consistent with the mass-value, so that in the entire containing set, energy is consistent with the 2-plane (circle) model of the space's mass. Thus, the system (either the solar-system or the galaxy) is fundamentally defined inertially on a plane, and (then) the energy is divided between the subspace containment-trees of the different dimensional—and different sized—"discrete hyperbolic shapes," which model the dimensional levels and their charged-material properties

The calculation, sum of E x dt defined on a time interval, is a math construct defined on hyperbolic space (or energy space) and its minimum (or extrema, eg also maximum) would be related to charges fitting into stable orbital systems which are characterized, in hyperbolic space, by the property of energy.

Furthermore, uniform stable system-types, ie the sets of orbital structures, which are allowed to exist by the containment set, implies that these orbital-systems form in a solvable and controllable (or causal) context for interactions, and this would allow energy to stay at a minimum for the allowed orbital system (energy) structures.

This can be contrasted with particle-physics where energy-values are not believed to remain conserved in a local context, where the system containment space is believed to become irrelevant, because point-particles destroy the local structure of space. This destruction of the idea of math containment is brought about due to the inconsistent construct of the models of point-particle-collisions, which are supposed to exist in a random context, this type of very specific geometry (a particle-collision) is not compatible with randomness.

Yet this math structure is made energy-invariant in regard to particle-states by unitary invariance.

That is, once a commutative set of operators diagonalizes the quantum system's function-space then the Hermitian-form (much like a scalar, or inner, product) preserves this diagonal structure, and the energy operator acts as a unitary operator which will preserve the Hermitian-form.

Thus, the energy operator will preserve the function space's spectral structure.

Thus, for an energy operator the occupancy of the diagonal energy-levels may be re-arranged but the system's energy structure stays in tack. This is similar to the stability of these energy states being modeled by as discrete hyperbolic shapes.

However, one can find the discrete hyperbolic shapes which solve the many-(but-few)-body problem whereas one cannot find wave-functions for general quantum systems, ie stab;le systems with definitive spectral-orbital properties.

Unfortunately, when a wave-function is given an internal particle-state structure, this new vector-structure, attached to the wave-function, will preserve the particle-state structure (when) in a unitary context, but now some (most) of these internal particle-states are unstable. Thus, the unitary operators acting on the particle-state structure are identifying ways in which the particle occupancy of the wave-function's energy states will disintegrate. That is, the math structure of particle-collision interactions is describing the dissipation of the system. [in fact, this is the only property which particle-physics genuinely claims to predict, ie the weak-force related beta-decay, as well as the dissipation of elementary-particles which exist in the debris of particle-collisions observed in particle-accelerators.]

Yet [even "in the average"] these (quantum) math model constructs of the reduction of material systems to point-particle (random) events in space and time (or in space-time) supposedly being related to a "stable" internal particle-state patterns of transition and/or decay, and where these transitions are unitary, is unrelate-able, in any convincing-way, to the specific stable spectral properties of general quantum systems from: nuclei, to general atoms, to molecules, to crystal properties, wherein the unitary property of an equation, whose containment space (ie its function space) is not well-defined, seems to provide no insight into understanding either spectral structures of stable physical systems, or in regard to the logical basis for the extrema of, sum of E x dt on a time interval, as being a valid math process for finding a system's relevant equations, rather it describes a decay process for the system's wave-function.

This more or less proves that the particle-collision model of quantum interactions is not related to understanding the stability of these system's energy properties (eg nuclei, general atoms, molecules, crystals etc).

Where an extrema of action depends on actions on wave-function by unitary operators, the math process, ie unitary invariance will maintain a quantum system's diagonalized structure, but this requires that material occupy stable energy-levels within a system, and not enter energy-structures where the energy-state decays.

In regard to the 2-plane model of mass for a high-dimensional containment set, can a C^2-space identify a higher-dimensional model of independent inertial sets, eg SU(2) x SU(2) ~SO(4), where inertial-C^2 is related to either SO(4,1), ie H^4, or 4-dimensional discrete hyperbolic shapes contained in H^5, in turn, related to either an SO(5,1) fiber group, or to an SO(5,2) fiber group, ie to SU(5,1) or SU(5,2) fiber groups for the spin rotations of metric-space states, where SO(5,1) and SO(5,2) are the fiber groups related to the two charge-types, or shapes which have hyperbolic-dimension-3, contained in H^4. Inertia modeled as a C^2 space would allow a 3-dimensional shape associated to the system's mass distributions, [eg perhaps this is necessary for the elliptic shaped galaxies, or gravitating satellite-systems which are not planar], where mass is dependent on the energy of the charged system-shapes and the occupation of the stable spectral-flows by charges, {so the spectral-flows (which the charged material can occupy) can have dimension greater than 1-dimension}.

Contrast the three ways in which a physical system's measurable properties are modeled:

1. New
2. Quantum
3. Classical

1. New, this context is used to both ensure that stable math patterns can be described and so that measuring is reliable.

   I.   One begins with metric-invariant metric-spaces.
   II.   Metric-spaces are given (physical-mathematical) properties

III. These properties are represented as pairs of opposite properties, and they identify metric-space states.

IV. Metric-invariant metric-spaces are (also) associated to shapes, the shapes which are stable have non-positive constant curvature, where the discrete hyperbolic shapes are the most stable.

V. Partition the subspaces and dimensional levels of an 11-dimensional hyperbolic metric-space by a finite set of discrete hyperbolic shapes, with different subspaces of both the same and different dimensions being given different sizes, and thus, subspace set-containment based on dimension identifies different sets of "trees-of-containment," within the 11-dimensional hyperbolic metric-space containment set.

VI. Whether a stable shape can exist in one of the appropriate-dimension metric-spaces of the 11-dimensional hyperbolic metric-space is determined by as to whether the stable shape resonates with the finite-set of spectra of the discrete hyperbolic shapes which partition this 11-dimensional hyperbolic metric-space (and has a size which can fit into one of the subspaces of the correct dimension, where an n-dimensional discrete hyperbolic shape would fit into an (n+1)-dimensional hyperbolic metric-space).

VII. Discrete operators whose definition depends on metric-space shapes and the geometry of the fiber group determine material interactions and the geometry of such interactions.

VIII. These interactions allow stable systems to form by the relation which the system's dynamic properties have to being resonant with the partitioning set of shapes of the 11-dimensional hyperbolic metric-space.

Note 1: The structure of the interactions hides the higher-dimensions from the lower-dimensional spaces.

Note 2: The interaction structure "applied to small systems" causes these small-systems to possess random properties, ie it causes the appearance of quantum randomness.

Note 3: The structure of the interaction causes the material interactions in 3-space to be spherically symmetric.

2. Quantum, this context describes randomness and it is used to try to identify stable spectral properties, but, in general, this attempt is a failure, thus, it is a math-method of indescribable randomness.

0. The domain space is either Euclidean space plus time (Schrodinger), or space-time (Dirac).

I. Material reduces to small components whose spectral-particle properties are random events defined in space and time.

II. A quantum system is to be modeled as a function space.

III. The physical properties of a quantum system are determined by sets of operators, so that there is a set of operators which diagonalizes the function-space so as to identify the quantum system's spectrum.

Note 1: The spectrum of general quantum systems, though observed, cannot be found by means of this method.

IV. A particle-state structure is imposed on the wave-function (where, in general, the wave-function cannot be found, so it is not clear there is any content in the statements of particle-physics)

V. A unitary connection-operator is applied to the particle-state structure of a wave-function, apparently so as to lead the particle-components of the quantum system to their pathways of decay, but it is claimed that this allows the energy-states of the wave-function to be perturbed and adjusted.

VI. But this model negates the idea of math-containment for the quantum-system, and thus, renormalization is defined which needs to re-adjust the description (by subtracting infinity from infinity so as to get a finite number) so that math-containment is re-gained.

3. Classical, this context assumes that measuring is reliable and only the solvable systems have stability.

0. The coordinate system is given a Galilean coordinate frame, amongst a set of equivalent coordinate frames, ie frames where the distant stars are either fixed or rotating.

I. The derivative is a model of local measuring of a function's graph, where the function identifies a point-particle's position in a geometric context, ie the function's domain is space.

II. Mass is defined in regard to the second-time-derivation of a particle's position function.

III. The geometry of material defines a force-field, in regard to relating local geometric-measures of a field-vector to material geometry, which, in turn, relates to a point-particle's geometric position, so as to form equations for the dynamics of the material system.

Note 1: Only a few of these (partial) differential equations are solvable, but they are highly controllable descriptions of the system's properties, and thus they are of great practical use. In these solvable cases, they provide sufficiently precise descriptions of the system's measurable properties so as to be reliable descriptions.

If not solvable then the context of the system's differential equations can allow-for feedback-systems, though the reliability of these feedback-systems is difficult to determine.

00. General relativity is based on the idea that measuring can never be reliable, and consequently general relativity cannot describe any patterns which are stable, other than the 1-body system with spherical symmetry.

When one compares the new ideas with the classical ideas one sees that they are quite similar operator structures, but classical cannot solve for a stable solar-system and electromagnetism cannot solve the atom, while the new ideas can solve these problems.

When one compare the new ideas with quantum ideas one see that there is assumed randomness, and an operator structure, and an added-on particle-state structure for the quantum viewpoint, but none of the general many-(but-few)-body quantum systems have solutions, ie order does not emerge from a context of randomness, while the new ideas require that states of material be a part of the metric-space structure, and in the new descriptive context, the general many-(but-few)-body (quantum) systems do have solutions, and it is done in a stable, geometric context, from which (apparent) randomness is derived.

35.

Banking vs. Useful knowledge

Banking empires vs. useful knowledge

The ideas expressed in this paper are ideas about existence in regard to human creativity ie they go to the heart of describing the creative relation that (human) life has to existence.

They are ideas which both take from the culture (they use authoritative ideas) but (and) they are expressed in the simplest of concepts, which are derived from the idea about reliable measuring and stable patterns and the relation of reliable measuring to quantitative sets

And

Why they are not considered by society is explained, basically because society is a narrowly defined empire ruled by a few.

The US is an extension of western civilization, and that civilization was (still is) the roman empire, where between 900 AD and 1400 AD the emperors passed the baton to the merchants and bankers, where the likes of the Medici family, and their aide Machiavelli, defined the ruling class in the 1400's in the urban city-states, and through the church had great influence over the other feudal kingdoms of Western Europe.

The US began as colonies of Western Europe which sought, and were "granted," freedom and equality (if they wanted), but being human products of primarily Roman civilization they expressed their freedom with violence and subsequent inequality, though there are exceptions, with the Quakers being the most notable exception. The Quakers remained equal "friends" with everyone, including with the native people, and the Quakers were the most energetic, creative, and thriving colony, especially, around the era of W Penn.

The American revolution was indeed about equality, where it was proclaimed in the Quaker colony and in the Quaker city of Philadelphia, and freedom was defined around knowledge and creativity, and the second amendment was about not having a standing army, that is, policing is not to be based on the violent advantages of armies, but nonetheless the expression of the "violent advantages of armies" model of a justice-system is what occurred anyway, quite often in the name of an arbitrary Puritan-like culture, and the avarice involved in the extermination of the native-peoples so as to steal their lands and plunder their riches.

Thus . . . , between (1) the violent armies wherein wage-slavery was imposed by a justice system (which was also an army) and (2) the dominance of the narrowly defined activities of banks . . . , the experts of knowledge were no longer capable of expressing truth, but rather they have competed so as to be able to service the technical instruments of a banker's empire.

That is, since WW II, and because of the incompetence of Truman, the US has whole-heartedly embraced the form of the Roman-empire, since that is what the arbitrarily defined structure of capitalism actually was, after the Bill of Rights has not been enforced, since without the Bill of Rights the US was then based on property rights and minority rule, as well as possessing a standing-army, ie the foundation of the Roman empire, and a merchant-banker's empire was significantly expressed by A Lincoln, who joined with the Northern corporations to defeat the south, the choice between Scylla and Charbydis, but, earlier, when president G Washington put-down "Shay's rebellion," a rebellion against the banks foreclosing on revolutionary-war veterans, the direction of the US leaders was fixed, there would be standing armies, and thus, a new Continental Congress was already needed to right this injustice.

The thoughts expressed in these papers, if others with the credentials of empire (defined later*) expressed 1/10,000 of the content of these expressions, they would be hailed by the media as the highest illuminati of the land, (*where the credential-of-empire are related to the fill-in-the-blank social-structure which relates "the allowed knowledge" . . . within the empire . . . . to the creative actions of society (the creative actions which support the power of the owners of society)).

So, perhaps, others (eg readers of these papers), who are seeking knowledge, should "try real hard" to understand the ideas expressed.

They are ideas about existence in regard to human creativity ie they go to the heart of describing the creative relation that (human) life has to existence.

They are ideas which both take from the culture (they use authoritative ideas) but (and) they are expressed in the simplest of concepts, which are derived from the idea about reliable measuring and stable patterns and the relation of reliable measuring to quantitative sets, where the quantitative sets have an axiomatic structure associated to themselves; concerning the operations of adding (counting) and multiplying (grouping); and they are also ideas about the subsequent (needed) ideas about local independence, in a many-dimension containing space, (ie they are the ideas about algebra, condensed and reduced to their simplest functioning forms, in regard to derivatives (or other function-operators), especially in regard to basing descriptions either on functions or function spaces [and associated sets of operators]).

It is clear that the mathematicians do not understand the simplest functioning forms which are applicable in regard to, quantity and shape, so that measuring is reliable and the patterns described are stable.

 A space which contains an observable pattern (ie a stable system) and the properties which the system can have, based on its measurable containment context, where such a containment set is used to describe the stable measurable properties of such an observable system.

That is, the formulas associated to operators . . . , where the operators include models of reliable local measuring, and their inverse (integral operators) . . . , defined in regard to local measures of functions, ie local slopes, and thus defined on both functions and (the inverse operators of local linear measuring are about a local model defined as) the sum of local products of domain small-intervals multiplying a linear {approximation of a} measure of a function's graph (or a function's properties).

Note: numbers are always types and thus depend on a context, upon which the definition of a function depends, since a function maps the system-containing coordinates onto the measurable properties of the system.

The simple idea is that a limit structure (related to determining the "local" slope of a function's graph) ensures a reliable local linear relation between function-values and domain-values, but such a reliable quantitatively consistent relation is a possibility only if there is global commutative, or independent, set of such relations for each local coordinate direction at all points of the system, in the system-containing coordinate-space.

The geometric sets, which properly model quantitative sets . . . , whose shapes allow for consistent quantitative relations . . . , are circles and lines ie circle-spaces and "cubes," but (where) the cubes are highly related to the circle spaces.

The traditionally developed ideas, which are consistent with this simple analysis, are the ideas of geometrization of Thurston-Perlman, which explicitly states that only a few shapes are stable, and where the most numerous type of stable shape are the circle-spaces, more explicitly the discrete hyperbolic shapes or hyperbolic space-forms, but the "discrete Euclidean shapes," ie the tori (or doughnut-shapes), are stable, but not discrete in regard to their allowable geometric shapes, ie they can have a continuous range of sizes.

Thus, the ideas of quantity and shape (the content of math) so that there is both reliable measuring and stable patterns which can be built-upon is to be about circle-spaces and their higher-dimensional structures, where by higher-dimensions one means higher-dimensions than the dimensional-structure implied by the idea of materialism.

Where do stable spectra come from?

Answer: They are defined by their resonance with a finite spectral set upon which existence, within a context of reliable measuring and stable patterns which can be reliably described, is defined.

The finite spectral set comes from the partition of an 11-dimensional hyperbolic metric-space by a finite set of discrete hyperbolic shapes so as to partition the different dimensional levels and the different subspaces of the same dimensions.

Why 11-dimensional hyperbolic space (or 12-dimensional space-time space)?

Answer: Because of the patterns identified by Coxeter concerning the discrete hyperbolic shapes of the various dimensions he expressed the idea that there do not exist discrete hyperbolic shapes which are 11-dimensions or higher, and that 5-hyperbolic-dimensions is the last discrete hyperbolic shape which is bounded (and closed).

Furthermore, the 2- and 3-dimensional discrete hyperbolic shapes are fairly numerous and varied in their genus and the size relations which can exist between their toral components, but the 4- and 5-dimensional discrete hyperbolic shapes are fairly limited in their allowable shapes.

The partitions of the 11-dimensional hyperbolic metric-spaces by a finite set of discrete hyperbolic shapes can have various size relations associated to the different dimensional levels, where the relative sizes of shapes defined between different dimensional levels is controlled by the multiplication by a constant factor which is defined between (adjacent) dimensional levels . . . , as is done in regard to similarity relations between triangles in the 2-plane.

Thus, by the size relations of the partition-discrete-hyperbolic-shapes . . . , which is based on a finite set of discrete hyperbolic shapes of various dimensions and sizes . . . , various types of containment trees can be defined in regard to sequences defined in regard to increases of dimension of the sets, in the containment-tree, within a containing 11-dimensional hyperbolic metric-space.

Life and mind

There is the further patterns of odd-dimension discrete hyperbolic shapes which also possess an odd-genus, where the genus is the number of holes in a circle-space shape where a torus a discrete Euclidean shape also thought of as a doughnut has 1-hole so it has a genus of one, when its stable spectral flows are occupied with charged components, then the shape has a natural charge imbalance, and it naturally begins to oscillate, so as to generate its own energy. This natural energy generating shape, is such that, each shape is associated to a fiber group, eg a unitary fiber group, so that these (Lie) fiber groups always possess a maximal torus, within which the spectral properties of an experience (sensation, or detection of form) can be formed, and stored, by resonances between the existing spectra and the spectra which the maximal torus can carry within itself, ie this is a simple model of a mind, and the oscillating system along with its associated mind is a simple high-dimensional model of life.

These ideas exist outside the confines of the idea of materialism, and in the new descriptive context, the first life-form could be a 3-dimensional discrete hyperbolic shape which is contained in a 4-dimensional hyperbolic metric-space.

The 4-dimensional Euclidean metric-space, ie the space wherein inertial properties are defined, has a fiber group of SU(2) x SU(2), and thus this fiber group would have a 2-dimensional maximal torus, which can carry 1-dimensional spectral properties, and it also causes a dynamical structure associated to two separate 3-sphere interaction geometries in 4-Euclidean-space, which could also be used to separate the stable space-form material interactions from the oscillating space-form material interactions, where the stable material is in regular 3-space and the oscillating space-form material would be in 4-space, which is different from the regular 3-space (this could be a model of a life-form's energy-body).

Why are these ideas not explored more energetically?

This is a bigger revolution of thought than was the revolution of Copernicus, and it relegates the ideas of the 20th and 21st century sciences to small, often irrelevant, sub-categories, while these new ideas expand the context of Newton's and Faraday's classical models of measuring and linear quantitative consistency, in particular the realm of the linear, solvable, and controllable physical systems described in the general context of classical physics are expanded, wherein stable geometry is the basic pattern about which precise descriptions depend.

It could be said that these ideas not explored more energetically because it "takes time to change, in order to process information," but this is clearly not true, in a society where information can be acquired and processed so fast.

rather

It is the result of the sultans of modern business-social control model of society

That model is based on materialism and propaganda, and controlling people through the institution of wage-slavery, and dividing people within society based on a narrow set of society's defining fundamental categories . . . . of allowed human thought within society . . . , in the language of opposites:

Science vs. religion
Intellectuals vs. working persons
Authority and discipline vs. weak-minded and incompetent
Equality vs. violence and inequality
Creativity vs. traditions

As well as holding society to stay within narrow ways of doing things, organized to serve the interests of the ruling class, and done with arbitrary-value defined for society by the propaganda system and assigning to these categories of high-social-value a very complicated technical language (used to identify the experts within a category of high-social-value), and wage-slavery, and violence. So by using the language of experts one is agreeing to a set of assumptions which require that one's viewpoint (about the social category within which people act) remain narrow, and if one tries to question the arbitrary value within which (upon which) an expert category is to be conducted within society, then one is excluded from being in the category

Furthermore, it is the "sultans of society" (the dominant few) who control, within the social construct of experts who are also wage-slaves, both sides of this set of dichotomies.

The funny thing is, it is the system under the control of the sultans which has consistently demonstrated incompetence and weak-mindedness, which characterizes the set of people expressing the ruling authorities of our society, but there are extremely violent generals and judges (in a violent justice system) who can support the current authority. The institutions keep rolling-on, since they are deemed, "to big to fail." and they define the narrow categorical way of organizing society so that the society is all about supporting the interests of the dominant few, who demand that "the world be viewed so narrowly" in a highly compartmentalized division and control of society.

That is, math departments and physics departments at public universities are already doing the things which best support the interests of the banking empire, so new ideas are not to be expressed or considered.

36.

Science is politics

My main point is that science and math, have a "real relation to things" which is most concerned about practical creativity.

What I write, is really in a very simple and a very educational style,

[it exists in this form after a lot of discipline]

All language is controlled by the owners of society, who are the paymasters of the wage-slaves, where these wage-slaves foolishly and arrogantly help sustain the entire delusional structure . . . , essentially stopping the wide range of creativity which mankind is capable of realizing.

Society is an example of self-sustaining system of illusion.

The ruling class harnesses many social forces in order to achieve such wide-spread delusional beliefs:

arbitrary high-social-value (propaganda-education),

money systems controlled by banks (who in turn define social-value by their investments), and

extreme violence, which is mostly subtle and institutional, it exists but it is hidden, its institutional nature requires that people internalize the terror which these institutions impose on them, the people.

The core problem is law, where the US society's law is based on property rights and minority rule (the legal basis of the Roman-Empire), though the US's historic documents proclaim a basis in equality and freedom,

Whereas the ruling class say:

people have an "equal opportunity" to compete for high-social-value within the language controlled by the ruling-class,

That is, the freedom of the people is defined by their willingness to help the ruling-class become ever more powerful, so as to help along the lines of high-value, which the ruling class is defining.
and
People are free to get loans, ie and be creative, but the ruling class must OK these loans (thus they [the bankers] are able to filter-out anything which might be of significant competition).

Thus, the agents of empire, those who are given a literary or research-position, are narrow and quite arrogant and coercive in regard to the ideas that they express, as well as the ideas which these people . . . , who are, effectively, "media agents," . . . believe to be true.

37.

Harmonic analysis

Introduction

The hyperbolic metric-space (or equivalently, the space-time metric-space) within which the equations of a material interacting system are to be (can be) defined, is (itself) a stable discrete hyperbolic shape which identifies an envelope of the orbital-spectral stability of the material components which compose the interacting system . . . (this is (or can be) the model of the stable solar-system (or a stable atom), a stable orbital geometric-structure which the equations of "material-component interaction" [or equations of statistical organization ***] cannot identify, since these equations are [either] non-linear {thus quantitatively inconsistent}

[*** or indefinably random equations, which are built upon harmonic functions, which are oscillating functions, which can effectively be defined as a set of functions defined on circles, since Fourier analysis is quite similar to the L^2-function space (where L^2-functions also allow such functions as "the oscillating part of cut-off polynomials," but this is, effectively, non-linear sets of oscillations, which are still related to shapes characterized by their genus-number {geometrization shows that these are, mostly, unstable shapes, which, often, evolve into the stable circle-spaces})]

{the energy structures and the angular-momentum structures, in turn, based on spherical symmetry, are not sufficient math structures which can identify the stable spectral properties of general systems, which this method, based on indefinable randomness, is trying to describe (or identify)}).

This inability to identify the stable spectral-orbital properties of observed systems has resulted in the idea of such system's (assumed to exist) containing spaces be first, nullified (or ignored), then corrected, and then re-introduced back into the original containment space.

These are the ideas of:

1. Feedback systems (chaotic action is forced back into the original quantitative setting by feedback processes) and
2. Hartree-Fock (and Slater-determinant) atoms (ie do an averaging process with a normal-distribution, for charged and spin-½ particles around the atom's nuclear-center, until one finds a data-fit, but such data-fits are not (or seldom) found), and
3. potential-wall for nuclear containment (where the potential-wall is introduced in both an ad hoc, and data fitting, manner, where the potential-wall is vaguely, related to the strong-force of particle-physics, but with no valid quantitative models for such a potential-wall) and
4. particle-physics most notably the incompatible "highly geometric particle-collision, but randomly based" descriptive structure (leading to the radical need for claiming "space breaks-down in the small regions identified by particles, and particle-collisions," and then renormalize along the parameter of mass so as to re-adjust the correcting perturbing (math) construct, then returning the system (the wave-function, but defined with particle-states) to the containing-space one originally left), and
5. string-theory, which is based on geometry and on an arbitrary incorporation of materialism, by making higher-dimensional shapes have extremely small micro-geometries, but it is a geometry which is still consistent with randomness, at least in the context of materialism, where a form of continuity, in regard to all material interactions in any dimensional structure, is assumed, but such an assumption of continuity has no basis . . . , it is an arbitrary assumption.

That is, in a discrete context (ie quantum physics), math operators which relate discretely (and discontinuously) between both (1) different dimensional-levels and (2) between small time-intervals, might be a much better model for the actual containment set (for existence).

Etc.

The new model identifies the:

1. cause of apparent randomness of micro-systems, and
2. the reason for the appearance of point-particle interactions, and
3. the reason for a spherically symmetric force-field's relation to inertial properties in 3-space,
4. etc.

The, above mentioned, existing stable metric-space shapes are in resonance with some subset of the set of spectral values of the finite spectral-set, in turn, defined by the "partition" of an 11-dimensional over-all containment hyperbolic metric-space.

The finite spectral set is defined by a partition which identifies each subspace of the 11-dimensional hyperbolic metric-space with a (largest) stable discrete hyperbolic shape, one such stable shape for each subspace, so that each subspace of each dimension has a largest shape, so that, that largest shape for a subspace identifies that particular metric-space shape (defined on a particular subspace with a definite dimension) as being a part of the partitioning-set of the 11-dimensional hyperbolic metric-space.

There are $2^{11}= 2048$ stable shapes in the partition of the 11-dimensional hyperbolic metric-space. However, due to the dimensional properties of discrete hyperbolic shapes only 5-dimensional discrete hyperbolic shapes can be bounded shapes and there are [11C1 + 11C2 + 11C3 + 11C4 + 11C5] = [11 + 55 + 165 + 330 + 462] = 1023 subspaces which are 5-dimensions or less (excluding the two subsets, which are 0-dimension, and the 11-dimension set, itself).

Now each stable discrete hyperbolic shape can have various numbers of holes in its shape (a number called the genus of the shape) and each hole can be related to different spectral values. This is especially true for 2- and 3-dimensional discrete hyperbolic shapes. The 4- and 5-dimensional discrete hyperbolic shapes might have other restrictions on their geometries, eg limits on their genus numbers (see D Coxeter).

This means that the finite spectral set, which defines what can exist, can be quite a bit larger than its number of subsets, though if the partition is defined to be only uniformly regular fundamental domains, so that each such uniform shape is only associated with one spectral value, then 1023 would be a lower bound on the finite spectral set.

Thus a hyperbolic metric-space with a certain size and shape can contain within itself any set of shapes which are:

(1) 1-dimensional less than the given metric-space shape,
(2) smaller than the given metric-space size, and
(3) their spectra are in resonance with some subset of the finite spectral set of the 11-dimensional over-all containment hyperbolic metric-space.

Each of these smaller, stable shapes are metric-spaces, and these metric-spaces can contain within themselves smaller and 1-dimension less and resonating metric-space shapes, too.

Inside of each metric-space the equations of material interactions can be defined. These equations are mostly non-linear, but stable constructs (of systems) are either orbitally constrained by their containing metric-space shape, or the collisions of material components can be related by "resonance and energy relations" to new stable, 1-dimensional greater, discrete hyperbolic shapes.

The smaller, lower-dimensional components, which a metric-space can contain, can either be (condensed) components, or they can fit (by resonance) into the faces, of the fundamental domain of the metric-space within which these components are contained, and thus be a spectral-flow of the shape.

Harmonic analysis, main part

If one compares harmonic-analysis with the model of a many-dimensional set of circle-spaces, related to the containing space by resonances, then one sees:

"on the circle-space side" a set of stable patterns, but it is a many-dimensional model, but wherein the idea of resonance allows for a wider-range of possibly relevant spectra, ie spectra which are made relevant by the idea of resonance of the dimensional level's relevant circle-space shapes, ie the resonance of the observed system, with the over-all containment set's finitely defined spectral set, while "on the harmonic-analysis side" the first level of harmonic analysis is also about circle-spaces, namely, a trig-series (or Fourier analysis) (but unrelated to a (stable) definite model of the circles, which the harmonic values are related), while at the second level one has the idea of an L^2-functions, ie functions which have a cut-off radius . . . , so this would include the trig-series (or Fourier series) type functions, as well as polynomial functions, where polynomial functions have the properties of oscillating functions in a bounded region around zero, where polynomials with such vaguely oscillating properties are the focus of algebraic geometry, . . . , where the radial cut-off of such (oscillating) functions is needed to ensure that the integral of "a function-squared" converges (or identifies a specific value).

One sees that, though polynomial oscillating functions may be relevant in the study of why non-linear feedback systems which seem to have ranges of relatively-stable applicability within a quantitative, or measurable, setting (context) [since polynomials have the structure of both numbers, and the structure of relevant orders of magnitude, where orders of magnitude are relevant to a measurable property {ie a function's values} of a system (or math pattern)] but their non-linearity actually makes them unstable, and quantitatively inconsistent (ie chaotic) and uncontrollable patterns of very limited practical usefulness.

The main issue of math description is distinguishing when a system (an observed measurable pattern), which is related to a measurable shape, has stable controllable properties, ie a description which can be used in a controllable and measurably reliable context (as opposed to feedback systems), and when its observed vague patterns are not relatable to control and subsequently not relatable to practical use.

This distinguishes between:

(1) controllable, and practically useful, patterns of (observed) existence
    vs
(2) unstable patterns, which cannot be used in a context of control (of a [stable] system's properties).

That is, one wants to use the observed stable properties of systems in (practically useful) creative new ways.

It seems that both the nucleus and the atom are (can be) 3-dimensional discrete hyperbolic shapes, where an oscillating, unstable nucleus is most likely a 3-dimensional discrete hyperbolic shape.

This would mean that our existence (characterized by inertial material interactions) is dominated by our being in a Euclidean 4-space. Thus, electromagnetism is the main force holding material system's together, but we observe primarily the inertial results of material interaction.

However, the fiber group structure of Euclidean 4-space is

SO(4) = SU(2) x SU(2)

Which can separate the two different materials which exist in R(4,1). Namely, the odd-genus and the even-genus circle-spaces shapes, and relate the two types of inertial interactions to R(4,0) and then separate them into two separate material-containing subspaces, so that we see the inertial properties of the even-genus charged-system shapes, in regard to the R(4,0) material properties, ie the material properties of inertia, ie material's relation to position and spatial displacements.

Furthermore, the tensor structures of geometric-measures, ie the differential-forms, have different arithmetic relations on the odd-dimensional vs. the even-dimensional containment metric-space sets, of a material system's description, so that the even-dimensional containment-spaces fit more naturally into an adjacent lower-dimensional spaces (of odd-dimension).

Furthermore, the fiber group structure of SO(4), associated with base space R(4,0), allows such a arithmetic relation of the differential-forms to be more noticeable (or more important) in R(4,0).

It is the infinite-extent space-forms (or unbounded discrete hyperbolic shapes) of dimensions 2-, 3-, 4-, and 5-dimensional hyperbolic spaces, and their relation to bounded (or semi-bounded) discrete hyperbolic shapes in the same dimensional spaces, which have a relation to various subspace and component properties, and their possible relation to other angular momentum systems in either different dimensional levels or in different 11-dimensional hyperbolic containment metric-spaces (each 11-dimensional space defined by its own finite spectral set).

Thus, there are issues about bounded-ness and unbounded-ness of components and their adjacent higher-dimensional containment metric-spaces, and the shapes (or bounded, or unbounded properties) of these containment metric-spaces.

That is, there are different types of questions which can be asked in different math models of containment, and the math models which are based on "stability of shape" seem to be the better spaces within which to frame questions in relation to descriptions of existence which can be practically useful.

Social aspects of intellectual activity

The current descriptive dogmas are designed to answer questions concerning the rate of nuclear reactions, and the relation of these rates to any possible model of particle-collision probabilities.

This is a very narrow viewpoint, concerning measurable descriptions and their relation to practical usefulness. It seems to be a viewpoint {of an abstract and wrong (deceptive) viewpoint [which has been captured by a use in regard to making bombs]} about linking high-social-value with narrow world-outlooks, which can only considered in a mental context of obsession, with no vision of reality, except an assumed absolute reality associated with a dogma.

This is exactly what the Godel's incompleteness theorem . . . , about the limitations (or inability) of precise language to identify patterns . . . , addresses.

Namely, the need to base language on new contexts assumptions and interpretations, etc, so that the descriptive patterns of a measurable context are practically useful.

Some techniques for manipulating social forces

Yet the accepted dogmas of obsessive experts, ie a community of carefully selected autistic types (who want to possess models of obsessive detail), is defended with the same fervor as the political-economic vision of social organization, with a viewpoint that mankind is fundamentally a sinful animal (ie an incapable, ignorant, violent, and selfish animal), best controlled by an enlightened few.

But in reality it is a society being controlled by the narrow limitations and extreme domination of a small set of ignorant few who are pleased to be selfish and who defend their social positions with psychopathic violence, ie it is far too narrow an outlook for the creative natures of human being . . . , (ie human beings whose creativity have been bent and distorted by the evil social constructs within which they are governed) . . . , and such narrowness and required violence leads to destruction (central planning for the selfish [destructive] interests of a few).

The institutions of propaganda and law (government, religion, and education) are all filled with the types of dominant individuals who are arrogant and murderous in their deceit (to follow the social analysis of the US provided by M Twain).

That is, the context of the corrupt emperors of Rome is still the context of western culture, but they have come to an end in regard to their gifts of engineering which they can provide to a community, ie the brick-laying (as preformed by the Roman military) are too limited in their vision, and now the current emperors, apparently, see only themselves, in the abstract language of delusion, ie an economy and a knowledge which is failed.

So that knowledge and creativity are all about people being controlled to serve the selfish interests of the few.

One cannot see anyone in the media and in the justice system and in the upper reaches of the education system who are not fatally flawed with their self-centered arrogance

The surveillance state's apparatus could be used to end the "failed top-rungs of the controlling social hierarchy," and instead control their violent and failed (intellectual) minions . . . (this could possibly be done by controlling a handful of people), . . . , so as to transition to a state where law is based on equality and the freedom to believe and to express beliefs with an eye towards creativity based on selflessness, where a truly free market could be realized, so that equal free-inquiry can lead to ever more creative capacities of a human society, so as to realize the original intent of the US revolutionary war.

The above discussions about math and science are about the fundamental math properties which are associated with the construction of measurable descriptions, wherein the patterns described are stable and practically useful.

38.

Angular momentum

Introduction to Angular momentum

A re these ideas useful?

The world is not determined by material existence . . . where your comment suggests that existence is based on materialism, . . .

However, these new notions about precise descriptive language are also about the limits of language . . . a language must always reflect on itself . . . .

One wants reliable measuring in regard to descriptive containment (of measures and derived properties, or functions) and one wants to use math patterns which are stable, and controllable, when placed into a context where measuring is reliable.

Then descriptive information can be used.

In fact, this may be a definition of meaning . . .

But the lack of usefulness is the main criticism of: quantum physics, particle-physics, general relativity and all theories derived from these math models of "a random, non-linear, material world," eg string-theory, that is, all of these theories are practically useless.

That is,

"Virtually all useful technologies are derived from the ideas of classical physics and its fundamental relation to geometry, and stability, and subsequent control."

Eg even micro-chips are made with both thermal processes and optical processes etc.

atomic bombs are based on 19th century models of chemical reactions, coupled to the property that there is a "critical mass" for radioactive substances.

These above mentioned, but accepted as authoritative, physical theories cannot describe the observed stable properties of: nuclei, general atoms, molecules, crystals, solar-systems, dark-matter etc . . .

. . . . . Whereas the new descriptive context can describe the stable properties of these systems in an (the) "active context of cataloging the finite spectral set of a (particular self-contained) model of existence."

In this age, the "capacity to describe vague patterns" is considered a use, but, it might be claimed that, this is similar to religion having a use.

Namely, religion is used to deceive people by the use of a descriptive language so that this language is backed-up by violent terror.

The Roman emperor Constantine linked violence and a hierarchical society to propaganda, ie the practically useless descriptive language of religion.

This deceptive use of a practically useless language and violent terror in a hierarchical society is, today (2013), what the media is all about, as the pinnacle of the social hierarchy, attacks the lower classes, all based on an arbitrary idea about high-social-value.

In 4-space (or 5-space-time) the resonances of naturally oscillating energy generating physical systems, such as radioactive material can be tapped-into as an energy source, by means of "circuits," in regard to ideas which are similar to the ideas of what Tesla may have considered, along these same lines of thought, planets and stars also depend on such oscillations as their energy sources, which can also be tapped-into, ie if done in regard to the earth's oscillating energy-dynamo, then this would provide clean, cheap energy sources.

The angular momentum links within a stable discrete shape, or equivalently, the shape's fundamental domain has relationships to its lower-dimensional faces, and these relations of an associated control over spectral-resonances can be used to model "a central macroscopic-system" "a command-control system with a hierarchy of controls," where this would be a good model for a living-system, so as to have such "central hierarchical control" over the living system's glands (subsystems) and molecular properties (spectral resonances).

The higher-dimensional models of life, for example, some life-forms (such as human life-forms) are 7- and 9-dimensional, and thus they are unbounded. Such systems are centers of light (not as much material positions), and these relatively stable systems link, by angular momentum paths, such a life-form to a wide range of spatial and temporal relationships (as well as relations between lower-dimensional shapes) so as to allow both "communication connections," as well as "travel connections (or links)," where such links are stable, linear, and controllable set of patterns.

These are the types of links, which the currently accepted, but un-useful, vision of science descriptions calls worm-holes, but the current math models of worm-holes are unstable and non-linear and quantitatively inconsistent, they are models of systems which cannot be used, because they are unstable and chaotic models.

The current viewpoint about science and materialism concerning highly complicated descriptive patterns, which are unstable and quantitatively inconsistent patterns, are used in a context of illusion, and absurdity, veiled in complication, and mired by their relation to illusions.

It is useful to try to extend knowledge beyond such a useless descriptive context.

However, the new descriptive context describes the most fundamental context of existence encompassing both life and material, and the context ranges over many-dimensions . . . , wherein size can have various dimensional and set-containment relationships . . . , so the main use of this new language is providing the simple, but rigid, context of all of existence, so that one is able to consider "how one can expand, by creative effort, the current state of a rigid, limited existence,"
[but an existence, to which we have lost such knowledge of our "true relation to existence"]

The natural goal of human creativity:

"how to expand existence in regard to its simplest and most fundamental properties?"

Angular momentum, (main part)

Both the radial equation, and other orbital-spectral shapes, of stable physical-systems are related to properties of angular momentum whose assumed spherical symmetry is disrupted by the shape of the metric-space within which the equations for the physical system's material components, and/or the system's angular momentum, are defined. That is, these physical systems also happen to be metric-spaces, which have shapes, and which are, themselves, contained in higher-dimensional hyperbolic metric-spaces.

The spectra of systems with stable orbital-spectral properties which are made of charges (or condensed material) depend on solutions to geometrically separable equations related to both a system's radial equation and the system's angular momentum properties, where both of these sets of equations, which are separated from one another, are (now to be) associated to (or defined within) the very stable circle-space shapes, which model the containing metric-space of the system's material components (where angular-momentum is not simply related to spherical symmetry), and these equations are defined in a many-dimensional context, where each subspace of each dimensional level is given a metric-space shape (where the higher-dimensional shapes can be either macroscopic or microscopic sizes), and the shapes are always the very stable patterns of circle-spaces (or discrete hyperbolic shapes). Without such a geometric structure associated to a dimensionally related set-containment tree of an 11-dimensional hyperbolic metric-space, the observed stable patterns of physical systems cannot be identified mathematically, but the (new) context of measurable description focuses on the envelopes of stable orbits, which are the set of metric-space shapes which are resonant with the given set of metric-space shapes, and these resonant metric-space shapes define these systems and their observed stable properties, and so that the confinement (or the bounding shapes) of such a system is (topologically) an open-closed metric-space, which define the envelops of orbital-spectral stability for the fundamental stable physical systems, eg nuclei, atoms, molecules, solar-systems, etc.

Both, relatively large (but bounded) and unbounded shapes of hyperbolic metric-spaces, and of higher-dimensions, allow angular-momentum properties to inter-connect both lower-dimension as well as higher-dimension metric-spaces (which model stable systems) and which possess stable shapes.

If a stable system, which conserves energy, is a discrete hyperbolic shape, where a discrete hyperbolic shape is a shape which is composed of toral components, and, in turn, these toral components are related to one another by various angular orientations (or the toral components angular relations to one another) where these angular relations are determined by discrete Weyl-angles___, where, in turn, Weyl-angles are defined to be related to a finite set of conjugate-actions which exist between maximal-tori of a fiber Lie group, . . . where these different maximal tori cover the Lie group, and each maximal torus can be transformed into one another by these Weyl-group (group conjugation) actions associated to Weyl-angles . . . the set of maximal tori cover the entire fiber Lie group (of a metric-invariant metric-space, where the metric-space is the base space of a fiber bundle, which has a fiber Lie group),

Note: The discrete isometry subgroups of the fiber groups of SO(s.t) (or SU(s,t)) identify lattices on the base space of the fiber bundle, ie a regular arrays of vertices defined on the metric-space (similar to the set of vertices on a checker-board), where these set of vertices, in turn, define the fundamental polygonal shape (similarly, these would be the squares on the checker-board) [or polyhedral shape, depending on dimension of the metric-space] which, in turn, determines the stable circle-space shape, which are the set of stable shapes naturally associated to a metric-invariant metric-space whose fiber group is a (classical) Lie group . . .

___, then the (inertial-dependent) angular momentum of such a system would be defined by the (axial) vector sum of the angular-momentum of each toral component, so as to define an axial rotation-vector for the system, (yes?), as well as the spin rotations between metric-space states (defined between orthogonal spectral flows which exist on the discrete hyperbolic shape).

One question is, "Can the set of spectral values of each of the toral components be determined from angular momentum properties, or are the Weyl-angles (of the system-defining discrete hyperbolic shape) set-up so as to define a single 2-plane orbital structure, which is similar to the Bohr-Sommerfeld H-atom (ie elliptic orbits defined within the toral-envelopes of orbital stability), but now with (what seem like) a set of extra non-orbital toral components (positioned between separate concentric toral shapes), which are needed to separate the different (approximately) circular sets [of a concentric-circular-shape of an orbital structure (or concentric toral shapes)] in regard to the system's discrete hyperbolic shape (where the concentric toral shapes [similar to concentric circles] define either the stable orbital envelopes, or the stable spectral-flows, which are occupied by a set of "correct sized charges," ie charges the size of the spectral-flows of the stable physical system)?"

Since angular momentum is an axial vector (anti-symmetric 2-tensor), wherein the different axes of distinct planes of rotation, eg associated with the orientation of the different toral components, are summed to get a single axial-vector of rotation (in turn, defining a single axial vector direction, which is associated to a single 2-plane of the system's conserved rotational properties) can the spectra of the various different toral plane orientations be recovered?

or

Does it become quantitatively inconsistent, in a non-commutative relation which exists between the different axial-vector-components, associated to separate toral components of the system's discrete hyperbolic shape, which are part of the vector-angular-momentum sum?

A definitive sum may require that the different toral components, which separate the concentric toral shapes of the system, be at right-angles to the concentric toral components.

On-the-other-hand a material system's moment of inertia is symmetric, thus diagonalizable (in regard to a local set of coordinate directions), thus defining a local coordinate-vector to which the rotation-axis can conform. Thus, the axis of rotation is best related to a moment-of-inertia tensor, which, in turn, is associated to a set of circles, wherein each circle represents a toral component of the system's discrete hyperbolic shape.

The underlying math-issue, in regard to this discussion, is the relation of function-spaces to sets of operators, and whether there actually exist sets of operators which have the capacity to diagonalize a function-space, (the usual quantum-system models of angular momentum). Apparently, if the assumed symmetry is rotational symmetry, as opposed to shapes related to circle-spaces, the answer is, No.

That is, should the operators really (do they need to) be a set of discrete hyperbolic shapes (ie circle-space shapes) and the descriptive context be contained in dimensionally-related "trees of set-containment," as well as (needing to know) the finite spectral sub-set, which defines the particular system (whose stable shape is a circle-space). The finite spectral sub-set of a system, modeled as a circle-space shape, depends on separation as well as both a solution to a "radial equation" and a valid geometric model of the system's moment-of-inertia, ie its angular momentum related to a simple, quantitatively consistent geometry, ie its circle-space shape.

Other issues in relation to a system's set-containment structure

However, there is also the issue of whether the system of concern is modeled as being contained within a particular dimensional level, from which the system's higher-dimensional structures, ie angular-momentum structures are difficult to determine. That is, the angular-momentum structures are difficult to determine from the properties of a system, when the system is modeled as being within a particular dimensional level, ie the higher-dimensional relations (of angular-momentum) are lost.

Furthermore, there are also the issues concerning the (hyperbolic space-form) model of the observer and its dimensional properties, and the relation of a material system to the dimensional structure of an observer, and a subsequent (limited) interpretation of a system by an observer, so as to identify the effects of an observer in regard to determining, both a containment set, and the set of set-containment trees, which are relevant to a system's description.

How is mass conserved?

If energy is conserved for systems (whose energy is invariant to time displacements), then the 4-space momentum-energy vector implies that mass is also conserved, eg mass = (k) x (energy), but these are independent spaces, ie Euclidean space and hyperbolic space, so what is the mechanism of mass conservation?

It must be related to the dependence of masses (inertia) on their toral shapes being in resonance with a stable discrete hyperbolic shape, a shape which best models a stable (energy-invariant) system based on charge.

However, From the above paragraph one must ask:
"Is this a planar Bohr-Sommerfeld-orbital shape?"
or
"Simply a sum of the energy of each toral component of the system's discrete hyperbolic shape, ie line integrals of force-fields for each toral component?" *?

or

"Can it be related to an axial vector, based on (built from) a non-commutative axial-vector sum, which is, nonetheless, related to a diagonal moment of inertia tensor?"

The answer seems to "need to be" a Bohr-Sommerfeld-type orbital system-structure, at least for atomic components, eg molecules would be more complicated (with several different nuclear positions in the geometry of a molecule, where each nuclear position in a molecule would be related to a 2-plane associated to that nuclei's atom and that atom's concentric toral components, but each 2-plane could have different angular orientations to their adjacent 2-planes [or nearest neighbor nuclei]).

However, the "non-orbital toral components," . . . which are needed to separate a concentric-circular-type orbital structure of the system's discrete hyperbolic shape . . . ., may add some variation to the energy and mass distribution within a charged system. These non-orbital toral components may have Weyl-angles which are not planar (in particular, they might have an orthogonal orientation to its associated atom's 2-plane orbital-structure), so as to discourage the existence of orbital instabilities associated to the non-orbital toral components.

Is the relation between inertia and energy ultimately to be found in relation to a regional 2-plane to which all inertial properties of the region are defined? Is this 2-plane either the galactic 2-plane, or the 2-plane of the solar-system, or the 2-plane of the moon's orbit, or the 2-plane of an atom etc?

Modeling atoms as 3-dimensional "discrete hyperbolic shapes" and the "unbounded" neutrino

The art of mathematics is about finding stable patterns which approximate a system's measurable properties and which can also guide the system to a state of very precise observable properties.

The nucleus and the electron-clouds are 2-dimensional discrete hyperbolic shapes, but they fit together into atoms, which seem to be 3-dimensional discrete hyperbolic shapes (also called space-forms). Furthermore, these 3-dimensional space-forms are smaller than the size of the discrete hyperbolic shape which defines such small shapes in our subspace, namely, the discrete hyperbolic shape which determines our stable solar-system properties. Thus, the 3-dimensional discrete hyperbolic shapes, into which atoms are formed, are in resonance with 3-dimensional discrete hyperbolic shapes, which are 3-dimensional subspaces which are different from the 3-dimensional subspace of our solar-system.

The stable spectral-flows of 3-dimensional hyperbolic space-forms are the 2-dimensional discrete hyperbolic shapes. Thus, the atom has the geometric-structure of orbiting-bodies of what are (like) condensed material in orbits, where the condensed material are (also) 2-dimensional hyperbolic space-forms, which are charged either as positive of negative charges, and these condensed 2-dimensional space-forms also possess the properties of having neutrons and neutrinos as a part of the charged composition of these 2-dimensional space-forms which are contained (as condensed material orbiting) within a stable orbital envelope of a 3-dimensional discrete hyperbolic shape (Note: The vertices of the infinite extent neutrinos would converge to the distinguished vertex of the [bounded] 3-dimensional hyperbolic space-form within which they are contained [ie within which their orbits are defined]). These 2-dimensional space-forms, which are also charged orbital components, would be forced (by the hyperbolic-coordinate geometry within the 3-dimensional hyperbolic space-form, of which they are a part), towards the stable 2-dimensional spectral-orbital-flows of the 3-dimensional hyperbolic space-form, and if they are not still the right size for these spectral-flow 2-dimensional shapes (ie they are not exactly [or not almost] resonant with the 3-space-form's 2-faces) then these condensed components would (again) be forced again by the hyperbolic-coordinate structure of the 2-dimensional discrete hyperbolic shape towards the stable 1-flows of these 2-dimensional hyperbolic space-forms, {where these 2-dimensional hyperbolic space-forms form the stable spectral 2-flows of the 3-dimensional hyperbolic space-form}, in turn, the set of 2-dimensional toral-components of the 2-dimensional hyperbolic space-form, {which form the stable 2-flows of the (original) 3-dimensional discrete hyperbolic shape}, would be folded into the concentric tori of the stable 2-dimensional hyperbolic space-form, which forms what would appear to be the stable orbital properties of an atom as observed in 3-dimensions. Thus, the folded 2-space-form, which is folded so as to be concentric toral 2-components, would form the envelop of orbital stability for the condensed 2-dimensional discrete hyperbolic shapes which are the orbiting charges of the atom.

This allows for

(1) elliptic orbits within the envelop of orbital stability, and
(2) the various toral components of such an orbital structure composed of folded concentric toral-components of a 2-dimensional discrete hyperbolic shape . . . , [which is, in turn, a part of a 3-dimensional discrete hyperbolic shapes within which the condensed 2-dimensional discrete hyperbolic shapes have their atomic orbits defined] . . . , allow for angular momentum structures for the atom to be defined.

In such a case, the condensed 2-dimensional discrete hyperbolic shapes which form the orbiting condensed material components of the atom would have to be smaller than the orbital envelopes of which the condensed orbiting-material is a part.

330

Both the actual elliptic orbits of the charges (which can be determined in the stable context of a 2-body orbital material-interaction with a spherically symmetric force-field), where the nucleus is defined on the inside toral component (which is either very small, or the nucleus is in actual resonance with a 2-flow of the 3-dimensional discrete hyperbolic shape about which are defined the concentric toral components of the electron-cloud where each toral component contains an electron-neutrino shape behaving as condensed material trapped in a toral-envelop of orbital stability), and the electron-cloud defined on the other (folded) toral components, and the structure of the angular orientations which the other toral components have to one another, and together these shapes can be used to determine the radial and angular momentum properties of the general atom, as it would be observed in 3-space.

Thus one sees both an approximate structure guiding orbital properties and the relation that these guiding shapes and an actual (relatively) stable model of a 2-body interaction (acting between condensed material-shapes) can together result in precise spectral-orbital properties of many-body systems.

Is the difference between a crystal and an atom that the atom requires a 2-plane geometry about which concentric 2-toral components are defined by folding, and that molecules define an intermediary geometric structure in which the 2-toral components of the various 2-flows of an 3-dimensional discrete hyperbolic shape can still have well defined 2-plane orientations to one another, while the 2-plane structure of a crystal comes to be defined by inertia, thus geometrically the crystals becomes Euclidean, while the crystal can still have the charged energy-levels of a hyperbolic space-form?

If the nucleus of an atom is a 2-face of the atom's 3-shape, then the vertex-at-infinity of an independent electron-neutrino shape would be associated to a vertex of a bounding 2-face of the atom's 3-dimensional fundamental domain (is a possibility).

Inertia

Inertial values can be defined as the diameter of planar circles . . . , due to the fact that the opposite pair of metric-space (inertial) states can be defined as a circle on a 2-plane (so that this circle of mass defines both its center, related to translational spatial displacements (or translational momentum)) . . . ., and the circle which both defines a mass's position in space, and is related to angular spatial displacements (associated to angular momentum), so the spin-rotation of opposite metric-space states is the circle mapped into itself, so that in a circle-representation of mass both the circle's center and "the circle itself" are always separated, whereas for charge there are two types of charge, and two types of shapes (bounded [ie nucleus and neutrons] and unbounded [ie electron-cloud and neutrinos]) where the opposite charges can combine on either of the two shapes, but the pair of metric-space states, for each type of charge, must remain orthogonal to one another, and this requires the stable shapes of the "discrete hyperbolic shapes" of 2-dimensions, whose shapes are contained in 3-space.

Thus, charge is associated to a standard 2-dimensional "discrete hyperbolic shape," while these charges can adjust to different sized orbits, where the different sized orbits (different from the standard charge size) add slightly to the energy of the orbital charge.

It is a fundamental question as to whether the hyperbolic 1-flow charges are of a fixed length, or if they can change their lengths based on the orbital properties of the discrete hyperbolic shapes within which these charges are contained (ie the charges occupy the stable orbits of the discrete hyperbolic shapes)? Clearly the electrons, of the infinite-extent electron-neutrino cloud, can have variations in regard to the electron's charge-size. Ways to consider data from Dehmelt's isolated electron might have a relation to the question as to the variation of charge-sizes for electrons, and an unbounded 3-dimensional discrete hyperbolic shape, which could be a possible model of an electron-cloud, is also the type of unbounded 3-shape, which Thurston considered in his expressions about bounded 3-dimensional discrete hyperbolic shapes.

Neutrinos (a connectable path to infinity, and back)

There is the unbounded neutrino, which is spin-½, and there is unbounded light, which is integer-spin (spin-1), so "Is the difference between these two types, that one can identify, an unbounded relation within a containing metric-space?" but "What if the metric-space is bounded?" . . . , where the bounded metric-space would be a circle-space . . . , so that within such a bounded metric-space, the integer spin-type identifies an unbounded metric-space subspace, whose unbounded-ness is related to (or characterized by) a sequence of ever larger spheres, while the spin-½-type becomes bounded by the metric-space, where the neutrino's vertices-at-infinity become identified with the (distinguished) vertex of the bounded metric-space?

Thus, one could consider one model of a neutrino . . . , modeled as an unbounded 2-dimensional discrete hyperbolic shape, and contained in a bounded 3-dimensional metric-space . . . , to be represented as an unbounded polygon in a 2-plane, so that the unbounded polygon is modeled as being contained in a circle, where the bounding circle, itself, represent points at infinity for a 2-dimensional hyperbolic metric-space modeled as the interior of a circle, as well as being contained in a different independent metric-space whose inertial region is defined by a 2-plane, (so that the unbounded polygon is modeled as being contained in a circle, and) so that this circle also becomes the neutrino's metric-space bound on that particular inertial 2-plane (for the Euclidean region). (??)

Alternatively, the unbounded polygon model of a 2-dimensional neutrino (whose shape would be identified in 3-space) which has vertices on the infinite distant circle (where the interior of the circle is modeling an unbounded 2-dimensional hyperbolic metric-space, as mentioned above) but then these vertices on the circle can, in turn, be defined to identify some of the separate vertices of a bounded 3-dimensional fundamental domain of the hyperbolic metric-space, within which the neutrino is contained, so that the distinguished point of the 3-dimensional fundamental domain of the metric-space, within which a neutrino is contained, identifies the model of infinity (a point at infinity) for the neutrino (this [bounding vertex] would be similar to the points at infinity for the circle representing points at infinity on the hyperbolic 2-plane, upon (within) which the unbounded 2-dimensional model of a neutrino is contained), while a similar unbounded discrete hyperbolic shape, used to model light, would pass through the bounds of the 3-dimensional fundamental domain of the neutrino containing metric-space.

It should also be noted that the shape of a 3-dimensional fundamental domain of a hyperbolic metric-space would be contained in 4-space.

If the neutrino is bounded by the metric-space within which it is contained, then there are the issues of:

"how the mass of the neutrino can be determined?"
or
"within the neutrino's containing metric-space, 'is the neutrino still mass-less?'"

2-planes and defining a system's inertia

It is observed that stable orbital (inertial) systems seem to form in the structure of 2-planes, eg the solar-system, many galaxies (spiral galaxies), and, apparently, the atoms too. That is, only folded discrete hyperbolic shapes can account for the stable properties of general atoms, ie the solution of the radial equation of a general atom requires an "envelope of orbital stability" which can be provided by a folded, discrete hyperbolic shape.

Note: (where the question is, are the orbits filled with charges, which more-or-less fill (or fit into) the atom's natural stable spectral-orbital flows, or are these orbital-flows filled with condensed charges?)

[The spin structure of the orbits of charges in atoms suggest that the charges "fit into" the stable spectral-orbital flows of the atom's discrete hyperbolic shape. However, the spin properties can also be related to an independent 2-dimensional discrete hyperbolic shape, which acts as condensed material, which is contained within an envelope of orbital stability.]

What would cause such a well defined axial rotation-vector in such general systems, such as solar systems galaxies and atoms?

That is, Why is a 2-plane so prominent in these systems?

What is causing such an limiting geometric 2-plane structure for inertial systems?

Is this a fundamental property, of existing in a 2-plane, in regard to the properties of inertia?

Apparently, it is, where the 2-plane relates to the two-opposite-inertial states (translation and rotation) being mapped into one another by mapping the circle into the circle, since the circle also identifies the circle's center as well as a rotating circle.

In this context, mass can be identified as being proportional to the circle's radius.

In turn, does this necessitate that an atom, a charged system with a definitive mass value, also have a planar orbital structure?

Consider, the case of condensed material forming, in what appears to be a non-linear many-body system, which also has properties of frictional, or tidal, forces acting on the many-bodies, "How can such a system be stable?"

But the geologic record seems to imply that the solar-system has been stable for 4-billion years.

This solar-system stability seems to only be understandable if the system also includes an orbital envelope of stability, composed of a folded discrete hyperbolic shape, where the shape of the orbital envelop of stability, itself, interacts with, and adjusts, the orbital properties of the condensed material it contains, eg the planes are contained within an orbital envelope (which is a 4-dimensional shape, ie if one focuses only on material and its interactions the envelope cannot be identified).

Thus, one has as a part of a system's model either a stable orbital envelope for condensed material, or a stable discrete hyperbolic shape within which charged material occupies the natural stable spectral-flows, so that the occupying charge is the same shape as the higher-dimensional shape's stable spectral-flows.

In both cases there is the possibility that either the condensed matter, or the geometry of the spectral flows, can be affected by other 2nd-order (partial) differential equations (associated to 2-body interactions, which the discrete hyperbolic shapes allow to be determined) which identify a dynamic structure either "in the envelop of stability" or in regard to stretching and wiggling (vibrating) "of the flows" which are occupied by charges.

It is the context of both . . . :

(1) a set of toral-components associated to a stable orbital structure, ie a discrete hyperbolic shape, within which some form of condensed material (or even charges which fit into the

spectral flows [of a (the) higher-dimension discrete hyperbolic shape, which is determining the orbits which are being considered, or observed] in a condition of resonance) is existing in stable orbits, so that this stability is related to resonances with a finite spectral set [defined by the finite partition {of an 11-dimensional hyperbolic metric-space} of the different dimensional subspaces {of the over-all containment metric-space} by discrete hyperbolic shapes], so that the toral components can be related to angular-momentum pathways (through the various toral components of the discrete hyperbolic shapes) of various dimensions and various sizes and various sets of trees-of-containment . . . , and

(2) the infinite-extent structures of discrete hyperbolic shapes which exist at all dimensional levels, and of which the neutrino is an important example, so that the set containment structure of an angular-momentum defined pathway between various sets of toral components, defined within various trees-of-set-containment, which allow for a wide-range of possible links (toral-component pathways, which connect between various sets of discrete hyperbolic shapes over a range of dimensions) between stable systems, which could allow for communication and travel which can be defined locally or they can be defined over great distances as well as being defined across what would appear to be disjoint sets, where each such disjoint set is ruled (determined) by a finite spectral set, ie displacements and connections between "different worlds."

These new and expanded viewpoints concerning the structure of angular momentum . . . , with both extended viewpoints concerning the relation of angular momentum to a stable system's toral components as well as the prevalence of infinite extent types of stable shapes so as to extend the range to which a property of angular momentum can be related . . . , allows the currently expressed idea (within the media) about "worm-holes," in regard to a controlled and stable physical-connection between various contexts of geometry, and containment, and position in space and time.

These ideas are opposed by both the owners of society, and by the, highly touted illuminati . . . , whom work for the owners of society . . . . , because these ideas oppose the notion of materialism, and they challenge the authority of the, so called, illuminati, ie the intellectual-class who dominant the authority of science but who follow the dogmas dictated to them by the owners of society, ie this is the knowledge which is used to uphold the creative-knowledge of the production industries which maintain the ruling-class.

Furthermore, they are new ideas which challenge the way in which instruments are best related to the properties of existence, eg see paragraph above.

The ideas expressed above contain the ideas (or word-quantitative constructs) expressed by today's illuminati (2013) as subsets, where the properties of these subsets are of minor importance in relation to the control of the properties of existence, eg materialism is a proper subset of the new descriptive context.

Are numbers real?

Only if they are used to describe a quantitatively consistent pattern.

Numbers, or measuring, are about "types of numbers" associated to measurable properties of a stable pattern.

The questions which these new ideas allow, ideas concerning issues of a new containment set, and the relation of the new containment-set to the creative activity of life, which is about creating existence itself, are the proclaimed properties of both containment and partitions, in regard to a set of (rigidly) fixed math properties, or are there (the newly described) properties, which are possessed by the new containment context, and which can be used to construct "what the fixed math properties allow," namely, new avenues of expansion of existence, so as to be able to change an existing set of (math) properties . . . , so as to maintain most of the old properties, ie the context of existence and the possibility of creativity is expanded (ie creative possibilities are not lost).

Life is an instrument of unfathomable capacities, whose structure may be the same structure as the outside spectral structure of containment, or nearly the same.

Comments: these new ideas are opposed by the ruling-class, since these new ideas are opposed to the idea of materialism, and they are ideas which place the center of human creativity, not in the material world, but rather well beyond the material world.

Though the ruling-class has a set of religious authorities, whom the rulers support and where the religious authorities serve the interests of the ruling-class, and where these religious authorities claim to believe in an "ideal world," where idealism is the opposite of materialism, they are really authorities who want to possess dominant social positions in the material world.

Thus they are religious authorities who support the ruling class, and in so doing they can maintain their domineering social positions.

The only strategy that this ruling social set-up can adopt is to either ignore these ideas, which is what they want to do, or to claim they are the ideas of an incompetent person, ie a person with no value. But explicit refutations do not exist.

The only refutations, that do exist, are those where the authorities claim that their dogma is correct, and that anybody who has social-value must accept the assumptions and dogmas of the authorities, and after adopting the assumptions of the authorities, then prove the new opposing ideas. But this is like requiring Copernicus to adopt the assumptions of the Ptolemaic system, and then prove his (Copernicus's) ideas within the context of Ptolemy's descriptive structures.

This is not possible, since the descriptive languages are from completely different contexts. That is,

The new ideas are based on a context within which there is reliable measuring, and the set of describable patterns descend from stable (math) patterns, where the basis for stable patterns are the stable geometries associated to geometrization of Thurston-Perelman, namely, predominantly the stable discrete hyperbolic shapes, and the relatively stable and flat (when contained, or observed, within) discrete Euclidean shapes. Note: within the currently accepted descriptive structures only the solvable systems have technical value, and solvability is based on the global commutativity of operators, where the operators are models of "local linear measuring."

Whereas the descriptive structures, of the current math and physics authorities, are based on both/either:

(1) indefinable randomness, wherein any vaguely distinguishable feature (it is assumed), regardless of the stability of the (observed) feature, can be used as a basis for the laws of probability . . . ,

note: many of the failed risk-models of the financial sector, were built by MIT PhD's following this math structure of indefinable randomness, and/or

(2) non-linearity, where most classical geometries and general relativity are descriptions of non-linear math patterns, but these patterns are not stable, and not solvable, . . . , but the properties of the most fundamental systems, eg nuclei and the solar-system, are stable, and this observed stability cannot be described using the properties of indefinable randomness and/or non-linearity.

Addenda:

A 1-shape circle-space (or a circle) is contained in a 2-plane, whose 2nd-order interaction equations are defined in 3-space, and have a geometric relation to 3-space, while a 2-shape (circle-space) is contained in 3-space whose interactions 2nd-order dynamic equations are defined in 4-space, but the differential 2-forms in 4-space, which are part of the 2nd-order differential equations of these physical systems, have an arithmetic relation to two dependent-sets of 3-spaces, where the dependence is related to the relations that the geometric measures of 4-space to the force-fields, but there is also the natural geometry of inertial 4-space, ie related to the fiber group SO(4), which, in turn, can be related to two sets of independent 3-spaces, ie one space for each of the two types of material defined by the 3-dimensional discrete hyperbolic shapes, which are contained in 4-hyperbolic metric-spaces, and so that each of these two material-types can be associated to inertial 4-space dynamics, thus each 3-space will have two sets of 3-space vector fields associated to it.

The differential 2-forms of 5-space, ie SO(5) fiber group, are related to one 5-dimensional vector-field defined on 5-space (which can be related to inertia), while the differential 2-forms of 6-space, ie SO(6) fiber group, are related to two sets of 5-spaces vectors, ie two vector-fields which are in 5-space, but these two (separate) 5-vectors are inter-related by the differential 2-form defined on 6-space.

For each of the 2-toral components . . . . of the facial structure of a 3-hyperbolic-space-form (contained in hyperbolic 4-space) . . . there can be defined an inertial 2-plane . . . for each of these 2-dimension toral components (of the 2-faces of the 3-shape) . . . so that in the inertial context of these 2-planes each of the component 2-tori can siphon either the condensed orbiting material (contained in this metric-space shape) or the 2-spectral-flows, by a "change of rate of the flow" (or of the orbiting material), of the charged material which occupies this 2-flow (of the discrete hyperbolic shape) where this "increase in rate of flow" is due to the negative curvature of the discrete hyperbolic shape. Thus, the energy of the radial equation is adjusted by both elliptic orbits of condensed material in the envelopes of orbital stability which the material containing metric-space causes as well as the angular momentum flows which are associated with inertia, ie energy, at Weyl-angles (or at right-angles) to the radial orbital shapes of the system (or equivalently the material containing metric-space).

Knowing the inertial 2-planes of a system's angular momentum can be related to the finite spectral properties of the containing 11-dimensional hyperbolic metric-space.

39.

Civilization and thought

Humans teach by creating in an independent manner

Humans can metamorphize into what they actually are, ie all of existence is central to the human essence

Civilization is mostly opposed to thought, and it mostly expresses a will to have ever more people survive.

The choice about "the role of the person in society," which civilization seems to collectively make, seems to be about:

Equality vs. inequality [Where western civilization has selected inequality and its attendant violence]
Or

Creativity and rational control-over changing language
vs.
dogmatic authority, which is related to increasing populations, built on very narrow (or constricted and limiting) traditional social structures

Or
Individualism vs. oligarchy (or collectivism).

That is, the idea of an oligarchy is about collectivism which supports a narrow viewpoint about how man fits into the world.

Is mankind a creator, or is mankind to be focused on survival in a narrow and constrained context, constrained both materially, in regard to how the material world is organized around the investments of a few in the ruling class, and constrained in regard to what mankind is capable of creating?

The pattern of social power is about the public serving a few, the few who form the ruling-class, where the ruling-class has won "the game of ownership, whose rules are based on violence," which is related to property rights and minority-rule, and these few investors control the basic attributes of the society, which are related to the society's living-style, both material-resources, and how these resources are related to creativity, which, in turn, is related to the actions of the society which allow for an easily managed large population, which needs to be a part of a civilization whose population grows and the structural-social-engine which supports this population also expands, but expands in a way in which the few are kept in power, ie the rules are designed to support the few rulers.

The large populations of western civilization have the technology of violence and destruction so as to be able to march-out to exploit the (basically) gentle, curious, and creative nature of mankind, so as to steal the riches of the creative, and to conquer the rest of the world, so as to expand the influence of some violently inclined investors to extend their control which is defined in a narrow context of violence and narrowly defined material uses.

For science in the age of Copernicus one finds that language (in regard to science) could not be changed since it was being used in a particular way as propaganda-tool for (of) the civilization. Science in our age is also "not a rational endeavor," again language cannot be changed, since a narrow viewpoint is being used in a particular way for military productions, and military advantages, which defines and maintains a narrow way of doing things, and only a narrow context for creativity is allowed, ie only certain types of creativity are defined as possessing high-social-value, in regard to the context within which the wage-slaves are allowed compete for high-wages.

When science is criticized today (2013) [which is either a rare occurrence, or if science interferes with traditional interests] the criticism is not based on rationality . . . , since the only people allowed to criticize science are the high-ranking scientific authorities, themselves, and they are well-indoctrinated (or funded to do research in particular ways) . . . , rather science is claimed to be irrelevant, and a bad joke is made about this irrelevance, causing science to have become: "a society (or exclusive club) of science-literary criticism,"

. . . , but this is really because science has become irrelevant to development, in regard to, practical creativity, and furthermore, science is already being used in the ways, which are needed to support and maintain the interests of the investment class.

There is nothing rational about modern science, it is all about irrational obsessions which marginally support the military interests.

Thus, rationality is considered to be (or is accepted as being) about using language in narrow ways (consistent with the propaganda system) so as to enhance the power of those in-power, where "in the context of the destruction . . . ,

which this narrow way in which society is organized (so as to be able to support a large-ever-expanding-population and an ever-expanding growth in the way in which "a narrow range of resources" are to be used) . . . ,

seems to have led," the very powerful investors . . . , ie those who possess so much money that they can totally control any aspect of the highly controlled and narrowly defined markets . . . , to now proclaim, that "the population of the world needs to be made smaller."

That is, the ruling minority are opposing the large-population social structure upon which their social power seems to most depend.

That is, the psychopathy . . . , which the propaganda system forces onto the thought-processes of its occupants . . . ,
. . . , has some obvious signs . . . ,
eg it leads to a solution to the problem of over-population . . . ,
. . . , where the oligarchy is organized to cause such an over-population "problem" . . . . ,
. . . , as being that of exterminating the population. [The first model of population extermination, due to science, was the atomic bomb.]

Apparently, these thoughts of "devising some highly organized systematic extermination" are based on trying to maintain the inequality upon which the system is built [protect the ruling minority, those illuminati who possess a deep mental capacity, to which the competitive wage-slaves, and their narrow visions of allowed avenues of creativity, aspire to realize such high-mentality (ie a mentality of obsession and delusion) (a high-valued mentality about which there is no valid definition {a valid definition of intelligence might be about the capacity to discern truth})].

In the world of propaganda, the social system . . . , in which a few people are the rulers of society, and they preside-over a very narrowly defined structure of thought and life-style, and it is these few who preside-over an ever larger populated civilization, ie empire, or oligarchy, or collectivism (it seems that Marx's contribution, to the dialog concerning western civilization, was to ask the question, "Why not make the hierarchical structure of western society responsive to community-social needs?"

(This is supposed to be the subject matter of a democracy, but western society has dismissed democracy [except as a propaganda slogan], so the answer (to Marx) seems to be, because of the selfish nature of the powerful few, but (nonetheless) these few have built vulnerable social-hierarchies based on stealing and violence, so their solutions will always be "more violence")), it is a very narrow western social-system of propaganda, . . . , which is, nonetheless, referred-to (by the propaganda system) in terms of the society being based on "the virtues of individual freedom," while equality, and true individualism, and creativity, are social attributes referred-to (within the propaganda system) in terms of:

collectivism,
being incorrect,
incompetent, and
inefficient,

but such criticisms seem to be more of a statement about self-reflection concerning the attributes of the ruling-few (rather than being rational statements).

The high-paid wage-slaves, the academic science, which forms the basis upon which the narrowly directed engineers apply their attentions, has become, since about 1900, a world of mental delusions, ie it is a complicated memorized construct which has no descriptive content.

It is a descriptive structure of a precise language which is trying to relate
Either
An "unreliable measuring context," ie non-linearity, to stable patterns, [the Thurston-Perelman geometrization has shown that such a non-linear (or unreliable measuring context) either disintegrates or evolves toward a small set of "stable shapes," of which the most numerous are the discrete hyperbolic shapes and the discrete Euclidean shapes)]
Or
Trying to use unstable random patterns in order to describe stable spectral patterns.
But
This is a game of delusion.

There are replies which the media puts-forth to these criticisms, that:

"random patterns associated to unstable properties cannot be used to describe stable properties, and that a context wherein measuring is unreliable, ie non-linearity, cannot be used to describe stable (geometric-spectral) properties."

Namely, the replies of the propaganda system are:

0. There are intelligent people who try to use non-linearity (1) in order to describe stable patterns, and (2) to design feedback systems. (But these attempts are about imposing new states on the original system, in an arbitrary way, which is designed to get the results which conform with what is observed, ie it is about suddenly discounting the description and its containment set, introducing new states into the old (but deposed) context, and then re-starting the old context, but now in a new state (which did not exist in the original containment set), ie it is an epicycle structure used to fit data.)

    [How can the calculation (of the precession of Mercury's perihelion) be taken seriously when Mercury has zero-mass, ie it is a one-body description where the one-body is acting on objects which have zero-mass?]

1. The idea of randomness means unstable

    But, This is not true, randomness is a selection process which must be related to a fixed and stable set of possible events (or outcomes), otherwise, the probabilities which are identified by some "counting of events process," in regard to an event identified by the property of distinguishable but which are not stable, cannot be relied upon as valid probabilities of vague unstable properties (which are vaguely discernable).

    Or equivalently,

2. There are discernable (but vague and unstable) patterns which can be counted, and thus statistics can be formed.

    However, vaguely discernable patterns, which are not stable, cannot be counted, since in such a context there is no valid relation between unstable patterns and counting, since the unstable, vaguely discernable, patterns cannot be a basis for a counting process, thus a valid number system cannot be determined in such a context (ie who knows when a vaguely discernable pattern will disintegrate [and, thus, is no longer, vaguely, discernable, ie what was vaguely discernable has disintegrated and thus it is not counted, etc]).

A string-of-symbols (ie a descriptive language) needs to reference some content, eg a stable pattern, that is, numbers are measures of a type of measurable properties which are associated to a stable system (stable pattern).

If one uses either math-methods or math-models, then the elementary properties of numbers need to be satisfied, if a valid number system cannot be built based on either unreliable measuring or based-on "counting vague and unstable patterns" then math-methods are not valid.

One should be appalled at the combination of ignorance, and narrowly conceived deceptiveness, which now (2013) passes for intellectual activity (centered on language), where descriptive knowledge should be related to practical creativity:

either

in regard to material systems,

or

in regard to "real" higher-dimensional manipulability and controllable descriptive contexts.

The math language now used (2013) is aimed at doing nothing, ie a math language which has no stable patterns also has no content.

whereas

Religion is now (2013) about "the propaganda of ignorance" so as to allow hierarchical social control by the few, whose focus is on exploiting reason so as to make; both their own individual fortunes, as well as (creating) very large populations, and to exclude the use of reason by the rest of the public, so that this exclusion is accomplished by a university-investment institutional control over "how knowledge is used for creative purposes" within society.

Science:

I. Materialism,

    A. Physical [instruments]

        1. Non-linear (feedback systems)
        2. Indefinably random (bombs, competition, chemical identification and models of correlations between chemicals)

    B. Life

        (indefinably random, DNA, molecular biology, chemical identifiers and correlations etc, evolution, competition, scarcity)

        but also the new context in which knowledge can be used to extend survival contexts so that there is not scarcity of resources, etc) [these ideas are marginalized as being "the concerns of survivalists," that is, the propaganda system is all about "making jokes" used to marginalize certain types of thought. This is done by the thieving owners of society, ie the joker works for the thief.

        Materialism and social Darwinism say that life is based on violence. If there is no valid model, which is different from material based science . . . , then social Darwinism is taken as a scientific fact . . . , which it is not.]

II. Economics-politics-law [property rights and minority rule; the same basis for the law as in the Roman Empire, defined in a narrow material context of products and efforts, where this narrowness is held in place by extreme violence]

This implies: inequality, selfishness, and extreme violence, justified by an arbitrary claim of possessing superior knowledge:

(where possessing superior knowledge can be used to justify extermination of others, as exemplified by the actions of Puritans, of Massachusetts, against the native peoples, or as the Catholic Church burned Bruno at the stake, for believing-in the ideas of Copernicus {the rights of life and death "over the lives of others," which it is assumed in western culture by those who claim to possess superior knowledge}, and a thoughtless (mercenary) militia can be motivated, or commanded, to carry-out the extreme violence), and it also allows for a unity of both the organization and the actions of a big institution.

How should society be organized to best use the nature of mankind?
(creativity and equality)
How should society be organized to best use resources so as to support society in a sustainable manner?

III. Control of language by the propaganda system.
[How language is used to manipulate science
and
How both language and narrow focus (eg identifying a narrow range of needed skills) are used to divide people,
and
How language is used to institutionalize violence]

Religion

New religious interpretations based on;
New viewpoints about life: its structure, and its origins, and its relation to existence

These views are the most rational expressions about the above, just, mentioned context of language and science in western culture.

This also requires a rational viewpoint concerning science and math, a viewpoint which is not allowed by the propaganda system ie where the propaganda system is about the expressions of society's arbitrary idea about what has high-social-value.

In regard to a valid rational viewpoint about science, consider that science is best modeled as classical physics where a surprisingly simple and relatively general set of laws are applied, namely, inertia is related to force-fields, and force-fields are a result local linear geometric measures associated to force-fields being related to material geometry of a spatial-position's surrounding material, as well as a similar law for thermal physics,

So that the descriptions (the solution function of the system's equation s) are: correct, ie can be observed, and to sufficient precision, and the description provides good information in regard to placing and controlling a solvable physical system, which is acting as a subsystem in a newly created system, eg an invention of a new instrument.

Whereas, the current descriptive structures of: either reducing material to its elementary components which are governed by a solution function to energy-wave-equations where the solution function has a relation to the probabilities of local spectral-particle random events in space and time, or identifying the large-scale macroscopic material properties by means of geometric properties whose solution functions which are non-linear, and thus both quantitatively inconsistent and, subsequently, chaotic.

Neither of these two techniques of physical description, either random or non-linear, can be solved for the most elementary of the fundamental systems, which are observed, and which are also observed to be stable.

Thus, neither of these descriptive techniques is capable of describing the measurable properties of general systems, which possess stable properties, in a correct manner, and neither of these descriptive techniques has any relation to practical uses.

(the rational thing to do; is to, consider many new ways in which to try to describe the observed sets of properties of fundamental stable physical systems, which have no valid descriptions.

Eg using the very stable discrete hyperbolic shapes as models of both metric-spaces and as models of material systems fitted into a many-dimensional containment context, ie a description which transcends the idea of materialism.

Then to consider the relations which the "new ideas" have to religious ideas, instead of promoting an idea of a "big bang" about an instance of creation which is related to an absolute idea about materialism)

In the new viewpoint about existence, a viewpoint which contains as a proper subset the current material-geometry and random based descriptive structures.

Our (material) experience is essentially contained in a metric-space, which has the size and shape of our solar-system (otherwise the "observed" stability of the solar-system has no explanation) so if we are life-forms which are oscillating 3-dimensional discrete hyperbolic shapes then we are contained in a 4-dimensional hyperbolic metric-space whose size is smaller than, the size of the solar system

(eg the chlorophyll molecule might also have "such properties of oscillation" as a molecule, when it interacts with light.)

But if we are an oscillating 5-dimensional discrete hyperbolic shapes, then in order for that 5-dimensional space to contain our lower-dimensional material composition, then that oscillating 5-dimensional discrete hyperbolic shapes will have to contain the 4-dimensional hyperbolic metric-space, whose size is greater than or equal to the size of the solar-system, which means that such a higher-dimensional model of a human-life form is extremely large, in regard to the size-scale of the solar-system.

That is the origins of human life might actually be the story of giant forms of life descending down to earth, so as to live in harmony with the life-forms, which can be gently used as food, so as to sustain the lower-dimensional aspect of a life-form's existence, eg and with an easily considered relation to a story of a metamorphosis.

Thus, the idea that we are created in the likeness of God, might (be interpreted to) mean we are formed in the likeness of giant life-forms, whereas the purpose of these life forms is to create existence, itself.

But the false version of the story of life states, "humans were formed from 'lumps of clay,' and humans have dominion over the earth."

But humans are also about creating existence, just as is "the character God," in relation to the story about "giant-Gods" inhabiting higher-dimensions.

In fact, the context of a very large existence, might be what is the correct context of our existence, which we should be able to perceive (to perceive the world as it really is, ie this is a religious saying of Buddhism) and use our life-form as an instrument through which we can travel, and communicate (where both activities are done) through stable angular-momentum channels [connecting different spatial places, {or connecting different life-forms, in regard to communications}] of the higher-dimensional shape (eg the large 5-dimensional shape) which we really are, wherein a life-form can control (by intent) the flow of energy through the stable angular-momentum channels of its own (large and higher-dimensional) properties of existence.

This is a model of life which allows for a channel through which a precise control over the living system's actions can be realized (or described in a measurable and stable context).

That is, the mythology of the ancient Greeks and Sumerians, concerning a race of giant (possibly higher-dimensional) life-forms who make man from a "lump of clay," can be easily related to the mythology of creation, (creation myths) as is described in the Judeo-Christian-Islam

Bible, the mythology of being related to the giants of mythology, as is the mythology of ancient Egypt in regard to the constellation of Orion (the giant), as well as the mythology of changing levels of existence as are many of the old myths of the Native American cultures, eg changing the dimensional context of awareness, etc. (apparently the Islam Bible [the Koran] tweaks this Greek story a bit, but apparently it stays close to the Jewish Bible [Old Testament], but also admits Christ was a prophet {whatever all this might mean?}).

That is, the Bible is not a description of a useful measurable truth, but rather a description of "what 'it is trying to claim'" as a truthful history of the human life-form, and its origins and its relation to existence, which entered western thought through the Sumerians and ancient Greeks, etc etc.

But, in fact, the human life-form might be a 7-dimensional or a 9-dimensional discrete hyperbolic shape which possesses an odd-genus number. In such a case, such a life-form is in fact unbounded (like-light) and possesses no position in space, yet its shape can be related to positions (or regions) in space, and if its lower-dimensional structure could also realize an unbounded shape (as the, apparent, unbounded shape-structure of neutrinos would allow) then such a high-dimension center of energy (existing in a region of space) would not need nourishment in its purpose of creativity. Thus life is about knowledge and creativity at the fundamental level of creating existence.

Each of us is an unbounded source of both light and creative energy, seeking to create existence itself, or seeking to crack the cosmic-egg, so that the new life-form emerges out "into a new context of existence" (which life intends to create).

Realizing this true state of existence, is what the Buddhists might call "waking-up."

This seems to be the context about which C Castaneda wrote, concerning the knowledge of the native people's of the Americas.

However, the western related cultural stories are all about battle and violence and treachery, in regard to the giant entities associated to the idea of creation, but they are powerful entities who seem to want to rest on their laurels, they are entities who seem to not want to have to consider the struggle of knowledge, and creativity, rather they seem to be concerned about being dominate in a fixed-existence, so as to behave in a petty, selfish manner (in regard to tradition and the status quo), and engage in violence and battle on a fixed level of materialism, which is related to violence and control.

The bible (or Koran) is organized so as to be about a hierarchy and the propaganda of an empire based on materialism.

Whereas:

The true basis for human activity is: not property rights and inequality, and such a context's attendant violence, but rather equality and creativity and selflessness.

The "lump of clay" is passive, to the selfish goings-on of "Gods," except not passive . . . in regard to "dominion over the (material) earth" where the "lump of clay" expresses its selfish material dominion, so as to follow the petty behaviors of the material-fixated "Gods."

The "lump of clay," picture of mankind, is to have the effect of making "ordinary" people subservient to superior, more powerful, entities, ie the violent rulers of civilizations, eg the Roman civilization.

Thus, one can see that the bible is, indeed, organized so as to be about a hierarchy and the propaganda of an empire.

Thus, as "lumps of clay" the people within a society look to the upper 10th-percentile group, on the normal curve, which can be defined in regard to some arbitrary but fixed state of mental and/or material existence.

When a skill is measured for any (arbitrary) identifiable skill, then measures of abilities of that skill can (be used to) define a normal curve, and the higher skill-levels distinguished, so that looking for an apparent better-skill capability [(for any arbitrary skill) which is defined in a narrow constrained social context] it is natural to choose from the upper-10th-percentile of the normal curve.

Do these normal curves have any meaning? No. & Are they based on stable properties? No.

However, this method of defining normal curves in regard to a narrowly defined set of skills is very effective in regard to the expressions about defining a false absolute-dogma, which has come to be a part of a highly-valued creative process, where the new instrument to be created might be "needed" for society, but where both high-value and "need" are invented (or determined) by a hierarchical civilization's elite ruling-class.

The ruling-class identifies the skills wanted, then a normal-curve is defined, and only the highest tested, who possess the skills which the ruling class wants, are selected (so as to form a self sustaining viewpoint about the possession of an absolute merit).

That is, the public takes these arbitrary measures as an absolute definition of personal worth, and this is because, that is how the propaganda system represents these arbitrary values (as being related to an absolute truth, ie an absolute measure of high-value) which the ruling-class identifies for the public and then expresses these needs in the propaganda system.

What is being discussed, when language essentially either has no meaning or its meaning is determined by propaganda and emotional contexts?

The major division in regard to social organization deals with the idea of:
Equality vs. inequality

This may be contrasted with the idea of:
Individualism vs. collectivism

Equality means individual creators who compose their own language, based on elementary concepts of assumption, interpretation, context etc, upon which each of their languages is built, where "the language's function" is in relation to creativity.

Whereas inequality is arbitrary, narrow, built on violence and maintained by the control of language so that there is a uniform agreement about what has high-value and how creativity should be controlled by the few in the investor class.

Yet if one asks the "highly controlled public," ie controlled by the propaganda system, then the idea of individual freedom essentially means "the freedom of the 'strong' to impose narrow language upon others so as to control the public for the selfish purpose of the strong," ie there is an implied inequality when the idea of individualism is expressed.

That is, individualism is associated with violence and domination, which, in turn, depends on collectivism and conformity.

These ideas of individual inequality are justified by the notions of materialism, wherein there is a scarcity of resources and all living creatures must compete for these resources so that the strongest (or the best) survive, ie it is social Darwinism, and, despite the rhetoric to the contrary, the idea of social Darwinism is the strongest affect which "science" has had on social structure, whereas classical physics has had a strong effect on engineering, but engineering is stifled in its creative range by the investor class.

Furthermore, when one expresses the idea of equality then there is an incorrect model of: "each person, in an equal society of, having to conform to a single idea, and each person must have equal parts of the material pie," . . . , instead, for an equal society, one needs what is necessary to survive and to create at an individual level, or (if so chosen) to form as a group working to create together, eg a corporation.

Thus one sees that it is the context of inequality in which there needs to be both mental and language conformity in order for "inequality to survive."

Inequality is about: materialism, control of language, arbitrariness, and the extreme violence needed to maintain the arbitrary hierarchy, a hierarchy which historically emerged from a narrow social structure originally imposed by violence.

That is, collectivism is more a result of an unequal social-order, rather than collectivism emanating from an equal society of equal creators.

40.

The choice

The choice society makes is between protecting everyone's individual freedom in order to learn and create, that is, equality in regard to both knowledge and creativity, (ie the declared point of the US revolutionary war, declared in the Declaration of Independence, wherein equality was declared to be the basis for US law) vs. protecting a few who administrate a narrow, collective, and traditional viewpoint of (or identified in the propaganda system as being) dogmatic-authority, violence, and expansion (ie the banker-Roman-Empire-(or Marx's)-viewpoint)

For new ways to consider the question "What exists?" in relation to both practical creativity and "How society should be organized, eg what words should mean" {or how many ways of organizing "what words mean" can be used in many (new) creative contexts.}

[religion and science both deal with the surprising, but true, newly identified structure of existence, ie the new structure of existence which transcends the constricted idea of materialism, where, as it now is, materialism is the basis for modern science, and subsequently, materialism is the cause of the failing of modern science, ie there is no valid descriptions (to sufficient precision) of the observed stable spectral-orbital structures of: nuclei, atoms, molecules, crystals, solar systems, and no explanation of dark matter, dark energy, etc, and furthermore, modern science has no relation to practical creativity, all of our engineering is still based on classical-physics, even the building of micro-chips, and the nuclear bomb is based on 19th century models of "rates of chemical reactions."]

[Note: The US revolutionary war was supposed to support everyone's individual freedoms, and not support the freedoms of the bankers to be tyrants, yet G Washington went out, while president, and put-down Shay's rebellion, a rebellion built out of a group of unpaid revolutionary war veterans, a rebellion against the tyrannical banks whom were foreclosing on the (broke) veterans' farms. That is, G Washington presided over maintaining European social-class structure in the new nation. That is, when material value was established in North America the banks rushed-in to take it over, and this allowed the already existent European class-structure of "the beginning US" to be maintained.]

A person is responsible for the survival of oneself, and is not responsible to administrators of a failed line of social pursuit. If there is a state then the state (claims) to take on responsibilities of survival of the individual, whereas the individual is responsible only for knowledge and creativity within the state (but the reality of knowledge is that it is not narrowly defined, as social institution now constrict knowledge).

The nature of being a person

People seek knowledge so as to be creative. One's creative responsibility is to challenge, correct, alter, and improve existing knowledge and to expand the creative context so as to give back a creative gift to the earth. If one's creations have results which are harming the earth, then these structures need to be corrected

Be responsible in regard to an attempt to "see the world as it really is."

The current social structure is based on minority rule, where the minority are chosen by their theft of both property and knowledge from both the people and the culture, and the subsequent violent ability of these (violent) few, who (apparently because of their violent capabilities, ie apparently, their self-importance gives them contempt for others) have high-social-standing, to stabilize their condition of being thieves and thugs within society. Their thieving-violent activities are claimed to be building the society. They twist the idea of a state so as to define the state as that which serves them, the violent thieves who own the state, ie law is based on property rights and minority rule.

Thus, there are two levels of justice within such a society, where justice is used by the owners of society to both protect themselves and to attack rival thieves and associated armies, ie justice is used to attack the public, but now the public is organized in a narrow condition (of thought), so that merit . . . , along the lines of narrowly defined pursuits . . . , is easily discerned (the normal curve for many sample-sets of "samples of the mean" for a vaguely identifiable property).

Thus, it is demanded that the failing ways of this type of a social system are not to be noticed by the public, ie the propaganda system cannot talk about all the failures of this system.

The current administrators . . . of the few guiding the collective expansive (ie expansive but narrow to the point of extreme destruction) and narrow vision which motivates society, ie the bankers (and oil-men) . . . . , need to accept their obvious failure . . . , and seek to serve new ideas . . . , because the authority and tradition of the banker-administrators of the society's collective motivational structure ie their authority and traditions have failed.

They (the banking-class) are destroying (poisoning) the world and their knowledge is not leading to any valid form of development.

That is, the model of expansion has failed (the limits of earth's people-population, upon a heavily-poisoned earth, for a people-population which focuses on the material-side of existence, where this condition [of the earth] was reached due to the expansion managed by the owners of society, ie they are people who are unaware of what their actions are causing (they only see themselves and their own interests).

The narrow set of knowledge upon which the current way in which society is organized is helpful in regard to easy administration of social institutions by a few Engineering is used to create an institutional order, and technical development facilitates a social order and it is used to help manage this order.

How the world is put-together is not of much interest in such a social context

However,

It is the state which has responsibility to the public (nourish, help people know and create, and protect the people from the destructive selfishness of a few); but in a tyranny (minority rule) that responsibility (of the state) is about protecting the process of violent theft.

The US law is based on property rights and minority rule.

And the value of knowledge is measured by how that knowledge contributes to the (management) power of the ruling few.

On the other hand, in regard to a descriptive knowledge concerning "perceiving the world as it really is,"

[Note: this is a religious saying of Buddhism, as well as Taoism]

The current social structure is based on the entrenched power (of narrowly defined ways in which to use material resources) which is associated with fixed ways of doing things in a context of materialism. It supports innovations which make these fixed material processes cheaper, or better designs for doing certain (narrowly defined) things, in a social structure attached to a materialist viewpoint about (scientific) knowledge. This is all managed by a very small set of powerful administrators (whom watch over this totalitarian vision for society).

Thus, the engineering of our society is stuck in "the use of 19th-century scientific knowledge," ie the knowledge of materialism, wherein (in regard to classical physics) the description has a math-structure, wherein the descriptive information is controllable, reliably measurable and practically useful.

Whereas, quantum physics, particle-physics, and general relativity and all the physical theories derived from these core sets of descriptive constructs, do not describe a context where measuring is reliable and the systems are not controllable.

Thus, either

One can choose the currently accepted authoritative dogma of basing measurable description on reduction to (local) point-spectral-particles and describing these assumed-to-exist point-particle entities to be ruled by indefinable randomness and non-linearity, or macroscopically by assuming that geometry of material interaction is to be based on a one-body system which possesses spherical symmetry, ie a non-linear geometry,
Whereas in both cases there are no valid descriptions of the observed stable spectral-orbital properties of nuclei, general atoms, molecules, crystals, as well as the stable orbital properties of the solar system.

All these stable systems go without valid descriptions, where the main statement within this dogma is that "these (relatively simple, yet very stable) systems are too complicated to describe."

This descriptive context is a failure.

Or

One can consider a new math structure (ie new assumptions, new contexts, and new interpretations of data [as the ideas of Copernicus defined a different context than the context of the Ptolemaic system]) . . . , wherein the descriptions of systems depend on a math context of reliable measuring so that the descriptions are, in turn, related to stable patterns so that the observed stable patterns of material systems can be both described and controlled . . . ,

. . . , a new way to view these stable material systems in regard to a many-dimensional containment set whose subspace structure is partitioned by a finite number of discrete hyperbolic shapes, which model metric-spaces, so that for sets of (these) different signature metric-spaces (including the inertial containing Euclidean metric-space) each type of metric-space is associated to a property and an associated material (or material property). So that all of the stable systems are in fact metric-spaces (or condensed sets of such metric-spaces) which are in resonance with this above mentioned finite set of (hyperbolic) metric-spaces so as to belong to some tree of subspace and metric-space size containment sets within the higher-dimensional containment set. That is, the controllable context for material systems is actually a descriptive construct which transcends the idea of materialism.

Thus, within each metric-space containment set the lower-dimensional material components which the metric-space contains have interactions defined by (partial) differential equations, both of an inertial type (ie dynamics) and of a force-field type, where the structure of these equations is dependent on the new sets of properties defined by the new context of set-containment, where these new sets of properties account for the same set of observed properties which are interpreted in different ways in the old (currently accepted (2013)) descriptive system.

Note: This point needs to be stressed in regard to the critics, ie this means that I am taking responsibility for a failed construct of knowledge (which defines a failed authority today (2013)), and correcting it.

This correction of knowledge is opposed by the bosses of the wage-slave critics of dogma (the propagandist-critics), since it would interfere with their investment structures, and subsequently, their social power, and the failed (narrowly defined) collectivism which their tyranny defines.

Note: The propagandist-critics Whom, apparently, these propagandist-critics like to "lick their boots (or caress their own paychecks)" but most likely (apparently) these propaganda-critics like "being a helper" to the powerful, they are the types who seek to believe the authorities and whom obediently follow the dogmas and traditions associated to the knowledge which upholds the narrow interests of the owners of society. Furthermore, these easily influenced critics (since they are followers and cannot take-on the responsibilities of challenging a failed authoritative context), like the advantages which supporting the owners of society provides to themselves (for essentially very little effort (the owners of society, and their propagandist wage-slaves, are some of the laziest and most irresponsible people in society) the status quo is allowed to reign supreme as an absolute blanket (of fog) over society. That is, the dogmatic authority of the social institutions (which have been built and guided by the owners of society) allow those who support the status quo to be in a state of "being not-responsible."

That is, it is assumed that the rest of society has to take responsibility to solve the problems caused by the owners of society.

This is the natural logic of a people who are wage-slaves serving the interests of the few, ie the few who destroy both the earth and the culture of the society.

[the Chinese communists and the capitalist-bankers are equivalent masters {Marxists-capitalists}, who reign over a collective state of arbitrary irrationality].

Even the high-priests of rational dissent, ie the few who express anemic social criticisms from academia, or who come from a journalistic-editor viewpoint, express ideas which are, essentially, consistent with the interests of the owners of the society, in regard to the assumptions of the rationality of the owners of society's institutional structures ie they have been brain-washed, ie they believe the useless dogmas (endless, repetitive baloney about: law, science, politics, economics, the organization of society, etc, where the assumptions of these descriptive structures, which, in turn, are associated to institutional operations, are never challenged). This leads to an intellectual waste-land, which exists because the authority of social institutions are not allowed to be challenged, rather the "myth of merit" which is defined by wage-slaves competing in regard to narrow dogmas, effectively states that other ideas have no merit, similar to how the European invaders claim that the "savage" native people were worthless and to be exterminated. Similarly, many of the first adherents to the ideas of Copernicus were exterminated, by the authorities of that age.

The best which these weak-critics have expressed is that "society is organized to serve the interest of the owners of society," where ownership is centered "in the banks."

This note applies both to the propaganda guards and the authoritative guards of academia, ie the protectors of authority and tradition, ie protectors of a narrowly defined tyranny (but everyone needs to look around and see the obvious failings of such a system).

A descriptive structure which transcends the idea of materialism

The stable shapes which such system structures can possess are the: cubes, cylinders, tori, and shapes built from toral-components, ie the cubes [or right rectangular shapes] and circle-spaces are the set of stable shapes.

Thus, the geometry of the charges, neutrons, neutrinos, nuclei, atoms, molecules, crystals, and solar systems are based on:

. . . cubes, circles, rings, cylinders, tori, and folded shapes of toral-components (where the allowed folds are related to the Weyl-angles defined in the material containing metric-space's fiber Lie group [easiest thought of as a unitary fiber Lie group]), concentric sets of toral components, etc . . . ,

. . . , within which the material and condensed-material-components-orbit in a relatively stable manner, and often in a neutrally-charged manner.

Of great interest is the relation between a set of low-dimensional components and relatively small components, of adjacent 1-higher-dimensional metric-spaces, which define a stable system, within which the lower-dimensional components occupy. There is the bounded nuclear charged structure, wherein neutrons and nuclei alike as well as electrons and neutrinos . . . , where each component is based separately on 2-dimensional shape, which are both either bounded or unbounded, and apparently . . . . , these 2-dimensional components come together in a bounded stable 3-dimensional shape which is a charged closed metric-space (or as an atom is a neutrally charged closed metric-space). This seems to have a shape of some form of concentric set toral components, and would be similar to an atom's orbital-spectral structure, but there seems to be an energy-gap which distinguishes the neutral charged structures, ie atoms and molecules, from charged, but stable, shapes of bounded nuclei. That is, nuclei are 3-dimensional shapes whose shape and structure is quite similar to that of an atom. The energy gap between atoms and nuclei would be about the relation of these shapes sizes (or spectral values) and the resonances which are defined in regard to the subspace partitions into a finite set of metric-space shapes associated to the 11-dimensional over-all containment hyperbolic metric-space. Or, the energy-gap may be related to the particular organization of the set-containment tree, related to our solar-system (and the shape properties of the metric-space which our stable solar-system defines) and/or the relation that this set-containment tree has to a higher-dimensional energy generating shape within this set-containment tree.

The particle-physics model of quarks and leptons is all about the unstable 3-dimensional spatial structures which form when the 3-dimensional shapes with their 2-dimensional components are broken apart due to high-energy collisions, and which are unitary since these metric-space shapes are a mixture of opposite metric-space states.

Furthermore, this description allows for the usual structures of dynamic systems to exist, ie their differential equations can be defined within the metric-spaces within which the material components are contained, that is, if these material components define "free" systems, ie well defined unattached (condensed) material component. Furthermore, these dynamic structures are still often related to non-linear states of dynamic change.

However, when material is a component of a higher-dimensional containment set, which itself possesses a stable shape, and if that shape also possesses a capacity to generate its own energy, then components, and "regions which contain components," within the over-all containment system, can have their properties (or states) changed, by the over-all containment system controlling energy-flows through the (higher-dimensional) angular-momentum channels (or regions, or faces) of the system.

These containment sets might be made more complicated by the size relationships that exist between the system's lower-dimensional face and contained-component structures (or relationships), but such size-relation might also determine more opportunity to act (in a wide range of contexts for existence, as well as for intentional actions [of life-forms]).

41.

Illusions of high-value

Society is based on a political-economic-military system . . . , which is an extension of the idea of civilization held within the (Holy)-Roman-Empire (where Holy is to be interpreted to be related to a propaganda system which controls language and the high-valued thought within society, where the media is thought of by the public as being the sole-voice of authority which provides to society an absolute and objective truth) . . . , and which defines, in a very (both) narrow and arbitrary manner, high-social-value, wherein the public is defined as wage-slaves, who seek high-paying positions working on the projects of interest to the owners of society.

The experiment of wage-slavery defined within a controlled market of products . . . , which fit into the interests of the owners of society and these products are to be chosen either by the wage-slaves or the government . . . , has been a failure, but it is called free-market, democratic capitalism, but it is a . . . controlled-market, totalitarian society so that there exists a society based on central-planning, in turn, based on selfish interests . . . , really a dictatorship.

Apparently the ideas of Marx were all about oligarchy, the same types of thoughts about social organization as those of the Roman civilization, but where the choice between social-organizing oligarchies was a choice concerning propaganda, where the goal was either exploitation to serve the few (called free-market democracy), or material exploitation to serve a propaganda system which is claiming to serve society, where the organization of this societal-serving system was also to be based on the decisions of a superior-few, who would administer such a system,

. . . , so the issue again becomes
"what is actually true, ie the truth as discerned by the (elite) few?"

"can the unequal few, actually determine the truth, since their choice as being the superior-few (essentially imposed by arbitrary military-violence [ie property rights and minority rule, ie the basis of Roman law]) is a choice made within a narrow viewpoint?"

Can a language incorporate both the propaganda of governing (by oligarchs) and the language of a useful and observed truth, ie the language of science and math, with such a scientific language's correct relation to truth, namely, a truth being related to practical creativity?
[Answer: No, this is why oligarchic society always fails]

The American-Quaker (Philadelphia) solution to the problem of social oppression was, to "base law on equality" and, in turn, relating equality to both freedom to believe and freedom to express beliefs . . . , but the object of such beliefs is in regard to practical creativity . . . , so that human creativity defines a truly free-market.

Within such a truly free-market, advertising expresses an invention's properties and its uses, where a free-market must be regulated so that no one can control (or dominate) the market (since without such control the market cannot be free).

Law must be administered in regard to the spirit of the law . . . , though different communities can express some forms of law intrinsic to the beliefs of their community, but people must always be allowed to leave such communities.

This solution to either the Roman or the feudal, or (now) the non-free-market banking system of social oppression, was provided by the Americans during the revolutionary war, well before Marx expressed his belief in oligarchy—related to both material and technological exploitation of the many by the few.

Unfortunately, the winning side (of the American Revolutionary War) was administered by the higher-social-classes of people, within the (new) US, which were comprised of an aristocratic community which, essentially, expressed the values of the upper-social-classes Europe (at that time), and such European viewpoints were put into the governing organizations of the US, where banks and debt were big issues at that time. But where such issues should not have been so important, since the US was then independent, and should have had confidence in its own inner strength.

Thus, by relating US social organization, ie by means of European-style economics, so as to relate US economics (then, 1780) to trade with Europe, so that this trade was organized by European banks, was an un-necessary illusion of economic dependence.

(where economics simple means organizing trade of created-products, but not organizing investment in any particularly established manner, so as to unduly restrict the social-knowledge structure of the (then, 1780) American society).

A hierarchical (language) system determines a circular structure of agreement, where the high-valued wage-slaves accept the underlying principles of the system . . . , since they are the ones who are succeeding in that social structure, . . . , and it is these successful types who are asked to play the role of experts for the media. Thus, these (successful) wage-slaves, who have relatively successful stories (within the confines of social-organization), have come to be in agreement with the underlying principles of the social-system, within which they achieve their (relative) successes.

This highly constrained "state of the state" can exist as a result of pathological interpretation and enforcement of law, which is based on property rights and minority rule, and this was the result of the ruling class administering the new American government after the revolutionary war. This can be re-visited and changed with a new Continental Congress, so as to base US law on the principle of equality.

The revolutionary war was fought for equality and the freedoms to believe and to express one's beliefs, with an idea to developing science and its relation to practical creativity where people were to live on self-sustaining farms and to create, based on selfless motivation, an ideal which B Franklin demonstrated.

Unfortunately, Franklin also developed the idea of a controlled press, where W R Hearst might be the legacy of the wise-cracking Franklin within the press.

The definition, by the propaganda system, of high-social-value is that upon which knowledge and creativity have been defined within society, where the propaganda system was eventually touted as being a sole absolute voice of truth. But, if this image were true, then Franklin would have, most likely, been forced to resign his press activities before he became rich.

Thus, knowledge and creativity has been organized within society in very narrow ways, so as to serve the interests of the owners of society. Yet, this narrowness is denied to exist, and instead it is claimed that the system is expansive, and all encompassing, and the system really is based on absolute truths, of which only a few people are capable of absorbing, and this is because it is such a complicated descriptive structure, which contains the absolute truths of society. But in reality it is a descriptive structure based on obsessive focus on irrelevant complications which in the end, possess no content (other than expressing the selfish interests of the owners of society)

But this is like the absolute truth "being about"

0. the fact that large corporations are the most efficient ways in which to use resources . . .
   . . . . but this is a great fallacy . . . it is the worst way to use resources;

This is because:

1. it narrows the way in which "changes in the use of resources" can be realized, and
2. it is simply the opposite, that is, it is very inefficient and
3. it was always the basis for the idea of "too big to fail" but
4. what its really social-function was, was a way in which to organize society, so that the few owners of society could better control society, by large investments, so that the stability of society was based around protecting these investments, and
5. the wage-slave fiction, which was placed into society by both propaganda system and by the selective enforcement of law, instituted by the justice system, effectively, destroys the creative capabilities of the public,

    The institution of wage-slavery, instituted to protect the investments of the very-rich, made this type of hierarchical social-control of the public very easy.

History of hierarchical social-control, within Europe, since the Roman times:

Craftsmen-guilds and banks form monopolies, for both products and life-styles, so that these monopolies are controlled by financing or market success, ie dominating and controlling a market.

It is narrow, and not expansive, but it can come to be appreciated by the public, due to the familiarity which the public has with such a social institution, as well as some of the conveniences it provides to society.

Thus, knowledge and creativity are to be irrelevant to the public, and creativity is considered to be related to monopolistic social interests. Thus, there is very narrowly defined ideas about high-value: art, science, engineering, but art and science are thought of as disciplines which can only be practiced by the experts.

And thus, people came to believe that:

The public could not distinguish between truth and falseness, so that only experts can determine truth within society.

Furthermore, The public could not judge "law cases" based on their consciences, but must follow the "letter-of-law instructions" given to them by a "corrupt and overly domineering," judge (ideas of both J Adams and T DeChristopher 5-24-13 B Moyers TV show on PBS).

High-social-moral-value, including truth, is to be determined by the ruling-class.

This dysfunctional way in which to set-up society, can only allow change which is based on the beliefs and control of the ruling class, and it is a society which acts on delusions and it cannot perceive either the failures of the society or the danger to which such a delusional society is subjecting both itself and the world (the destruction of the earth by poisoning [pollution] and decimation, eg over-fishing, mass extinctions, etc)

When a story unfolds, the public, both right (those who take sides with the boss) and left (those who follow either the golden-rule or the intellectual-class), go along with the (absolutely authoritative) government's explanation, ie the propaganda system is the sole voice of an absolute truth, where the violent arm of the government works for the rich (ie the interests of the right-wing) and the left-wing works for the intellectual, or academic language-side of the propaganda system, so as to defend the high-value, which upholds the interests of the ruling-few (and the high-value defined by the ruling-few), while the right stands for tradition and a need for violent control (so as to protect the interests of the rich [and the high-social-value which they have (supposedly) built and maintained]).

Science, or the social position of any expert, is all about the politics which is associated with the manipulation of the language within the media, so that the determination about the subject-matter of science can only be done by the owners of society. That is, the scientist is a pawn within the institutions which are controlled by the owners of society.

This is because everyone is a wage-slave, and thus people (as wage-slaves) seek to compete along the narrow lines of thought, which define "expertness" within institutions, or they are simply willing to "have a job," so as to exist within the social constructs of conveniences and familiar-ways, which are provided by (the owners of) society, where these societal conveniences are manipulated in a social-context whose range is far too narrow.

Ultimately, this narrowing . . . , of the activities of people in society . . . , was brought about by the narrow conditions by which civilization formed (in the west) when it was developed by the Romans, where the Romans required such narrowness, for both their armies and their state, so as to function as an expansive civilization, which functions to steal the wealth of other (nearby) societies (by means of violence), but also providing engineering development, which, in fact, is the edifice which supported the empire, where the Romans imposed . . . , their engineering infra-structures of their Empire by force . . . , on the societies which the Romans conquered.

In turn, these narrow social conditions, in regard to what the Romans built, eg building and water-systems and an easily organized agricultural construct, were used as the basis for a banking empire, where bankers are a less personalized and more personally remote than were the very public-group of Roman emperors, ie the bankers were insulated by a wall of language and institutional positions.

That is, the bankers are hidden (veiled) by an investment social infra-structure, which can be presented (within the propaganda system) as being both expansively directed, and being helpful to society (they provide engineering projects and/or they provide art-cultural gifts to society, but these projects mostly serve the interests of the bankers).

Thus, craftsmen-guilds (in narrowly defined sets of social activities) in partnership with bankers, created monopolies, which defined fixed traditions, as well as defining an authoritative knowledge within society. Where the main intellectual authority had to do with religion, but where the knowledge of building (engineering) was consistent with the interests of bankers. The institution of "professional, academic, science" was all about bankers controlling the creative capabilities related to new knowledge and this knowledge was narrow and was developed incrementally (except for Newton and Faraday).

Thus, what is considered to be true . . . , in the categories into which the society is divided . . . , is a very limited truth, yet it is only the successful experts who are allowed to express "what is considered to be true" within the media, where the media is a carefully-gated communication system, wherein people are led to consider irrelevancies, as being of central importance, in regard to both life and knowledge, and the subsequent very narrow relation creativity (based on investment) has to knowledge, . . .

. . . the experts who have internalized the narrow dogmas of such a truth, which is necessary in order to compete for high-social-positions, where these narrow authoritative dogmas are developed and fed to them by a propaganda-education system (where the education-system is masquerading as a propaganda system), [that is, this internalization of narrow dogmas happens because of the success of those who do internalize these dogmas, because the dogmas are institutionalized and controlled, which means that the experts have to internalize these dogmas, in order to compete for social positions associated with institutions, and creative projects, based on banking-investments]

That is, knowledge is limited so as to be used to create and produce the instruments and products which enhance the power of the owners of society, ie it is a craftsmen-guild and bankers monopoly which is based on the control of knowledge and the organization of resources.

What are the categories of thought and activity into which a society has come to be rigidly divided?

It is based on the banker-oil-military business interests and the markets which have been developed, where a market is a relation between knowledge resources related to investment and associated to the productivity of products, and the relation of this narrow monopolistic scenario to the needs and life-styles which are organized and controlled by the media around this narrow investment scenario.

Basically it is the knowledge of science and/or carefully monitored statistical models (of distinguishable features of interest) which is related to product, and how such products (or instruments) are used in society. Namely, they are used to maintain the power of the owners of society.

The problem is, that such a viewpoint about knowledge leads to large amounts of error, since most of this construct "called knowledge" is not valid, it is not true. Rather it is only vaguely true in a limited context, but "the point is," that the information about which these constructs, or contexts of markets and products (or the contexts within which an instrument is functioning), are made, in turn, defines a context which is monitored so much, that adjustments can be made to try to make the context remain (relatively) stable, for both markets and instruments.

Only the high-salaried experts who work on projects which are in the interests of the owners of society can discern truth.

But the only truth which is allowed to be discerned is to be defined within the narrow context which defines the interests of the owners. But this is the only type of truth which the experts are trained to identify.

Only the authoritative and learned experts of the church can discern truth, and that truth is to support their interests, and it is demonstrating an expertness in this dogma which identifies one as an authority. That is, "the way truth was used in the age of Copernicus."

In the highly controlled media a media which is even more controlled that the government regulated market, where the market is highly regulated by the government because the government takes its orders from the dominating rulers of society, and the rulers of society want the market controlled to their liking. Ie central planning by selfish interests.

But look to the actual development of technology . . . , the development is all based on 19th century science, . . . , but most of this development was undertaken by people on the margins of society . . . , not the ruling class, nor their never-wrong overly authoritative minions (academic authorities) who are ware-housed (and made ineffective) in the academic institutions (managed by the ruling-class), . . . , rather the actual development of technology is done by marginalized people like: Faraday, Tesla, and the Wright Brothers, while all the domineering authorities have a surprisingly modest contribution to technical developments, though they are good at adjusting and maintaining instruments already in use. Yet the investment class is quite "quick to steal the new ideas developed by others (eg mostly the small adjustments)" for their own exclusive investment purposes.

What type of creativity is wanted (defined by the owners of society)?
and

How is it to be organized and used in society (also defined by the owners of society)?

Answer: It is used to serve the interests of the owners of society either through the economy or by a military-justice-system strategy used against the people, who the owners fear, or through the media, ie using propaganda to: define, secure, and maintain a narrow viewpoint about knowledge and how language is to be used within society.

"What all" can one person (or a small central planning committee) control through their control over language?

When one controls a communication system, which the public views as providing to them the sole voice of an absolute truth, then one controls both . . . "how language is used by the public" and "how the public thinks."

Controlling language allows one to control, by means of the media:

Science

(the control of science has been accomplished by controlling what is considered to be the authoritative dogma of what is presented to be an absolute truth which is associated with science and the process of observed verifications (but science always pretends modesty, claiming only that, so far no other models of truth make sense (though they judge other models of existence based on their dogmas, since other models based on other sets of assumptions cannot be judged as to their truth-value by the experts who believe-in different dogmas {this is the same as the example of Copernicus vs. Ptolemy}), but their model of truth is really in relation to how the owners of society both invest-in and use such narrow truths (of a dogmatic science) to create the things which support the interests of they themselves, the owners of society, eg their investments in instruments),

Note that . . . :

social organization, (institutions, administration, justice system, education, propaganda, celebrity, inequality)

social forces, (talking primarily about issues which divide people, and calling this free speech)

Economics

Markets

. . . , use electronic communication Technological devices, so that . . . (how) . . . they are used in society . . . , (note: especially after excluding the delivery of written letters [end the post office])

. . . , so that all communication is forced into a set of electronic instruments, and then (the owners of society) use these electronic communication channels to spy on all the information sent through the electronic communication channels,

. . . , so as to use this information for selfish purposes, (purposes of the domineering central planning committee), but, the owners of society, do not use these communication channels to improve knowledge expand the discussion about useful (descriptive, measurable) knowledge, and thus, expand creativity, as well as expanding the contexts of creativity, eg electromagnetism was a different context (within which to create) than was the existing contexts of mechanics and thermal physics.

Furthermore;

Law (which is used to attack the public),

personal behaviors,

beliefs in high-value,

psychological types of people whom deal with social forces by means of different types of identifiable behaviors (where this is possible if the social forces are held stable),

manipulating psychological types (eg autism, psychopaths, etc)

Control religion to express certain types of arbitrary, but "absolute moralities,"

eg the native people do not believe the things which we (the European derived thoughts) believe,

so these native people are inferior,

so they can be exterminated,

ie this is about manipulating ideas and thoughts of the public,

ie ideas can also be exterminated . . .

All of these social and intellectual institutions have their languages tied to (tethered to) the arbitrarily identified high-valued interests (identified by the propaganda system) of the owners of society, so as to make their investment-bets less risky, ie sure-bets, and thus the ruling-class can maintain an absolute control over:

1. society,
   as well as
2. its faithful meritorious wage-slave adherents
   (where their merit is determined by their faithfulness and competitiveness within the narrow dogma, which the propaganda-controlled institutions define for the wage-slaves),
   as well as
3. the disappointed wage-slaves
   (the losers, in the game which is all-about serving the interests of the owners of society)

The narrow way in which this viewpoint about knowledge is structured
(and how it all works together to advance the interests of the owners of society):

42.

Intellectual Revolutions

Scientific revolutions: can lead to social revolutions (or scientific revolutions can exist only after social revolutions, so as to make it clear how the propaganda system is manipulated, and even the experts are duped by this manipulation)

Godel's incompleteness theorem is about the limitations in regard to the practical usefulness of precise (measurable) language, and it requires that the development of precise, practically useful descriptive knowledge exist in an equal society, wherein equal free-inquiry exists. This is the idea of equality which is the fundamental teaching of all religions, ie equality related to both the development of knowledge and the relation of knowledge to practical creativity, and to new contexts for practical creativity.

Descriptive and measurably verifiable knowledge develops at the (elementary) level of assumption, context, interpretation, etc, and only (small) adjustments to existing uses of language are related to an overly authoritative context which is narrowly defined and it is a descriptive language which is held rigidly fixed, so as to become so complicated it is practically useless.

The propaganda system is considered by the public to be, "the single, and only trusted, voice of an absolute truth" which is provided to the public by (the intellectual authorities, who work for) the owners of society through the media, ie the instruments of communication within the society, which are controlled by the owners of society, and which are used to set-up and determine institutional and societal goals, so that the "absolute truth," which is so expressed by the media, is the truth which best fits, in regard to serving the (business) interests of the owners of society. Ownership, business, knowledge, and practical creativity, as well as deceptive use of language, are the main forces of society which are used by the owners of society to control society.

Where it should be noted that business and economics is not about the quantitative structure of money being governed by quantitative rules about the economy (since the large players in the game can make the economic rules invalid by how the large players use their money), but rather economics is the arbitrary social-institutional set-up, in regard to the culture's knowledge, and how it is used in the practically creative processes (which are narrowly defined by the owners of society so as to maintain their social power) within society. Thus, the communication system is used to maintain the social power of the owners of society, and to control the language and the thoughts of the public, so that their beliefs and actions will be channeled and used to maintain the interests of the owners of society.

The owners of society have been able to manipulate social institutions so that the owners of society effectively play the role of God within society.

(This was done mostly through their absolute control over the propaganda system (ie society's communication system) and because politics is a subset of the propaganda system, so politicians are "vetted (chosen)" by business interest)

This was also do-able because of the corruption of religion, a corruption which characterizes the Protestant relation to Calvinism, where Calvinism is about discerning that a person has "a relation to God" (and that a person is virtuous) primarily if they prosper in the material world.

The owners of society are, in fact, the hidden-hand, which is defined by A Smith in his descriptions of capitalist-economics, but the deception is formed so that the hidden hand in economics is the hand-of-God, a God who controls economics (and their wage-slave minions).

That is, (macroscopic) economics is not a reliable quantitative descriptive structure which can be used to determine cause-and-effect. This is because there exist the owners of society, who have enough wealth and institutional influence so that they can control markets in an arbitrary manner, ie they play God, and they represent the hidden-hand of God within economics, the hidden-hand which is supposed to cause economics "to work" based on quantitative economic laws, this is a derivative of Calvinist thought. That is, this is all a bunch of baloney because economics is controlled by the owners of society, ie the ruling class, the bankers and oilmen who the military serves. Note that the standing-army is the backbone of the effective tyrannical emperor-bankers who rule society, just as was done in the Roman-Empire.

There is a relation in the language of the propaganda system in which (so that) "materialism" is linked to:

1. Calvinism, and to
2. Darwin's evolution (see below), and to
3. A Smith's descriptions of economics, as well as to
4. Marx's, amusing belief in a, "help-the-public oligarchic economic system,"

where Marx's system of oligarchy is incorrectly attributed to be about social-collectivism. This viewpoint is incorrect since the nature of western civilization, at least since the Roman-Empire, has been defined as a collectivism, a collectivism which is forced onto western civilization by the extreme violence of either emperors, or now (2013) bankers.

Note: The history of a slight amount of equal-freedom of the British-US society, supposedly, related to the Magna Charta, has been toppled by both the extreme violence of Empire and the wresting of science-freedom (the freedom to re-define technical language) by an overly dogmatic and authoritarian science and math professional community, which is associated with professional university science and math institutions, as well as being related to the need of the owners of society to have their important technical instruments adjusted and maintained (by the professional scientists).

The development of new science has, mostly, come from the margins of the academic system.

Note that new ideas about science and math interferes with the stability of investments in regard to technical activities (eg for corporations based on a given technology).

Thus "if" one sees that "what is being done by institutions is failing" then one has to oppose the authoritative commands of the propaganda system, and subsequently use new ways in which to "build and use language" so as to describe a new structure for language within which the failure, in regard to what is currently being done, can be stopped, within both the propaganda system, and within the society's (academic) institutions.

In particular, technology is failing to develop new contexts within which practical creativity can expand.

So if one sees that the current way in which society's institutions are being used is destroying the earth, by a form of poisoning of the environment (and thus the earth will not be able to sustain the life-forms which now exist upon it), then one needs to provide and alternative set (an alternative descriptive construct) in regard to how society should be organized.

For example, the use of oil by society is poisoning the earth and though the current knowledge of physics and math can provide clean relatively-cheap renewable energies, these energy sources are not developed, but the current knowledge of physics and math cannot provide to society a more economical (cheap), clean energy source based on new ideas about math and science, where science and math have dogmatic ideas concerning what is assumed and how the descriptive language is organized, in regard to:

1. Reduction to smaller components, and assume both
2. Randomness, and

3. Non-linearity,

which have been the dominant ideas in the intellectual communities of science and math for over 100 years.

If . . . . , one sees that the institutions of math and science are failing in that both

(1) the descriptions of fundamental stable physical systems cannot describe to sufficient precision the observed properties of these fundamental systems (listed above), and
(2) the currently used descriptive constructs are not remotely related to the further development of practical (technical) creativity, . . . , Then . . . , one needs to provide and alternative set (an alternative descriptive construct) in regard to how the descriptive constructs of both math and science should be organized.

But to be able to provide a new construct . . . , and to begin a dialog based on the acknowledged failures of current practices in regard to social and (technical) ways in which language is organized . . . , one needs to deal with the propaganda system.

The saying about free-speech, provided by the propaganda system, in the US society is . . . , One can personally have free-speech if one owns a significant part of the propaganda system, ... , but to acquire an expensive part of the communication system one must gain one's money by adjusting one's viewpoints to be consistent with the absolute truths proclaimed by the media (this must be done in order to make money). Thus, in order to get rich one must be rewarded within the context of the given absolute truth of the propaganda system, ie the same truth which defines business interests in which investment is allowed, and after being rewarded, so that one becomes rich, then one has lost touch with alternative ideas (which one wanted to express in the first place), and one can be sure that it is easier to continue one's success by remaining consistent with the authoritative dogmas of the "owners of society."

That is, authority is defined by the (very narrowly defined and highly controlled), so called, free-market, but this identifies an even more narrow a definition of authority than was the definition of authority determined by the Catholic Church in the age of Copernicus. This current practice of determining narrow authority within society is really an expression of Calvinism, where "a person having 'grace with God'" is determined by worldly successes of the person (one cannot help but see the hand of JD Rockefeller in this conniving propaganda maneuver), ie success in business (in Calvinism, worldly success shows that a person has self-evident virtues in the eyes of God, ie this was believed by the New England Protestantism, at least from the 1600's to the 1700's, but it is also the (hidden) theme of the propaganda system today (2013) and is the real basis for the "moral authority" of the right, where the political-right are the political factions which support the owners of society).

Because both

(1) science and math are failing to develop a new context for practical creativity, and
(2) that because of the way business is done today, namely, the earth is being irreparably damaged, by it being poisoned, and this is done because it (burning oil and poisoning the earth) provides the owners of society with great profits based on the resources which they currently control, and thus it allows them to continue to control society, . . . , so that the society needs to be re-configured.

One way in which to do this, to help both problems, is to have law is based on equality (and not property-rights and minority rule, ie the basis of law in the Roman-Empire). This idea of equality is immediately (and wrongly) categorized as a society in which the people are required to believe-in a collective society, however, this is wrong (both the ideas of A Smith and K Marx express ideas about the oligarchical structure of society imposed by Roman civilization, where oligarchy is based on society having a collectivist social structure), consider the fact that the propaganda system remains so narrow, in regard to "what it allows to be expressed within society," and thus the propaganda system defines a society which is a collective-society, which focuses on a narrow range of ideas, and this means that such a society is not an equal society, ie it is not now (2013) a society of equal individual creators.

This might be a societal law, that "a society can only be considered to be an equal-society if its people are allowed to be independent, equal creators."

A retort which the propaganda system would provide, in regard to the idea of "a society of equals," is that society is based on the survival-of-the-fittest (apparently where food and water are scarce, a viewpoint which is not true "in general") and that people naturally form into hierarchies based on their own natural unequal abilities.

But inequality, in regard to ability, can only be measured in regard to a narrow definition of what a valid human activity should be. However, the development of knowledge requires that narrow definitions be allowed to be changed, ie that narrow definitions are not to be allowed to define a rigid authoritative context.

One sees that each individual should define what has value in their own lives, and not forced to personally value "what the-owners-of-society value."

A narrowly defined, and fixed, set of ideas is used to determine a system of extermination of both thought and life based on arbitrary high-value, ie the value defined by the emperor (the value defined by the investments which banks make).

But the high-value concerning the idea of survival-of-the-fittest is claimed to be a property identified by science, as a "law of biology."

However, mutation and survival-of-the-fittest, ie the basis of Darwinian evolution, is not a descriptive structure which can actually describe the way in which life-forms, and the characteristics of life-forms, come into being, ie the simple systems cannot be shown to develop into more complex systems based on mutation and survival-of-the-fittest.

This is a failing of this "type of math-construct," a math-construct which is based on "what is best described as indefinable randomness," ie it is a failed math-construct which is a result of defining probability in an improperly manner. Where it needs to be noted that "probability based descriptions" must be based on fixed sets of stable events, not on events which are unstable, ... , and not on events which cannot be defined (or identified).

"How can complicated constructs be defined by constructs of simple patterns without an implied reason to seek a state of "being a different construct" (ie the simple functions "as is," so "why should the simple seek to be different?")?" The claimed answer is mutation and competition for scarce resources. But mutations statistically are destructive, thus mutations lead to simpler systems, while new complex systems of a new life-form require internal intent in order to be driven towards such complexity, and not driven by acquisitiveness and brutal battle, in a context which begins within a similar simple context (see below for a more mathematical explanatory context).

The only way in which evolution makes sense is if things began in their most complicated context, in regard to DNA structure, and then destructive mutations allowed for complexity to develop, but the complexity would have to be driven by intent, not randomness.

In the current authoritarian propaganda-system, or "style of social communication," where only authorities and important personalities are allowed to express ideas within the media (ie publications and broadcasting, etc), and the public is to be "taught by the university authorities" so as to only be allowed to express narrow dogmas which support the interests of the owners of society.

This is not a valid model of knowledge.

The professional scientist is a social construct used to control knowledge, so that knowledge will only be used to support the interests of the owners of society (the professional scientist, mainly adjusts the instruments, which are used to support the interests of the owners of society), and the relation of the authority of professional scientists to teaching and subsequently related to a narrow vision of practical creativity, and it results in a narrow focus of science.

It might be noted that experts are people who do not believe in themselves (they do not believe in their own ability to discern truth) rather they believe in, and follow, institutional authority.

Note: In the age of Copernicus the Catholic Church was the only institution which was allowed to discern truth, today truth is primarily determined by the owners of society, ie the materially successful people, where material success is the basis for believing in the superiority of these people, ie a Calvinistic (and very materialistic) viewpoint about "how a society determines personal-worth."

Science and math are failing (too),
The nature of physical science

Assume that an observable pattern of material (or energy) is in a quantitative and geometric containment space, and the physical system exists in a descriptive (containment-set) context where measuring is reliable and patterns are stable (or, at least, some of the most fundamental patterns are stable).

However, in physical science,
Observable properties (qualities) are "no longer" related to a . . . . , precisely determined (geometric)-(material-position) descriptive containment-set whose patterns are reliably measurable, and which are used to describe stable (controllable) and observable patterns.

This is because, in the descriptions of physical systems, it is now (2013) assumed by an overly authoritative science and math (institution, eg universities) that,

Either

Existence is the reduction of physically observed properties to indefinably-random, local, spectral-particle-point events in space and time (the domain space of a "reduced to a point and observed" measurable property) influenced by system-defined measurable properties (or operators acting on wave-functions), apparently, defined in bounded regions of the domain space.

In this context, measurable properties, (or, equivalently, operators acting on function-spaces), are most intuitively defined by their actions on sets of harmonic functions, where these sets of harmonic functions can, be taken together to represent a physical system, whose fundamental observable properties are random, and where these harmonic functions have the form of waves, $f(x,t)=Ae^{i(Et-px)}$, upon which local linear operators can act, so as to represent (or determine, after they operate on the quantum system's wave-function) a set of measurable properties, so that the measured-values of these properties are spectral-values (in regard to the spectral properties which the (partial) differential equations of quantum physics defines, and which these sets of operators might define).
(ie Quantum Physics and particle-physics)
However, in this descriptive context either the stable observed events cannot be calculated by this method for general (quantum) systems, whose properties this method is trying to describe,

or because the events (or observed patterns) themselves are unstable (which is definitely true in particle-physics) so either way . . . , either incalculable eigenvalues, or trying to determine probabilities of a system's spectral properties based on unstable patterns . . . , this precise descriptive construct of a probability related description of measured eigenvalues for the system does not have a proper definition (ie is not well-defined) in the context of these math constructs associated with the "reduction of a material system to random particle-spectral events in space and time," and thus this context cannot be used to identify valid measures of probability, for these (always assumed to be) reduced random events (which are defined as local spectral-particle events). Note: This is also the basic, and invalid, math structure of indefinable randomness upon which Darwin's evolution is also based, ie observed properties (in the fossil record) of new complex subsystems introduced to new life-forms are not calculable, and the probabilities are based on unstable (simple) events, ie both the mutations and the events of survival-of-the-fittest, do not define stable patterns upon which a probability can be based in a valid manner. The probabilities of an organism developing along a particular path of evolutionary development cannot be calculated in a reliable context.

Or

Existence is assumed to be about non-linear, local, (supposedly) measurable properties of space and time . . . , (the geometric measures defined on the domain space of a system, or pattern, which, supposedly, possesses a measurable property) . . . ,
. . . , influenced by system-defined measurable (material-geometric)-(energetic) properties. This is about trying to identify measurable, stable patterns, observed for physical systems, based on math methods defined on unstable and non-measurable math-patterns.

In classical physics there is the differential equation of F=ma, where a is acceleration, and a is related to spatial displacements, and where F is (quite often) a force-field, so that the force-field is a 2nd-order differential-form (ie a function related to local geometric-measures) which is determinable as a solution to a differential equation if this differential-form has been differentiated two-times by an exterior derivative (and its dual), ie a second-order partial differential equation, and then related, by the partial differential equation, to the geometric distribution of charges (or masses) and currents in the domain-space of the force-field solution function, ie in space-time (or space and time). This equation can (most) often be a non-linear equation, while the only useful and controllable solution functions (to these types of inertial-force-field equations) are related to linear, continuously commutative (independent in each local coordinate direction), and metric-invariant context for solution functions to a classical system's partial differential equation.

Whereas in general relativity a 2nd-order partial differential equation, is defined in regard to a connection, so as to define, in regard to the containing (curvilinear) coordinates, a non-commutative local operator relation (though diagonalizable at a single point), which acts on a metric-function (or metric-2nd-order-tensor) and then (this differential equation defined on the metric-function) is related, by the partial differential equation, to the geometric distribution of either masses and/or energy-density distribution of the system in the domain-space of the force-field (solution) function, ie in space-time (or space and time).

[Note: According to E Noether's symmetry relations, inertia is to be defined in Euclidean space, while energy is to be defined in a hyperbolic metric-space, ie (equivalently) a space-time metric-space, where it should also be noted that the energy-time symmetry relation of Emily Noether is the (real) basis for the Einstein relation of "mass equals energy."]

Once the metric-function is solved (found), by solving this non-linear partial differential equation, then the dynamics of the system's energy-distribution-inertia properties are supposed to be determined by the containing coordinate system's geodesics, ie not by F=ma, rather inertia follows geodesics.

However, since this is non-linear these general systems cannot be solved (nor controlled).

That is, it is a descriptive context which has no practical value in regard to practical creativity.

It should also be noted that, one wants to stay in the metric-invariant context, and not base physical description on general (non-linear) metric-functions.

(ie this is the descriptive context of General Relativity and the (above mentioned, in a previous paragraph) set of non-linear systems of classical physics)

However, very stable properties are observed for the fundamental systems, where these fundamental systems, in turn, provide the building-blocks for our experience, where in our experience, "measuring is reliable," where the fundamental stable physical systems are: nuclei, general atoms, molecules, crystals, and the solar system, etc, and where none of these systems has a valid description which is based on physical law so that the descriptions are sufficiently precise, in regard to some of the very precise, stable properties which many of these systems are observed to possess.

That is, science and math are failing.

Either one accepts the language of the authoritative institutions, in which case only traditional (institutional) authority is considered to be correct,

Or

Ones sees that "what is being done, ie what language is being used" is wrong, ie the authoritative language is not valid, and thus one must present an alternative set of ideas.

Are precise descriptions containable in a measurably reliable geometric (or event containing) context, so that the descriptions are based on stable patterns?

In order to understand both stability of material systems, and to have these systems be contained within a reliably-measurable descriptive context, in regard to the domain space (or domain metric-space), one can consider an array of many hyperbolic metric-spaces (but whose inertial properties are relatable to a Euclidean space), modeled as discrete hyperbolic shapes (ie following the "time is related to energy" properties, determined by E Noether's symmetries), which possess both many-sizes and many-dimensions . . . , [ultimately, all, contained in an 11-dimensional hyperbolic metric-space, so as to define a finite set of both subspaces and spectra] . . . , so that both material and space are the same type of thing. Namely, they are both hyperbolic metric-space constructs, but they identify different adjacent dimensional levels, while force-fields and inertial properties are defined in, yet, a higher-dimensional metric-space, which is adjacent to the material-containing metric-space.

That is, material is no longer modeled as a constant (or as a spectral-massive-point), but now is geometric. However, material-interactions are not (in the new descriptive construct) contained in the material-containing metric-space, but rather the material-interaction is defined in yet a higher-dimensional metric-space, which is adjacent (in regard to dimension) to the material-containing metric-space.

Nonetheless, material-interactions are also (still) defined as (partial) differential equations defined on metric-spaces, as is true for both classical physics and quantum physics (and particle-physics), but the metric-spaces are metric-invariant, and the stable shapes (of both material and metric-spaces) are linear and continuously commutative.

Thus, either the partial differential equations related to material interactions . . . , or the shape of space . . . , acts on the material components contained in a metric-space-shape . . . , and the number of material components within a metric-space can vary depending on the size of the metric-space-shape and the relation that the material components (which are contained in the metric-space) have to the structure, eg spectral-flow-structure (or, equivalently, sub-face structure of the shape's fundamental domain), of the containing metric-space. That is, there may only be enough material components to occupy the spectral-orbits which exist in the metric-space, or there may be many material components which can condense to form what are essentially independent-free material bodies, which can be related to the elliptic, parabolic, and hyperbolic second order partial differential equations, which identify the local measurable structure of material interactions, and its geometric relation of spatial displacement [or discrete velocity change] which exists between the local measurable properties and the fiber group {which causes the discrete changes, which, in turn, are defined in discrete time intervals}.

Note: Particle-physics is both indefinably random and non-linear, ie it is a descriptive structure with a very limited range of practical use, and it seems to only be useful in regard to determining the cross-sections of elementary-particles, ie cross-sections are related to determining probabilities of particle-collisions, and thus related to the rates of nuclear reactions (used in bombs).

The new context for physical description can easily account for the stability of material systems, as well as the structure of complex material-interactions, eg material component collisions, which, mysteriously, result in new stable systems, depending on both energy and resonances, but it is a description which can also account for the apparent random properties which small systems are observed to possess, as well as the point-like properties of their observed interactions.

That is, the new descriptive structure is consistent with observed properties of material systems contained in metric-space so as to account for both randomness and stability, and it is a new context which possesses relations with widely diverse new math constructs, related to the widely diverse properties of:

1. material-systems,
2. angular-momentum,
3. life,
4. mind,
5. religion,
6. cosmology, and
7. it identifies a new context for practical creativity.

That is, the new descriptive language both transcends the idea of materialism, and it contains the material-world as a proper subset.

Furthermore, it identifies a context in which "the shape of space" defines envelopes of orbital-spectral stability, ie the context of general relativity, wherein the shapes of interest, in regard to these envelopes of orbital stability, have the properties of being: linear, continuously-commutative, and metric-invariant; so that it is a context which can be related to angular momentum, ie it provides general relativity with a practically-useful context, wherein the often expressed idea of (non-linear) worm-holes, associated with general relativity, are now relatable to the linear, continuously commutative, and metric-invariant structures of angular momentum structure of very large and higher-dimensional shapes of metric-spaces, ie related to the toral-components of these very stable discrete hyperbolic shapes of these metric-spaces.

43.

Higher dimensions

Why are (the) higher dimensions (hypothesized by the new ideas about physical description) not observed?

There are several reasons for the difficulty in perceiving higher-dimensional properties of existence:

0. In an 11-dimensional over-all containing (hyperbolic) metric-space an n-dimensional fundamental domain may fit into any of (11-n) different independent (n+1)-subspaces. That is, when a fundamental domain takes on a shape then it also chooses the (n+1)-dimensional subspace into which it is contained. Thus an 11-dimensional containment set which is partitioned into different dimensional subspaces which are represented as (discrete hyperbolic) shapes (where the shapes are based on being fundamental domains), and so that each subspace of each dimensional level has a largest shape associated to itself . . . , (and also a smallest size shape which the metric-space contains, where the need for a smallest is derivable from there existing a largest shape for each subspace, ie a finite number of largest shapes in a partition, where the partition is a finite set, and the requirement that all shapes of metric-spaces contained in the 11-dimensional containment set are resonant with a finite spectral-set defined by the partition of the 11-dimensional space into shapes) . . . . , identifies sets of set-containment trees based on the sizes of the shapes which are associated to each of the given dimensional subspaces.

Is it needed that all of the smaller shapes whose dimension is less than or equal to the dimension of the containing shape (in the containment tree) be resonant to some spectral value of a face of the largest shape which contains the lower dimension or smaller shapes?

Must the smaller shapes, whose fundamental domains have the same dimension as the largest fundamental domain, in that particular dimension subspace, also be contained in the shape of the subspace's largest fundamental domain? (Yes, otherwise the properties of these same dimension shapes, associated with these smaller fundamental domains, would not be stable within the containing metric-space (where the containing metric-space possesses a shape). And (Yes, in regard to higher-dimensional interaction properties of these smaller shapes, ie if the properties of these smaller shapes are not stable then their associated stable properties would not be observed, where observation depends on interactions.)

1. Apparently, we perceive in 3-space, but the 2-dimensional discrete hyperbolic shapes . . . , which define charged systems . . . , can be used as faces to define a "discrete hyperbolic 3-shape" contained in a particular 4-subspace . . . where these 3-dimensional shapes contained in a particular 4-subspace are charge-interaction shapes that are contained in 4-space, where the metric-space referenced here is a hyperbolic metric-space.

   However, the 3-subspace, wherein we perceive, is (also) a discrete hyperbolic shape which is the size of the solar-system, where this solar-system-sized 3-shape is (also) contained in a particular 4-subspace, so the natural size of material, which would be perceived, so as to appear to interact in such a 4-metric-space would be the 3-shapes (which are contained in a 4-space) which would be the size of the solar-system.

   Thus, in the 3-subspace within which we focus our attention, we perceive the smaller condensed material, whose 3-shapes are smaller than the solar-system, and within this context we perceive only the inertia of the 2-plane (inertial) circles of some [ie only the 2-planes contained in the particular 3-subspace] of the 2-toral components of the 2-dimensional facial structures of which these 3-shapes are composed.

2. The containment of material in metric-spaces, where the metric-space can be a fundamental domain which, in turn, can possess a discrete hyperbolic shape (contained in an adjacent higher-dimensional subspace, also restricted to being a fundamental domain), so that dimensional changes are not continuous, but rather dimensional changes are discrete changes in metric-space containment and material interaction structures, ie descriptive structures constructs on higher-dimensional coordinate containment spaces have different properties (as well as different shapes) due to changes in dimension (and their various relations to higher-dimensional containment subspaces of these shapes), and these discontinuous changes of structures can include different sized materials contained in a new containment metric-space-shape, where all these new properties are identified between adjacent dimensional levels, so that these differences in size determine different size-scales within which material interactions are observed, eg the relative sizes of the

interacting material-components depend on the particular subspace within which the (interacting) material-components are contained.

3. The nature of inertia, where inertia is defined as a circle-size defined in a 2-plane, and our focus on inertial properties of material.

4. The dimensional structure of material interactions, and the discontinuous changes in the dimensional and size nature of material and its interactions, where these interactions (and the properties of material) are re-defined between adjacent dimensional levels.

A dimensional structure is only continuous for a given dimensional subspace-metric-space in regard to its own lower dimension subspaces, and because . . . . of both the dimensional properties of interactions, and size and dimension properties of the metric-space, within which both the planets (upon which we exist), and our (learned) perceptions are contained,

Where . . . . we interpret our perceptions as being material on earth . . . . , and where the dimension of the metric-space which determines the stable properties of our solar system is both 3-spatial-dimensions and 3-hyperbolic-dimensions, (*) so that . . . , thus, we only can perceive the 2-dimensional faces of any higher-dimensional shapes which are contained in such a 3-space.

However, the effects of higher-dimensional metric-space structures, ie higher-dimension than 3-space, are observed;

such as

1. the stability of the solar-system, which implies the 3-shape interior of the orbital 2-shape of the solar-system (ie concentric 2-tori) . . . , where this 3-shape (whose fundamental domain is "bounded" by 2-faces) defines both the metric-space of our perception, and the structure of the solar system . . . , but this 3-shape is contained in a particular 4-subspace, which, in turn, also defines a fundamental domain so as to possess a 4-shape (or a 4-metric-space), which, in turn, is contained in a particular 5-space, etc.

    as well as

2. the other stable spectral-orbital structures associated with some interacting material systems, which have the property of being stable-patterns,

    such as

2a. nuclei,

2b. general atoms,

2c. molecules (and some molecular shapes),

2d. crystals,

2e. the solar-system, etc.

That is, the new context associated with the partition of dimensional subspaces by metric-space shapes so that each subspace has a largest (observable) such partitioning shape, seems to be the only context within which stable metric-space structures of various dimensions can both, exist, and possess various interpretations in regard to these shapes playing the role of either metric-spaces or material-components whose properties are stable.

The physical systems of: as nuclei, atoms, molecules, crystals, the solar-system, are all either 2-shapes or 3-shapes, though 2-shapes are mostly associated to being models of explicitly charged systems (as well as the phrase following (*) above, which means that we cannot perceive these 3-shapes in 3-space, rather only (some of) their 2-face related shapes (which are consistent with the particular dimensional subspace structure of the set-containing subspaces).

Identifying our perceptions with 3-space means that our perceptive abilities are cut-off from perceiving the 4-space within which our 3-dimensional-metric-space is contained.

These 3-shapes are contained in 4-space within which there are defined interactions of charged material components, due to force-fields, which in 4-space define a pair of 3-dimensional fields which are contained in 4-space, ie in both 4-Euclidean space and 4-hyperbolic space. The pair of 3-fields in 4-space can be used to define a pair of 3-fields in 3-space by identifying $z + w$, and $z-w$, as the 3rd vector or similarly identifying w with t (time). These different 3-fields can be further distinguished as elements of particular 3-spaces due to the group structure of $SO(4) = SO(3)$ x $SO(3)$, where each $SO(3)$ is related to a 3-space, (x,y,z) which in turn, is contained in 4-space, (x,y,z,w).

Conjecture: In order to observe a 3-shape in 4-space one needs an inertial interaction, but the material involved in such an interaction, for the 3-subspace of our perception, would have to be the size of the solar system, except for the slight interactions of Van der Waals forces between molecules, etc. That is, the resonance-size hyperbolic metric-space (or 3-fundamental-domain) is the size of the solar-system, where this 3-shape, in 4-space, including the earth's orbit, which is related to the 2-face subspace-shapes of the 3-shape.

Note: There are a pair of 5-fields in 5-space, but there are three 5-fields in 6-space, etc.
(note: these field constructs were incorrectly stated in a recent previous paper)

The 3-flow (of a 4-shape) . . . , which defines the 3-subspace within which we perceive material properties . . . , is the size of the solar-system.

Basically we perceive inertial material interactions.

Inertia is defined by discrete Euclidean shapes which resonate with certain 2-toral components of the discrete hyperbolic 2-shapes which identify stable material systems (or components) which are perceivable in hyperbolic 3-space, where the inertia of these 2-torus's is determined by the sizes of circles on 2-planes, so that inertial interactions define 2-tori contained in 3-space (3-Euclidean space).

Thus, we also only perceive 3-space, since we focus on inertial changes in regard to material properties of interaction, and these properties are restricted to 3-space.

Complicated life-forms, such as human-life, can be a: "relatively-small shape, which is both oscillating (so that, such an oscillating shape generates its own energy), and 3-dimensional 'discrete hyperbolic shape,'" where such a 3-shape system would be smaller than the size of the solar-system.

If the life-form (which is related to the 3-subspace of existence within which we perceive) is higher-dimensional (than 3-dimensions) then it would have to be a size which is bigger than our solar system.

This 3-shape would be contained in 4-space and its material (inertial) interactions would be associated with the fiber Lie group SO(4) = SO(3) x SO(3), where one SO(3) is related to stable material and the other SO(3) is associated to oscillating material. Furthermore each SO(3) can also be associated to a pair of 3-dimensional force-fields. In regard to the 3-subspaces (of 4-space) upon which the SO(3) groups would act, consider that 4-space would be (x,y,z,w), [in regard to the 3-subspaces (x,y,z) and (x,y,w)] by associating the (z,w), or z + w, and z-w, with each of the two 3-subspace upon which the SO(3)'s act, so that in one of these, say, the w-coordinate, is associated to time, while in the other 3-subspace the z-coordinate is associated with time, so that the two types of material each have separate 3-subspaces within which to exist, and each type of material would also have a pair of 3-force-fields, ie the electric-field and the magnetic-field, associated to their material interaction structure, but in each such space-time, and for each force-field, the material interactions would be similar to the material interactions of 3-force-fields in 3-space but now there are a pair of these 3-force-fields where one is associated to position interactions and the other force-field is associated to interactions associated to moving charges, since the time coordinate can be related to velocity properties.

44.

Property rights vs. the commons

The model of civilization provided by the Roman-Empire is the same model of civilization used today.

Namely, one acquires new property and new riches by violence, then, supposedly, one uses knowledge possessed by the culture, and related to practical creativity, to build structures, eg water-supplies and sewers and public buildings etc, which, it is claimed, improves the condition of life for those just conquered peoples, so that the conquered population becomes "used to" the built-structures, which are associated with these structures being used within society, in a practical (but very narrow) manner, by the people, where this public use would occur during the time-period of an oppressive occupation, an occupation which is based on violence, so that a way-of-life becomes familiar to the people, (a way-of-life associated to the social-physical structures of a technically identified social context), and the violent-imposition of oppression is administered by social institutions, within the new social-technical context [Does this sound familiar?], so as to define a social hierarchy.

Subdue a people (and their culture) by violence (steal what these peoples have, which has "value"), occupy based on both oppression and, so called, technical civic improvements, [ie provide free-lunches (or public works)], so that within such social context an institutional hierarchy is implemented, then take away the free-lunch (by instituting taxes) after a hierarchical social-oppression is institutionalized. Then within this narrowly defined but easily managed social context, swell the population so as to define a growth, based on narrowness, which is organized to be managed by only a few in the ruling class.

That is, the commons have been (come to be) redefined as the public-works, which have become a part of hierarchical institutions. The society has been privatized, based on a controlling social hierarchy. The commons can be ran-sacked at any time based on property rights and

384

minority rule. This expansion was based on violence, and "the commons" were where there was no apparent value, and, thus, it was the place upon which the low-levels of society depended for their survival.

The issue is about law being based on equality, or based on property rights and minority rule.

In fact, one sees that water-supplies and sewers are unnecessary, when one considers the lives of the native people's of the Americas, their needs for food, water and shelter were easily met, simply by living in harmony with the world's environments, whereas, sewers get emptied into the water-ways and then the wastes polluted the village down-river.

If the water from the streams is not drinkable then it is being polluted by some improperly considered technology.

However, people should easily adapt to changes by moving elsewhere, rather than staying fixed, so as to admire their public-buildings.

That is, the great-value [related to the, so called, practical creativity (or to civic engineering)] of a created context results more from "a circularity and repetitiveness of a communication system," which exists within a hierarchical social system. That is, a clean water-supply is highly valued by the population only after the river has been polluted by a sewer-system (which is located somewhere).

Today developing water-filters to combat water-pollution is important, as is combating-disease, and building sanitary conditions, eg using the new pit-toilets, and developing energy through the re-new-able energy sources (including thermal-solar) are all great uses of practical knowledge.

But tinkering with genetic chemistry, wherein the relation of "DNA to the enzyme systems of life-forms" is not understood, and thus such tinkering is total insanity, since the relation of the chemical properties of DNA to the life-form and to the web-of-life is simply not understood.

Such chemical-tinkering is based on ignorance, an insufficient knowledge, and this use of ignorance is allowed, based on decisions within a (corrupt) justice system, since big corporations are using this incomplete-knowledge as a way to use a hierarchical system of knowledge to control a market (and the population), so the justice system is upholding this insanity and this great ignorance.

The earth supports the life which exists on earth, while practical knowledge is used to "improve one's living conditions," or make one's living conditions more conducive to developing practically knowledge related to creative (and selfless) works.

The basic question is about existence, and the realm (context and range) within which the "practical creativity" of humans is defined. Is it only the material world, or is there a natural context of existence and "practical doing" which is not confined to a material existence? This is where both science-math and religion have failed, and they have become managed by corrupt and/ or ignorant people who are locked-in to a very narrow narrative, based on materialism and selfish gain, and who are afraid to seek to realize what they are within existence, frightened to either consider or experience how they reflect all of existence.

That is, "it is the violent destruction for selfish purposes of acquiring riches and property" which, of course, implies the idea of property-rights, which results in such a social construct being associated to a failed, but narrow and

fixed, way of doing-things.

It is the relative (or controlled) implementation of already developed practical knowledge (of the society which has been over-run by violence), and which determines if a "violent take-over of property" can be used to (or justified by) institute public works, which "improve the living conditions of the newly conquered society (people)," so as to then be able to institutionalize a social hierarchy based on an external knowledge, in turn based-on the selfishly motivated view of a material existence which is held by the conquering forces which seek to institute a social hierarchy, primarily built for the selfish-gain of a few.

In the US it is a question as to whether it is the social model of either the "violent Puritans" of New England (who represented the Roman model of violent take-over of the native people's land, but the so called better-life was only better for the colonizing Europeans) (this leads one to ask, Was the Roman-Empire also based on the same type of self-referential culture based on colonization?)

Or

The friendly Quakers, whose social model was to lived in harmony with the existing native populations and to interact with the other native cultures, whereas the native people's did not have the acquisitive life style of Europeans, and it is the native people who lived happier, more self-confident lives and who possessed a very deep knowledge both of the material environment of the native peoples (their lives were not all that difficult [they easily had food water and shelter], and they traveled and traded over great ranges) and

Knowledge about what existed between the material and the spiritual worlds, or the verbally-spiritual world, where the verbal-spiritual was the propagandistic model of religion which the Europeans knew about religion, (that is, the native peoples used the knowledge which they possessed and that knowledge existed between both the material and the living nature of existence).

The Quakers were the more energetic and more thriving community while the Puritans seemed to relish torturing and subjugating the people in their community, who deviated from a norm, where this norm was identified by a few people who comprised the ruling class [This should also sound familiar, ie What "norm" do today's surveillance-mangers serve?].

For example, B Franklin ran-away from Puritan Boston so as to thrive in the Quaker communities, but Franklin was arrogant and did not sufficiently praise and support the equality of the Quakers of which he needed so much in order for his life to thrive.

That is, the spirit for the Europeans was a form of propaganda about possessing high-value, but they attached themselves only to a narrow vision about possessing riches in an exclusively material-world.

Today, the "ancient aliens" TV show on History-channel, is still grappling with the ideas of ancient religions concerning the powers of giant peoples (eg so many religious stories of creation deal with the giant origin of the first humans, ie emerging from the constellation Orion (the giant)), which the religions represent as being both narrowly confined in their views and selfish in that narrow outlook as to the existence-properties of the Gods, ie the difficult challenge of continually transcending the creation of existence by the natural living entities which are fundamental to existence,

So that the "ancient aliens" TV show, attributes these mysteries to the existence of superior-intelligences who can better deal with the material context, so as to provide the lowly mankind with technical gifts concerning the material-world, ie a myth of inequality and the diminution of the powers of individuals of the human species, ie a very narrow context concerning the nature of existence.

The amazing building capabilities of the ancient world was a world where knowledge was about both (1) the material world and (2) "the living world and a world which transcends the material world," and this knowledge has been demonstrated by the skillful work in very old cities, which are over 12,000 year old (perhaps greater than 30,000 years old) wherein structures of great size were built so that the work demonstrates deep technical knowledge of material and a capacity to technically create in a material context at a high-level, even beyond our technical understanding today.

These engineering marvels of ancient mankind are shown on the "Ancient Alien" TV shows on History-channel, but they are improperly interpreted to show that mankind is inferior except for the few elites.

But the correct interpretation is that;

This was possible for mankind who possessed a deeper knowledge, which transcends the idea that there only exists "a material world," and the absurd idea, based on violence, of "an arbitrary high-value being represented by an elite few," who are to be allowed to "dominate society by violence."

The current basis for science, ie the knowledge needed for practical creativity, first it is assumed by science that "the world can only be a material world" and that different regimes of differential equations define measurable properties of physical systems

1. Classical solvable and based on the existence of measurable properties of relatively stable material systems, this is the set of precise descriptive structures upon which a great deal of all the practical technical development in the western civilization is based. Note: "western" means society based on the Roman-Empire, or whose main religions are Judeo-Christian-Islam.

2. Classical, non-linear these are quantitatively inconsistent descriptive structures, used to describe unstable material system patterns, which can only be controlled based on the properties of the system's differential equation itself, ie not the (solved) properties of the system, where the information derived from the properties of the system's differential equation can be used in feedback systems, ie this means that the system is contained in a discontinuous set of descriptive structures, between which the system's description is changed where these changes are based on the observed properties of the system being fed-back so this information is used to alter the description of the system, but the system still being related to the differential equation (the changing conditions are discontinuously and totally outside the context of the system's descriptive structures)

3. General relativity also non-linear and un-relatable to practical use, since the observed properties of the solar system are stable, but the descriptive structures available to describe the solar-system can only describe a non-linear quantitatively-inconsistent, and thus chaotic, context.

Furthermore, there are the very stable nuclei, general atoms, molecules, molecular shapes, and crystal properties, which all go without valid quantitative descriptions.

Thus there are

4. Quantum physics randomness, ie function spaces of harmonic-functions, and operators, which represent measurable properties, together are used to define a statistical mess, which *cannot* be used to identify the observed stable spectral-orbital properties of general quantum systems, because the observed spectral properties of general, but very fundamental, quantum systems cannot be found to sufficient precision with this method, it has become a method of statistical (or probabilistic) manipulation of non-physically motivated models of these general quantum systems. Within this failed context, of not being able to identify by physical law the spectral properties of general quantum systems to sufficient precision, it can be defined as a method of indefinable randomness.

It is, more or less a method which tries to identify macroscopically physically measurable properties with sets of operators, which, in turn, act on sets (or spaces) of harmonic functions, ie functions of the form, Ae^i(Et-px), that is operators which act on sets of functions whose domain spaces are defined on sets of circles.

That it, quantum theory is a description which tries to use the properties of probability to fit the descriptive structure to the observed data.

Yet many of the reasons for abandoning classical, or geometrical-and-measurable based, description are not really resolved by the quantum context, such as accelerating charges giving-off electromagnetic radiation (so as to give-off energy) . . . , [whereas a bounded system composed on individual, free charges moving in an unknown, yet, nonetheless, bounded context, in turn, requires these free charges to possess various types of charge-accelerations, and, thus, causing these charges to give-off electromagnetic radiation], ... , so as to make the system unstable.

An alternative

Whereas, If one identifies the descriptive context so that the (quantum, or physical, charged) systems and their interactions exist in a set of open-closed metric-space structures, whose geometry is that of circle-spaces, so that the description depends on processes defined in a discrete manner, with discrete relations existing between the separate descriptive structures (or separate metric-spaces, which are involved in the descriptive structure), then this is a descriptive context which has many similarities to a harmonically-based quantum description, but would account for a closed system, which, when left undisturbed, would not give-off electromagnetic radiation.

and

5. Particle-physics, whose properties are only related to calculating elementary-particle cross-sections, "Where is a precise description of a general nuclei's stable spectral properties, where the description is based on the laws of particle-physics?" (Answer: No such description exists, ie it is a physically useless theory since it is based both on indefinable randomness, ie it is mathematically structured to adjust the wave-function of a regular quantum system, and non-linearity, ie its is both chaotic and quantitatively inconsistent)

Particle-physics is based on data which can be interpreted to mean that there are dimensional levels in regard to U(1) x SU(2) x SU(3), which are unitary, ie based on pairs of opposite metric-space states, where these levels have stable patterns associated to themselves, ie the particle-collision patterns, but whose interpretation is confined to the idea of materialism, but the particle-collision patterns are not stable properties, but the patterns of this suggests the idea of grand-unification which is an idea related to SU(n), and this does imply the material world is but a single dimensional-level within a many-dimensional context where the many-dimension model implies a transcendence of the idea of materialism. There is a new idea which builds the higher-dimensions with stable geometric shapes, rather than trying to build it with the ideas of indefinable randomness and non-linearity.

and

6. Derived theories, eg string-theory, derived from both particle-physics and a model of gravity consistent with general relativity, it is a geometric theory consistent with indefinable randomness, non-linearity and the idea of materialism.

Where all of the above descriptive structures demand the idea of materialism, and all of these descriptive structures cannot be used to describe the observed stable properties of the above mentioned fundamental physical systems. The basis for the current math models of physical systems is both indefinable randomness and non-linearity, and these descriptive structures cannot be used to describe the observed stable properties of (general) physical systems, which exist at all size-scales.

45.

The formal language of mathematics

These ideas go to the heart of oppression . . . . , and at the heart of oppression is arrogance . . . . , and a blinding confusion (the media, including Portland indy, is filled with great concern about complicated, irrelevant issues . . . the issue is equality and creativity . . . , the rest is a bunch of mind-distortion brought on by the media . . . . where the media is especially about "the control of technical issues").

Perhaps the professional math community could have the correct interpretation (or are correct), that the experts have superior intelligence which ordinary people cannot understand.

However, formally fixing the axioms of math and fixing its context, was an agenda imposed on the math community, which (at the time) this agenda was greatly opposed by some of the "top" mathematicians at the time. Namely, Brouwer and Klein, etc, who expressed the ideas, such as the formal axiomatic language of math not being able to express any useful ideas, which my comments have also expressed, but I also provide a complete alternative to the failed context of math today.

The paper (upon which you are commenting) is about a book review, and the book (Plato's Ghost: . . .) was about the transition of classical math (with its more intuitive basis) to the axiomatic context for math, ie formalism vs. intuition is how it was framed in the book, where it was stated in the book that Brouwer . . . "aimed at nothing less than overturning Hilbert's formalized axiomatic methods, an approach which Brouwer regarded as content-less."

These are not my words, rather this apparently was expressed by Brouwer, approximately in the year 1900, but this belief that "formalized axiomatic math" will be "without any content," or "the patterns which it describes are not useful," is exactly what the record of technical development suggests.

One must look at what is being accomplished in regard to development related to the idea expressed by math and science in the modern era, ie since 1910, and it is not much, other than a lot of hyped-up literary-expression about modern math and modern physics, but these disciplines have virtually no relation to the development of practical creative efforts, in regard to building new instruments, or in regard to using new processes (whose affects due to the actions of the process within the context of the process are not catastrophically destructive, eg tinkering with DNA when the model of life, as a material entity, is such a primitive model, ie a process done within the context of virtually complete ignorance, and thus could result in a destructive context which is not understood by those tinkering with the DNA).

Essentially, the modern ideas of math and physical science, in regard to physical instruments, are only related to the bomb and the laser; while other technical developments, since 1910, are about coupling quantum properties to classical systems.

However, classical physics provided the basis for a wide range of technical developments around the 1900's, and the vast majority 99.5% of our technical devices, which we have today, including computers and microchips, are based on 19th century classical physics, while 20th century science has contributed virtually nothing to this development.

Though, in 1950, it was promised that cheap, clean fusion-energy would be online by 1955, but they still claim it is only "two years away," even today.

More recently there is the claim that "there will be quantum computers," again this was made around 1980, and today it is claimed they will be developed in about two years. Does this claim sound familiar?

This lack of relation to practical development is at the heart of Brouwer's criticisms, about math formalized about a fixed set of axioms which are to be applied in all contexts of measurable description, subsequently, the measurable descriptive contexts are no longer of any practical value.

If one tries to hold a technical and precise language fixed, held fixed to axioms and certain contexts and certain interpretations, then it rather quickly loses its relevance in regard to practical usefulness.

What the media does, is that it expresses the idea of inequality being the natural law of human existence (evolution, business competition, history [as told by the winning side], etc), and it provides rational for what the ruling-class does, and it develops a narrow model of knowledge, which is the knowledge needed to support "what the ruling-class does," and it expresses these ideas by continually repeating these ideas.

Subsequently, high-social-value comes to be defined by traditions and narrowly considered ideas, all of which serve the interests of the ruling-class, and which the public internalizes, due to its constant repetition, and due to the fact that they are wage-slaves, so that the ruling-class and their minions . . . , (ie those who have internalized these ideas of inequality and successfully competed for high-valued jobs) . . . , possess disdain for the ignorant-herd, ie the public, but, in turn, these high-valued minions support (in a thoughtless manner) the ignorant constructs of the ruling class, and the so called superior intelligence's whom work for the ruling-class, ie the scientists, the journalists, the religious leaders (ie the experts of the narrow dogmas and their supporting propagandists), so as to either divide society into squabbling "arrogant" factions claiming superiority over one-another, or into the "general public," whom are attacked by all the higher factions of society. Inequality can only be maintained with extreme violence, and in an amoral culture.

What the media does, is that it makes claims about the superior intelligence of its experts, so that the small-intelligence of the public cannot be accepted as possessing any type of valid thought.

However, if one considers the definition of intelligence, one sees it is mostly about rates of acquiring cultural knowledge, ie it is about acquiring the knowledge which the ruling class wants its experts to consider in regard to developing ideas which will improve the power of the ruling class.

That is, intelligence is vague idea with no valid definition which is measurable, except in the context of the statistics of indefinable randomness eg the speed with which one acquires culture which is (believed by the ruling class to be) of importance to the ruling class.

On-the-other-hand, intelligence could be tested in regard to a young person identifying: their own house, and finding ones way to school (or to the bus stop, to school), and identifying information about one's favorite TV show, etc. Then it would be found that all people have equal intelligences in regard to their rate of learning measured against learning things which are important to them.

Is it not convenient that axiomatic math also defines indefinable randomness? However, indefinable randomness is not valid . . . , "as the failure of the financial risk calculations, which was a large part of the cause of the 2008 economic crash," . . . , because the invalid risk calculations were quite wrong. That is, monkeys throwing darts to calculate risks, do better than the MIT PhD math and physics people did at calculating economic risks for the (deregulated, gambling-casino) banks.

Apparently "the thoughts of the public" are at least as valid, if not more valid than the failed ideas of the experts, where cataclysmic failure was expressed by the experts failed knowledge concerning risk calculations, where this axiomatic knowledge was rigorously proven to be true, but it is only true within the formal language of the math experts, as usual, it is not true within a context of an actual existence.

In the same Bulletin of AMS journal, there were articles about differential forms, where the context was about non-linearity, perhaps this does have some relation to the issue concerning the range a validity of a system's (non-linear) partial differential equation, and thus to the structure of limit cycles associated to the non-linear partial differential equation, but it is describing a math context where nearly all the math patterns, related to solution functions, are unstable and chaotic. That is, it is useless information in relation to the about 50 page effort given to the discussion.

There is another article (book review) about discrete probability, where randomness is categorized into pure randomness and randomness with a pattern. Now the numerical properties of random sets (or processes) can sometimes themselves be related to a pattern, eg if one has the math structure to calculate a mean and a standard deviation, defined on a random set of events, then there are rules about the number of data points with so many standard deviations from the mean, and there can be external reasons for the existence of patterns in data, for example, the measurable context might be related to a non-linear partial differential equation, and thus the data has limit cycle patterns in its (time dependent) convergence patterns, but there are reasons why the category which possesses patterns may not be able to determine a valid probability for such systems. The process may possess "discrete properties," in which case the discrete process might perform a "discrete jump" which avoids the (convergence) structure of the detected pattern, eg a discrete jump to a different region of limit-cycle convergence for the partial differential equation, or the external structure may suddenly become irrelevant to the random process, eg the system's partial differential equation may suddenly becomes irrelevant, where this is the sort of scenario in which the above mentioned article, about the differential forms of non-linear systems, may or may-not have some information to add to these considerations about what are mostly indefinably random contexts, ie based on vaguely distinguishable patterns, ie patterns which may be unstable, and thus they are random events cannot be counted in a reliable manner.

Now this indefinably random context is used with some success, but this is not because of the structure of randomness rather it is a measure of an underlying pattern of behavior, most often related to the obsessive and continual repetition by the media in regard to certain ways of using language (by the media), so that what is measured is a stable structure of the language of the media, and not the "free behaviors" in regard to the context of the events being counted.

Basically, the context in which random properties can be used, is when there are many components which compose the system whose (component) properties can be related to average values, which can be measured in regard to the large system, and which, in turn, are related to a causal, measurable system. Otherwise randomness is mostly about determining probabilities, but it is only in a very limited context . . .:

"of the probability being relatable to finite sets of stable, well-defined random events"
. . . , wherein these probability calculations should be considered reliable.

On-the-other-hand, discrete structures can easily be related to randomness, in regard to discrete interaction processes which possess a random property.

Formal language vs. language built around separate sets of contexts and interpretations etc

Can the formal language of math be related to the existence which we perceive?

Answer: Basically, no, since there are too many limitations as to the patterns which a formal language can describe, and subsequently, a formal language can be virtually unrelated to any pattern of existence, but once formal language of math is entrenched in the professional math community, the formal language becomes both ever more complicated and relatable only to the patterns of other math literature, with nothing (or very little) to say about the existence we experience.

Science has also become equally formal and mathematical (using the same formal, complicated math language) with its math structures improperly interpreted in regard to the relation these math patterns have to observed patterns, ie if the nuclei cannot be described using the laws of particle-physics then this gap between math patterns and observed patterns needs to be taken as an indication that the formal math patterns are not relatable to the perceived properties of our existence.

That is, in both math and physics there is an endless stream of literature . . . , which only has a valid logical and formal relation to other math and science literature . . . , whose structures and laws cannot be used to accurately describe "what is observed" by applying, in a both a wide-ranging and sufficiently precise way, set of rules, which allows math descriptions to precisely describe the observed properties of general systems.

By the failing of math methods to describe the stable properties of physical systems, this means that the formal structures of math and science are inconsistent with what actually exists, ie the world's as well as life's containment set.

There now is an alternative containment set, interpretative context for "what it is that we observe" so that assumptions, axioms, contexts, interpretations, and containment sets, etc are re-invented so as to be able to describe the observed patterns of the physical world, but it is a description which transcends the idea of materialism. That is, it is an example of the intuitive method which works better than does the formal methods now used in math and science.

Is the context for physical description to be: differential equations defined for functions or functions spaces whose domain space is a metric-space,

or

Is the shape of a metric-space the fundamental property of the physical structure, where the metric-spaces (which have shapes) are, in turn, contained in a domain-space which is another (higher-dimensional) metric-space, which is also a shape; and so-on; and where derivatives are defined in a discrete manner on stable metric-space shapes, in a context in which there are different, and separate (because of the shapes of the metric-spaces), dimensional levels?

This is actually, the correct context which allows for stable physical properties to exist in a stable manner and to be described in a measurable context.

In general relativity, the question about the metric-space properties is central to describing physical properties, but it is framed in a non-linear context so that measuring is not reliable, and the context is quantitatively inconsistent, ie it is non-linear, so that no physical properties are describable, ie it is useless in regard to practical use.

Modern formal math is based on the idea that one can consider any context, which is vaguely related to the math structures of quantity and shape, and calculate with formal math patterns in a consistent manner, but this is not true.

Most contexts (of existence) are not consistent with the elementary properties of quantitative patterns, ie the formal math patterns are quantitatively inconsistent patterns and most shapes described by formal math patterns are not stable.

So the formalism of math allows for a great range for math literature but it is mostly quantitatively inconsistent nonsense, which can neither identify (stable) patterns, nor can the properties of formal math descriptions be consistently measured, so it is unrelated to a measurable context within which stable patterns exist.

Formal math does not have any (or it has very little) practical value.

Can math formalism be applied to any quantitative context? or Must math descriptions of math patterns always adhere closely to the properties of elementary arithmetic? The latter.

Must the patterns described in a math context, which is quantitatively consistent, be stable or can they be unstable? They must be stable.

Is a derivative a local linear model of measuring or can it be non-linear? (It must be linear.)

If many-dimensional then must the local matrix properties of a derivative, which are associated with the local linear vector structure, which the derivative gives to the local coordinates of the function's domain space, be continuously commutative, in order for the math description to remain quantitatively consistent, and so as to remain consistent with the properties of elementary arithmetic,

or

Is non-commutativeness a valid property of formal math descriptions? (Quantitatively consistent patterns require continuous commutative relations on matrices, as well as on function spaces.)

Are derivatives operators which act on function spaces so as to allow non-commutaviety between derivative operators,

Or

Are derivatives about local measurable relations which exist in a discrete manner defined between stable metric-space shapes? (the latter is a better context for the description of existence)

In order for measuring to remain quantitatively consistent, must the domain space be metric-invariant, and be defined on a local linear context, so that all the local linear geometric measures are consistent with metric-invariance in a continuous manner, and so that the matrix associated to the metric-function's symmetric 2-tensor structure is continuously commutative? Yes.

Is a derivative best thought of as a local linear measure which is defined in a discrete manner in regard to the changes in the properties, such as position, to which its local measures are related, ie discrete changes associated to a derivative which is also defined discretely? Yes.

In this discrete definition there are also discrete relations which exist between different adjacent dimensional levels, and are related to discrete shapes of either the interaction structure, or discretely related to a stable shape for a metric-space.

Within a given metric-space the discrete structure of the derivative can be approximated so as to define the usual (usually second-order) sets of differential equations of physics, mostly classical physics, but the new description also accounts for the property of quantum randomness, and so differential equations associated to random properties do make sense, but they are very limited in regard to their practical value, since the stability of physical systems actually comes from their stable metric-space structures.

That is, formal math structure can exist but its relevance is very limited.

Are formal math structures to be fixed and proclaimed for all-time, and then applied in a formal way, where the validity of the math patterns is assumed to always be true? No.

This is a paper related to a book review in Bulletin of the American Math Society July 2013 concerning the book "Plato's Ghost: . . . , by J Gray, and Reviewed by D E Rowe. This is a review whose focus is on the conflict between the formalists (determine a fixed language for math) and the intuitionists (relate math patterns to the properties of existence so that math patterns are stable and quantitatively consistent) which occurred in math, in a time period around 1900.

Rowe's review apparently is more sympathetic to the intuitionists view (where the author of this paper is an intuitionist) than was Gray's outlook.

The main issue of the book is about how modern math has come to primarily be the domain of the formalists.

But formalism has led to some significant failures, since rigorous formalism can define a formal truth, but many, if not all, of the applications of the formal patterns of modern math fail to provide valid answers when applied to the real world, ie existence and the formal truths of mathematics are inconsistent with one another.

There exists a deeper criticism of the formal language of math, where the alternative to formal math is to continually re-building language so that it remain both valid and practically useful.

Where the formal language of math, though rigorous, (the language) is neither accurate (in regard to what, about the world, it is trying to describe) nor is the formal language of math practically useful, ie it is not the correct context in which to describe a physical system's measurable properties, however, formalism deals with extremely complicated patterns which delve into a very general context, but the math information about this general context seems to be un-relatable to practical uses or to actual (observed) properties.

Namely

Either, in probability

trying to define risk (or a quantum-system's spectral properties) based patterns

either defined by identifying vaguely distinguishable, but unstable, patterns

or on sets of spectral values (related to a function space) which the descriptive structures cannot identify,

Or, in geometry

Trying to describe an unstable geometric pattern by means of a quantitatively inconsistent descriptive (or containment) context, eg using a non-linear context to describe unstable shapes (or a chaotic context).

This is the realm of formal math.

That is, formal language claims to be capable of describing a system's measurable properties, by applying the formal language structures of math, based on formal properties of an assumed containment in regard to either coordinate domain spaces or function spaces, but it is either not the correct context, or the ideas are inconsistent with the actual patterns which (that) are being observed, yet the formal quantitative structure claims to be the correct containment space for the observed patterns of, for example,

eg trying to identify the probabilities of financial patterns, but the patterns can never be sufficiently defined, ie the patterns are unstable and thus not properly relatable to a probability containment context.

However, there are many examples where formal language is trying to relate patterns to a quantitative context but it is failing, and this is done "even though" there already exists an alternative set of: assumptions, contexts, and interpretations concerning pattern containment, within which one can choose many different quantitatively descriptive contexts, and there exists an already developed new "quantitatively based" language, which exists as an entire descriptive structure, whose logical basis is more intuitive and more consistent with the properties of existence, and which can account for the observed patterns, which the formal language is trying to describe.

An example of an alternative math language is based upon using the ideas of Thurston-Perelman geometrization in a many-dimensional context, in order to describe the stable spectral-orbital properties of physical systems which are observed at all size-scales.

Instead of defining derivatives on either general metric-spaces or general function spaces, rather define the descriptive context to be a discretely determined set of metric-spaces which posses stable shapes and are associated to specific dimensional levels and subspaces contained within an 11-dimensional hyperbolic metric-space.

This math structure, as well as the criticism of the accepted context of formal language used by the professional math and science communities, is presented in several books "A new Copernican Revolution" B Bash, P Coatimundi Trafford publishing, "The Authority of Material vs. the Spirit" D Hunter, Trafford publishing, and more than four books by M Concoyle.

In regard to social forces, formal math structures are about a formal power relation between math and the state, where the state wants a hierarchical structure imposed on the math community, and in this formalism the state can manipulate control of the subjects of math by placing obsessive competitive and aggressive types within such a community ie controlling a subject by the use of borderline autistic people. It is about control by requiring narrow vision in regard to a subject from which power, which can challenge their (the corporate-state's) arbitrary social constructs of high-social-value, can emanate. This is about the cultural knowledge remaining consistent with the collective actions of a society which collectively supports the interests of the ruling class.

46.

Natural containment

Media language usage vs. technical language usage

Technical thought is (carefully) controlled by society, whereas politics is simply a part of the propaganda system.

The social context of controlling thought

In science and math, just as in the media, it is difficult to express ideas which are different from the authoritative dogmas, (or in regard to politics, difficulty in expressing ideas different from) of the political interests of those (owners of society), who both control the media, and who determine the authoritative dogmas of the expert scientists, since science is about the knowledge used in the society for practically creative purposes, which the ruling class wants to control, but where the word "creativity" is most often about the technical-skill of a narrowly defined artistic medium.

That is, creativity is not discussed concerning engineering, where in regard to engineering, the ruling class must control what is being created.

(They do not want competition of new practical creative products or processes, which might come from new knowledge. The banks had a difficult time surrounding and controlling the late 1800 technological explosion [a process done through copy-right law], which was brought on by the development of the ideas of electromagnetism and thermal physics.)

Issues about "the actual structure of social power" and the "nature of mankind" in regard to "how such a social structure should be organized" (civics, or governmental processes to be used to institute changes do not work, since the media is controlled by the ruling-class, where politicians are propaganda-people selected by the ruling class) . . . . , as well as issues about the actual structure of the math and scientific languages and the nature of math and science in regard to how the language structure of math and science should be organized in order to realize the widest range of practical creative possibility . . . ,

. . . . , are not allowed to be expressed.

When new ideas, which are precisely expressed in a logically consistent context, on some of the, so called, "free speech" outlets, the only responses are a few who defend authority {since elitism, domination, and inequality are the non-stop repetitive messages of the media (and the ruling-class)} otherwise an organized attack (composed of many people), done either directly by the editors, or by a prevalent organized set of people who seem to support the designs of the corporate-government interests, ie the interests of the bankers and oilmen and military. Perhaps these commenter's . . . , who clearly do not read the posts, but who continually comment, nonetheless . . . , are orchestrated by the media moguls themselves, eg G Beck, or perhaps G Beck has written a book "How to make-fun of any ideas which the bankers and oil-men oppose" and it is put to use by the "G Beck ditto-heads" (lol), who may be organized by powerful corporate-government interests.

That is, those who claim to want to stop big-oil from collapsing the ecosystem, need not seek to control all of social organization (which appears to be their plan, based on the way in which their demands fall onto deaf-ears of those who possess social-power), but rather they may need (or find an ideological friend in) new ideas about physics, which can be used to find clean, cheap energy sources, which, in turn, can depose big-oil outright, due to superior technology.

But nonetheless, the faithful-opposition apparently, only believe in the ideas which are supported by big-oil.

Apparently, they do not want to be "made a fool" (they apparently see themselves as being "the intellectually superior types," who do not want to jeopardize this image of themselves, which they present to the world, ie they do not want to tarnish their badge of intellectual superiority in the eyes of others), and thus, they will not support (or will not show any interest-in) any ideas which challenge the authority of physics, an authority which has been used to stifle technology different from oil, and nuclear, and it has been used to build the military-state, which upholds the intellectual superiority of the big-bankers, whose investments determine "what is created within society." It is to these bankers to whom the "intellectually superior types" also demonstrate their intellectual subservience, which is demonstrated by their not being interested in new knowledge (only following a regimented authority, which is narrowly expressed in the media), ie the (so called) intellectually superior types must believe the same ideas about physics and math "in which the banker believe."

This again is understood in relation to the repetitive nature of the media, which only gives voice to the ideas which the ruling class wants expressed, where a journalist (or a professional, so called, peer-reviewed scientist) will not be allowed to express ideas in the media which challenge the very narrow authority upon which bankers base their investment actions.

Practical creativity, and the knowledge base upon which the narrow creative efforts of the society, is determined by the investments in equipment (ie in complicated instruments) and research directions supported by banking investments (in turn, the government investments follow the banker's lead).

Yet, despite this obvious relation between knowledge and practical creativity, eg engineering, being controlled by banker's investments (which, in turn, guide the investments of a government which serves "big industry") the continual repetition by the media of the narrow dogmas of "science," ie the science constructs which are being followed by the bankers, in turn, seems to be the reason that the public, or a "voice of dissent," will not challenge the so called educated authorities (who serve the (investment) interests of banks).

That is, the dissenters are deceived by the media.

They want to save the environment, but they also support the intellectual framework, associated to university math and physics, which, in turn, supports the knowledge and practically creative structures of the banker-oilmen, who are intent on destroying the environment for their selfish interests.

The language of science and math has come to be without any content, due to its "axiomatic formalism" structure.

The control of technical language by society (the ruling class), controlling the language of the experts

This is a paper related to the debate concerning "axiomatic formalism" vs. "intuitiveness," in regard to the application of math patterns to measurable physical descriptions, so as to result in both practically useful precise descriptive information and a valid context in regard to it being consistent with the structures of existence (the debate concerning "axiomatic formalism" vs. "intuitiveness," were ideas expressed in a "book review," see (1) below).

If the axiomatic formalism, does in fact, lead to descriptions of patterns which have no content, as claimed by Brouwder (where this claim was made in the past [around 1900], in regard to the above mentioned debate, and where if one critically looks at the evidence, one sees a lot of evidence that this is, in fact, true), then what is an intuitive alternative?

Answer: One must focus on the (stable, measurable) patterns, which one is trying to describe, and then create a complete descriptive structure which is to be used in the descriptions of the measurable patterns, where the new descriptive context deals with (or are concerned with) one's creative interests, so as to build an entire descriptive construct in an intuitive manner which is of practical value, so as to provide some sense about (or some explicit expressions concerning): assumptions, contexts, interpretations, and set-containment, etc, in regard to the new descriptive construct.

Note: There may be some universal characteristics of math (or precisely measurable) descriptions, such as:

   I.   math is about quantities and shapes, and
   II. a precise mathematical description is only (practically) useful if the basis of description is both "measurably reliable" and the "patterns are stable," (where measurable patterns are both the basis of description and they are the purpose for [or goal of] a mathematical description).

However, in regard to axiomatic formalism, it is assumed that a quantitative structure exists in an absolute abstract form, where the cause of this truth is put into effect by the proclamation that "such and such an axiomatic quantitative structure 'does exist,'" and are assumed to be true even if [in a general math setting (under axiomatic formalism)]:

   1. measuring may not be reliable (there is not a context of quantitative consistency, eg non-linearity and unstable patterns, eg the metric-function is not reliable model for measuring),
   2. there may exist chaotic dynamic events (which one is trying to describe, ie the patterns are not stable),
   3. patterns and measurable-values of "vaguely distinguishable patterns" determine a condition of indefinable randomness, so that either
  3a. the patterns of either (1) "vaguely distinguishable events" are not stable, or (2) observable, stable patterns are not determinable by math methods,
     or
  3b. geometric patterns (which one is trying to describe) are not stable;
       yet the axioms, of the quantitative sets which are arbitrarily placed into this unstable and chaotic context, are still considered to (continue to) be (or to remain) true. That is, the patterns and the math structures of what one is trying to describe are not consistent with the axiomatic formalism, yet the axiomatic formalism is still considered to be relevant to the descriptive construct, anyway.

However, the only descriptions of math patterns . . . , which are:

1. "measurably reliable,"
2. controllable, and
3. stable

. . . , come from the math context of (partial) differential equation models of (physical) systems which are:

1. linear,
2. metric-invariant, and
3. continuously commutative everywhere.

But for quantum systems this is not possible in the (insufficiently determinable) context of sets of (Hermitian) operators acting on a (probability based) harmonic function-space [where harmonic functions are defined on sets of circles].

The failing aspect of modern descriptions based on axiomatic formalism is that they fail to be able to describe the observed stable properties of physical systems (such as being able to find the spectral values of a general quantum from calculations based on the laws of quantum physics).

The basic pattern of physical science has been "material contained in metric-spaces and differential equations," . . . . , where the properties of their solution functions . . . , which are about the material-system properties, which, in turn, are either geometric or "the random (system related) spectral properties" (of random material point-particle events in space) in nature . . . , are identified . . . , and whose solution functions have the same metric-space as their domain spaces.

But this containment structure has not been able to describe the stable orbital-spectral properties observed at all size-scales, since the differential equation, as well as the descriptive context, is most often both/either non-linear (ie quantitatively inconsistent) and/or effectively indefinably random ie randomness associated to unstable and incalculable contexts concerning a probabilities elementary event spaces.

That is, the metric-function of a system-containing domain space, and the model of local linear measuring associated to the measurable properties modeled as a (solution) function, as well as the very limiting idea of materialism (where, in usage, material is modeled as a point, where a scalar-value is defined), are (all together) not sufficient constructs, concerning the nature of existence, which can be used in order to describe the observed properties of stability which exist in a measurably-reliable (system) containment context.

There is a new (intuitive) descriptive context which is based on the geometrization properties of Thurston-Perelman . . . , ie the limited types of stable shapes which exist in the different dimensional levels . . . , where (in the new context) the material and its (adjacent higher-dimensional) containing metric-space both possess stable shapes, so that differential equations (concerning the dynamics of material interactions defined by differential equations) are also defined in the new context (and in a similar manner as before, but now in a discrete context [or in a discrete process]) but these differential equations play only an intermediary role "of temporary dynamical properties of material components" so that stability is determined by the stable shapes of the metric-spaces which exist in the [forms of] various dimensional-levels . . . , {and in various subspaces of the same dimensional level within an over-all containing 11-dimensional hyperbolic metric-space} . . . , so that each metric-space-dimensional-level possesses a stable shape, which, in turn, can contain lower-dimensional metric-space shapes, which are considered to be material-components, which exist within the higher-adjacent-dimension containing metric-space.

The model of a "material-point possessing scalar-value" is changed into a "model of material which is (usually) a stable discrete metric-space shape," but further, "all metric-spaces, of all dimensions (and for all subspaces), also possess a stable shape."

That is, material shapes (which are metric-spaces) fit into adjacent higher-dimensional metric-spaces, which also possess a stable shape. This fits into an 11-dimensional hyperbolic over-all containment metric-space, so that trees of both "subspace and metric-space shape's" containment properties are determined, [so that the metric-space shapes fit into a containment tree, which is determined by both subspaces, dimension, and metric-space sizes].

Descriptive context (for describing the measurable properties of physical systems)

There is a natural underlying stable structure of existence, which is directly associated with the idea of existence being contained in a metric-space . . . ,

Which is of non-positive constant curvature, and can be of various dimensions and of various (metric-function) signatures, ie related to R(s,t) or C(s,t)) . . . . , where "for there to exist the properties of both reliable measuring and stable patterns upon which to base precise descriptions" the context of description, eg a physical system's (partial) differential equation, is (or must be):

1. linear,
2. metric-invariant, and
3. continuously commutative everywhere,
   (for both (a) the matrix of the local elements of a linear (partial) differential equation, and (b) the metric-function's local matrix).

The metric-space is the base-space of a principle fiber bundle, in which the fiber groups are the natural classical Lie groups associated to both

(1) the various types of signature, and
(2) of various dimensional, metric-spaces.

Namely, the SO(s,t) and SU(s,t) Lie groups, where, s, is the spatial subspace-dimension, so that s is greater than or equal to 1, and, t, is the temporal subspace dimension, so that t is greater than or equal to 0, so that, s + t = n, where, n, is the dimension of the metric-space base-space.

The metric-spaces also have properties attached to themselves, as well as possessing opposite metric-space states associated to these metric-space properties, eg Euclidean space is associated to spatial position and inertia, while space-time or equivalently hyperbolic space has time, energy, and charge associated to itself.

The opposite metric-space states associated to Euclidean space are the fixed stars and the rotating stars.

The opposite pairs of metric-space states for hyperbolic space ([generalized] or space-time) are positive time and negative time.

Spin-rotation is the spin-rotation between these opposite metric-space states defined for each metric-space which is involved in physical description (or, possibly, involved in mathematical descriptions)

Geometrization

The underlying stable (metric-space) structures as well as the underlying (associated) stable properties are contained in a many-dimensional context, where the metric-spaces associated to the different dimensional subspaces are the natural stable shapes associated to the metric-space by means of the metric-space's fiber (isometry, unitary) group. Namely, the discrete (isometry or Hermitian invariant) subgroups (in SO(s,t) or SU(s,t) respectively).

In particular, in a hyperbolic or Euclidean metric-space, these discrete subgroups are, essentially, the stable discrete hyperbolic of Euclidean shapes, which model the system-containing (or material component containing) metric-spaces, where each discrete hyperbolic shape also has a discrete Euclidean shape associated to the discrete hyperbolic shape through resonance.

The stable shapes of the hyperbolic metric-spaces can (be used to) define a partition, wherein each subspace of each separate dimensional level is associated with a largest shape, where all metric-space shapes need to be resonant with some shape in the partition.

Note: These stable shapes, as well as the discrete shapes of Euclidean space, are the natural shapes of these metric-spaces since they are based on the discrete subgroups of each metric-space's fiber groups.

Material component interaction (partial differential equations)

Within each dimensional level, ie within each stable metric-space, which has a shape, there can be contained lower-dimension metric-space components whose spectra are in resonance with the finite partition set of discrete hyperbolic shapes {defined for each dimensional level and for each subspace of each dimensional level} . . . , whose (material component) properties are measurable within the component's containing metric-space . . . , so that these components are interpreted to be material components, which in Euclidean space possess spatial positions, and upon which partial differential equations can be defined.

However, these components also define either (when lower-dimensional) the shapes of a material system or (for the higher-dimensional metric-space shapes) the shape of the material containing metric-space, where the shape of the (material-component containing) metric-space can affect the dynamics of the stable orbits for the material dynamics defined by partial differential equations, which are defined within the metric-space.

Thus, these metric-space shapes defined on different dimensional levels define either the stable spectral structures of quantum systems, or the stable orbital geometries for the dynamics of interacting condensed material.

That is, stability is not controlled by dynamics (though the dynamic energy of the system is important), rather it is controlled by the natural (stable) shapes of the metric-space containment scheme for existence, so as to define either stable resonances (which can occur during component collisions which possess the "correct" energy) or stable orbital envelopes (elliptic partial differential equations) for interacting material systems (so that the energy of the dynamic system fits within ranges which will allow a, resonating, stable structure to become dominant in regard to the system's organization [either as a component or as an orbit])

(where these interactions can be either collisions, or orbits, as well as semi-free, [or parabolic-equation], systems).

The current descriptive structures used for material systems

One might note, in contrast; the order of classical and quantum physics comes from the idea of material defining a particular dimension metric-space and laws concerning the definition of a system's defining differential equation are followed by the descriptive process . . . , where in classical physics, in regard to material interactions, the local measuring properties of material position is related to the force-field properties, which, in turn, is determined by the surrounding material geometry (whose local measurable properties are determined by differential forms) so that the force-field is defined locally [either as a vector, or as a differential-form] at the same point in space where the material's position properties are being locally measured, so an inertial equation can be both defined and solved so that the solution function forms (becomes) a global-set of information about the system, within the material-component containing metric-space.

whereas in quantum physics, the random properties of local spectral-point-particle events are given a harmonic context, ie a function space, so that sets of operators associated to measurable properties, are selected so as to form a set of measuring values which are (hopefully) related to the local (random) spectral-point-particle events, ie random spectral-point-material events in space and time define the system's identifying set of local random-event values, which emerge from the (quantum) system. However, in the operator and function-space (with metric-space domain spaces for the functions) descriptive context of quantum physics, the math patterns cannot avoid non-commutative relationships, where this is due to the complicated geometric structures of the general functions which compose the relatively general viewpoint of a function-space.

Note:
However, if the harmonic functions (of a function-space) are related to circle-space shapes then commutative relations are determinable by the relationship of these (circle-space) shapes to the set of linear, continuously commutative, discrete hyperbolic shapes.

In classical physics, models of local measures are used to find global information about system properties, while in quantum physics, models of local measures are used to determine local information about local spectral-point properties (of quantum systems) which are measurable. That is the models of local measures are used in opposite ways in classical descriptions (where they determine global patterns) vs. quantum descriptions (where they determine the set of local spectral-properties) of a quantum system.

That is, it is either operators or differential equations, which are defined within a domain space, in a context of materialism, so as to either find global information about the system, or they are used to find the event space spectral-values for the set of local random point-particle events in space, which are assumed to be related to a (quantum) system.

Summary contrasting the two descriptive contexts (formal vs. intuitive)

It is . . . , either the local (measurable) spatial structures of material interactions or the local spectral values emanating "as spectra" from a material system, . . . , from which the description tries to derive an ordered set which can be used to describe a system's measurable and stable properties. [but both descriptive contexts have fundamental flaws in regard to being able to describe some fundamental stable properties either general quantum systems or the stable solar-system (which classically is non-linear)]

In both cases it is, "use materialism to define a differential equation of material systems contained in a metric-space."
The new idea is that material and the containing metric-space either bound (from below) or contain (from above), where above and below are defined in regard to dimensional values, whereas the differential equation is a result of new interaction shapes defined in a new discrete context, it is a new model of interactions but which is similar to the classical material interaction.

However, in the new context, it is the spatial structures of the containing space which provides most of the order for systems . . . , (but this new structure can also determine both the local measurable spatial structures of material interactions [macroscopic orbits], and where the spatial structures (shapes) of a material-component [which is a lower-dimensional metric-space than is the material containing metric-space] determines the spectral-values of a quantum material system) . . . , and (thus) the new structure of both space and material is quite different (than) from the idea of materialism.

To contrast these new ideas with general relativity, general relativity tries to make the shape (of a material containing) space be the cause for inertial material interactions, but mass is difficult to model, since it possesses shape (and thus distorts the shape of space in a non-linear manner), where the description is essentially non-linear and, thus, indeterminable (though discrete hyperbolic shapes are linear, and provide a linear model, in regard to the shape of space) and it has proven practically useless, in regard to general shapes of a metric-space's coordinate structures, with the only solution being a 1-body system which possesses spherical symmetry, a very unrealistic physical model.
Whereas in the new model material and the metric-space within which the material component is contained are each separate metric-spaces which are separate from each other where inertia (in 3-space) is given a simple geometric structure in Euclidean space (a circle-size [proportional to mass] defined in a 2-plane), and the shape of space is determined by the natural shapes of the (hyperbolic) metric-spaces, which either define material (which is associated

with a Euclidean mass, through resonances) or they contain material, depending on the relative difference in dimension of the two types of shapes.

These ideas slip between the two opposing sets of ideas of "geometry based classical physics" and "probability based (or harmonic) quantum physics." The new descriptive construct which is associated to material-component interactions are close to the classical ideas, while the math structures, which model both material-components and metric-spaces, are circle-space shapes which are shapes which are very close to the ideas of harmonic functions. However, the choice as to the basis for the descriptive construct is, "stable geometries in a context where measuring is reliable," ie this is similar to the solvable part of the classical viewpoint. However, the interaction structure assures that the apparent random conditions of small material components continue to be present, which is the basis for quantum randomness, but particle-collision data . . . , based on a consistent set of patterns which are related to unstable elementary particles . . . , is now interpreted as being about the unstable decay patterns of the natural set of 2-dimensional and 3-dimensional discrete hyperbolic shapes, which are a part of the decay process when the vertex of the high-speed component collides with the vertex of the target component so as to break apart these space-form shapes, where these shapes subsequently decompose into unstable quarks and leptons which are sort-of-like vector components which are associated to the original shapes which existed before the collision.

(1) This is referring to a book review in Bulletin of the American Math Society July 2013 concerning the book "Plato's Ghost: . . . , by J Gray, and Reviewed by D E Rowe. In this "book review" the focus is on the conflict between the axiomatic formalists (who determine a fixed language for math) and the intuitionists (who relate math patterns to the properties of existence, so that math patterns are stable and quantitatively consistent) which occurred in the history of math, in a time period (of plus of minus 20 years) around 1900. Axiomatic formalism was instituted for universities around 1920. This formalism, along with peer-reviewed professional publishing, does lead to intellectual tyranny based on narrow authoritative dogma.

Spying (who is doing it, who do they work-for? what do they hate? what hate group are they a part? Since hate and class warfare is the characteristic of the justice system)

As for spying, the issue is . . . , that (within the US society) there is already a proven amoral:

1. administration of the law, and
2. formulation of laws, and
3. the media (with its narrow one-sided focus), and
4. this amorality is often centered around economic manipulation . . . ,

. . . , so that without very strict control in regard to having access to such (spying) information . . . ,

Then the control of this way of collecting information will be (and one can be assured it already is being) used by amoral people for their selfish advantage, with control of this domain already in the hands of selfish psychopathic, domineering types (eg organized by people of the type as the amoral JE Hoover), so that the tyranny of the amoral emperors of Rome will be re-instituted, so that the expressions of even narrower viewpoints and more narrow petty personal interests are to be expressed, in terms of even greater authoritative intolerance which serve the selfish interests of a few [the administration of the law will be directed by the spy's] will be realized.

So

Who will composes "the army of spy's," will it be the Aryan Nation, A Bremmer (? spelling) (Norway massacre), J Laughner (Tucson Massacre) . . . , the handlers of LH Oswald, or the handlers of JE Ray (MLK assassin), etc . . . , etc . . . , or will it be even worse . . . , will it be the politicians, who lie, and who criminally collude with the ruling-class, eg use spying to do insider-trading in a rapid-time sequence (which high-tech spying allows) . . . . and (the politicians) who continually attack the lower classes?

47.

Class

The propaganda system is essential for the structure and maintenance of social-class

The formalized technical languages of peer-review, can and do fail, but there is an alternative, ie similar to the ideas of Copernicus challenging the authorities of his day, and this demonstrates that there is not any aspect of the highly controlled channels of communication which is capable of presenting any truth what-so-ever, but it only serves the interests of the ruling-few of a world empire. The down-fall of this empire can be accomplished by an intellectual revolution, throw-out the intellectual basis of the empire (materialism and fundamental randomness), and transcending the idea that science must be based on the idea of materialism.

See m concoyle's talk at 2013 San Diego math conference

Summary

Much of propaganda depends on people perceiving the media as carrying a reliable truth, this is allowed by peer-review, ie the news may be manipulated but peer-review is sure to be true, especially in regard to physics and math, wherein truth (since the 1910's) is based on formalized axiomatizing of a technical language, so as to "not allow:" new contexts, or new interpretations, or new ways in which to organize a measurable descriptive language; but this means that that the formalized axioms eventually become irrelevant, and then the technical language possesses no capacity for content, ie a model of an absolute truth but a truth which has no content.

That is, do not count-on any aspect of the highly controlled channels of communication to be capable of presenting any truth, when it serves the ruling-few of an empire.

But there does exist an alternative to the current non-functional science and math formal languages, ie similar to the ideas of Copernicus challenging the authorities of his day.

The western society does what the rich (ruling-class) want.

Value is not considered to be knowledge and (practical) creativity, rather it is what the ruling-class proclaims "what it is that has value," ie value is arbitrary.

This proclamation of value is done through the propaganda system which fills the communication channels and the education system, wherein the truth that is studied, is the high-valued knowledge most useful to the ruling-class.

The societal organization of Western civilization was determined by the Roman emperors, where actions and knowledge are narrowly confined (knowledge of the world is limited, while abstract hierarchical symbolic use is the main focus of the most learned members of the culture), and practical knowledge is associated to "certain uses" of practical creativity, ie civil works, about which society is organized. This narrow society is formed and maintained by extreme violence, and because of its narrow definition (or confinement) it is a society which is in need of the consuming processes of exploitation, destruction, and expansion, so that populations grow and more (usually material) "value" confiscated from other societies.

This social hierarchy was taken over by bankers, so as to be based on investments, which are determined within a stable, narrowly defined market. (narrowing allows "educational efficiencies," easier organizations, and violent expansion and exploitation, and an expanding population who possess the same narrow viewpoint, motivated by selfish interests in a context of materialism)

Though the most energetic and prosperous early British colony was the Quaker colony and it was the same Quaker concern for equality, which was stated in the US Declaration of Independence, as the cause for revolution; but it was the rich colonists, who were most-close to the trade practices of the bankers of Europe, and it was the rich who led the new US, so as to define freedom as "the freedom of bankers to invest," or as the Puritans state it, "the freedom of the morally superior people to continue to prosper."

That is, the elitist Puritan (Calvinist) ideal was used by the propaganda system rather than the egalitarian Quaker ideal shortly after the US revolution. (The main point of (Protestant) Calvinism is that the spiritually superior people are identified by God by their material success, ie the rich do not need to buy indulgences from the church, ie one of the main points of M Luther's reformation)

From the start the US was a society highly dependent on propaganda distorting the ideas of:

Equality and "the freedom to know and create" based on a selfless belief of helping other independent people, who are joined together by life and the commons within which we all live and are nourished, (we are all equal creators),

And

Changing this ideal to:

Inequality, controlling society in very narrow ways, associated to investment, which serve the few, and applying all the necessary violence needed in order to achieve this goal, ie the model of the European-banker-Roman civilization.

In the US there are two sides to a social discussion within the propaganda system, each serving different aspects of the same master. That is, the pairs of "Hegelian dialectic opposites" which, together, define a whole context of an artificially-limited realm of discussion. The point is, that the entire realm of discussion . . . , in regard to all the pairs of opposites considered . . . , is far too narrow a realm into which knowledge and (practical) creativity can be confined, so it is confined by violence which is perpetrated against the public and which is mostly implemented social institutions (law, education, investment).

The rational side (the liberal), who believe deeply in scientific truth and progress, . . . , where since the middle 1960's the changes are small; and focused on the telephone, the TV, the computer, the car, nuclear weapons, carbon fuels, and civil engineering similar to the Romans, ie progress and change is mostly an illusion . . . . , and this side believes in helping others, claiming a "believed to be" novel idea of a collective community . . . , but an Empire already is a collective endeavor, in which the public must support the rich ruling-class . . . , and the narrow focus of this endeavor is maintained by means of violence, both directly and through institutions eg education, the justice system, the political system, and a careful control over the communication channels of the society.

And

The-other-side (the second side, the conservative) which supports the ruling-class, where the ruling-class is attributed . . . , by Calvinistic religious views . . . , to a condition of being superior people, both morally and intellectually, and certainly above the law, where the harsh laws only apply to the public, and this second side also supports the extreme violence, which always must be directed towards both the public and to those who are outside of the Empire, this violence is necessary because of these other people's inferior human tendencies.

That is, this second side has the belief that "if one is favored by God then they will become rich."

Thus, the main form of the society is based on the opposite sides of Calvinism vs. Communism, both materialistic and collectivist viewpoints, but one believing in a divinely ordained superiority of the ruling-class, and their violent opposition to the inferior public, and the other side believing in the golden-rule, and the absolute truth of rational science, and subsequently, a belief-in competition and survival of the fittest.

That is, both of these sides support the ruling-class, and the high-value which they (the ruling-class) proclaim to society.

The scientific viewpoint of materialism is the most damaging idea to society, and to truth, and this harm is possible since the public does not see how the language of science is manipulated.

There is a belief, by the public (since the propaganda system told them that it is true), that peer-reviewed science is (essentially) an absolute truth, but many of the main problems with language (and the idea of a descriptive truth) related to peer-review is that peer-review is a "formalized fixed structure for a precise language" which is associated to the science community, and thus it has limits as to the patterns which it can describe. That is, Copernicus cannot prove his ideas by beginning with the assumptions of the Ptolemaic system.

The knowledge of science, which the ruling-class needs, is related to an ability to adjust and maintain complicated instruments into which the ruling-class has invested.

Investments in knowledge are the most risky, since knowledge can change, and a fixed, formalized axiomatic scientific language tends to lack content, so abstract complications are added to the scientific body of literature, but the knowledge only remains relevant to the (old) knowledge needed for the adjustments of instruments, ie the knowledge content within an axiomatic formalized language stays fixed.

Today (2013) the fixed structure of language is:
In physical science there is:

1. Material geometry (classical) or a function-space and a model of fundamental randomness (quantum), and
2. a derivative (classical) or a set of operators (quantum) (and differential equations (both)) and
3. a domain space for the solution function (classical) or "functions of the function-space" (quantum) within which the material-spectral system is contained (both), where the domain space is a metric-space whose dimension is determined by the idea of materialism (both).

The only useful part is (classical):

1. Material geometry,
2. the derivative (and a differential equation), and
3. the domain space of the solution function, and

4. only the differential equations which are solvable provide the knowledge which is needed to adjust complicated instruments whose controllability is reliable (there are also feedback systems).

{Note: The solvable equations of some very simple quantum systems (H-atom, confinement to a box, etc) can also provide some information, but the quantum methods cannot be generalized, as they can be generalized in classical.}

The rest is irrelevant complications, presented in a formalized axiomatic (or fixed) language (held fixed by peer-review), a language which has (come to have) no content.

The main very big problem, which is so artfully ignored

But there are very stable physical systems, which fit into the general categories of: nuclei, general atoms molecules crystals; which when their components are the same, then in their "process of formation" these quantum systems (which possess the same numbers of similar components) come to possess precisely the same spectral properties . . . . , so this means that the process of their formation is controlled . . . , and thus it must be a math model which is geometric and solvable.

Fundamental randomness of the (irrelevant) quantum context is wrong.

One can be in a fundamental agreement with Einstein that "God does not throw dice," ie quantum physics and particle-physics are both (all) wrong.

But, furthermore, general relativity is wrong, though metric-invariance is correct, when a metric-function is based on a linear descriptive context it is a statement about reliable measuring, and this requires that a metric-function be fixed, but when a metric-function is based on a non-linear descriptive context it is a statement about un-reliable measuring, ie determining a general (non-linear) metric-function is not about an invariant form for a differential equation as the idea which is the basis for physical law, and this is because local measuring is not smooth, rather the correct model of a local measuring process is that of a discrete process of change.

Thus technical language is thrown into a context of irrelevance by means of maintaining a fixed formalism, and this is done, at the convenience of the ruling-class, to stabilize the technical context of their investments, ie so that "new knowledge" will not make their instruments irrelevant (and thus bad investments).

Note: axiomatic formalism for the higher education system began around 1910-1920 and this is about the time when controllable-system technologies all became based on electric circuitry and electromagnetism, or on thermal physics, even the rate of (nuclear) reaction is based on 19th century ideas about statistical physics. This is also when general relativity became the standard formal truth for peer-review journals, ie journals published by publishing-houses controlled by the very rich

Do people ever wonder why science stays so irrelevant to issues concerning both

1. clean, cheap energy sources; and
2. many-dimensional models of existence?

Why, in the current many-dimensional models of either elementary-particles or string-theories, do the experts always adjust their model so as to include (or maintain) the idea of materialism, as well as to make their descriptive contexts relevant to large explosions of material (ie nuclear weapons technology)? This is because their interests (the expert's interests) are about being peer-reviewed and they are competitively concerned about overly authoritative ideas, or more accurately, their ideas are narrowly defined dogmas which are of the greatest interest to the ruling-class, in regard to the business investments of the ruling-class.

Why is modern science only concerned about finding the cross-sections for elementary-particle collisions . . . and idea which primarily relates to "rates of nuclear reaction" in systems which are carefully set-up to be used to induce nuclear reactions . . . , yet the cheap, clean energy systems, which were promised long-ago (1950), never materialize . . . , be it fusion or thermal-solar . . . ,
　　or
Why is modern science only concerned about unifying models concerning rates of particle-collisions (or reaction rates) with models concerning models of gravitational spherical-symmetry singularity-centers (in a context of materialism)? . . . , (Is the answer: So as to produce an explosion to rival the formation of a black-hole.)

Can math models of a "many-dimensional existence" transcend the idea of materialism? Yes.
　In such a more realistic (non-materialistic and many-dimensional) model of existence, can both the creative range of mankind be extended, as well as a more realistic basis for developing clean, cheap energy sources be determined? (Yes.)

Why are there very stable quantum systems which exist, and which are consistently similar in their spectral-properties when they possess the same number of quantum components, such as in the case of nuclei and general atoms?
　These properties suggest both a geometric structure (for these systems) as well as a formation process which exists in a context of being controllable (a solvable, geometric context, which is controlled by initial or boundary conditions).

Yet, the best statement which modern physics and math (which are being based on indefinable randomness and non-linearity) is supplying about these stable and consistent systems is that:

"these systems are too complicated to describe using the (very complicated) laws of physics concerning either quantum physics or particle-physics (laws based on random collisions [or random events] of elementary-particles)."

Then perhaps the laws of physics (and math) are wrong, and the current laws of physics are only focusing on a small area of concern, namely, the relation of small (apparently) random component behaviors to properties which only relate to nuclear reactions.

Why does science and math conform more to (almost exclusively to) business interests, than to independent representations about how precise math methods and various (new) interpretations of scientific data can be used to describe the properties of existence as "it actually is" so that, that existence is not based on the idea of materialism?

Part of this focus on narrow sets of ideas has to do with the complications which get associated to this narrow set of ideas and the ability to behaviorally manipulate people within the public education system, principally through (using a fraudulent idea of) "intelligence-testing" so that autistic-types are distinguished by the test, where these personality types are related to obsessions over memorized complicated strings of symbols. That is, these people are obsessed with narrowly defined complication and they do not consider the obvious way in which the narrow dogmatic and fixed language structures are failing to provide satisfactory answers concerning observed patterns. That is, whole classes of observed very stable spectra associated to these wide sets (nuclei, general atoms, molecules crystals) of fundamental physical systems, a condition which implies both geometry and solvability

(note: string-theory is sort-of a pathetic joke of endless and incomprehensible complications, trying to (do the impossible) couple geometry to random processes)

So that the best statement of modern science is that "these systems are too complicated to describe" . . . ,

. . . , this certainly suggests that the, so called, "leading intellectuals" are leading us to their own obsessions, concerning complications about the illusions which these "experts" (autistic-types) have been induced to consider, induced by illusions of high-value associated to these types of goals (eg using a language based on both geometry and probability) in a fixed and subsequently irrelevant formal language.

Now that there is an alternative math model for both materialism and the apparent fundamentally random properties of existence . . . . , namely a many-dimensional containment space of highest-hyperbolic-dimension-11 which is partitioned into a finite set of stable discrete hyperbolic shapes of metric-spaces of various "spatial-dimension" and metric-function signatures with each different signature associated to different physical properties, so that material is a stable metric-space shape which is contained in an adjacent higher-dimensional similar metric-space, but the containing metric-space also possesses a stable shape, . . . , so that the main issue before mankind is about determining . . . in a language context of different assumptions, new contexts, new interpretations of patterns, and new ways in which to consider containment, . . . , the relation between (1) a math context of "reliable measuring based on and describing stable patterns" and (2) a practically useful truth, associated to practical creativity (even in a non-,materialistic and many-dimensional context concerning a descriptive language of existence).

In regard to this fundamental issue facing mankind "the relation that knowledge has to practical creativity" that, now (2013) the social structures of mankind are based on elitism and this elitism is based on an absolute idea about truth and this social structure is opposed to people expressing alternative views and presenting these ideas in an intuitive, but carefully presented, manner.

Currently there are the orthodoxies of quantum physics which has very little relation to practical creative development and particle-physics and general relativity combined in string-theory so as to only be related to an assumed spherically symmetric core of a grand nuclear (or particle) reactions.

That is, modern physics and math are only focused on nuclear weapons.

But this is possible since the ruling-class controls the communication channels so that the education system filters into the top levels autistic types who see reality as memorized sets of complicated related symbols, ie apparently they are not capable of assessing the value which their deep thoughtful considerations have within society, so that these autistic types deal with knowledge in a context of formalized axiomatic structures, ie fixed symbolic structures, which are not capable of describing patterns that have any real content, ideas that can be used for practical creative development, except in the most narrow of nuclear weapon contexts.

The usual narrow context of increasing the range to which a fixed instrument can be applied, ie not developing new contexts within which practical creativity might lead to much more practical creative variety.

The ruling class are in charge of language within an oppressive, "collective society" whose collective purpose is to serve the interests of the ruling-class.

Note: The success of statistics . . . , in regard to its (so called, former) great accuracy at which a political-election is predictable after a few votes are counted . . . , is really expressing the idea that in the very limited range of choices are possible in regard to the limited range of symbolism (or language) is allowed to be used within society. Thus, the amount of variation of how people . . . , who are forced into a narrow viewpoint about life . . . , adapt to such a limited language usage, ie the vote, is quickly measured by a few samples. The validity of the statistical methods is related to the narrow variation of symbolic (or language) usage within society, since people are trapped in a language which is violently imposed on them (as is their condition of being wage-slaves).

The professional scientists are those who both adjust and gradually advance the range of use of complicated instruments (into which the ruling class has invested), and they have no relation to truth. That is, there is no room for a scientist to identify new ideas . . . . , such as Copernicus, to challenge the scientific dogmas (used by the ruling class for their narrow interests) . . . . , in regard to how information, of all types, is controlled in a very narrow manner by the ruling class.

Academic institutions are organized around sets of narrowly defined languages, in turn, based in (for math and science) axiomatic formalization, so this narrowness requires that the focus of professional science people also be narrow, so the symbolism of knowledge is narrow, and the efforts of a material based knowledge can also be narrowly channeled into the (material) instrumental needs of the ruling class, since material is controllable by violence and by law.

Thus, intelligence was defined to be about obsessive focus on symbolic knowledge.

That is, curiosity, inventiveness, new contexts, new interpretations etc are not considered to be a part of new knowledge.

Rather new knowledge must be derived from the fixed absolute scientific truths held so "dear to the heart" by the professional scientists, who, apparently, love so dearly to support the interests of the ruling class, and the truth the ruling-class provides to society through their control of society's information channels, and controlling an manipulating personality-types within society by institutions, be it education, prisons, mental health, law, as well as peer-review publishing.

That is, Copernicus must accept the assumptions of the Ptolemaic system and then prove that "the earth travels around the sun" based-on the opposite viewpoint (or opposite assumption).

The notion that trade, and narrowly defined commerce, can be the basis for social organization is nonsense. The trade structures can exist, but they can only exist in a context of equal, free creativity, not in a context of monopolistic domination in regard to a narrowly defined belief in (arbitrary) value, which must be held in place by both propaganda and extreme violence.

Though creativity is the core of what it is to be human, the building and trading of material related: objects, resources, and instruments, is bound to lead to failure, since it is (has come to be) defined so narrowly, so that resource extraction is very fixed and destructive.

The reason for this narrow definition is that monopolistic (and easily controlled social) institutions, eg corporations and the governments which they control, depend for their social power on their narrow relation to resources, and, subsequently, to narrowly defined knowledge, eg oil, and nuclear weapons and electrical circuits, etc.

Thus, control is about maintaining traditions and narrowness, yet promoting an illusion of constant newness and great diversity, eg new circuits or new weapons etc.

Whereas, ancient mankind was more aware of their personal power, and the relation that their life-forces have to existence, and they had a greater diversity of creative avenues to explore.

The point of western civilization is the focus and central theme of the need for extreme violence in both armies and in social institutions, and an extreme control over language and thought within the society, so as to oppress people whose creative directions would interfere with the "collective intent of a civilization" which is ruled (or directed) by a few, so that the conquests and associated trade markets are required to expand, and violence of the society's social institutions drives the people into a collective viewpoint.

Marx was limited in his viewpoint about the nature of the human life, apparently he believed in inequality (since he believed in oligarchic societies), and he believed in materialism and bought-into the baloney about efficiency of markets and large scale monopolies. Efficiencies are baloney, it has mostly been a story of too-big-to-fail. That is, narrowness is not efficient, rather it concentrates social power.

That is, the war with Marxism is a sham, since Marxism and capitalism are both oligarchical and collective enterprises.

It is like the wars between Christian sects, slight differences in dogma allow new leaders in the oligarchy, and that is the point of the wars (to put in charge a new boss).

The creative range of human life is not confined to the limited idea of materialism.

Yet the religions of the west, ie the Judeo-Christian-Islam (the Abrahamic) religions, are about belief in the material world, and the propaganda needed to focus on living together in a social hierarchy, and to have faith in one's own community needed to over-throw other material-based tyrants, who rule hierarchical societies, ie the Abrahamic religions are similar to Marxism, in that they believed in materialism and they believe in rule by the few, but they simply want to have a different set of tyrants in place.

Thus, the propaganda was about a particular community possessing a superior God, whereas for Marx it was about a hierarchical society whose propaganda was based on "helping the people."

When Russian communism collapsed, this meant, in capitalistic societies, that the wage-slave social position of the worker could easily be exploited (or ignored).

China quickly changed its propaganda from "helping people" to narrowly defined trade and controlling markets, but maintaining the same ruling hierarchy.

To look at the native people of the Americas in pre-Columbian times, one sees some cultures where the creativity of mankind is not focused on the material world, there is no need for such a focus on materialism, and this is because "living is easy," when the people live in a harmonious manner with an abundant earth.

The reproductive patterns of the people revolve around the female, and the society is mostly egalitarian.

But there was also a scientific-religious side to their knowledge (perceiving the world as it really is, in regard to the, true, many-dimensional structure of existence), so that creativity was channeled into the relation that life has to existence, ie life creates existence.

This knowledge (other than the idea of materialism) allows one to understand that "more than 10,000 years in the past" great stone monuments were built, with great precision, but without any apparent relation to material technology.

But one does see a similar history . . . , since over 10,000 years ago, also in the Americas . . . , of these communities becoming more violent and more material directed.

Yet, a deep science-spiritual knowledge about existence was maintained . . . , as demonstrated by the formation and disappearance of communities in the US southwest so as to not leave a trace, . . . , ie a reach of their knowledge which still extended outside the material world.

This shows (or can be so interpreted) that the existence of a (delicate) knowledge beyond the idea of materialism can serve to keep the nature of a social structure more harmonious with a deeper truth and a better society within which to exist.

Also

The living account of "perceiving the world as it really is" and an intricate cosmology of the ancient American cultures, as expressed in the memoirs of C Castaneda, which expresses this science-religious viewpoint about existence of the ancient American cultures. Eg lines-of-the-world (models of both neutrinos and light . . . , [in a math model of existence based on the primary role of stable shapes in regard to understanding the stable order which we experience, or which allows us to experience], . . . , in the world) experiencing many-worlds communicating with other (intelligent) life-forms, etc.

In the west, the mythology about many gods, who possessed extraordinary powers, with a vague account of their beginnings, and man's diminutive relation to these Gods, seems to have a similar history in the Americas, since over 10,000 years ago, where the gods of the west were depicted as being domineering and selfish and violent, rather than being creative.

In fact, "experiencing the many-dimensional world, as it really is" also explains that each of us has a relation to existing as "a giant-being," of the stars, who can move between many-worlds.

Perhaps best depicted as a "giant tortoise living on air, freely dispensing creative gifts onto mankind."

The small community of C Castaneda might have missed this central aspect of existence, which they may have experienced, but mis-interpreted.

Science and math

Science is based on:

Classical description is based on:

1. materialism,
2. local linear measures of properties, represented as functions (whose domain spaces contain the material system),
3. where coordinate containment is within a metric-space (metric-invariant), within which there is an external material geometry, which is separated from the material component upon which one is focused (and measuring in a local manner). This relation of local components to distant geometry is realized through force-fields . . . , (second-order differential-forms [ie mass modeled as a 2-form in 3-space, while charge is modeled as a 3-form in 4-space]) . . . , and local inertial properties, ie the laws of Galileo-Newton.

{Thermal physics is about representing energy and entropy functions with their domain spaces (coordinate containment spaces) defined by the thermal measurement properties of: p, T, N, V; in regard to a closed thermal system (in equilibrium, the question is, "How can equilibrium be achieved so quickly?" the answer seems to be about the rigidity of the actual physical constructs physical associated to physical descriptions, and its relation of the operators, eg derivative operators, (newly found) to discrete properties, or discrete processes), and expressing local measures of these functions (energy or entropy) as first-order differential-forms (along with the three laws of thermal physics).}

Quantum description is based on:

1. Random material-particle-spectral events in space and time
   (whose probabilities are associated to spectral-values, so that both of these properties are to be determined (or carried) by a wave-function [but for general quantum systems, this

dual quantitative structure associated with a wave-function (both probability and spectral-values) seems to not be possible]),

2. local linear measures of spectral properties (ie sets of local operators), where the spectral-random structures are represented as a function-space of global wave-functions (whose domain spaces contain the material system),

3. where coordinate containment (defined on the wave-function's domain-space) is within a metric-space (metric-invariant), within which there is an external material geometry, which is separated from the material component {upon which one is focused (ie the random spectral-particle event)} . . . , and where the local spectral-property is being measured in a local manner, . . . , where this external geometry is represented (in quantum) as a potential-energy function (ie local operators are both derivatives and multiplicative operators [acting on a wave-function as an energy-operator]).

But this type of intrinsic containment, which is necessitated by the idea of materialism, ie existence is determined solely by material properties, (ie local material and distant material properties mediated by local differential operators) is not sufficient to describe the observed stable structures of many-but-few-body systems . . . , which are either general (non-commutative) quantum systems, or non-linear global systems of classical physics . . . , both (non-linear classical and random quantum) of whose observed properties are very stable.

One needs a containment hierarchy of separate, independent metric-spaces (so that within a metric-space one has an open-closed topology, ie external spatial [or metric-space] structures are unnecessary, and difficult to ascertain (or difficult to observe), thus, leading to the idea of materialism) [of many different dimensions and different metric-function signatures] so that either a metric-space's shape can influence material orbits (of material contained in the metric-space) or in lower-dimensions a metric-space's stable spectral properties (or stable shape) can determine the spectral properties of a stable material component.

Both of these math constructs, as well as Einstein's axiomatic, non-linear characterization of inertia, are either non-linear, or non-commutative, or indefinably random, and subsequently they do not fit into a quantitatively consistent pattern to which math can be applied in a valid manner, ie the descriptions of these unstable and indefinable patterns do not conform to a measurably reliable context, and the patterns described are unstable, and this implies that such descriptions are without any meaning, and thus, without any relation to practical creative developments.

This non-relationship with practical creativity is the proof that these non-linear, indefinably random descriptive contexts are not valid.

That is, there exist fundamental, stable spectral-orbital physical systems which exist at all size-scales but which go without valid descriptions based on physical law: nuclei, general atoms, molecules, crystals, the stable solar system, and now dark matter, etc. The main statement made by the overly authoritative science and math communities is that "these systems are too complicated to describe," yet they are stable, and they also, apparently, form in a controllable context, since the systems which are of the same type also have the same spectral properties, which are precisely distinguishable.

In regard to the development of practically creative contexts, there is a non-productive aspect of a carefully guarded knowledge, the science-math knowledge of our society, wherein peer-review is supposed to guarantee a reliable truth-value of these descriptions, which are presented to the culture by the professional scientists, a quality-control is placed on truth, but the control is determined by an authoritative dogma, and where these wage-slave scientists compete for the best paying jobs, which in fact are associated to adjusting complicated instruments, ie getting research grants from the corporate-government.

That is, science does not need to be protected from intellectual deceit, rather peer-review is identifying an authoritative dogma within which the investor class wants the professionals to compete so that they will fit into the narrow knowledge-based interests of the investor class.

Integrity in regard to truthfulness is a great hoax, in regard to the nature of the US communication systems.

Consider that the media is considered to be, by the public, the sole authoritative voice of the society, yet all the media does is:

1. mis-represent and
2. mis-direct, and
3. Deceive, and
4. Collude with the fraud perpetrated by our corporate-government, and then
5. Participate in covering-up this fraud, so as to deceive, the public.

The only truth being protected is the truth into which the investor class is basing its investments, and these issues are about narrowness and domineering monopolies, ie it is about traditional fixedness, eg oil has been fixed as a monopoly for over 100 years, and coupled to coal 200 years, while nuclear is all about nuclear weapons and a willingness to poison the earth for selfish interests.

When a narrow technical level, associated to a set of product types, eg circuitry nuclear weapons etc, is attained and monopolies are built around such products or processes, and then this technical level can be given an axiomatic formalization so as to fix knowledge, by the use of a formal language, which has a limited range of applicability.

Yet the axiomatic formalization is a statement that this viewpoint is to be applied universally.

This is the viewpoint of truth which is derived from a narrow context, and it is the viewpoint which best supports a stable context for an already developed social structure for investment, ie it is an effort to lower investment risk in regard to the development of new knowledge.

Thus, it is a viewpoint which opposes any new technology, where the investors want to oppose any new creative context, since they have already invested in particular types of instruments.

This is noticeable in regard to oil, whose poisonous effects on the earth need to be curtailed.

However, in a justice system . . . , where law is based on property rights and minority rule . . . , this means that the oilmen will always be able to block any attempts (by the public) to curtail the use of oil as an energy source.

The technical languages (of the narrow truths which stabilize the investment context) are given an axiomatic formalization, which, in turn, is the basis for the dogmatic authority of a narrow viewpoint of science, despite claims to the contrary, where science is both protected by peer-review and it defines a professional wage-slave scientist.

But these formalized axiomatic truths are very limited in their range of applicability and when they are applied to greater ranges of descriptive contexts they always fail.

This narrow viewpoint about a "descriptive truth" is maintainable, since the education system is so easily manipulability, wherein intelligence is used to manipulate students . . . , where intelligence is defined as a personality who is border-line autistic and who obsesses over their own memorized symbolic structures so as to get entrapped by the complications of the generalized application of narrow axiomatic symbolic formalism to a context in which the language does not fit . . . , yet because the axioms define the "rules of the game," which have led to a great amount of irrelevant complications ensue.

The non-productive nature of these peer-reviewed formalized axiomatic descriptions has been born-out by the lack of any relation of the ideas (descriptive patterns) of quantum physics (or particle-physics) for general quantum systems to provide accurate descriptions (or accurate lists) of spectra of the system to sufficient precision, and the utter lack of these (quantum) descriptive structures to any practical creative developments.

Some observed quantum properties can be coupled to classical systems, which are controllable, but only the laser was developed based solely on quantum properties. Micro-chips are made in the context of thermal systems and classical optics.

Non-linear classical descriptions can be related to feedback systems, but this implies that the containment space "for such descriptions" need to always be changed, when a feedback induced system-adjustment is made. The feedback-action is outside of the descriptive structure.

Furthermore, the validity of the system-defining differential equation is always to be questioned, ie the focus is not on a chaotic solution-function, but rather on a system's differential equation (and its associated limit cycles related to dynamic convergences of regions in the system's domain space), but the chaotic nature, or unreliable measuring context, of the system as a whole, means that the range of applicability of the differential equation cannot be known, ie the feedback mechanism can suddenly no longer be relevant, ie the system (to which the differential equation is applied) has suddenly dissipated.

What is needed is a stable, solvable, controllable descriptive pattern as well as new contexts for creativity. This requires that math language not be formalized, rather, that description stay close to changes in assumptions, always adjusting interpretations of data or of patterns, the use of stable patterns as the basis for description, new contexts identified, and containment sets always re-considered.

The press (the propaganda system) are the enemies of both the public and scientific knowledge.

They view the public with contempt, while they view scientists, whose ideas have little (if any) relation to developing new practical technologies (this is true, despite the hype to the contrary) or to developing a new context for practical creativity (eg a context beyond the idea of materialism) so that their intellectual achievements, other than their irrelevant peer-reviewed literary achievements, are essentially non-existent. Yet the in the media these experts are considered to be an intellectual aristocracy, whose social status is of the highest ranking, they are close to the social-ranking of the ruling-class, but with a more limited domain of power, ie the domain given to them by the ruling-class of adjusting and maintaining the complicated instruments into which the ruling class has invested.

By worshipping a very limited form of knowledge thus causes communication channels to close-down to new ideas. That is, new ideas about science must enter public scrutiny through peer-reviewed journals, but peer-review is built around the authoritative dogmas of a formalized axiomatic language and as such it will exclude any ideas whose assumptions, interpretations, contexts, etc, are not consistent with the viewpoints of the peer-review institutions. That is, Copernicus must assume the beliefs of the Ptolemaic system (the sun goes around the earth) and then prove that the earth goes around the sun. Thus, no one is allowed to challenge the authority of the reigning dogmas of science, and (or because) these dogmas are consistent with the interests of the ruling-class.

This is all based on an endeavor to form an illusion, deceiving the public in an even more substantial way, namely, that the authoritative experts possess an absolute truth, which, because of their superior intellects (Calvinism keeps emerging into the language, ie substitute morality for intellects) these experts must also possess a superior truth, so that possessing such a truth is reason-enough to dominate and destroy both all other people and all other ideas (or thoughts), ie people are essentially their ideas and beliefs.

The role of propaganda is to dominate all language usage and all thoughts of the public, so as to conform to the collective will imposed on the public by the ruling-class.

This is all upheld in a circular and self-serving manner by peer-review, which determines the dogmas of science which in turn determines the content of technical language, ie descriptions which are measurable.

That is, the only way in which the public can determine an absolute truth is through peer-reviewed publications.

But

The media is the sole authoritative voice for all of society. That is, peer-review gets its high-value (in the minds of the public) by the way in which the general media treats peer-reviewed truths and its associated set of authorities.

Thus the following illusion is being presented to the world, Abstract mathematics has mastered the techniques of descriptions of random systems, random systems which often have underlying order, and that order can only be discerned and then used to understand the observed order of the system. This is clearly not the case, as the state of quantum descriptions clearly shows, namely, the order of general quantum systems cannot be explained, rather it is claimed that "the observed order of these random systems is too complicated to describe."

Nonetheless, this is the claim; so that

1. the relation of DNA to the epi-genome will come to be understood, through the method of being able to distinguish certain molecules and then finding the correlations, which can be used to determine the system's underlying order (this is far too big of a set to be able to describe in any comprehensible manner)
2. The patterns of random collisions of unstable elementary-particles and their non-linear, covariantly invariant, representations can be coupled with (a covariantly invariant) string-theory, to lead to the descriptions and control of all material systems (despite the fact that thus is far too complicated of a descriptive structure to be able to describe the observed properties of material-spectral-random systems).

That is, all aspects of the wage-slave nature of the society requires that any wage-slave view the public as a set of grossly inferior beings, and this can only lead to great injustices.

428

The social set-up of US goes something like this:

The top 500 individuals (But it is really, likely, the top five [maybe 10] families)
The 3% psychopaths (manipulated for purposes of extreme violence)
The needed 4% high-level personnel (top experts, and top managers, propagandists)
The 25-30% working in security

So there are 67-70% composing the inferior public, of which all but, say, 2% are deluded

However, there is an alternative, it is carefully thought out, it changes the context of human creativity. The changing of the context of creativity is central to any intellectual revolution of mind.

Does your loyalty side with equal freedom and personal knowledge and creativity, or must you conform to the oppressive world that the tyrants provide for you, where value and truth is well defined for you, ideas to which you must deform your soul, and you must "fight to be liked (or valued) by the ruling families of the world," the model of heaven-on-earth (which is provided to us by the media).

48.

Mis-use of math

Alternatives to both formalized axiomatic math and to physical descriptions which are based either on indefinable randomness or non-linearity

The failure of the entire social system . . . , a system in which all of society focuses on "serving the needs of the ruling class," a similar social structure as was the Roman-Empire, where Roman civilization was designed to serve the Roman emperor and his associated ruling-class in Rome, . . . , system failures such as the failure of (as the media story goes) the money lending between banks (since the banks all know that financial transactions are now (at least since 2008) all based on fraud), which was associated to failed investments [or the failure of the society to stop the destruction and poisoning of the earth, where this failure to stop the earth's destruction is done by the political system in order to serve the selfish interests of the few bankers-oilmen-military people who own and rule society] . . . ,

where the bank failures was, to some large-degree, brought about by failed calculations of risk, where this risk was determined by using the math methods developed concerning the random context of quantum physics, a method associated with indefinable randomness, where the distinguished states, whose probabilities are to be determined, are patterns which are either not stable (eg elementary-particles) or not calculable (eg the spectra of a general nucleus).

Now this failed math construct, concerning risk (or randomness), is becoming the basis for a (failed) probabilistic model of a future "crime perpetrator," who is to be pre-emotively exterminated by the state, based on an easily manipulated model of an enemy, a model which can be easily manipulated by adjusting data or adjusting categories to which events belong, ie by adjusting an arbitrary definition (or arbitrary context) of an indefinable and unstable pattern, eg where an enemy might be defined to be "a person opposed to the interests of banking-oil ruling-class" (where the protection of the ruling-class has defined the national security system).

Note: Because math has become based on axiomatic formalizations, much in modern math deals with descriptive contexts in which either measuring is not reliable, eg non-linear and/or chaotic contexts associated to the measuring constructs upon which the descriptions of patterns are based, as well as where the patterns being described are not stable patterns, eg from non-linear shapes to elementary-particles, or the basic events of random patterns are either not stable or not well-defined, ie no known math methods can be associated to a calculation which leads to the identification of an observed system's events . . . , which are either not stable events or are not definable . . . . , or to the determination of their probabilities,

. . . , (the [so called] failure of the money lending between banks, which was associated to failed investments, though more likely, the bank failures were associated to failed calculations of risks) where this set of failed investments was turned into an opportunity to . . . ,
    lie to . . . ,
    de-fraud . . . ,
    steal from . . . ,
    and then oppress . . . ,
    . . . , the public,
so that the ruling class was saved from ruin by their failed investments . . . , where this opportunity to save the ruling-class from its own induced failures was facilitated (caused, or set-up) by the media (or propaganda system) in the media-governing-military system in which we live, which is designed to serve the needs of the ruling-class.

This (same) Roman-model of civilization has repeatedly proven itself to be a failure.

Yet, this failed social structure has come to control the knowledge and creativity of the society.

This was the main point of power, which was central to Roman civilization, namely, the ruling class was served by the military, and the military was also where the central repository of knowledge and the basis upon which practical-creativity was instituted, ie the Roman legions built the water systems and sewers and roads for the towns in their newly acquired lands.

The narrow models of cities, agriculture, and finance (stable money) of the Roman civilization was built and organized around the engineering structures which the Roman armies built.

One sees that this is the same type of narrow context requiring ever more expansion and exploitation and destruction about which the banking-oil-military defined ruling-class also depends today.

Similarly, today, technological engineering is funded and designed to support the ruling class.

That is, today, the ruling class control the information channels and the (university) centers of learning within society, so that these structures are constructed and designed to serve the interests of the ruling class.

The failures of math and physics . . . , [most notably that quantum physics and particle-physics both of which have no relation to practical creative development within society, note: micro-chips are built based on knowledge of thermal physics] . . . , are related to a formalized axiomatic language, which was made fixed by the decree's of the ruling-class (though instituted by the experts who serve the interests of the ruling class), this was around 1910, so that the descriptive constructs of math and physics are based on rigid axiomatic formalism, and as a result the described patterns of the professional mathematicians have come to be without any content, ie they cannot describe the observed patterns of general fundamental systems to sufficient precision (both in general quantum systems and in regard to the stable properties of the solar system) and they are unrelated to any form of practical creativity (except the solvable parts of classical physics, which are still the basis for nearly all of our "modern" technology).

The math statements made within the context of a formalized axiomatic structure form a formalized language which can be used to describe properties, which belong only to an illusionary world, and thus, these ideas (or patterns) can be manipulated so as to form a quantitative structure, which, if carefully constructed, can, nonetheless, be used to support arbitrary ideas, so that these arbitrary ideas appear to be based in an objective "quantitative set of measurements," eg statistics being based on vague and unstable patterns used to define a random event space.

This manufactured quantitative relation to an illusionary world of false images is the result of math having come to be about expressions (of a system its containment and its measures) about relations which exist based on a formalized axiomatic basis, which it is believed "can be assumed to be present in regard to any measurable description, in regard to any (vague) pattern, and in a context where measuring is not reliable," and, subsequently, this is a descriptive context, which has no practically useful content, other than creating illusions of measurable objectivity which are used in the media (to deceive the public). Nonetheless, it is a descriptive construct which allows for rigorous treatment concerning any invented illusionary world, so as to be based on the (this) formalized axiomatic basis (for a too technical language of formal mathematics).

New ideas come from people who are at the margins of society rather than coming from the class of peer-reviewed, professional science and math people. Referring to peer-review as if this makes an authentic truth, is similar to look to the media to describe the relationship which the public has to the actions and intents of the governing institutions, as the practically useless and failed math-physics constructs of: quantum physics, particle-physics, string theory, and non-linear based general relativity have shown.

The professionals are good at adjusting complicated instruments and gradually extending an instruments range of uses.

Furthermore, a professional math and science person, is a carefully selected particular personality type, and would be a person who is at a prestigious university, ie aggressive and competitive, who might best be characterized as being obsessed with (their own) memorized dogmas, which is mostly symbolic, and these symbolic dogmas would be without content, in regard to what the technical language (of such professionals) is trying to describe, eg particle-physics is not capable of describing the stable spectral properties of a general nucleus, but that is what it is supposed to be describing.

New ideas which are practically useful, come from people on the margins of society, where such people as: Faraday, Copernicus, Kepler, Tesla, the Wright brothers, Einstein (at first), etc existed (but Einstein was raised-up to become an "illuminati," while the much more substantial, Tesla, was pushed aside [maybe because Tesla did not talk with academics, but rather talked to investors]).

On-the-other-hand, within the academic world Einstein axiomatized electromagnetism and special relativity, and he borrowed from E Noether, the symmetries between time and energy, to find [Energy = mass], but his general relativity, ie trying to axiomatize inertia, has been a disaster, ie trying to quantize a non-linear context where measuring is not reliable, and these efforts fit into the newly begun effort to formalize the axiomatic structure of math, eg the efforts of the Bourbaki professional mathematician's program of axiomatizing math.

Math as a technical descriptive language

But math is about:

1. Quantity and shape
   And
2. Where, there is reliable measuring, and there are stable patterns, where these stable patterns are both (a) to be described, and (b) there are stable patterns which are a part of the descriptive constructs.

These general math properties are more necessary, in regard to descriptions of stable measurable patterns, than are the frames defined in physics, where frames are concerned with the relative motions of the (physical) system-containing coordinate space, within which measuring is assumed to occur.

The frames of physics relate to both metric-stability, or metric-invariance, and the relation that these frames of coordinates have to the properties of a system defining (partial) differential equation, ie the laws of physics.

The idea that a physical system's defining set of (partial) differential equations which are covariantly invariant . . . , ie related to differential-forms (which should also, in turn, be related to an invariant definition of the containing space's coordinates metric-function, ie not a general metric-function which defines an unreliable measuring context)

. . . , is not sufficient (ie Einstein's principle of general relativity is not sufficient) . . . ,

. . . , one also needs the math property of these equations being continuously commutative everywhere.

Note: An alternative containment context where "inertia might be defined due to the shape of the material containing space." The geodesics defined in a metric-invariant context can be related to inertial dynamics, but only in the energy-space of a hyperbolic metric-space, if the material is contained within a discrete hyperbolic shape (which is also a metric-space).

A physical system's defining set of (partial) differential equations which are covariantly invariant is modeled after the structure of the classical force-fields, F, of classical physics, where such force-fields are differential-forms which determine relations (by means of exterior derivatives) between material geometry and local linear geometric measures . . . , (ie general covariance, but in classical physics there is metric-invariance), . . . , which, when solved, can be placed into an inertial differential equation of, ma = F, which, in turn, when solvable, is both accurate to sufficient precision, and its description is stable, and the system can be controlled, which allows for practical usefulness in regard to this information's use.

General relativity assumes that only local linear derivatives are fundamental, and a (general) metric-function for a coordinate space can be found in a context of general covariance, ie a partial differential equation pertaining to differential-forms, so that a solution to such an equation can determine either the coordinate space's natural local linear measure for length, ie the coordinate space's general metric-function, or the sets of local linear geometric measures consistent with the general metric-function.

In turn, the geodesics, which are determinable from the metric-function, determine the inertial properties of the system.

This is a circular relation between local coordinates and the model of the derivative as a "local linear measure" which cannot be unraveled.

The problem is that measuring requires both a metric-function and local linear derivatives to exist as stable patterns within the descriptive construct.

Otherwise the partial differential equation is non-linear and defines a descriptive context (or a set of containing coordinates) which are found from a non-linear differential equation, ie this is an unreliable context for measuring.

Thus, general relativity has only solved one problem; the one-body system which is assumed to possess spherical symmetry, a continuously commutative shape almost everywhere. But, try to put another piece of material into such a one-body construct, and an ensuing attempt to find a stable shape (eg a stable solar system) can never be resolved, the context does not allow measuring to be reliable, and the math patterns will always be chaotic. So using this descriptive context to describe the stable solar system is unthinkable.

There are two basic categories for physical description (having just dispensed with general relativity) classical description and quantum descriptions.

Classical physics

Classical physics is based on geometry and measuring and its laws about a physical system's partial differential equation apply to general classical systems, when solvable they identify accurate measurable patterns for the system defined to sufficient precision (to be both verifiable and practically useful) and they are practically useful descriptions of the physical system's measurable patterns. The descriptive structure goes from local measuring to a global solution function, and it depends on:

1. materialism,
2. a differential equation, and
3. a system-containing (metric-invariant) metric-space.

Non-linear systems of classical physics

However, most partial differential equations, which are related to the definition of classical systems (based on classical laws of physics) of a material system's defining (partial) differential equation (which has been defined within a containment set and associated sets of measurable properties), are non-linear, and thus they are not (in general) solvable. However, the critical points of a non-linear (partial) differential equation . . . , [ie zeros of the homogeneous non-linear differential equation, and boundary points, and points where the differential equation is not defined] {of such a non-linear partial differential equation} . . . , determine different regions of the domain space where the dynamical properties of the interacting material (which composes the system) determine convergence properties, convergences to limit cycle boundaries of the different regions of the domain space.

Furthermore, some solutions to non-linear partial differential equations are related to geometric properties, eg the lower-dimensional boundary shapes associated to higher-dimensional geometries, ie cobordism. But if the shapes are non-linear then they are almost always unstable, and thus, they define a context where measuring is unreliable, and the patterns are not stable.

However, in the context of general (or non-linear) shapes, it has been shown that there do exist sets of very stable shapes, ie the discrete Euclidean shapes and the discrete hyperbolic shapes, but (even though they were derived from a general context concerning shapes) these stable shapes are: linear, metric-invariant, and continuously commutative everywhere (except, perhaps, at one point). This is the geometrization theorem of Thurston-Perlman.

But elementary considerations concerning "reliable measuring and requiring a descriptive basis in stable patterns" are sufficient to see that a: linear, metric-invariant, continuously commutative everywhere, context for measurable descriptions, wherein the stable reliable measuring shapes are: the line, the circle, and cubes, and circle-spaces, which all together form the proper context within which to consider these ideas about "reliable measuring and the existence of stable patterns" as being central to math descriptions concerning patterns about quantity and shape.

Quantum physics

Quantum physics is based on randomness (indefinable randomness) and a stable spectral set, which are concepts (which are supposedly) defined by both a quantum system's function-space and a set of operators, which represent the quantum system's measurable properties, but these sets of operators cannot be found for general quantum systems, (ie quantum system's whose stable spectra can n be measured), and if this description is formulated, the description is neither accurate nor determined to sufficient precision in regard to the quantum system's observed very stable, and distinctively precise, spectral properties, and, furthermore, the descriptive patterns have virtually no practical value, since it is usually a many-but-few-body quantum system which is described by probabilities (ie it is not a description which can be controlled). The descriptive flow, for the logic of quantum description, is from a set of local operators to a local solution function, which is a descriptive context which, in turn, implies that "measuring in the lab is sufficient for determining the quantum system's properties."

So the question is, "Why even try to describe its properties by means of physical laws of quantum systems?"

Is the main problem with quantum physics either,

(1) that "it is a probability based description"

or,

(2) it is a descriptive context which is local, and it [ie this "local to local" math construct] leads to (other) information which is also local, so the math process cannot lead to any further information about the quantum system (that cannot be collected in a local context)?

However, the quantum system is (can be) provided with a new model [eg the quantum system is a stable metric-space shape].

The quantum description also depends on the ideas of:

1. materialism (random material-point-particle-spectral events in space and time),
2. a set of spectral differential equations (or sets of spectral-operators defined on a function-space), where
3. the system containing domain space is a (metric-invariant) metric-space.

But in this function-space context the math structure is seldom continuously commutative everywhere, thus, the stable measurable properties of general quantum systems can (virtually) never be found (only a handful of exceptions, eg the main exception being the H-atom [a two-body system reduced to being a one-body system which is (in this one physical case): linear, metric-invariant, and continuously commutative, ie solvable]).

The only context in which either of these "descriptive constructs" is solvable (either classical or quantum) is when the partial differential equations are:

1. Linear,
2. Metric-invariant metric-space,
3. Continuously commutative everywhere,

The shapes which allow this set of conditions are the shapes which are also quantitatively consistent, and, in turn, these quantitatively consistent shapes are the lines and circles; and then the associated shapes of:

cubes,
cylinders, and
circle-spaces, eg tori and discrete hyperbolic shapes (which are composed of toral components attached to one another).

That is, the metric-invariant metric-spaces whose metric-functions have constant coefficients, and whose curvatures are constant and non-positive.

The descriptive context of:

1. materialism,
2. a differential equation, and
3. a system-containing (metric-invariant) metric-space.

Can be changed to:

1. Material equals an open-closed (within itself) metric-space shape
2. a differential equation, and
3. a system-containing (metric-invariant) metric-space, is an adjacent higher-dimension metric-space than the metric-space-shape model of material, and the material-component containing metric-space, now, also has a shape.

The descriptive context of quantum physics:

1. materialism (random material-point-particle-spectral events in space and time),
2. a set of spectral differential equations (or sets of spectral-operators defined on a function-space), where
3. the system containing domain space is a (metric-invariant) metric-space.

Can be changed to:

1. Material equals an open-closed (within itself) metric-space shape
2. a set of spectral differential equations (or sets of spectral-operators defined on a function-space but now the functions in the function-space are discrete hyperbolic shapes which model material), and
3. a system-containing (metric-invariant) metric-space, is an adjacent higher-dimension metric-space than the metric-space-shape model of material, and the material-component containing metric-space also has a shape.

Note: Even the math structure of particle-physics, 2-dimensional and 3-dimensional shapes in complex-coordinates (which are related, in particle-physics, to quarks and leptons), can be related to the (small) discrete hyperbolic and Euclidean shapes (of the same dimensions, just listed), upon which a new descriptive context for physical containment can be modeled.

In fact, these small shapes, in complex-coordinates, might be related to the energy partitions of a quantum system's energy space, ie its relation to the quantum system's quantum numbers (in regard to an energy-wave operator model of the quantum system).

This partition of the energy space's volume by small fundamental domains of discrete hyperbolic shapes (which are metric-space shapes related to an energy space), into which charges naturally fit, is related to the entropy properties of thermal systems.

In the new structure of:

1. Material equals an open-closed (within itself) metric-space shape
2. a differential equation, or sets of spectral-differential operators acting on sets of discrete hyperbolic shapes which are functions (ie differential-forms), and
3. a system-containing (metric-invariant) metric-space, is an adjacent higher-dimension metric-space than the metric-space-shape model of material, and the material-component containing metric-space, now, also has a shape.
   and
4. This (stable) pattern of lower-dimensional material being contained in an adjacent higher-dimensional metric-space, can continue in this dimensional-containment pattern up into higher (hyperbolic) dimensions, where the last known discrete hyperbolic shape has a dimension of ten, and thus this pattern of shapes of "a particular dimension being contained in an adjacent higher-dimensional metric-space" can be continued up to an 11-dimensional hyperbolic metric-space, which is an over-all containment space, so that each dimensional level is to be a discrete hyperbolic shape, which is an open-closed topology, and thus each dimensional level is separated, or is discontinuous, from the higher-dimensional metric-spaces. Within a metric-space of a given dimension the properties of the higher-dimensional metric-spaces are excluded, ie either not observed or difficult to understand how to observe these higher-dimensional properties. The relative sizes of adjacent different dimension metric-spaces can also affect what can be observed.

In this new way in which to organize physical description (I) new stable material components can form during collisions (due to resonance), and (II) stable orbits can be determined by condensed material, which is both interacting within a metric-space, but also being contained in a metric-space-shape, where the shape (and its geodesics) can define an envelop of orbital stability for the interacting condensed material components of the system (within a relatively large-sized metric-space shape), eg the planets interacting with the sun so as to form the stable solar system (where all the material components are contained in a metric-space shape).

I. Within this new structure of description the collision-dynamical-system (defined by a dynamic partial differential equation of material components contained in a metric-space) can result in a new stable discrete hyperbolic shape as a product of the collision, if the collision has both the right range of energy, which allows a new material component to form (where it forms due to resonance), and if the original discrete hyperbolic shapes (which are colliding) and the final (new) discrete hyperbolic shape, of the lower dimensional material component shapes, have either the correct range of spectral values, or have the correct spectral values, so that the new (lower-dimension) discrete hyperbolic shape resonates with the finite spectral set of the over-all containment hyperbolic metric-space . . . , which is defined within an 11-dimensional hyperbolic metric-space [which, in turn, is partitioned into discrete hyperbolic shapes so that each subspace of each dimensional-value has a largest discrete hyperbolic shape associated to itself, so that this set of "largest shapes" (associated to all the different dimension subspaces) all together define the finite spectral set associated with the over-all containing 11-dimensional hyperbolic metric-space].

II. While on-the-other-hand the condensed material, which is contained in a metric-space [which possesses a shape], can dynamically (defined by a dynamic partial differential equation of material components contained in a metric-space) define an orbital structure which is related to an envelope of orbital stability, which the shape of the material-containing metric-space defines.

The dynamic structures associated to either the classical differential equations, or, in quantum description, to sets of differential operators (which are now acting on sets of functions which are differential-forms associated to discrete hyperbolic shapes) do not have enough of a relation to stable shapes so as to allow for their solution functions to be able to define the stable properties which are observed for such fundamental physical systems as: nuclei, general atoms, molecules, crystals, or the solar system (as well as dark matter), and thus new containment spaces need o be considered, for all of existence (ie and not simply maintaining the idea of materialism), but it also helps to use consistent math patterns which are measurably reliable and so that there exist patterns in the descriptive context which are stable.

49.

Commons and equality

Equality and the commons

The commons

The commons can only . . . , exist, and be maintained, and be protected . . . , if law is based on equality.

Knowledge can only be developed if it remains a part of the commons, so that big business does not co-opt what is considered to be a scientific truth, ie a measurable verifiable description which is useful in regard to practical creativity, so as to destroy the need for equal free-inquiry which is associated to developing new scientific knowledge. Peer-review most serves corporate interests, in regard to their investments in expensive instruments, in new contexts, instruments might quickly become out-of-date.

But the fluid relation that a precise language has to a "descriptive truth" requires that scientific dogmas are to always be challenged by a scientific community. Yet corporate interests want knowledge to be secret and difficult to acquire, so as to protect the trade-secrets and complicated instruments associated to their monopolistic market controls.

The commons

The commons are the places which have no obvious, easily exploited value, in regard to a very narrow definition of arbitrary-value defined by both (1) the organization of society's social power-structure and (2) the market-place (which conforms to the structures of social power), around the ideas of materialism and rare-elements in narrowly confined and highly controlled markets, in an investment-based world, markets define regions of power (in regard to resources and their use), ie control of a material or a process, or way, of organizing and selling material based creativity (ie both material resources and the knowledge which is used within society are defined by the highly controlled market system of distribution, capitalism is central planning based on selfish interests, in a social collective whose purpose is to support the ruling-class).

The commons are those places (or regions) to which the public retreats when it is being attacked by the ruling-class.

When the commons are "homesteaded," and subsequently acquire value, then the ruling class confiscates them, or taxes them, so as to extract their value. Furthermore, there is also the taxes paid to the rich within the context of a highly controlled and narrowly defined "free-market" wherein the public also gets taxed (by the rich) by means of the private market, which is called the free-market (but is really the natural system of taxation into which the collective nature of society must pay).

That is, the commons can only be defended if law is based on equality, but if law is based on property rights and minority rule, eg a republic which serves the interests of a minority, then as the commons acquire material (or organizational) value, the ruling class will take the commons from the public by means of an amoral law processes, which involves using both the letter-of-the-law and writing laws by a legislature whose "member's primary purpose" is to fit into the propaganda system which is controlled by the ruling-class.

However, if law is based on equality and a requirement for the government to promote the common welfare . . . , in regard to knowledge and practical creativity, wherein the people (the equal creators) selflessly create for others (for one another) . . . , is the duty of the government, ie a people-determined governing body (law based on equality keeps majority-rule from unjustly treating individuals) [where representation can be determined by individual means of mass communication], then "all" would be a part of the commons, and the issues of "homesteading" and subsequent temporary ownership could be sorted-out, though moving from place-to-place may become the standard life-style.

That is, law based on equality is a basis for a society which supports individual freedoms and as people naturally acquire knowledge which is properly associated to their creative interests and subsequently create in a selfless manner then this is a valid model of a free-market, but a market where there is very limited advertising, eg the spec's of the created instrument, and there is no very large producers, if production needs to be large then a cooperative structure which responds to protecting individual creative freedoms is the better organization so as to not become "too big to fail" which is the real basis for our current monopolistic business structures, where banks are simply monopolistic businesses.

There would also have to be limits to resource extraction, so that production activities can remain harmonious with the earth.

Knowledge and the commons

The knowledge which is developed within a culture is also part of the commons, yet the knowledge which is useful to the ruling-class, in regard to either artistic-cultural markets (narrowly defined, and thus susceptible to very judgmental standards and to hierarchies of vapid value), or knowledge about material properties which can be used in practical ways, in order to make material products for the market, is controlled by means of property rights (and/or copyrights) either by the distribution process into the market (of highly regimented art) or by slight adjustments to specific aspects of material instruments (patent laws), where an instrument works based on (culturally common) knowledge about material properties, claiming the adjustment is a newly created aspect (a new patent) of the culture's knowledge, and a newly created aspect of the culture's (ie culture controlled by the ruling-class) ability to create at a practical level.

That is, slightly different meanings of words are manipulated in the context of the letter-of-the-law, so as to steal (through the institution of justice), and thus control, the parts of the cultural knowledge which are used by the ruling-class in its market-propaganda system of organizing society. Which is all held together in the coercive and highly competitive structure of wage-slavery, which serves the ruling-class.

The main strategy for controlling practical knowledge of material is to create a class of professional (and academic) scientists and mathematicians who consider knowledge in a formal complicated language, eg formalized axiomatic language structures, whose fundamental assumptions and contexts and interpretations are all closely related to particular types of instruments, which are of great usefulness to the ruling-class, so that making those instruments becomes complicated, both because of the history of the instrument, and by the complicated formalized language which is developed around the instruments uses, and then these professional experts are supposed to "slowly expand the range of applications" of these complicated instruments. This is what formalized languages and peer-review are all about.

Thus, the important knowledge of science, used by the ruling-class to enhance their own social power, is veiled by a complicated language, and an difficult "to deal with" hierarchical educational institutions traditions, and feuding turfs-of-authority, ie petty selfish arrogances, but totally blind to the fact that their descriptive language structures are fixed, and thus, such a fixed and precise language can no longer be relevant in regard to describing the observed, measurable stable patterns of the world.

The authority granted by corporate interests encroach on how cultural knowledge is related to the commons

Nonetheless, apparently, it is imperative that the academic-authorities go on-and-on with great authority about irrelevant models concerning irrelevant, or imaginative (ie speculative), systems, which deal with both the extremely-small and the extremely-large regions in space associated to the structure of these imagined material systems.

That is, their speculative focus is on models of regions in space where observation is limited to a small range of spectral signals, so that their interpretations about their speculative models of these regions, which is based-on limited information about cross-sections and masses of colliding particles, are no more valid than is any other imaginative interpretation about these remote aspects concerning speculative constructs, that anyone might have, about these limited contexts (or limited numbers of identifiable properties associated to very remote parts of existence).

Thus, if anyone wants to provide a precisely considered speculative model then one wants for any such interpretive model a set of descriptive patterns where (1) measuring is reliable and (2) the descriptive patterns are stable, so the descriptive properties possess content, ie are measurable and possess a practical useful capability).

Note: That both particle-physics and general relativity are non-linear, ie their model for the derivative is a connection. Thus, these patterns will be both unstable and quantitatively inconsistent. It is this context, where math properties are questionable, about which overly authoritative speculative descriptive properties are being provided. Furthermore, it is done in a very authoritative manner, and in a way which excludes any other speculations (so this arbitrary authority is protected and required by peer-review).

This is the structure of a religion, a community of those faithful to a dogma, (wherein only the high-priests of the religion can interpret the religion to the society) not the informative communication structures of a science, where science is about challenging authority, so as to not having to kowtow to those (monstrous) people who want to use physics for the exclusive purpose of making bombs, which is how physics is now (2013) construed.

The overly protected, and insulated from reality, academic people compete over intellectual models which do not have any (or only very limited, as expressed here, in the paragraph below) content.

They may as well debate the number of angels which can fit on a pin-head.

These models are mostly only relevant to both (1) gravitational singularities of spherically symmetric geometries and (2) the cross-sections of elementary-particle collisions, so as to have both models related to reaction rates in a very high-energy explosion at the (assumed to be) spherically symmetric center of the blast.

Obsessions, autism, and the psychopathic

One cannot help but believe that this was the direction given to university physics departments by E Teller or some even more monstrous person. Such psychopaths thrive in totalitarian, and essentially lawless, states such as was the case for the Roman empire.

One would expect that psychopathy is a form of autism, ie obsessive behavior in regard to a delusional mental model of the world.

It seems that the top social (high-level management) positions, in our collective-based society, are filled with people whose lives revolve around their obsessive behavior patterns concerning a "mental construct" which is more of an illusion than a valid model of existence.

We are guided more by illusions, in our society, than by a valid assessment of existence and concerning the value of others.

This set of illusions has been provided for us by the propaganda system.

Freedom is followed by more oppression

Only a very limited set of viewpoints are allowed to be expressed. This is the result of control over publishing and broadcasting (with the editors and their peer-review demands, ie gate-keepers of a limited vision of truth, which are associated with these publishing institutions).

But with the internet, new ideas are being expressed and accessible to others, and it is these new ideas which the owners society need to control by means of blanket-spying.

Thus the true amoral nature of the controlling forces of society are coming to light.

The owners of society want politics and publishing to be 100% propaganda, so as to have ideas, which are allowed to be expressed within society, to have only a very limited relation to truth . . . , a truth as believed by the owners of society, and which is a part of their model of social control . . . , so that this limited set of ideas are the only ideas which should be allowed to be expressed within society.

That is, any story on the media . . . , including any web-sites which are well organized associated to many sources, and thus they must be well funded . . . , is about presenting false models for everything, science religion, economics, psychology, sociology, politics, the structure of social power, etc etc.

The image of science and math

The image of science presented by the media is that a physicist is a scientist who considers equations of systems, either based on the laws of physics, or to determine the laws of physics, and then these equations are solved, so all information concerning all realms of physical description have been considered, by the physicist, for all physical systems ranging from the macroscopic to the microscopic, and the microscopic has been reduced further to elementary-particles, and string-theory unifies particle-physics to gravity, so all the solutions to the equations (concerning the laws of physics) provide all relevant information about all conceivable physical systems. However, the physicist must proclaim that for most fundamental physical systems which are: general, stable, and precisely identifiable physical systems, the standard set of information which the physicist provides, is that "the system is too complicated to describe."

But "what is the context of their descriptions (anyway)?" Is it all physical systems large and small?

Answer: Basically quantum description . . . , (which, it is believed, encompasses all physical phenomenon, since even gravity, it is now assumed, has been reduced to particle-physics by string-theory) . . . , is a locally based descriptive math structure, which is being used to determine local particle-spectral properties, so why not simply go to the lab and get the data on these systems (this is, essentially, the true practice of quantum theory, then ways to attach quantum properties to classical systems are considered).

The systems are built from many-but-few components, and the description is probabilistic, so it is unrelated-able to either a definitive prediction of an event or to control over the system.

Furthermore, for general, but very fundamental, systems within this descriptive context "the equations are not known" and only "approximate equations" are provided, which are supposedly consistent with the laws of quantum physics, and are "almost always" non-commutative, so as to define a quantitatively inconsistent descriptive context, ie any information provided by such a model is unreliable.

That is, this descriptive context is not capable of identifying . . . , through their own physical laws (the laws of quantum theory) . . . , the stable properties of a general quantum system. Such a condition for physical description, based on physical law, provides a good definition of an indefinable random descriptive context.

When the description, or wave-function (which is not, in general, determinable), is adjusted by the math constructs of particle-physics . . . , ie "the crown of the physics community's physical model" (according to the propaganda system) . . . , the equations become non-linear, and so these equations (math relations) are describing a context, ie a system's equation, {and associated solution function (which never exists)}, which is indefinably random, and non-linear. That is, it is an equation which is describing a physical system which is chaotic, and there are no math stable patterns which are describable in such a (math) context, ie it is an indefinably random, quantitatively inconsistent, description of a system in which none of the patterns it describes are stable (if it describes any valid patterns at all).

Physical law based on either dynamics or a harmonic structure defined around a system's geometry of energy, has failed (the intermediary dynamic processes which exist between stable [and neutral] states of material cannot be used as a basis for stable patterns)

The problem of the math models of physics

It should be clear that trying to describe the properties of physical systems "based on physical law, in turn, defined by (partial) differential equations," has a very limited range of applicability.

Namely, to such a descriptive range's solvable context.

Yet it is an accurate description, in regard to the mostly observed chaotic and unstable dynamic patterns which exist, and are a big part of our experience (at least, in regard to observing dynamic systems).

However, it is a descriptive context which has no value in regard to trying to describe the observed stable patterns of material systems of all size scales, since it is an unstable and quantitatively unreliable math context.

It is the stable patterns and quantitatively consistent patterns, associated to math constructs from which useful and accurate math descriptions emerge, in regard to the stable properties of physical systems which need to be described in a practically useful context, where these stable physical patterns are the most fundamental aspects of our experience, and it is such stable and quantitatively consistent math structures within which useful, applications of physical descriptions are going to exist. That is, it is from stable patterns and reliable contexts for measuring that an ability to describe the stable patterns of the fundamental material-components . . . , which compose most of existence, eg atoms, nuclei, etc . . . , is going to emerge (in a mathematical manner).

Bomb engineering is about the chaotic transition from one stable state of material to another, different but also, stable state of material, so this seems to be the motivation for the physics community to focus on the chaotic and unstable aspects of physical description of dynamic properties.

Thus, there is the focus of differential equations as being the correct focus for the laws of physics, since it is the chaotic processes where the rates of reaction (and thus the power of the bomb) are determined.

It is dynamic properties which physical law is trying to relate, by the laws of physics, to stable physical structures.

The main models for this are the successes of:

Newton's two-body model of planetary orbits, and
Schrodinger's model of the H-atom, where the H-atom is also a two-body system.

But this context of physical description is not working. But the interests of big business which win-out.

Nonetheless, particle-physics is a descriptive context in which cross-sections of elementary-particle collisions (relatable to rates of reactions) can be determined, and the string-theory linkage (or correlation) between rates of elementary-particle collisions and a gravitational singularity (of general relativity) is being modeled and related to ideas about nuclear weaponry.

It is clear what the owner of society want from the physics community, and the autistic structures of human mental properties are being manipulated to get obsessive types, who are mostly separated form a valid viewpoint of reality (a mental state in which most of the citizenry are also in), to only consider such models so as to serve the interests of the military big businesses.

Economics and the hidden-hand

The central focus on economics within society is the crowning achievement of both propaganda and the Calvinist rendering of Christianity, and that Calvinist interpretation is, that the very rich are not only "favored by God," but, effectively, they have come to be "God on earth," where all social institutions serve the economic titans, ie the very rich.

The dregs of our intellectual personnel are the stars of the propaganda system

The pathetic intellectual structure of the people who make the propaganda prevent them from questioning, in any meaningful way, and then their distorted mental constructs also prevents them from possessing the courage to see through the fake "dialectic of opposites" which is used in the propaganda system to divide the population. It is a fundamental part of the language used in the propaganda system. They do not report on the manipulation of personality—and mental-types done through public (and private) institutions: the justice system, the education system, the propaganda system, etc. and the autistic-psychopathy of the owners of society and the brutal way in which they terrorize and exploit the population, apparently, a necessity for an extremely hierarchical society (but, alas, they seem to be equivalent of the Roman Emperors). Apparently, the end-game of this form of social organization is both the destruction of the earth and an ever-growing and expanding population.

What story is true concerning Tesla?

Was Tesla highly influenced by his humanitarian beliefs in that creativity should be directed at helping all the people? or Was Tesla highly influenced by Christianity, and was his construct of existence radically different from the academic physics community? Did most of his ideas actually fail, but along the way he created the basic structures of modern electrical engineering? He had to be well educated about the electromagnetic laws of Faraday. Did he get influenced in a negative way, ie was he manipulated by bankers and investors, into changing his humanitarian beliefs so as to build his electrical-particle ray-gun in which he was inventing a weapon to, supposedly, "end all wars," or more likely to please the interests of the investors with whom he hung-around to a significant amount?

What was true of Tesla? I suspect he had a model of "the harmony of the spheres" (whatever) which gave him geometric models of things, to which he applied electromagnetism and determined a geometry's spectral dependence, a geometry to which currents could be linked.

There are other enigmas, like Tesla's personality is an enigma, which the propaganda system destroys for the culture.

Namely, once one could look to the stars as a symbol of a great mystery, but now some jackass will tell the overly authoritative stories of the stars presented by the science community and the propaganda system, so that these stories say that the stars are no longer a mystery, that arrogant mankind has figured-out the entire material world, and the people must grovel at the feet of these great intellects who supposedly understand these mysteries and are pushing such superior knowledge forward, ie ever more baloney. The truth which the experts recite to us is all a bunch of baloney, the main truth which these experts convey, is the truth about inequality, as the rule of society, and our intellectual endeavors must serve their dogmas.

The justice system is all about property rights and inequality

The justice system is in the business of both forcing the public to conform to the collective interests of a society run by a handful of people (they enforce wage-slavery), and of manipulating mental-types so as to terrorize the public, where the propaganda system is a partner in these efforts to terrorize and to divide the public. This is because inequality requires that the lower social classes be attacked by the ruling-class, since that is essentially what has happened within society, in regard to owning property.

This has been clear since the un-denied stories of a very manipulative and murderous JE Hoover came forth. Why would Hoover's organization change from these practices when he left?

The distortion of science comes about because:

The state of current math and science (physics in particular, but also biology) knowledge is all about the veil of complication which formalized axiomatic's has introduced into these disciplines, and essentially made them completely dysfunctional, and it has made science more like a religion, which worships nuclear weapons, than such behaviors being related to being a science. The single mindedness of pursuing the development of only one instrument (of destruction).

There are stable and relatively easily identifiable spectral properties associated to many fundamental quantum systems (this implies that these systems are geometric, and solvable, and controllable), but whose spectral properties cannot be identified to sufficient precision by the methods associated to the descriptive laws of quantum physics, eg an atom with more than five charged-components or any nucleus with more than five particle-components, nor can the (apparent) stability of the solar system be described based on currently accepted physical law (though this paper provides an alternative math model in which these observed properties can be described, based on the rules of the new descriptive structure).

Furthermore, most of the physical systems of classical physics define non-linear (partial) differential equations, which cannot be solved in a global context, and the local, numerical solutions are chaotic, so that no patterns are identifiable.

Yet, the many-body non-linear solar-system, apparently, has been stable for billions of years.

Why? Its stable properties fit into folded (by Weyl transformations) discrete hyperbolic shape, which, in turn, models our containing metric-space.

General relativity is also non-linear, and it has only been solved for one problem, and that is the one-body problem in which the geometry around that one-body is spherically symmetric. But this limited context of descriptive properties (for general relativity) has no relation to being able to describe a stable solar-system, which is a many-body system.

Furthermore, action-at-a-distance, or the property of non-locality, has been demonstrated in certain types of quantum systems, ie there is action-at-a-distance. This means that smoothness which is associated to non-linear models of physical law and containment sets is not a necessary property of a descriptive structure.

This is a weird issue since quantum theory claims to be discrete, but it is discrete only at certain points in an otherwise smooth descriptive structure.

This means that it is a good idea to model operators which model a local measuring property as being non-local, or discretely non-local, as opposed to differential operators, ie operators which model a local measuring property, which are only smooth (all derivative exist and are continuous) [or at least C^2].

Consider math

Math is about:

1. Quantity and shape
   And
2. Where, there is reliable measuring, and there are stable patterns, where these stable patterns are both (a) to be described, and (b) there are stable patterns which are a part of the descriptive constructs.

This leads to the questions:

3. What math structures, or math contexts, result in quantitative structures which are reliable?

   Answer: Being able to identify stable, simple number relations in a descriptive context, eg being able to count and identifying the set (or a subset) of integers, properties which must be consistent with an assumed measuring context of the math patterns being described, and there are conditions of the pattern being described in such a way in which changing size-scales between two quantitative sets is defined to be a reliable math process, ie linearity and metric-invariance and local properties of coordinates which are continuously commutative everywhere.

4. What shapes are stable?

   Answer: The stable shapes are the circle-space eg the torus (or the discrete Euclidean shapes), and the discrete hyperbolic shapes,or more generally

The shapes which allow this set of conditions . . . , (linear, metric-invariant, and continuously commutative everywhere quantitative patterns, which fit into a coordinate system which is assumed to contain the system whose properties one is trying to describe) . . . , are the shapes which are also quantitatively consistent, and, in turn, these quantitatively consistent shapes are the lines and circles; and then the associated shapes of:

1. cubes,

2. cylinders, and
3. circle-spaces, eg tori and discrete hyperbolic shapes (where discrete hyperbolic shapes are composed of toral components attached to one another, and in turn, these circle space shapes are associated to "cubical" simplex shapes (or fundamental domains)).

Circle-spaces can determine both (1) envelopes for stable (circular) orbits deformed slightly to be ellipses by material interactions, and (2) faces which can define both the material structure and the geodesics, or minimal-spatial-occupying-spectral properties in a continuously commutative shape (where this commutative property allows for 1-dimensional spectra to be associated with any minimal-regional-facial related spectral structures associated to these stable shapes (where for example a 3-cube has six faces, or facial regions)

These four general math properties are more necessary, in regard to descriptions of stable measurable patterns, than are the frames defined in physics, where frames are concerned with the relative motions of the (physical) system-containing coordinate space, within which measuring is assumed to occur.

The frames of physics relate to both metric-stability, or metric-invariance, and the relation that these frames of coordinates have to the properties of a system defining (partial) differential equation, ie the laws of physics.

But the model of system-defining differential equations associated to dynamic properties has not led to an understanding of the stable properties of physical systems.

A new context, where the shapes of metric-spaces are central to understanding both material organization and spatial properties and a metric-space shape's relation to dynamics.

Metric-space shapes determine stable material properties, but now existence is contained in a many-dimensional containment set, where each dimensional level is defined by a metric-space shape, whose topology is open-closed, and thus the dimensional levels are related to one another in a discontinuous manner, so that once inside a metric-space the (other) higher-dimensions are difficult to detect. In this context, differential operators become discretely defined on discontinuous time intervals, so as to be defined between different adjacent dimensional levels in an action-at-a-distance model of material interactions which result in dynamic changes.

The metric-space subspaces, ie either metric-space fundamental domains or metric-space shapes, can be associated to a frame, but a more inclusive viewpoint shows that it is the entire set of metric-space shapes which is the frame (or viewpoint) which allows a greater understanding concerning the properties of material systems as well as a greater understanding of metric-spaces (and their shapes).

The idea that a physical system's defining set of (partial) differential equations which are covariantly invariant . . . , ie related to differential-forms (which should also, in turn, be related to an invariant definition of the containing space's coordinates metric-function (a metric-function which has constant coefficients), ie not a general metric-function, which can be non-linear, and thus it would define an unreliable measuring context where there are no stable patterns)

. . . , is not sufficient (ie Einstein's principle of general relativity is not sufficient, though, perhaps, a natural focus of a dynamic centered descriptive construct) . . . ,

. . . , one also needs the math property of these equations being continuously commutative everywhere, in order to be in a context in which measuring is reliable and that the patterns being described are stable.

Note: An alternative containment context where "inertia might also be defined due to the shape of the material containing space (as in the case of general relativity)." The geodesics defined in a metric-invariant context can be related to inertial dynamics, but only in the energy-space of a hyperbolic metric-space, if the material is contained within a discrete hyperbolic shape (which is also a metric-space).

General relativity assumes that only local connection derivatives (ie non-linear math structures) are fundamental, and a (general) metric-function for a coordinate space can be found in a context of general covariance, ie a non-linear partial differential equation is pertaining to differential-forms, so that a solution to such an equation can determine either the coordinate space's natural local linear (but chaotic) measure for length, ie the coordinate space's general metric-function, or the sets of local non-linear geometric measures consistent with the general metric-function.

In turn, the geodesics, which are determinable from the metric-function, determine the inertial properties of the system.

This is a circular relation between the metric-function of local coordinates and the model of the derivative as a "local non-linear measure" which cannot be unraveled.

The problem is that measuring requires both a metric-function and local linear derivatives to exist as stable patterns within the descriptive construct (where non-linear models of local measuring do not allow for measuring reliability and they do not allow for stable patterns to be identified).

Otherwise the partial differential equation is non-linear and defines a descriptive context (or a set of containing coordinates) which are found from a non-linear differential equation, ie this is an unreliable context for measuring.

Thus, general relativity has only solved one problem; the one-body system which is assumed to possess spherical symmetry, where a perfect sphere is a continuously commutative shape almost everywhere. But, try to put another piece of material into such a one-body construct, and an ensuing attempt to find a stable shape (eg a stable solar system) can never be resolved.

The context does not allow measuring to be reliable, and the math patterns will always be chaotic.

So using this descriptive context to describe the stable solar-system is unthinkable.

In order to understand the order which is associated to the stable orbits of planets (or a quantum system's precisely defined stable spectral properties) one cannot use the local properties of measuring (modeled in a C^2 math context, where C^2 means that the second order partial derivatives are continuous), for both motion and for models of force-fields, in order to find this stable structure, since such a structure does not supply enough restrictions on the measured properties to obtain the stable patterns which observations say exist.

However, stable orbits (as well as stable, and precise, spectra) can be related to the stable shape of our (material-containing) metric-space, and where stable metric-space shapes (or metric-spaces) are associated to either material components, or the containing metric-space of the many (atomic) material components of the world we experience.

These stable patterns will be "discrete hyperbolic shapes" and such "discrete hyperbolic shapes" can be of many different dimensions (they exist in dimensions one through ten; inclusive) and they are differential-forms of solvable-shapes defined by cubical simplexes (which are related to continuously commutative shapes [when moded-out to form into an adjacent higher-dimensional shape, eg a rectangle can be moded-out to form a torus, defining the topology of the newly moded-out shape by an equivalence relation, ie or identifying opposite sides of the rectangle with one another]).

For condensed material contained within a stable metric-space shape, the stable orbits (planetary orbits) of this condensed material can be related to geodesics of a 3-dimensional discrete hyperbolic shape, ie solvable.

Furthermore, the spectral order of quantum systems is best modeled as discrete hyperbolic shapes (of course smaller than the discrete hyperbolic shape of the solar-system), perhaps of lower-dimension than their containing metric-space (although this is not necessary, but in our metric-space we would naturally see material interactions for material which is the size of the solar system.

Modeling material-interactions as dynamic math constructs

The differential equations of transitioning material systems (material components contained in a metric-space), ie systems of materials and their interactions, are discrete operators or models of local measuring of properties of a system which is contained in the domain space (or coordinate measuring space) [ie domain space of the function whose values define the property of interest], discretely defined (in a discontinuous manner) between both (1) discrete time intervals and (2) the different dimensional-levels of the system's existence, where the different dimensional levels are stable discrete hyperbolic shapes.

The metric-space shapes determine both the material components and the stable orbital envelopes for condensed material, ie condensed material components which are contained in the fairly large metric-space shape.

To re-iterate, the differential equation model of physical systems has many, sever, limitations

Neither differential equations identified by models of local measuring of properties of material systems whose quantitative pattern is not known . . . ,

Nor sets of operators (each representing physical properties) applied to function-spaces with each function representing harmonic distributions (in space) associated with (or of) spectral sets (local particle-spectral properties [seen as random events in space]) . . . ,

. . . , can be applied to either material or energy systems which are contained in a metric-space in order to determine either interaction dynamics (of a stable system) . . . , or determine commutative sets of operators acting on function-spaces so as to be able to determine or confine or constrain the relevant properties of a system . . . ,

. . . , so as to identify the order into which material systems composed of material components organize themselves.

That is,

Differential operators cannot be used to confine or restrain the properties (one is trying to describe) being described, and thus, they cannot be used to find the order, which is observed in physical systems.

Thus it is assumed that it is the geometry of either material or energy which needs to be the restraining cause (force), or which confines the pattern of a material system so that the system takes on the properties of order and stability, but this is not happening in the current formalized axiomatic which peer-review defines.

The confining context of existence's actual containing set, ie sets of stable metric-space shapes defined in many-dimensions

Thus, an alternative is to have these differential operators act in a discrete manner so as to inter-relate different adjacent dimensional levels of metric-space shapes, so that this is done in discrete time-intervals, where this time-interval would be defined by "the time it takes light to traverse the diameter of an H-atom" (this would also be the period of the spin-rotation of metric-space states, see next note).

Note: Metric-space states are derived from the fact that metric-spaces of particular metric-function signatures are associated to particular physical properties, so that these properties have two opposite states, eg hyperbolic space is related to the two-opposite states of (+t) and (-t). Spin-rotation, rotates between these two opposite states, which are properties (or opposite states) of each such metric-space.

It is the geometry of both the material and the metric-space, as well as the finite range of spectra which the containing space allows to be a part of an existing "material" system, within such a high-dimensional over-all containing 11-dimensional hyperbolic metric-space.

Thus, in a many-dimensional structure of metric-space shapes, there are both (1) material metric-space shapes and (2) material containing metric-space shapes, whose metric-space shape can define a stable orbit for the condensed material which the metric-space contains.

The finite spectral set which defines existence within any particular 11-dimensional hyperbolic metric-space, ie the correct containment set for existence. It is a construct which transcends the idea of materialism.

A finite set of these stable metric-space shapes can be identified by identifying the largest such stable shapes, which are defined for each subspace of each dimensional level, and then define the possible spectral-orbital values which can exist in any dimensional level and in any subspace by means of resonances which can be defined between any of the different subspaces, and this is because all such subspaces (identified as stable shapes of particular dimensions) are contained within an over-all containment hyperbolic metric-space, which is an 11-dimensional hyperbolic metric-space. That is, all the lower-dimension metric-space shapes which are contained in a metric-space of some given dimension must be in resonance with some largest metric-space shape defined for some subspace of the correct dimension, ie the dimension of the material component, (which exists because it has this resonance property).

Dynamics still affects the structure of stable material patterns

These metric-space shapes are not directly observable, except they are the cause of stable orbital properties for the planets in the solar system, and they are the cause for the stable spectral-orbital properties of material components (or quantum systems), where along with the 2-body models of gravitational interactions such dynamic structures can distort the shape of the stable-orbit within the geodesic structure of the shape of "the planet containing metric-space" so as to define ellipses which are slightly deformed from being circles (ie the case of the elliptic equation).

But for quantum systems it is the differential equations associated to point-models of small material systems colliding, wherein during the collision a new small material component might form, due to resonances . . . , of the new system's stable and small discrete hyperbolic shape . . . , with the finite spectra of the over-all high-dimension containing hyperbolic metric-space, ie an 11-dimensional hyperbolic metric-space (which, itself, has no discrete hyperbolic shape) (the case of the hyperbolic equation).

There is the possibility that the material components are not simply shapes and lengths of material metric-space shapes which fit into the facial or orbital envelopes of a quantum system, rather there are small material components also confined to envelopes of orbital stability within the quantum system's metric-space shape, so that in this case within a quantum system there are also elliptic equations to consider. This would be the model of A Somerfeld's elliptic adjustments to Bohr's model of the H-atom (and which fit the observed data on the H-atom very well).

The parabolic differential equation model of harmonic averages existing around the system's energy properties, ie the wave-equation of quantum physics, is valid to the extent that small material components interact in a continually changing set of geometries from one time interval to the next time interval and thus possess Brownian motions whose over-all effect is exactly the same as the material components of a quantum system appearing to be fundamentally random (this effect identified by E Nelson, 1957 Princeton), rather they are random until they form into material components (during collisions), where these stable components have a stable metric-space shape, whose spectra is consistent with (and resonate with) the finite spectra of the over-all containing hyperbolic metric-space.

The basic math pattern which is being used in the new math construct

Thus, it is the structure of the containing metric-spaces and their local coordinate actions (or the natural metric-invariant group actions) and this set-structure's relation to the stable geometric structures, which are naturally associated to such a metric-space . . . . , ie the discrete isometry subgroups, of the metric-space of the local coordinate transformation isometry group, ie a fiber group in a principle fiber-bundle . . . , which identify the order, or stable geometric patterns, (possessed by material systems) in a measurably reliable context, concerning the descriptions of the stable properties of material systems, which are assumed to be contained in these metric-spaces.

This same structure of natural stable geometries can be applied to various dimensional levels, and for metric-functions with constant coefficients, and for various signatures for these metric-functions, where all of these metric-spaces are associated to isometry groups, which, in turn, have discrete isometry subgroups and associated fundamental domains in their metric-space base-spaces (where these fundamental domains, in turn, are defined as "cubical" simplexes). This is the context from which stable properties of material systems, which are contained in metric-invariant metric-spaces, and which allow for both the existence of the stable order (stable geometric-spectral patterns), and also form the confining structures, which require this stable order to exist within material systems, where material has now come to be modeled as a stable metric-space shape.

The different signatures apply to metric-spaces whose material components, which these metric-spaces contain, can have various different relations to time, ie R(s,t) where, s, is the dimension of the spatial subspace, and, t, is the dimension of the temporal subspace, so that s + t = n, and n is the over-all dimension of the R(s,t) rectangular coordinate space.

Thus there is both reliable measuring and the existence of stable geometric-spectral patterns within such a descriptive context.

How stable microscopic systems form

In this new way, in which to organize physical description, (I) new stable material components can form during collisions (due to resonance, and if the energy of the collision is within the "correct" energy-range), and (II) stable orbits can be determined by condensed material, which is both interacting within a metric-space's shape and geodesic structures (in an open-closed topology, within the space), but also being contained in a metric-space-shape, where the shape (and its geodesics) can define an envelop of orbital stability for the interacting condensed material components of the system (within a relatively large-sized metric-space shape), eg the planets interacting with the sun, so as to form slightly elliptic orbits within the envelopes of orbital stability defined by the big material-space's shape, in the stable solar-system (where all the material components are contained in a metric-space shape).

I. Within this new structure of description the collision-dynamical-system (defined by a dynamic partial differential equation of material components contained in a metric-space) can result in a new stable discrete hyperbolic shape as a product of the collision, if the collision has both the right range of energy, which allows a new material component to form (where it forms due to resonance), and if the original discrete hyperbolic shapes (which are colliding) and the final (new) discrete hyperbolic shape, of the lower dimensional material component shapes, have either the correct range of spectral values, or have the correct spectral values, so that the new (lower-dimension) discrete hyperbolic shape resonates with the finite spectral set of the over-all containment hyperbolic metric-space . . . , which is defined within an 11-dimensional hyperbolic metric-space [which, in turn, is partitioned into discrete hyperbolic shapes so that each subspace of each dimensional-value has a largest discrete hyperbolic shape associated to itself, so that this set of "largest shapes" (associated to all the different dimension subspaces) all together define the finite spectral set associated with the over-all containing 11-dimensional hyperbolic metric-space].

II. While on-the-other-hand the condensed material, which is contained in a metric-space [which possesses a shape], can dynamically (defined by a dynamic partial differential equation of material components contained in a metric-space) define an orbital structure which is related to an envelope of orbital stability, which the shape of the material-containing metric-space defines.

Again the problems with the current viewpoint of the structure of set containment for existence

The dynamic structures associated to either the classical differential equations, or, in quantum description, to sets of differential operators (which are now acting on sets of functions which are differential-forms associated to discrete hyperbolic shapes) do not have enough of a relation to stable shapes so as to allow for their solution functions to be able to define the stable properties which are observed for such fundamental physical systems as: nuclei, general atoms, molecules, crystals, or the solar system (as well as dark matter), and thus new containment spaces need to be considered, for all of existence (ie and not simply maintaining the idea of materialism), but it also helps to use consistent math patterns which are measurably reliable and so that there exist patterns in the descriptive context which are stable.

Particle-physics and the new math context

Note: Even the math structure of particle-physics, 2-dimensional and 3-dimensional shapes in complex-coordinates (which are related, in particle-physics, to quarks and leptons), can be related to the (small) discrete hyperbolic and Euclidean shapes (of the same dimensions, just listed), upon which a new descriptive context for physical containment can be modeled.

In fact, these small shapes, in complex-coordinates, might be related to the energy partitions of a quantum system's energy space, ie its relation to the quantum system's quantum numbers (in regard to an energy-wave operator model of the quantum system).

This partition of the energy space's volume by small fundamental domains of discrete hyperbolic shapes (which are metric-space shapes related to an energy space), into which charges naturally fit, is related to the entropy properties of thermal systems.

In the new structure of existence:

1. Material equals an open-closed (within itself) metric-space shape
2. a differential equation, or sets of spectral-differential operators acting on sets of discrete hyperbolic shapes which are functions (ie differential-forms), and
3. a system-containing (metric-invariant) metric-space, is an adjacent higher-dimension metric-space than the metric-space-shape model of material, and the material-component containing metric-space, now, also has a shape.
   and
4. This (stable) pattern of lower-dimensional material being contained in an adjacent higher-dimensional metric-space, can continue in this dimensional-containment pattern up into higher (hyperbolic) dimensions, where the last known discrete hyperbolic shape has a dimension of ten, and thus this pattern of shapes of "a particular dimension being contained in an adjacent higher-dimensional metric-space" can be continued up to an 11-dimensional hyperbolic metric-space, which is an over-all containment space, so that each dimensional level is to be a discrete hyperbolic shape, which is an open-closed topology, and thus each dimensional level is separated, or is discontinuous, from the higher-dimensional metric-spaces. Within a metric-space of a given dimension the properties of the higher-dimensional metric-spaces are excluded, ie either not observed or difficult to understand how to observe these higher-dimensional properties. The relative sizes of adjacent different dimension metric-spaces can also affect what can be observed.

50.

Education 2

Education and the corruption of society

It is materialism, and the math structures associated to materialism, which are not capable of identifying the cause of the observed stable properties of many, fundamental, material systems. Yet, it is the model of particle-collisions in a context of materialism which both defines modern physics, and is the basis for nuclear weapons technology. This structure for knowledge exists because these are the wishes of the ruling-class. Yet, the failures, of such a model for knowledge, are the causes of the social collapse, due to not enough practical creative expansion, in turn, due to the act of constraining knowledge and restricting its relation to practical creativity.

If one has rules and principles which apply to machinery and instruments so that the instruments function within the descriptive context of these narrow rules, then one can set-up a means to distinguish if another person, who is learning these rules, has understood these principles and been able to apply these principles and rules when maintaining particular instruments, then this is narrow fixed model of information acquisition is determinable, and perhaps testable, but when there is observed order, and there are no valid precise measurable descriptions which apply to, or describe the cause, of this observed order, then there is no reason to believe the ideas . . . . , that are provided by any "experts" who claim to have "the best way" in which to try to understand this material-order which is related to (other than that) the ideas of the experts (whom are dedicated to the notion that the ordered systems are reducible and the cause of the order is the interactions which are defined between the particle-components by particle-collisions to which the experts believe their instruments [or particle-accelerators] are reducing these ordered systems) which, in turn, are ideas which are related to devising particular types of war instruments, eg nuclear weapons . . . . , but this reduction scenario is not even remotely close to identifying the math patterns which best apply to both describing, and then using the ordered patterns which are observed for material systems.

For example, these stable and highly ordered systems are some of the most basic, and fundamental, stable spectral-orbital properties of material systems which exists at all size scales, eg nuclei, general atoms, molecules, crystals, condensed matter, the stability of the solar system, and the motions of stars in galaxies, the apparent motions of galaxies in space and time, etc,

while on-the-other-hand, the idea of rates of particle-collisions, ie particle cross-sections, used by the designers of nuclear weapons, is useful for designing for a greater power of a nuclear explosion, but it (particle-physics) is simply a property of broken-apart components and their random collisions in a chaotic transition process, which occurs between relatively stable states of material, which exists both before and after the chaotic transition process, ie before the chaotic transition process begins there exist stable, neutral physical systems, and thus, "the main question is about" "why there are stable states in regard to material-systems." Namely, "On what principles is the order and stability . . . . of the fundamental, stable material-systems which make-up most of our experience . . . based (ie what causes stable structures and neutral material systems to exist)?"

The experts who have faith in reductionism, within the context of materialism, are not providing valid descriptions of the order of these fundamental stable systems, they claim that "the spectral properties these stable spectral-systems are too complicated to describe by trying to calculate these spectral-properties based on 'what they claim to be the physical laws of quantum physics.'"

Thus, their opinions and speculations are no more valid than anyone else's speculations concerning both these systems and the true nature of physical law (since they cannot calculate to sufficient precision the observed stable spectral-properties of these fundamental physical systems). Yet "why is it that these material based reductionists are in total and dominant control over the physics departments of public higher educational institutions?"

It is because the government partners more with big business interests than it is concerned about the culture developing new knowledge based on equal free-inquiry?

Yes, that is the correct conclusion, and this is one of the main causes for the failings of the society. Even the financial failure was due, in part, to believing the experts associated to the failed quantum-reductionist, random math models, which were subsequently applied to determine business risks.

But these math structures are not valid, they are false when applied to events in the actually observed patterns of the world.

This is an intellectual failing, to which such overly dominant behavior . . . , of experts held fixed upon a single narrowly defined dogma, which is based on a formalized axiomatic structure for language . . . , leads.

It leads to illusions (which are only relevant to a fictitious world) and this being the leading edge of technical development, ie an extension of the idea of adjusting existing instruments (that is, adjusting instruments, but not developing new contexts for creative development), then the society, and its delusional basis, has to be propped-up with more dogma and ever more domination by the few, and this all depends on this social model being maintained by extreme violence.

Technically this is the context of language in which a descriptive structure is trapped within a fixed context,

Namely:

Within an open-closed topology (where topology deals with the set structure of a continuum, in regard to measurable properties and their functional relation to the domain spaces) of a material containing metric-space, the (local) constructs of measuring, and the containment sets for these measured properties of material systems . . . , ie both the system and the external geometry of which the system is a part are contained in the metric-space, . . . , so that the measurable properties of these systems and their external geometries . . . , (note: this set structure of containment and measuring [where the math structures {eg math operations} associated to measured-values deals with algebra]) . . . , cannot be used to identify the stable patterns defined in (and observed within) a reliable measuring context, where these stable measurable patterns are observed to be associated to the system, and furthermore, this cannot be done either in regard to geometric dynamics, or in regard to harmonic averages in a random descriptive context defined around a system's energy (or measurable) math structures.

That is, a system's observed stable patterns cannot be identified in regard to the containment of measured properties defined by the idea of materialism (ie an open-closed topology associated to a material containing metric-space).

In particle-physics the higher-dimensions are internal particle-states, while string-theory extends this idea so as to distinguish a material containing 4-dimensional subspace so that the other higher-dimensions must be curled-up as very small geometries defined about the material containing subspace's points.

These math constructs adhere to a context associated to rates of particle-collisions so as to be applicable to nuclear weapons development.

However, there are other math models of higher-dimensions, where the higher-dimensions can be macroscopic, ie not internal particle-states, and these higher-dimensions are still difficult to detect.

That is, without the possession of a valid knowledge, then there is no way in which to be able to judge the validity and merits of any one's attempts to describe the cause of the order and stability which material systems most often possess, or want to (or are quick to) realize.

That is, the university, the seat of higher learning, cannot be based on sets of authorities and, so called, experts, who judge "the merits of other's ideas" when the point of a learning-center is to search for knowledge which, apparently, no one possesses.

This, of course, is an example of the mantra of "choosing winners" which the government is not supposed to do, since it interferes with the so called free-market, but the government's job, given to it by the owners of society, has been to always interfere with markets and society, so as to help maintain and grow the social power of the owners of society.

The republicans, those who openly support the interests of the owners of society, get away with:

Choosing market-place winners (funding research needed by big business),
Creating immense government (military expansion),
Regulating small businesses out of the market, and
Effectively taxing the public to death, but where the taxes are on medical, and energy, and communication systems, etc, expenses, which are associated with the privatization of necessary public services,
And
They support the counter-insurgency actions against the public, which is organized in the justice system, such as wage-slavery, and which depends on a network of people who thrive on acts of coercive domination perpetrated against the public, eg racists, hate-groups, and mobsters, etc.

And yet it it's the republicans whom rail the loudest (in the media) against the very acts they faithfully perform for their pay-master overlords.

But the democrats are "in-on-this too," since it is a set of socialization processes all of which are being orchestrated by the propaganda system, to which the politicians are "cogs in the wheels of the propaganda system," where the propaganda system is owned and controlled by the owners of society, eg it was the Carter administration who passed laws of "bankruptcy being associated to the raiding of pensions by the managerial class."

That is, there cannot be a hierarchy when one is searching to unravel (fundamental) mysteries, since no truth is known.

Yet, because by the ideas of Calvinism "the sign that a person is chosen by God is if they are rich" instead of a church which represents God, as was the case in the age of Copernicus, now (2013) it is the very rich who have become God, and the absolute truth . . . , which they dictate to the politicians and to the research communities . . . , is that all language and thought must be designed to serve the business interests of these Gods-on-earth.

The Gods-on-earth are imposing absolute truths into the cultural context which is related to the process of developing truth and developing new contexts of creativity.

That is, education is about the fundamental mysteries, and it cannot be based on narrow dogmas and formalized axiomatic language structures (used in both the propaganda system and in the narrow dogmas of many expert languages), which are dressed-up by a false image, namely, that there are experts (who are forced to serve big business interests) who possess an absolute truth, but rather education needs to be exactly about the fundamental mysteries, where everyone is allowed "a voice" in a discussion which allows various and wide-ranging different viewpoints (or different categories of considerations), since actual knowledge does not exist. For example that the stable spectral properties of atoms can be used to distinguish atomic-types implies a descriptive context for these atoms which is linear and controllable. That is, there should exist a general set of principles which allows one to identify each of any particular atomic spectrum to sufficient precision.

That is, one wants an educational institution to be based on equal free-inquiry, along with an idea about creating something where both of these purposes are the front and center of the discussion about knowledge. Otherwise one will get either partial truths, or distinguishable features of a system which the experts might claim to be related to an absolute truth, such as reduction to particle-components, which has an all encompassing range of application . . . , but which cannot answer the simplest of questions,

eg
1. DNA cannot describe how an embryo forms, and thus they should not have any special position in a discussion about knowledge (ie a described: measurable and practically useful; truth).
2. The DNA model of life cannot describe "how life began," or equivalently "what life is."
3. What causes so many material forms . . . built from many-but-few-components, to be so stable in regard to precisely identifiable spectral patterns, which are related to these stable material systems?

Modern physics cannot answer these questions.

The answer which modern physics now provides about these systems is that "these systems are too complicated to describe."

This is not a valid answer for such stable and precisely identifiable systems, since these (just mentioned) properties imply that the properties of these systems are both describable to sufficient precision and controllable.

Furthermore, one wants this knowledge to be based on laws, or principles, which are general, so as to be applicable to all the different types of stable material systems, so as to provide sufficiently precise descriptions, which are deduced from these principles, and so that what is described is very useful, in regard to practical creative efforts.

There cannot exist a "free-market" unless everyone in society is considered to be an equal creator, ie everyone is equal, but this also requires that the justice system focus on personality types who follow ideas about personal domination and selfishness, and the justice system must oppose these personality types, whereas, now these personality types are an integral part of the coercive terror which a justice-system counterinsurgency action is aimed at the public, and managed by the government, for the purpose of supporting the owners of society.

The so called free-market-democracy, which now exists, is a social structure designed to serve all the wants of the ruling-class, a more efficient expression (and a better disguised model) of the Roman-Empire, which was designed to serve the needs of the Roman Emperor.

Unfortunately, the ruling class is selfish and ignorant, and the social machine they have created is most efficient at the destruction of the earth, and the destruction of human society, than at providing for the needs of society, where these needs are (or can be) defined by a truly free-market, but it is a market which needs to be regulated so as to stop destructive practices to which it might become a part (simply the possibility of destruction due to market forces needs to be sufficient to cause changes in the market. That is, the idea which needs to be a part of a societal belief is that the society is creative and resilient and can easily cope with any changes). Rather than, the social power of a company depends on a particular resource and the social power of the company must be maintained at all costs.

The experts cannot identify the observed stable properties of physical systems . . . , both large and small, where these physical systems are composed of many-but-few-components . . . , by describing either the dynamics, or the averages of material-interactions, either associated with geometry, or based on averaging about the energy properties (or energy operator) of the [small] system, where the small system is modeled by means of a function-space . . . , [Note: Energy operators applied to function-spaces which are composed of sets of harmonic-functions] . . . , where, in either of these math constructs, now being used to model these stable systems, the descriptive process is about identifying and solving a general system's (partial) differential equation, and this implies that the properties (of the system contained in the domain-space) are smooth.

One can say, with great confidence, that "the experts cannot do this." This is because it has not been done so far (2013), so that after over hundred-years the experts should "throw-in-the-towel" and say, let other ideas be considered, (where the experts mostly work on ideas about nuclear weapons, but in all likely-hood these experts are unaware of this "state of physics," and this "state of the physics departments of "public" universities," which are solely and single-mindedly serving the desires of the (military) owners of society).

Furthermore, this formalized axiomatic has served the interests of the owners of society well, but it has destroyed the knowledge basis of an entire culture.

However, in the quantum context, which the experts are using, a local smooth construct defined on function-spaces is used so as to unravel the properties of observed, local-values which are discrete, but this seems to be a backwards-way in which to do this descriptive process, whose focus is (should be) about the order possessed by small, stable, discrete systems.

That is, discrete physical structures might be best related to a new way in which to define, in mathematics, discrete operators.

The current attempts (or the current math methods) lead to too many math complications, where the complications of the construct lead to unstable (both physical and math) patterns (which are being described) and to quantitative inconsistencies, eg non-commutativity and chaos.

The experts try to navigate these complications by using algebraic formalism, and a continuum-modeled containing-space, eg set structures, limits, convergences, etc, but the context is non-commutative (leading to both quantitative inconsistencies, and to the descriptions of unstable patterns) and containment is defined in an invalid manner, eg using unstable math patterns; and quantitative-inconsistencies; and vaguely defined ideas about set-containment; and the sets used are "too big," so as to not be able to remain within a logically consistent context (a continuum is not definable).

The results of the math methods . . . associated to such an invalid definition of set-containment and instabilities . . . . are re-adjusted, by introducing, in a discontinuous manner, a new containment set, which, in turn, possesses new quantitative contexts, ie containment and quantities seem to be indefinable parts of the discussion, which is part of a, so called, math discussion, where math is supposed to be characterized by being able to define these properties (for a system, or a math pattern) explicitly.

This describes the indefinable character of particle-physics eg re-normalization, and such a discontinuous change of containment sets are the central feature in descriptions of feedback systems (of classical physics), etc.

The motions and geometric-field structures {or [partial] differential equations} and containment-sets with material components represented as constants, which are associated to an equation, eg mass and charge etc, which assumes a containment-space defined by materialism, and the motions of these materials {or which assumes, random spectral-particle events in space} so that material defines an open-closed topology for a "material containing metric-space," whose dimension is fixed by the idea of materialism.

In fact, this scenario of either material motion in a geometric context, or material averaging about an energy-operator, does not define material confinement, in regard to: identifying a precisely defined order, and stability, and charge-neutrality of most stable states of material systems, ie material systems which possess both order and stability.

That is, these properties of the existence of a stable set of spectral values, actually, imply a controlled, geometric, and many-dimensional context for the system, but where both motion and randomness only play-out, in a mostly irrelevant manner, when considered as (the material) existing only within one particular dimensional level, whose topology is open-closed, and is related to, apparently, smooth math structures.

That is, since differential equations and the idea of materialism and force fields and dynamics, or energy operators applied to function-space models of quantum systems, (where both) cannot be used to define the fundamental order of observed, stable material systems, to sufficient precision so as to be based on general principles [note: though the solvable systems of classical physics are: based on general principles, and sufficiently precise, and related to practical usefulness], then this can be interpreted to mean that the containment set is separated from the idea of stable math patterns, and separated away from math contexts where measuring is reliable, eg the system is to be contained in a metric-invariant metric-space.

Using a local smooth math construct defined on function spaces to unravel the properties of the observed local-values, which are discrete . . . ,

. . . , seems to be a backwards way to try to describe such systems, where the focus of the descriptive process is about the observed order of the small, stable, and discrete systems. Nonetheless, the focus is on randomness, instead of the order. The algebra of operators is supposed to supply "the cause" of the observed order for small, stable discrete systems, where these systems are often composed of many-but-few-material-components. Yet, the algebra and limit structures of the containment set become very complicated, and there are also the quantitatively inconsistent, non-commutative sets of algebraic structures, as well as an ambiguous construct for containment, where the continuum is disintegrated (destroyed) in the realm of the very small, where all of these difficult (and essentially un-resolvable) math issues become the characteristic math properties of such a math construct.

For macroscopic systems there is an apparent and effective smoothness (in a very limited sense), while for quantum systems the math constructs for quantum-discreteness, which are used now (2013), have completely missed the correct (or observed) discrete structures. Namely, it is the individual functions of a function space which are being used to model the quantum system's discreteness, and then a (hidden) particle-state structure is being added to the wave-functions, where a wave-function represents a quantum system as a mix of all the individually discrete spectral functions. But, if the function-space is given a particle-state structure (where particle-states are supposed to account for the discrete-nature of the small, local (particle-event) properties of the containment space), then this is about the disintegration of the continuum of the system-containing domain space.

Namely, the domain space for each function of the function space, ie a collapse of the idea of math containment, where material systems are assumed to be contained in a continuum, so as to allow for a system's locally measurable properties (which are assumed to be smooth), which is implied by the notion of materialism, mass is defined by its resistance to changes in motion, ie a smooth construct.

But, in quantum description, materialism is changed to a model of explicit geometries of particle-collisions, which are defined in a non-linear, random-event math structure . . . (with an explicit set-structure [or algebraic structure] of the hidden particle-states) . . . , but this context is in a descriptive structure which is based on the "geometrically incompatible" idea of randomness.

That is to say, there is no valid notion of a containment space in quantum description, but there are also no solvable models of "general many-but-few-component quantum systems."

The model of randomness and point-particles go hand-in-hand . . . , (and can be easily modeled [by a new math containment model of existence] if the math operators are discrete and the material components are formed into discrete-shapes, which possess distinguished points, and whose interactions in-the-small make the behavior of these small-components, effectively, appear to be random. Yet, the new description is based on the stable discrete models of a geometry associated to the small material components and a model of discrete operators, ie not local smooth operators) . . . , while a quantum system's model "as a set of harmonic functions (or a basis for randomness)" is supposed to provide a wide-range for the system's spectra, but using these math structures within an open-closed metric-space, in which the idea of smoothness and local measuring (ie these smooth operators act on function spaces), is kept, but the idea of math containment is abandoned, due to a geometric model of point-particles which mysteriously emerge out of a chaotic model of the new idea about a system-containing domain-space (?), where this domain space seems to still possess a memory of an ill-defined particle's (mysterious) momentum properties, ie a part of the quantum system's smooth math structure. Yet, the description is based on the mysterious emergences of particle-states (then why is there a memory of momentum for some of these particles?), which, in turn, are assumed to be associated to the smooth quantum system (defined by sets of differential operators); but this math construct does not lead to a valid descriptive structure for the observed properties of these quantum systems, eg the stable spectral properties.

In fact, it is simply incoherent babble, which is designed to fit data, similar to the Ptolemaic models of epicycles, but the data-fitting is really the data associated to bomb-engineering, it is completely irrelevant to describing stable spectral properties (ie it has no general relation to this endeavor, maybe a handful of overly precise examples, which amounts to a bunch of baloney).

This is not valid mathematics at all, but it emerges from the US educational traditions of "the traditions of the authorities," . . . who are to guide education, and yet it is these authorities who serve those "who really do guide society, in a totalitarian manner (ie the owners of society)" . . . , this invalid math construct is authoritative babble, which is cloaked as rigorous math (in formalized axiomatic), but it is really based on the social authority of the personalities in these high educational and military institutions which now define the moral-and-intellectual-rot which now represents western society, these personalities only express the idea of extreme violence.

In the days of Copernicus the leaders wanted earth in the center, so as to form their propaganda around such an idea, whereas today (2013) the leaders want science to serve military and big business interests, ie different visions of propaganda, ie the church vs. today's big businesses.

This baloney, which passes for physics, should have been uncovered, as baloney, by the 1950's.
But, instead, endless "highly complicated" efforts are pushed upon a education system, which filters special autistic types, who are pushed into these pointless efforts.

In fact, this is the blue-print for the endless fraud, based on intellectual complexity, which is now the foundation of western culture, a culture which is essentially the same as the culture of the totalitarian Roman-Empire, but with the emperors now cloaked as anonymous bankers, where science is now based on obsessive autism (and not on the earth being the center of the universe, as in the era of Copernicus).

In this social context the academic system still services the complicated instruments for the rich, but intellectually, in regard to their forward looking outlook for knowledge, the academic authorities are spinning their wheels over irrelevancies, so as to let concerns about nuclear-explosion-processes still be the central focus, where, apparently, the autistic-obsessive authorities (in the higher-levels of the education system, ie at research universities) are unaware of the social institutional uses of this baloney, which their useless intellectual efforts are actually supporting.

That is, fraud is OK, at the highest educational institutional levels.
This seems to be the blue-print for the media construct; that fraud is OK everywhere in the upper ruling social classes.

The ideas in this paragraph (and the following five paragraphs) are about adjusting the math idea of physical containment, which is not based on the idea of materialism.

To have a consistent reliable measuring model of existence constructed then a local linear representation of a measurable property must be consistent with a metric-invariant structure. Furthermore, what is also needed is for metric-spaces to have metric-functions which have constant coefficients. That is, the non-positive constant curvature metric-spaces, where, in this context, discrete isometry subgroups (associated to metric-invariant metric-spaces) determine the stable (linear) shapes, which can be contained in the metric-space, which, in turn, must be used in reliably measurable descriptions, which model physical systems.

Note: consistency of a description, based on the physical system being contained in a coordinate space, along with the coordinate space's metric-function, implies a continuously commutative everywhere pattern to (or must) be associated to local models of linear measures of properties (ie functions) defined on the system-containing coordinate space.

The new discrete structure of the derivative:

Local linear models of measuring are discontinuously defined (ie discretely defined) between both

(1) adjacent dimensional levels, and
(2) discrete time-intervals, which are defined between discrete discontinuous metric-space states of discontinuous changes (during the discrete time intervals) in spatial positions (or values of velocities); in a descriptive context in which both

(1) geometry of position in space (Euclidean), and
(2) changes in energy (hyperbolic) are properties defined in separate spaces, but where the main containment feature would be the stable geometric patterns defined by discrete isometry shapes.

However, within a particular spatial-dimension metric-space, the discrete changes in spatial position can be affected by the shape of the containing energy-metric-space, so that the shape of the energy-space defines an envelope of orbital stability for condensed matter, where this orbital-envelop is based on geodesics of the energy-space's shape. This is for large size scale, so that the small time-intervals (of discontinuity) appear to define a smooth differential equation (or dynamics).

Whereas lower-dimensional faces (of smaller energy shapes) can discontinuously "jump" to new energy-spectral configurations of both shape and spectra, based on their resonances of the new shape with entire containing space of an 11-dimensional (hyperbolic).

The "jump changes" are a natural part of discrete context of interactions, while orbital envelopes are natural parts of condensed material contained in a large stable shaped metric-space which has geodesic structures, to which discrete (apparently-smooth) local measuring structures adhere.

Properties of observed physical systems are consistent with a context wherein measuring is reliable, as can be deduced from our own experience of being able to precisely spectrally-identify different: nuclear, atomic, and molecular material systems, wherein each identified system does possess the same number of quantum-components which compose the system, which, in turn, possesses the same spectral properties.

The current practices of the academic and media-world:

All intellectual efforts, which are allowed to be expressed, are used to both form and maintain a few sets of categories of thought, or activity, upon which the empire (and its few owners) depends for the maintenance of its power.

This means that a few sets of allowed opposites are the main way in which language, within society, is allowed to be used.

That is, language is used in a very controlled manner, by the media (and education system) in which the very limited set of categories of interest to the owners of society are the only topics of interest within the media (and in the education system).

This leads to a language of experts, but it is a language of fixed narrow formalism, where such axiomatic formalism is the basis for assuming that math patterns can be imposed on any measurable context without consideration about whether the context allows reliable measuring or if stable patterns are the basis for the description, or result from the description.

That is, the thoughts of the experts are allowed to be arbitrary and capricious, so that this has become the basis for truth within society, and this began in the 1950's.

It actually began in 1900 with the advent of quantum theory and it was fiercely contested by Brouwer and F Klein who supported the intuitive method where math development associated to math applications to the world, depended on assumptions contexts interpretations and containment structures but D Hilbert won and his minion H Cartan (his father E Cartan was the better mathematician) instituted formalized axiomatics, eg the Borbaki program, around 1910 or 1920. Subsequently, it has been believed that any math construct can be applied to any "applications context," once a formalized context is established. That is, unstable patterns and quantitative inconsistencies became the central topics of the formalized math structures of Borbaki, leading to rigorous patterns only associated to an illusional world.

A math context which is great in regard to a desire to "fit data to whatever quantitative model one wants."

People should be made aware that either racism or elitism . . . etc, . . . , is a necessary part of a society based on inequality.

This elitism can be based on defining educational communities, in turn, based on the narrowly defined, fixed authoritative dogmas, which are espoused in the peer-reviewed journals:

be it a, so called, DNA theory of life, or a nuclear-bomb-engineering theory of physics,
Etc.

These dogmas fit into the limited number of categories, which are defined by the monopolistic big businesses, which run and control society.

These (above) examples are simply expressions, cloaked in a context of an elite set of people, who are expressing an arbitrary value, which is being based on, and upheld by-means-of, both institutional and personal violence, which is all organized around the justice system, and a rule of law based on property rights and minority rule.

This is an elitist-based form of "the principle" that "might-makes-right," which, in turn, is justified by the material based biological science based on "the survival of the fittest."

Furthermore, the act of manipulating the 2% of people (or so . . .) who are inclined to seeking domination by means of violence (or by some form of narrowly defined competition, as are the autistic-types) is crucial to a state being organized around elitism, in its many varieties within society, ie clubs and communities, but where these violent types are used to keep the social structure of the culture very narrow and controllable, in an institutional sense, where these acts of social narrowing are ultimately based-on violence.

One can look to the journalists:

The objective reporter, in a society, where fraud and injustice is easy to find and report, whereas these reports are used by the politicians to re-write laws so as to legalize this fraud and injustice, where similarly in theoretical physics and mathematics the authority of personality is used to legitimize the intellectual fraud in the physics and math communities.

C Hedges remains consistent with the narrow set of dialectical opposites which define the accepted categories of society, the set of categories about which journalists and successful writers must adhere in regard to what they may write and subsequently be published.

Hedges rails against the popularization of the viewpoint that "a person makes their own reality through their mental constructs," where this is represented, by Hedges, as "one can believe that one day they will be rich and famous, if they have merit" (as the value, within which "one's merit" can be judged, is narrowly defined for the public by the owners of society), but in the dialect of opposites, the opposite to this "belief in one's self" is homelessness.

The set of false opposites is a way in which to construct an artificial language (a way mentioned by Hegel, and his dialectic of opposites) wherein the list of opposites also defines the list of acceptable categories of discussion, so that the use of language is always done in a way which supports the interests of the owners of society, since it is the owners of society who control all publishing and broadcasting, and as they try to control the internet by spying, so that the content of the language, which is used by the people, can only be related to the narrow interests of the owners of society, and these interests are defined by the categories associated to sets of opposites about which social discussion is allowed to be based. On-the-other-hand, these categories of opposites are mediated by experts, ie the person who has merit, so that no one except experts are supposedly qualified to talk about the subject matter of the categories in which discussion is allowed.

But the ideas of the experts, associated to the owners of society, have run their course, and will stay around as institutions of a life-style . . . , [just as the life-style structures of the Roman empire stayed around so that the Holy-Roman-Empire based on kingdoms took over these life-style structures] . . . , and by the behavior (of these owners of society) of over-domination, subsequently, life and its creative interests are exterminated, so as to squelch competition, so the owners of society do not have to relinquish any social power, but their markets and expansion capability dies with the loss (extermination) of creativity, since knowledge is no longer relevant to new creative contexts, and truly new products, indeed the natural place to go intellectually, in such a failed social context, is to reject the notion of materialism.

The main set of opposites, in this contrived language, is not a false opposite, but it has an obvious correct side (as do all of these limited numbers of categories of opposites), and that correct side must be fought against.

This pair of opposites is:

Equal vs. unequal,

or

An intuitive basis for precise language (the side of equal creators)

vs.

Formalized axiomatic (peer-review, ie content controlled by editors), . . . , Where equal is the correct side, as expressed by Socrates, as he represented equal free-inquiry as a basic human activity, and as a precursor to practical creativity.

But inequality is the rule of the land, and in choosing inequality the society must also be arbitrary and it must use extreme violence.

Thus, there are false opposites, which define many other categories, and some of these are:

Science vs. religion
Particle-physics vs. general relativity
DNA models of life vs. to seek a more inclusive model of life's functions
Smart vs. stupid
Expert vs. non-expert
Absolute truths vs. developing ideas about the mysteries of existence
Moralism vs. hedonism
Hard working vs. lazy
Detail oriented vs. careless

Freedom vs. being controlled
Rich vs. impoverished
Survival of the fittest vs. those who want hand-outs
Right vs. left
Society vs. nature

Then there is the mantra:
Our great technology, provided to us by the monopolies, and how we cannot live without this technology, and the great efficiencies of monopolies, etc
vs.
Alternative ideas

Civics vs. propaganda

Experts vs. the non-expert

The categories which cannot be talked-about

Oil vs. renewable energy sources
DNA manipulated farming vs. natural based farming
Property rights vs. equality (as the basis for law)

Truth based on expert opinion
vs.
Truth of a member of the public's equal free-inquiry, a person who is, by definition, a non-expert

Behavior modification in an imposed social structure
vs.
creativity as the main attribute of people

Wage-slaves vs. a free-people
Banking vs. experimentation by non-experts

And in these categories, everything is considered to be a war, wherein there must always be both "a winner and a loser," and those people who are outside of this short list a categories (determined by business interests) are attacked by society.

Basically the false dichotomy is:

The word of the experts within a category (and in the context of materialism)
vs.
The word of the intuitionist non-expert, where the expert must always win, since the expert represents the categories of interest to big business.

But to look at the main categories of the market . . . the social structure through which the owners of society derive most of their social power . . . . ,

military,
oil,
narrowly defined agriculture,
banking,
education,
housing,
transportation,
the media,
medical development,
the psychology of manipulating people, and
the counter-insurgency tactics of the justice system against the people
. . . , this narrow and fixed way of social organization is failing;
it is failing to develop new products, it has not provided new clean cheap energy sources through knowledge, it measures medical successes at 10% effectiveness,
it calls . . . totalitarian social structures and markets built around fraud . . . freedom,
it is failing to preserve the earth and its environment,
it is failing to expand based on inventiveness,
its model of markets are models of war and social control, so as to preserve society's inequality

R Nader remains a stalwart believer in civics, but he does not analyze how "control of communication channels" allows discussion to be framed by those politicians whom the corporate masters, want the people to "vote for."

The "ancient alien" journalism is a very strange set-up for journalism, it is about discussing a wide range of ideas concerning:

1. archeological anomalies, of very ancient ruins, (ancient finds which are not discussed by academics), and

2. the strong possibility of the existence of many different life-forms, since life on earth began almost as soon as the earth cooled, ie within 5- or 10-million years after the earth cooled, then life must either be a property of existence itself, or "randomly forming life-forms" are a relatively common event, as well as

3. ideas about higher-dimensions, but when higher-dimensions are mentioned, they are either referenced to such standard ideas as string-theory, ie they are referenced back to materialism, or they are described as categories of either thought or emotional thinking, ie apparently, ideas and feelings are the higher-dimensions, and

4. in regard to many different life-forms, there is the idea of being able to communicate with the alien higher-mental life-forms.

The diverse possibility for life should have many different attributes, including superior technology, and superior intelligence, where intelligence seems to mean, to the ancient alien journalists, a superior social position.

This discussion, supposedly, is all done outside the realm of experts, yet the journalists quite often defer to the material based ideas of the experts, so as to make it appear that these are serious things to consider so they consider these ideas along the lines of thought provided by the experts concerning the material world, though nearly everything the ancient alien journalists discuss are ideas opposed to the idea of materialism.

The main attribute which the ancient alien journalists express is that human life is dependent on other superior life-forms which, apparently, are far away from us.

It is a discussion about a set of diverse ideas but which nonetheless are pinned-to either ideas of the experts or to superior life-forms . . . , ie it is expressed so that it is still an attack on the public.

It OK's the right of the ruling-class to experiment on the public, or exterminate the public, as the ruling-class sees fit, due to the superior knowledge which is controlled by the ruling-class.

In fact, the new alternative math model of physical descriptions based on many-dimensions which are filled with both macroscopic and microscopic stable geometric shapes, which both allows for an understanding of the stable order which material systems (including living systems) can possess, while also identifying a geometric relation to material, life, and mind. Yet, the ancient aliens journalists are not interested. As usual it is the selfish interests which dominate expressions within the media of what are now vague ideas but which could be deep relations which affect our ideas about what we are, which seems to be the real point of the interest in ancient aliens, but the new ideas are not elitist, we are (each individual is) the center of the miracle of creativity (not some remote superior: either intelligence or technology, etc).

51.

Science and religion

Note: The following use of language is outside of the language structure used by the propaganda system for both academic and public discourse about science and religion, where the usual context of the language of the propaganda system is a pathetic excuse for intellectual discourse.

Thus, buttons in the minds of everyone will be pushed in a negative manner, by reading this paper, for all who have previously entered into such discussions concerning either science or religion.

Nonetheless new ideas need to be expressed, even though everyone believes in a childish viewpoint concerning rationality, where, supposedly, intellectuals only deal with absolute truths, or they only deal with a consensus, where science should be exactly the opposite of a consensus. Namely, science needs to challenge authority at every turn, and new ideas need to always be considered, especially, since (for example) the spectra of atoms with more that 4-components, eg a nucleus and some electrons, cannot be found to sufficient precision by direct calculation by using the laws of quantum physics etc, etc, eg similarly in regard to identifying the spectra of general nuclei to sufficient precision, etc. etc.

It is materialism, and the math structures associated to materialism, which are not capable of identifying the cause of the observed stable properties of many, fundamental, material systems. Yet, it is the model of particle-collisions in a context of materialism which both defines modern physics, and is the basis for nuclear weapons technology. This structure for knowledge exists because these are the wishes of the ruling-class. Yet, the failures, of such a model for knowledge, are the causes of the social collapse, due to not enough practical creative expansion, in turn, due to the act of constraining knowledge and restricting its relation to practical creativity.

However, this short discussion depends on a new math construct for "science" where a model for both stable material systems and living systems are based on these systems being stable shapes, which can be of various dimensional types, so that these new types of models of material systems possess stable spectral-orbital properties which allow the stable spectral properties of these systems to be determinable to sufficient precision, and it will show that this new math construct for "science" (where science is a quantitative and measurable descriptive language, which provides a model for interpreting the world's experiences) is a language which transcends the categories and stereotypical-beliefs concerning both science (ie materialism) and religion (the spirit (whatever the spirit is?)), a separation of subject-content which the propaganda system tries so desperately to instill into the culture, where its success at this time (2013) is phenomenal. And thus, the very limited range of thought about these topics within the culture.

The material events we experience appear to remain consistent with the idea of materialism, since the appearances determined by close (or nearby) events, ie the construct of (local) convergence, do remain within the material-containing metric-space, ie the material containing metric-space has an open-closed topology, but the stable and/or functioning properties of both material systems and living systems (as well as math operators) cannot be understood without reference to a higher-dimensional math structures (ie the observed, stable properties of many fundamental material systems cannot be described, in a valid manner, based only on the idea of materialism), and having a relation between sets of discontinuous structures requires that the operators, which mediate between the dimensional levels, be discrete, in regard to both discrete time-intervals and between discontinuously separated different dimensional shapes (where these shapes model different metric-spaces).

The existence of stable patterns, which can be consistently identified in one's own experience, ie can be measured, is the math context which is equivalent to (or can contain, without contradiction) a passive-creator model (of religion), but this new math context . . . , where in the new alternative viewpoint, math constructs are associated to sets of discrete hyperbolic shapes, defined in a many-dimensional context, where there are many metric-function "signatures" (see below) for the different dimensional levels (where odd-dimensional levels (or shapes) are contained in even-dimensional levels, and vise-versa) . . . , also allows (or must allow) for an set of energy-generating, and self-reflective, stable geometric patterns to exist in various high-dimensions, including the 3-dimensional level, ie a 3-dimensional hyperbolic metric-space, modeled as a stable discrete hyperbolic shape, which, in turn, is contained in a 4-dimensional hyperbolic metric-space, where the related metric-function has a "signature" related to the coordinate space R(4,2) (see below).

This is an active-creator model of existence, but the creator is, actually, a life-form, in fact, it could be ourselves.

The new alternative ideas are about a measurable descriptive context, which depends on a math model of a (local) measuring-construct, ie a derivative, in turn, based on a new discrete structure for the (local) operators, where these discrete operators are a part of (or contained in) a "many-dimensional and many-time-interval" context, where the multiplicity of time-intervals, which are (or can be) associated to various metric-spaces, and are also associated to different "signature" metric-functions, eg R(4,2) has 4-space dimensions and 2-temporal dimensions [and whose metric-function "signature" is, 4-2=2]. Thus this model of existence can be about a 3-dimensional discrete hyperbolic shape, contained in four spatial dimensions . . . , (a shape of odd-genus, ie odd-number of holes in its shape, eg a doughnut (or torus) has one-hole in its shape), . . . , which can generate its own energy (due to a charge imbalance, due, in turn, to the shape's geometry), and it is contained in a 4-dimensional hyperbolic metric-space, and this accounts for the interest in the R(4,2) metric-space, which has "two dimensions of time" associated to itself, one time-dimension associated to regular material motions and the other time-dimension associated to the motions of energy-generating "materials."

In other words, the high-dimensional energy-generating shapes, associated to certain odd-dimensional discrete hyperbolic shapes (whose genus are odd) . . . . , and defined on generalized non-positive constant curvature metric-spaces, where the metric-functions must have constant coefficients (so as to be relatable to a "continuously commutative everywhere shape" [ie a circle-space shape]) . . . . , is the natural context for both science and religion, where religion is about practical creativity of life-forms which fit within the true math containment structure (or math model) for existence, ie many-dimensional and very deeply associated to the very stable discrete hyperbolic shapes, where these stable shapes are needed so as to be able to account for the stable spectral properties of quantum systems, so that the spectra of these systems can be identified to sufficient precision.

These stable shapes, placed in a metric-invariant containment space, provide models both for stable material systems and the higher-dimensional structures which allow living, energy-generating, shapes to exist. Thus this new math construct allows for (or provides) a model for life, whose living functions come from a set of stable shapes which exists in higher-dimensions than the dimensions, which the idea of materialism define.

If the "creationists" want a math model which includes ideas about stability, measurability, and control, and of a "repeat-ably measurable" descriptive context . . . , ie a model of our experience, and which is associated to higher-dimensions, ie transcending the idea of materialism, ie being able to mathematically model an existence which is beyond the material world . . . , then they need to consider these math patterns very carefully.

This math construct models "the spirit" as an actual existing structure in a higher-dimensional space, where "spirit" now means; existence in higher-dimensions than the dimensions of material containment (where material containment is usually defined by the idea of materialism).

But the idea of a "deep-creator" is not external, rather it is within each and everyone of us.

This is also why law should be based on equality, and not based on property rights and minority rule.

Basically people become used to the "abuse by society," and its subjective hierarchy, because they like the public buildings and the sewer systems which their emperors built (eg in a similar manner as the Romans), but such (civic) creativity does not need to be attached to the idea of an oppressive emperor (or, today, a modern big banker, or oilman, or weapons supplier).

Part V
Book X 2
New material

52.

Comments on AMS

Comment about the focus of the American Math Society

It is better to model of physical system by stable geometries of simple shapes, than it is to try to identify the properties of physical systems by formulating and (trying) to solve physical systems by means of non-linear (partial) differential equation models of these systems.

One needs a valid model of complex physical structures so that the model is based on a relatively small set of rather simple stable geometric shapes, thus, the description is (can be) based on stable shapes and the context is reliably measurable, and one does not need,

A bunch of "not understood" sets of (partial differential) equations, which provide no valid model of anything.

That is, the equations, upon which the math community focuses, are not understood, since neither the form of the equation associated to a physical system nor the solutions to these non-linear (partial) differential equations can be found, and placed into a context where the information provided by this descriptive context are either accurate over a wide array of systems or practically useful, in regard to practical creativity.

Rather (for accuracy and practically useful information)

One needs a descriptive context, where measuring is reliable, and the patterns which are being described are (or upon which the description is based is) stable, so as to be associated to a set of stable spectral-orbital properties which model sets of material systems which exist at all size scales:

galaxies,
solar-systems,

crystals,

molecules,

atoms,

nuclei, and

neutrons, and

neutrinos,

and

this stable descriptive context is needed because the main types of observed patterns of the material world are these very stable properties of material systems . . . , so that these physical systems are measured in a context where measuring is reliable and quantitative relations are consistent with the numbers defined by the measuring units . . . , and this stability implies a descriptive context of math where:

1. local measuring is linear,
2. the containment space is metric-invariant, and
3. the local relations between the natural coordinates of a system's shape and local measuring-changes are continuously commutative (almost everywhere), so that
4. the equations are linear and defined and the equations are solvable and controllable.

Thus, the need to provide a model within a context wherein there exist stable metric-space shapes so that there is an answer to the question about how math can be organized so as to provide a model of material systems in the context of stable geometric shapes which possess stable spectral-orbital properties.

One can Either try to provide useful answers to necessary questions concerning the nature of existence, ... , similar as how Copernicus and Kepler provided the correct answers while Galileo and Newton provided general ideas about what set of properties and how local measuring relates to material geometry so as to extend (or solve) this descriptive context so as to derive (or contain) or to be consistent with the already provided answers of Copernicus and Kepler, [However, Copernicus's answer has not yet been provided with a general descriptive context associated to their observed properties of stable orbits, unless one considers describing the properties of the planets within a context of sets of stable metric-space shapes.]

Or

one can try to formulate and solve a sets of many (partial) differential equations, which model sets of physical systems, and then try solving these mostly non-linear equations, which are essentially unsolvable, . . . , since the idea of materialism has come to mean that the descriptions of material systems' properties are to be based on solutions to (partial) differential equations, which, in turn, are supposed to model the physical laws of a material world. But this viewpoint is not providing any type of practically useful descriptions, neither accurate descriptions, nor widely applicable, nor related to practical useful contexts for practical creativity.

That is, the idea of Einstein that physical law . . . , [of a descriptive context which is based on materialism, means that all the ideas about the nature of set-containment of a measurable description emanate from the idea of materialism] . . . , must be based on general, and invariant (to arbitrary changes in coordinate frames) types of (partial) differential equations, has not been a practically useful idea.

Thus, today the professional math and physics communities are trying to both identify and solve the, supposedly, needed (partial) differential equations so that the properties of a material existence can be rigorously developed (for material systems) by means of identifying and solving (partial) differential equations which, in turn, determine physical law, where the forms of these (partial) differential equations has come to be dictated by Einstein's vision of a coordinate frame's relation to physical law.

But Einstein was wrong as the results of A Aspects experiments concerning non-locally have already shown, the ideas of general relativity cannot be used to describe the stability of the solar-system, etc etc.

The non-linear context of Einstein's form-invariant equations in regard to arbitrary coordinate frames, which are supposed to represent physical law for material systems, is in opposition to the observed stable patterns for both spectral-systems and orbital systems.

That is, the observed properties of non-locality and action-at-a-distance suggest that coordinate frames are not the main issue in regard to physical description, while the existence of stability of spectral-orbital properties of physical systems which exist at all size-scales implies that the solvable context is more important than is the context of (partial) differential equations in regard to identifying the main structures of existence which allow for wide ranging and practically useful descriptions of physical properties.

What is a geometric alternative to a wave-equation defined within a general metric-space as stipulated by Einstein (or wave-model for oscillating wave-actions) either free-propagating and material-interacting-(electromagnetic)-waves (ie classical), or waves which are averaged around an energy construct (ie quantum)?

There is action-at-a-distance in Euclidean space, but charged-components can relate to infinity by means of their (natural) relation to sets of unbounded discrete hyperbolic shapes, where in an odd-dimensional hyperbolic metric-space shape, these unbounded discrete hyperbolic shapes can be oscillating (if the shape has an odd-genus), so that the energy-packets (propagating along the unbounded shape) are not reflected by the (a) discrete boundary which is defined with each discrete time-interval. Rather, with each discrete time-interval the energy-packets are only displaced toward infinity, and away from the source in regard to the odd-genus and odd-dimension unbounded discrete hyperbolic shapes which both defines the particular hyperbolic subspace of the over-all high-dimensional containing space and defines the path-context (or orbital-context) of an electromagnetic-wave. The energy-packets (propagating along the unbounded space-time shape) are toral shapes, which emerge from the toral components of a discrete hyperbolic shape (which is the source of this wave-packet energy) so that these energy-packets are (to be) centered about (or over) the lines-to-infinity,* which are defined by the unbounded discrete hyperbolic shape of the sub-space "to which the (original) energy-amount belongs."

The energy-amount would be determined from the difference between both the energy of the original toral component of the discrete hyperbolic shape from which the energy-emerged, and the lower-energy of the toral component to which the (lower-energy) charged component then went to occupy (at a lower-energy), and then there is also a proportional opposite-state energy as the 2-state shape of the toral component from which the energy emerged, thus forming into an energy-packet shape, which is similar to the shape of the toral component from which the energy-emerged, but in the opposite metric-space state the energy of the wave energy-packet is returning.

The unbounded discrete hyperbolic shapes connect (or relate) the energy-values associated to a position of the light-energy source in space, to the entire sub-space, so that the motion of the energy-packets is from the source out to infinity. However, the discrete time-interval structure of the motions of these energy-packets can either be un-reflected, if the unbounded sub-space shape is oscillating (where the oscillation causes destructive wave-amplitude interference on the reflected energy-packet), or reflected if the unbounded sub-space shape is not oscillating.

That is, the energy flowing into the hyperbolic subspace would be in resonance with the spectra of the over-all high-dimension containing space, if the spectra of the over-all space is defined by differences between the energies of concentric toral-components of an atomic (or molecular) source, and this energy-propagation can be modeled as energy-packets propagating on unbounded discrete hyperbolic shapes which are stable metric-space shapes and which identify a particular subspace within which the material components are contained.

When the toral energy-packets of an (electromagnetic) wave interact, by means of resonances, with similar shaped toral components of other discrete hyperbolic shapes, at other positions in space, then all the other toral-shaped energy-packets of the wave-action perform an action-at-a-distance disintegration, so as to conserve energy in the subspace.

Note: Discrete Euclidean shapes can act by means of action-at-a-distance but discrete hyperbolic shapes cannot possess a property of action-at-a-distance, but rather possess the property of possessing an (instantaneous) unbounded shape.

The lines-to-infinity*

That is, toral-shaped energy-packets which are centered-about the lines (of unbounded discrete hyperbolic shapes) which go out to infinity, where these infinite-extended "lines" are defined by the unbounded discrete hyperbolic shape, which, in turn, defines the subspace within which the energy (of component-occupation which is defined within a bounded discrete hyperbolic shape) is contained.

Thus, in odd-dimensional spatial subspaces an electromagnetic wave is a wave-front "shell" of energy which moves away from its source, while in an even-dimensional spatial subspace (or an even-dimensional hyperbolic metric-space), an electromagnetic wave is an enlarging-sphere which is filled with (reflected) light-energy. Thus, the even-dimensional model of electromagnetic-waves, the electromagnetic-waves are more likely to disintegrate sooner.

The wave-motion is from infinity back to the source, when within the opposite-time hyperbolic metric-space state, where each opposite-state identifies the other (opposite) state of wave-motion in a local manner.

Then there are the crystalline shapes associated to material organization, where the building-blocks of these crystals, are assumed to be, atoms and molecules, but the context is charge-neutrality, where the material-components are put-together in (charge-neutral) Euclidean translational symmetry patterns. However, charge-neutrality can only be understood if the atomic and molecular shape are being determined by closed metric-space shapes, so that the electromagnetic-fields are trapped (or confined) within a closed hyperbolic metric-space.

Nonetheless, after the neutral atomic (or molecular) components arrange themselves in a Euclidean symmetric manner then a charge structure re-emerges within the crystal.

There seems to be both a Euclidean (or inertial) symmetry to the crystal, which x-rays can identify, as well as a new discrete hyperbolic shape associated to the crystal's charged properties, whose shape might again be hypothesized to be, a folded circle-space shape, or perhaps a (new-type of) concentric-cylindrical discrete "hyperbolic" shape-forms, due to the physical-geometric-inertial constraints of the crystal's bounding surfaces, so as to not allow the moding-out process in one-direction, so as to not form toral-shapes, but rather only cylindrical-shapes.

These cylindrical shapes might be associated (at their top-faces and bottom-faces) to 5-, 6-, and 7-sided polygons (or polyhedrons) whose additional (but difficult to relate as circle-spaces) set of sides can be related to cylinders attached to toral components. These types of 5-, 6-, and 7-sided polygons can be observed for rocks which have formed into crystalline structures, such as seen at Devil's Post-pile located near Mammoth-Lake in CA.

For a molecule one would need both a (not necessarily concentric) system-unifying (discrete hyperbolic) shape, and atomic sub-shapes (which could be lower-dimensional shapes which are contained in the toral-components [or cylindrical-component] of a higher-dimensional unifying shape of the molecule).

Going in the other, smaller, direction, in regard to small-scale material components, one needs to consider nuclei, neutrons, and neutrinos, which are material components whose properties (of stable shape) are related to other issues about charge-neutrality and metric-space shapes.

The quantitatively compatible shapes are: the lines, line-segments, and circles, so that the stable shapes are rectangular shapes, cylinders, and circle-spaces whose fundamental domains are "rectangular," or the fundamental domains are moded-out to form a continuously commutative (circle-space) shape (almost everywhere). The circle-space shapes, which form from rectangular fundamental domains, are metric-space shapes. (Can cylindrical shapes also be used to represent an open-closed (within itself) metric-space shape?)

Consider that the key processes associated to math operations might best be described in a context of discrete and discontinuous ways of describing inter-related material properties of change, either of the component's shape or of the component's spatial position, where discrete discontinuities can be identified for:

Time intervals,
Between dimensional levels, and
Between folds by Weyl-angles, for folded discrete hyperbolic shapes.

Furthermore, each of these discontinuities (except for between time-intervals) there can also be defined a constant scalar factor, so as to change the geometric-sizes between the discontinuities.

Can valid lattices be built from such a discrete context, whose shapes are to stay within the shape-context (or quantitative-context) of:

lines,
circles, and
line-segments,
where scale changes, as well as discrete discontinuous folds in the lattice, can be defined at "the parts of the descriptive process associated to shape" where there are discrete discontinuous changes (where these scale changes are associated to either defining discrete scaling factors, or to discrete spatial displacements), and still be related to a variety of (different) numbers of faces, which define a (star-shaped) polyhedron (for a fundamental domain of a metric-space shape)?

In this context, the question seems to be:

Must these discretely discontinuous constant factors be defined so as to be consistent with the allowed fundamental domains of discrete hyperbolic shapes, or can they be used to form sets of new shapes, which are not necessarily related to the sets of allowable discrete hyperbolic shapes' fundamental domains?

Book reviews in the AMS Bulletin,

ie mathematicians seem to pay attention to irrelevancies, which are associated to authoritative laws, ie traditions of mathematical authority, ie an authority which also serves the banking-oil-military monopolistic business interests, but which do not lead to (or which lead to very few) practically useful developments of measurable descriptions.

Such blind following, and adherence to, dogmatic authority (which is represented as peer-review for professional journals) is similar to, and as damaging as, the arbitrary hierarchical social structures of racism.

There are 5, or so, book review articles, in the October 2013 AMS Bulletin, about complicated non-linear and non-commutative math structures, associated to both analysis and (harmonic) waves, apparently, motivated by an attempt to define either the wave-equation in a geometrically non-linear context, or defining transformations [often of (partial) differential equations] associated to geometric properties (so as to solve the (partial) differential equations, so as to, in turn, either uncover hidden geometric structures of a material system, or to define algebraically (with polynomials) a prescribed geometric-shape.

(but, as is always true in math, the math constructs, which are considered, are those constructs which are motivated by the adherence to traditions and authority by the professional math literature, but these constructs only have relations to math literature, and not to practical development of measurable descriptive contexts).

These topics are:

1. generalized sets of waves,
2. integral-geometry (or integral-geometric-transforms, eg like Fourier transforms),
3. algebraic geometry,
4. doing analysis by defining functional-distributions which possess singularities, and reverting to "generalized sets of waves" so as to be "used to make a simple partition of an energy-space by 'cubes' into a rigorous idea," and a book-review about
5. how knots are naturally associated with Lie algebras, where knots are often used in a context of identifying boundaries of complicated shapes, whereas Lie algebras also depend

on the simpler toral partitions of a Lie group where the toral components can be seen as linked (or knotted) shapes etc.

Note: T Tao, in his book review about "generalized wave sets," actually describes how waves can be solutions to 2-dimensional discrete hyperbolic shapes, [(perhaps) at least he tried to be relevant].

Except for integral-geometry, these are very difficult math-constructs, which (because they are non-linear and/or non-solvable) have only a limited relation to practically useful information, about an implied geometric context.

Integral geometry, if it is actually invertible (so as to provide an accurate image), can be useful. Nonetheless, in regard to models of the physical world:

"a guessed-at model of properties of the geometries of the physical world based on stable discrete geometries might be an equivalent-way of determining actual geometric properties of physical systems,"

while the way of identifying and solving non-linear (partial) differential equations, for physical systems, is not providing any definitive models (or properties) for such physical systems, so systematically probing a physical system by means of either light signals and/or by scattering processes and then doing integral geometry, perhaps is more valuable than is the information which is obtained from trying to identify and solve (partial) differential equations for physical systems.

So, relatively speaking, integral geometry has more practical value than do discussions about mostly irrelevant types of (partial) differential equations.

These mathematical ideas (in these five book-reviews) cannot have very much meaning, if such quantitatively inconsistent math models are not related to a set of stable geometric-patterns, to which a wave-model must conform.

That is, such non-linear models of waves (or geometry) cannot possess within their wave-function (or in their (partial) differential equation) structure any (meaningful) content, since the geometric patterns about which waves are to be oscillating, as well as the non-linear models of wave-properties, are, in general, not stable, ie either no valid shape exists about which the wave-oscillations would have meaning, or the non-linear wave-motions, themselves, disintegrate (or decay), ie non-linear shapes disintegrate in a mostly indefinable manner.

To focus on (partial) differential equations, as being the main model for the laws of physical properties of material interactions, results in a great difficulty to model physical properties, since the solvable aspect of (partial) differential equations are very limited, and because in the subject of (partial) differential equations, the set of (partial) differential equations are mostly composed of non-linear (partial) differential equations.

On the other hand:

To consider the bounding attributes of existence, as being directly related to solvable shapes, provides great clarity about how the properties of existence might be constructed. Rather than trying to both identify, and then solve, (partial) differential equations in order to be consistent with the, so called, "correctly derived" properties of physical descriptions, so as to be "correctly" based on the (so called) physical laws.

First one needs a context which possess properties of geometric stability, then non-linear constructs can be considered within such an outline of stable math patterns, so that the critical points of a non-linear set of equations can be related to limit-cycles defined within a scaffolding of stable geometric patterns.

The above context (of stable geometric patterns) which defines "stable geometric relations about which waves can be defined to oscillate" provides a "first" context of stability, about which other, perhaps more subtle (and defined at a lower order of energy), math relations can be considered.

Science must try to identify some set of fundamental patterns about which practically useful descriptions can be constructed. Apparently, that fundamental pattern is not the (partial) differential equation, but rather some scaffolding of stable patterns (which do have significant relationships to (partial) differential equations), within which measuring becomes reliable, and thus, the descriptions (even if higher-dimensional) can be practically useful, if the stable patterns (of physical existence) are accurately identified.

In higher-dimensional structures (partial) differential equations have many individual dimensional-level contexts, but they also have a more inclusive and inter-related structures, eg high-dimensional models of complex (eg living) systems, which are both properly related to stable patterns and they properly relate the stable patterns and inter-dimensional relations to individual systems contained in each dimensional level (the natural discontinuously discrete structure of material interactions), and provide new contexts for considering systems as stable, higher-dimensional shapes, eg (partial) differential equations can be used in each of the many particular dimensional metric-spaces (which is being modeled) so as to identify perturbation motions (or properties) within an envelop of orbital-spectral stability.

Furthermore, these many-dimensional stable shapes can provide a new set of models of set-containment (within higher-dimensions) so as to be able to define integrated systems which are very complicated.

Consider:

The models of either a nucleus as a set of concentric "rings" of charge, so that the neutron structure (with an electron-center) can also be considered, so that in the nucleus there can be concentric 'rings" of charge with the electron forming a central ring

    or

The positive charges are compact so as to relate to position, in the context of Euclidean space's definition of opposite metric-space states (in Euclidean space these opposites are either position vs. rotation; or position vs. unbounded; where rotational frames need a relation of the subspace to an infinite extent). Thus, one sees that the sizes of "hyperbolic-shapes" are the properties which are related to the properties associated to Euclidean-space properties.

Note: The neutrino has a concentric set of "rings" of charge with a positive-charge center and electrons on the outside the these sets of concentric "rings," and the model extends to infinity so that a subspace is identified.

Consider math objects which can provide answers, as Copernicus provided an answer, so as to be relatable to practical creative development, where practicality can be defined beyond the context of materialism.

Note: Formulating electromagnetism as differential-forms (a locally linear metric-invariant context) provided very useful information. Thus, the extension of a locally linear metric-invariant context to a continuously commutative (almost) everywhere context which allows for the stable properties of shapes, might well be the next context to consider, but now in a many-dimensional context, in order to continue to establish answers, which lead to practical development, as the answers of Copernicus (also) did.

53.

Sets of Opposites

The American experience, and an individual's (or a nation's) attempt to "put the colossus right," where "the colossus" first emerged as the Roman-Empire, and has defined western culture ever since, it is a colossus of social hierarchy which is based on extreme violence, and maintained by extreme violence as well as by using engineering in order to organize society to be ever more easily ruled by (and dependent on) a social hierarchy.

The western colossus is a collective effort to define a useless social hierarchy, where it is useless since it is based on materialism, which has already been proven to be wrong math models based on the idea of materialism cannot be found which can be used to accurately describe and identify the very stable and precise properties of the wide array of very stable spectral-orbital physical systems which exist at all size-scales. Furthermore, the data from particle-accelerators can be interpreted to mean that materialism is wrong, since in this context material reduces to unstable elementary-particles which exist in a higher-dimensional context of existence than the idea of materialism defines (ie higher-dimensions of material containment) and these higher-dimensional patterns are related to finite-dimensional unitary Lie groups. Thus, the failing colossus, of the western collective hierarchical society, has become painfully corrupt, whose apparent agenda is now the extermination of the world's population, and it is now destructive of the entire earth.

The corruption of the colossus is well known. Essentially, it steals both property and (narrows the scope of the use of) cultural knowledge, in order to maintain social inequality, but its violence and control over property, and the engineered instruments used to organize society, and, subsequently, the language which society uses, defies (or thwarts) any attempt to change the properties it imposes on society.

It is a colossus of inequality and extreme violence, justified by a propaganda-education system which serves the narrow investment interests, where actions and high-value are defined within society by means of investment and propaganda.

Whereas, the US originally tried to define a society based on equality.

Language can have a relation to Hegel's so called "dialectic of opposites" Language can be partitioned into sets of fixed easily managed categories by imposing sets of these "dialectic of opposites" upon thc language by the propaganda system. This seems to be a Roman invention, where Constantine's Bible might be a good example of a propaganda-manual for a hierarchical empire.

Consider the, supposed (sometimes assumed), opposites of:

Law based on equality and Law based on property
Creativity and regimentation of society
High-value and low-value
Good and bad
Male-domination and female submission
Intelligent and not-intelligent
Liberal and conservative,
Science and religion,
Materialism and "life and mind"
Creativity and materialism
Randomness and "stable geometric-shapes,"
Indeterminable randomness (quantitative inconsistency) and stable patterns

Math (quantitative patterns of words)
and
Physics (quantitative patterns of words associated to the material world, or to all of existence),
Etc

Male-domination and female submission

The extreme violence which characterizes western culture can be related to the idea of male-domination, and female submission, which is all about instilling a pathologically violent behavioral relationship between people, the violent behavior pattern of domination and a subsequent viewpoint of earning social-rank associated to the relationship of domination over others.

This behavior pattern is supported by, and developed from male-dominant social institutions.

Domination is rewarded and sought after so as to express a person's social-value, and where dominance emanates from violence. Thus, domination of the male over the female is expected to be expressed in relations between the sexes, rather than a set of behaviors between the sexes of: mystery, equality, and sensual experience, which is to (or might) also be associated with a desire to partner together so as to make children.

Conservatives and liberals

The conservative is for traditions and established authority

While the liberal is supposed to be for newness and developing new viewpoints or new contexts for creativity.

But this is not the, "actual very manipulated" relation which exists between the, so called, "liberals" and "conservatives"

The liberal is for intellectual authoritativeness, while the conservative is the self-righteous proponent for all aspects of established (social) power.

The conservative is for traditions and established authority (note: in a context of a highly controlled "market place," determining the truth is tricky), the conservative is for social-hierarchy which is equated with a moral-hierarchy, and an unequal society. The hierarchical social-order (observed in society), for a conservative, is based on individual capabilities, and their idea about society is that, "in a world based on inequality" only superior individuals can determine authoritative truths, and determine the course of society, ... ie the rich are examples of god-on-earth.

Whereas;

The liberal is, supposedly, for new things (but new things in a monopolistic social context are, supposedly, developed by the authoritative experts so the quality of the product is, supposedly, reliable, ie the need for authorities of particular trade associations defined in narrow contexts), the liberal is for the meritorious person defined in an authoritative context. That is, the liberal is not interested in new ideas, rather a liberal's behavior is consistent with the monopolistic trade-dominated market-place, but they (do) define an (the) activist context, namely, the act of doing "good works."

That is, the liberal is not for new things, and the liberals are not for equality, rather, they are for developing the narrow structures which serve the interests of both traditions and established authority.

The liberal has the role of "the intellects whom develop the authority (or social class) for the ideas associated with the elite monopolists," and this is the same authority which is being protected by the conservatives. That is, the liberals and conservatives are working together in order to support the interests of the emperors.

The liberal will state that there is a superior culture (mental-image), while the conservative will organize to exterminate any inferior cultures, or peoples, which get in the way of the expansion of the superior culture.

The liberal intellectual institutions, wherein the people which possess the, so called, mental-property of "true intellectual merit" are identified, but where this "identification of merit" is done in a narrow, authoritative intellectual structure, which is consistent with the interests of the bankers, military-contractors, and the oil-company monopolistic business institutions. That is, intellectual merit is defined in the context of investment interests.

The process of identifying "intellectual merit" is done (acted on within academic institutions) in terms of: intellectual fads, which are associated to both intellectual competitions, and investment interests, where intellectual institutions are controlled the funding processes of university research by the government, in regard to the product assembly-line relation that the "intellectual constructs of university research departments" have to the (new) products, which are of interest to the monopolistic business institutions, eg mostly the military contractors and the oil interests (in chemistry and agriculture).

That is, MIT is not doing research to find a "valid truth," rather MIT is doing research, which is both funded and consistent-with the engineering interests of the big monopolies.

The result is that the engineering-research is related to 19th century science, but whereas the, so called, scientific truths of: quantum physics, particle-physics, general relativity, and the derived theories of string-theory etc, are all about the interests of the military and oil monopolies, eg engineering a very large bomb explosion whose blast capability is associated to both particle-physics and general relativity, ie spherically symmetric gravitational singularities as the site for very high-energy, but random, elementary-particle-collisions.

Opposites

Science is about the material world, while religion is about the spiritual world, the world of "life and mind," ie the idea of creation and intent and "what happens to the energy of a living system when it dies?" etc.

When one looks at the western religions of Judeo-Christian-Islam, as well as the very socially hierarchical Hinduism and Confucianism, one sees an expression of a belief in a hierarchy of arbitrary value, where high-value is essentially defined by the act of creativity, as well as the act of controlling (with violence) one's fellow-man.

(whereas Buddhism's main statement is to "perceive the world as it really is,").

Yet, the main lesson taught by these religions is equality of (or between) people, ie a set of equal people being oppressed by a superior (external) foe, and the story of divine help needed in over-coming the oppressor.

Nonetheless, male domination over the female is also expressed by these religions, where these religions also express a natural hierarchical relation between a person and nature, as well as between a person and society.

That is, there is also the idea in these religions of hierarchy (moral value, or virtue), where these religions are based on the existence of a natural inequality amongst people within society, where people are to worship an all powerful creator.

But this apparent social inequality, which people see within society, is a result of the specialized narrowness which exists within society, which allows for tests to be devised so as to determine the absolute merit of a person, "their merit in regard to certain types of narrowly defined tasks," in regard to "comparing how different people perform a similar task" etc. That is, religion provides a handbook for designing a hierarchical society in which those with "moral value" rule the society.

The main idea of religion . . . , ie the religious act of worshipping of the creator . . . , where this religious idea gets re-worked as the administrators of the business institutions being the creators whom the public is supposed to worship . . . ,

which provide society with its creations, are associated with either the products (or things or processes) which "make a lot of money"

. . . , are to be given "all the social power."

All the social institutions . . . , including, the so called, free-markets . . . , are supposed to serve the interests of these (morally virtuous) administrators and managers of: product or process and propaganda, which make so much money, and thus, prove their worth in the market place.

Furthermore, it is only these "very-big business institutions" which are allowed to make a lot of money.

This becomes a narrowly defined society in which there is domination of the many by the few (the few who make so much money), and it is about this basic social structure in which the authoritative-intellectual (and good-works) liberal, and the self-righteous and dominant conservative identify their roles in society.

Thus, there is a large disconnect between how words are defined in a dictionary, and how they are used in the social context, wherein the, so called, liberals still claim to be for new things, but this is only within the self-righteous propaganda structure of creativity, which is defined within an "overly controlled and dominated" market-place, but it is the needs of this narrow market-place which determine the content and language of university departments.

This control of the intellectuals, as well as the subsequent control of university research institutions has a huge impact on science, and "what it is" that the propaganda-education system proclaims to be the scientific truths for society.

Consider:

Randomness and "stable geometric-shapes,"

Math (quantitative patterns of words)
and
physics (quantitative patterns of words associated to the material world, or to all of existence),

Materialism and "life and mind"
Indeterminable randomness (quantitative inconsistency) and stable patterns

Using convergence and limits to force quantitative patterns into a space within which the pattern cannot be "properly contained" and
Defining set-containment by using stable patterns (though, eg, in a many-dimensional context, where higher-dimensions are hidden, where they are hidden because they are discontinuously separated from both material and the material-processes of our experience, which we have defined (by overly authoritative physics, which serves investment interests) in a particular dimension [and at a particular size-scale]).

How many ways can one consider higher-dimensions and ways in which higher-dimensional stable (math) patterns can be organized, so that the material-world is a proper subset of such a newly given organization to a set of math-language patterns?

Using words (including nouns) contained in a sentence, where the sentence implies a context, and then forcing a (quantitative) relationship (eg by counting the number of the observed nouns) between a particular noun and the context of the sentence, eg "oatmeal can lower cholesterol" eg correlations between two nouns, eg the concentration of the molecule cholesterol (in blood) and oatmeal intake.

[However, identifying the molecule "cholesterol" has little value in sentences whose nouns, such as the word cholesterol, are being forced into containment sets wherein the nouns do not properly belong (in the context of the sentence). That is, the chemistry of life needs to be clarified (understood) before correlations between nouns (ie identifiable molecular types) in sentences, about living systems, have much meaning, especially, since the molecular context of life is so complex . . . ,

(with no over-all unifying model for life (or for the chemistry of life), but with only the very incomplete idea which relates the molecule DNA to all properties of a living system. Yet, there is no valid chemical model of the relation of enzymes to chemical reactions, yet, the main context of DNA chemistry is about enzymes.)

. . . , and the ability to identify (only a relatively few numbers of) molecules is quite limited.]

Eg consider "Earth as the center of the universe" and "the noun 'motions of planets'" (ie the sentence context of the Ptolemaic system).

Nonetheless, identifiable nouns, which exist in somewhat arbitrary sentence structures, are being used to try to identify quantitative relationships in (or quantitative patterns are being forced within) an improper containment context, which is implied by a (the) sentence.

This is often being done in a quantitative structure of convergences in containment-sets which are "too big," incorrect models of "system containment" are used, and overly general relations (eg correlations between containment spaces and sets of operators [or auxiliary sets of (quantitative) system properties]) are identified algebraically, and convergences (defined within containment-sets which are too-big) are used to force this ad hoc pattern into its assumed containment set (namely, the context identified by a somewhat arbitrary sentence, which is usually attributed with far too much authority (this limitation of authoritative language is possible through the relation of investment to the propaganda-education system, which exists within society)).

However, a particular sentence may define a containment context within which an (the) observed set of properties are being used.

For example, in particle-physics the assumed (to be useful) descriptive context of material-interactions is the context of material point-particle collisions, ie where the (so called) useful part of the sentence is defined in regard to particles in space (and related to a critical amount of a particular types of material), so that these particle-possess collision-probabilities (with other particles which compose the chaotic system) are related to rates of reactions,

(of the particular reaction process which is defined by the probabilities for the collisions of these point-particles, and, in turn, (these particle-collisions are) associated to a process of system-change, in which a reaction is defined between two relatively stable-states of material (ie the critical amount of a particular type and the resulting material into which it changes through the collision process), where within these two relatively stable systems, it is assumed, that the point-particles are components.

However, exactly how these sets of chaotically colliding particles . . . , which exist within the chaotic transition process . . . , (actually) fit into these two relatively stable system structures has never been clarified).

That is, though the sentence about the rate of a reaction defines a context of practical use, nonetheless, the context leads to no development of further (practical) knowledge concerning stable material systems.

This is the main criticism of science as practiced in a society today (2013) where investment-interests get to define what "language structures" are given authority, and how, so called, scientific language is to be used within society. Where it needs to be noted that it is a society defined by wage-slavery, which exists in a very narrow propaganda-education inter-related culture, which, in turn, is based on:

inequality,
theft, and
violence,

. . . , where the subsequent regimentation of culture leads to the systematic expansion of both the social-material-system of investment (which exists) and the population.

However, it is a society which shields itself (ie the society) from the development of both knowledge and a set of new contexts for creativity, which would be associated to the development of knowledge, where if knowledge to develop then language is to focus on assumption, context, containment, types of properties, and organization of math patterns into new types of (measurable) descriptive languages. But instead, knowledge is authoritative and dogmatically fixed to suit the interests of the investment-social-class.

{In the particle-physics model of the material world, the material system is contained in the domain space of the function space, but the system's set of measurable properties (both its internal and external states [or external wave-properties]) are determined by sets of operators, whose dimensional structure is greater than the dimensional structure allowed by the material system's domain space. Thus, the system's containment space is the domain space (ie where the material system exists) and the system is modeled as a function-space, along with the set of operators which are supposed to define both the full set of the system's external properties, and the further "(internal) states of the system" which are needed to define this set of operators, where all of these math structures all-together identify a containment set for the descriptive context.

Yet the internal-states (of the point-particles) are defined (not on a geometry, but rather) on a point-particle, even though the internal-states are vectors which suggest a further geometric-structure, (ie thus, the interest in string-theory).

However, the internal-particle-states are unstable, and their defining (further) local geometry is non-linear (based on a connection-derivative), ie and this means the geometric relations are quantitatively inconsistent (this quantitatively inconsistent property of non-linearity is the basis for the proof of Thurston-Perelman's idea of geometrization). That is, it is not clear if this descriptive context has any content, other than the relation of particle-collisions to rates of reactions (and this is not clear, especially, since the military de-funded the super-collider). That is, the rates of reactions (of a system of particle-collisions) are defined in a context, which is not quantitative, rather it is defined in a chaotic process of transition, which occurs between two stable contexts, but where word-meaning (really) only exists in a stable context, wherein one can talk within a context where patterns are stable and measuring is reliable.

Furthermore, a chaotic transition processes is not a context within which useful information can be provided, concerning the stable structures within which a sentence and its subjects (or nouns) are provided with a partially valid context (eg the initial and final states of material after the reaction), ie the context of the sentence and its nouns have a valid and/or stable relation to one another.

Note that "the sentence context" of Copernicus led to the development of knowledge while the sentence context defined by the Ptolemaic system, though fairly precise, did not lead to the development of ever more practically useful knowledge, ie practically useful information which is consistent with observed (measurable) properties (or observed patterns).}

That is, arbitrary relationships between nouns in sentences, whose context exists within a structure of very incomplete knowledge, are not reliable.

Often, what is accepted as science is about trying to place a quantitative structure which is associated to the nouns which are defined within their use in a vague language structure, eg cholesterol levels in blood correlated with intake of oatmeal, but these arbitrary and vague uses of math structure so as to identify a quantitative relation between nouns used in vague sentences are seldom (if not never) valid. Yet, these types of vacuous correspondences are claimed to be developed in a rigorous mathematical manner, by the professional math community, because professional mathematicians have been required to base their mathematical patterns mainly on the ideas of indefinable-randomness and non-linearity. Thus, vague sentences can be endowed with a vague epicycle (math) structure, which is designed to fit data, so that arbitrary relations become the basis for "convincing" arguments in the propaganda of persuasion.

One must consider the context, in regard to both stability of pattern, and reliable measuring, and not place elaborate math structures onto arbitrary contexts, even if the context of a sentence might have one relationship to a (the) (eg chaotic) process's further use (note: a process can be defined by a system, so it may be considered a noun), so that the idea seems to have practical value, but such a "practical value" will always be very limited.

Science and practical creativity (the material world)

Science is about a precise cataloging of observed patterns, which are placed into categories and types, or in particular, in physical science it (science) is about descriptions which are about the precise measuring of observable patterns, along with (as in the case of physical science of) an attempt to try to identify a single math pattern . . . , eg a particular type of an equation, . . . , which may be used to organize and/or realize (or predict) a wide array of sets of different observable contexts whose behaviors are precisely relatable to one approximate unifying math pattern, which, in turn, is (precisely) related to measurable properties.

The small set of general math patterns which are used to describe the underlying patterns of physically measurable properties, where these general math patterns are given an abstract math structure so that these math patterns can be related to a wide array of physical systems, which possess precisely describable properties (which are observed and measurable properties of physical systems) and where these properties can be (or are supposed to be) predicted by the solutions to the (a) system's (property) defining (partial) differential equations.

To get an idea about the relatively simple math ideas which need to be considered in regard to prediction by solution functions to a physical system's defining set of (partial) differential equations, then consider . . . :

{Eg they are (locally) linear, metric-invariant (thus related to differential-forms), continuously commutative (almost) everywhere, and single-valued, ie both the geometric structure of the system's containing space, and the initial or boundary conditions of the system determine a single solution function, or a single-valued solution function. There is also the multiple-valued function where the many-values are separated by a fixed interval (or a finite set of fixed-intervals), where these types of functions are the result of the function being defined on a space which has holes within the structure of its shape, where the number of holes in a shape is called the shape's genus-number. This type of many-valued function can also be related to a solvable set of (partial) differential equations, or, equivalently, can be related to both stable shapes, and/or to a zero-genus shape when considered (or described) in the correct dimensional-level, ie the dimension of the shape itself, so as to be considered as a fundamental-domain ie a "rectangular" simplex (or "rectangular" polyhedral) shape which fits into a lattice structure (or checker-board-type partition) of the metric-space, which has the same dimension as the dimension of the fundamental domain. Note: These stable shapes are (solvable) differential-forms.}

. . . for linear, solvable systems (observed and measurable patterns of interrelated material components, ie inter-related by a general math pattern) of classical physics . . . , ie mechanical, gravitational, electromagnetic, and thermal, and well as classical-statistical, physical systems (statistically based on stable, identifiable, random events, of a fixed number of material components, which are defined within a closed system),

That is, if physical systems are defined in the context of: materialism, linearity, metric-invariance, and both the dynamics of components and their associated geometric material context whose properties are defined by either differential-forms (which are locally measurable relation associated to geometric measures, eg length area, etc), or generalized metric-invariant differential operators, and the geometric relations of single-valued-ness of all of a system's measurable properties, which are associated to both a physical system's containment-space and its (local equation) system-defining relationships, or any such geometries of systems, which are without holes, so that, these many domain-space shapes, which do not have holes in these (many-space) shapes, so as to allow for both single-valued-ness of a measurable property of a physical system, and continuously commutative (almost) everywhere shapes, then the system-defining equations are (would be) solvable.

This same math structure (linear, solvable, and single-valued) can be "moved" to a more general, but closely related, context of both higher-dimensions and which contain new material-types.

{It is the second order, metric-invariant differential operators, often associated to second order (degree?) differential-forms, for the various differential structures defined in the context of . . . ,

(for either a one-patch [or global] or a two-patch {or, more generally, a compact} covering atlas for the material components, or system shapes, contained in metric-spaces, associated with)

. . . , the classical, finite-dimensional set of Lie groups . . . ,

[which, in turn, are associated to various R(s,t) metric-spaces, with spatial and temporal subspaces, ie s, t, so s + t = n, but it is, s, which is associated to a Euclidean R(s,0) metric-space, in turn, associated to a dynamic material interaction context] . . . , which, in turn, relate the geometry of a material system to its dynamic geometric properties of the system's geometric-material-changes, ie force-fields being related to the laws of inertia,

. . . , which allow some physical systems to be both modeled and solved and controlled by controlling either initial or boundary conditions of the system.}

Note: It is also the differential-forms which relate measurable thermal properties to equations which represent thermal laws for closed systems, where in thermal physics the first-order differential-forms are prominent.

The idea of science, being defined as math patterns used as widely applicable "pattern of prediction," . . .

ie used for determining precisely measurable and controllable properties of physical systems, where these math descriptions are based on their being applied to cause-and-effect (or linear and solvable) types of physical systems,

. . . , was a fine definition, which the math structures of classical physics determined, in regard to linear, solvable classical physical systems.

However, further considerations about the general laws of physics led to the realization that most classical system's are non-linear and not controllable

That is, most applications of classical rules, ie defining a system's equations based on both materialism and a general rule (or a law of physics), leads to patterns of systems which define non-linear systems.

Furthermore, general relativity is also based on non-linear patterns, yet it appears to be based on very general attributes concerning the relation between frames of a coordinate reference for a system and the equations (or physical laws) of any particular [a (that)] physical system. Unfortunately, A Aspect's experiment has shown that the spatial properties of physical systems can be non-local, ie action-at-a-distance (so Einstein is wrong)

But (or furthermore) non-linear math patterns are quantitatively inconsistent, and lead to the multi-valued-ness of the math models of physical properties, ie multiple solution functions exist, or bifurcations exist) and subsequently there exists both indefinably-random and chaotic behaviors, ie they are not causal and predictable systems.

506

That is, non-linearity is both quantitatively inconsistent and indefinably random, ie there is no set of stable events (or stable patterns) to which either a non-linear geometry or even a valid probability structure can be associated, ie the counting of such a probabilistic-system's events is not well defined.

Yet (or nonetheless) the critical points of many of these non-linear (differential) equations can be related to a partial-control (in regard to a dynamic component in the system) by using feedback information within the system, ie the quantitative pattern is not valid but the context of the system's classical (partial) differential equation can sometimes still be relevant, ie the surrounding geometry which is related to the system's defining (partial) differential equation is most often more relevant than is a system-component's motion.

For a new viewpoint, which leads to a linear, solvable (or causal) descriptive context

At this point one needs to consider the idea about stable, solvable, geometric shapes, in a context of material-containing metric-spaces of the general type of R(s,t)-metric-spaces, which contain new types of material . . . , where these new metric-space shapes are identified by the linear, metric-invariant, with non-positive constant curvature for metric-functions, which only have constant coefficients in the metric-function, and for geometric patterns associated to the system's geometry (or the system's containing space) which are continuously commutative (almost) everywhere, ie the linear and solvable math patterns, . . . , as the proper (underlying) context of measurable (and controllable) descriptions, in regard to describing the properties of stable systems and the containment-set structure for physical properties, and/or physical existence.

These stable shapes are also differential-forms, where the basis for differential-forms in an arbitrary dimensional-level are the polyhedrons, eg cubes and squares and rectangles etc, whose properties of adjacent angles [(to other polyhedrons in a checkerboard type lattice) defined at the vertices of the polyhedron's moded-out circle-space shapes] allow these polyhedrons to form into lattices, or checkerboard patterns associated with a metric-space's partition into polyhedra which cover the metric-space, (where the "rectangular" polyheda which compose the lattice are all of the same shape).

The boundary (of a polyhedron's lower-dimensional, facial) structure of polyhedrons identify a decreasing dimensional sequence of different dimensional polyhedral-faces which can be locally associated to geometric-measures (eg length, area etc) which are consistent with the dimensional properties of the local geometric measure structures of differential-forms, and a "cubical" (or "rectangular") polyhedral structure is also associated with a continuously commutative local coordinate transformation matrix of a shape's tangent directions (of the fiber group acting on the tangent directions of the shape) for either a 1-patch or a 2-patch covering atlas, for the stable shape associated to a (solvable) circle-space shape, ie a linear-and-solvable differential-form, where these linear, solvable shapes (or differential-forms) possess holes in their shapes, when viewed in the adjacent one-higher-dimensional (natural) shape-containing metric-space (circle-space shapes), so that this allows these shapes (or material components) to possess stable spectral-orbital properties (in the shape's containing metric-space).

"Rectangular" polyhedrons can be moded-out to identify stable circle-space shapes, where in the moding-out process the polyhedron's "pairs of 'opposite faces" are identified with one another, by a circular path which brings the opposite faces together, ie to be identified with one another. For example, the line segment identifies its opposite 0-faces to form a circle, the rectangle's opposite 1-faces are identified to form a torus (or doughnut shape), etc.

That is, it is the math structures of the linear, solvable classical context, which is associated to controllable physical systems, which, in turn, define the types of stable math patterns into which existence can be modeled as sets of different dimensional metric-spaces, ie metric-spaces which possess the property of being stable shapes, so as to extend the idea of a containment-set which exists in a context which extends beyond the idea of materialism.

These stable, circle-space-shapes are differential-forms which are defined by the discrete isometry subgroups, and associated discrete unitary subgroups, which, in turn, are associated to the metric-invariant (and Hermitian-form-invariant) contexts of the classical Lie groups.

They are circle-space shapes, ie tori and "strings of toral-components" which are put-together, but which can be folded in a constrained, but in a consistent, set of discrete ways (ie group transformations, or lattice folds), ie folded by means of discrete Weyl-angles.

It is the holes in the very stable discrete hyperbolic shapes which are central to understanding the stable, discrete spectra of quantum systems.

One losses single-valued-ness, when one observes the shape within the metric-space within which the shape is contained, but gains sets of discrete spectral-orbital properties for physical systems, whose dimension is one-less than the dimension of the (material-shape-containing) metric-space.

The circle-space shapes can be used to solve the many-(but-few)-body problems of both Newton and Schrodinger, but it is done in a new context for containment. The new context allows for stable two-body elliptic paths to be a part of an envelop of orbital stability defined by the folded stable metric-space shape (see many books by m concoyle).

This might provide a causal math process which might be used to understand the elliptic-path (or elliptic-curve) structures of a flat Calabi-Yau manifold, especially, when considered in the context of pairs of opposite metric-space states (which implies a unitary context).

The primary math structures of the new, many-dimensional, quantitative-containment context is simpler than the classical physics math structure.

These new math structure for physics define both

(1) the set containment envelopes for orbits for both stable material-components and/or condensed material (objects) and

(2) the finite spectral-set which determines the set of stable shapes which can exist (based on their resonance with the finite spectral-set) in a particular many-dimensional context for a higher-dimensional context which is defined up to (and including) an 11-dimensional hyperbolic metric-space.

Within this new context, for condensed material objects, and for some stable material component-shapes, the math of classical physics adjusts the system's spectral-orbital properties, and for (very) small, stable material components, the properties of randomness are also natural. However, randomness is not the mathematical feature which determines their stable properties.

Angular-momentum properties also apply to the circle-space shapes.

However, the math community, between 1890 and 1910, went forward with an ill-conceived plan [so as] to embrace quantitative inconsistency, as well as indefinable randomness, by assuming that one can place math patterns onto any context where numbers might be defined, so that this was done by using axiomatic formalization, so as to relegate shape to vague abstractions concerning: numbers, indefinable randomness, and non-linear quantitatively inconsistent sets, which are claimed to be geometric patterns.

But this is not true, these non-linear geometries are unstable, and thus the patterns have no definite content, ie their properties can dissipate.

That is, both shape, and "stable shapes, in particular," are the main issues to consider, in regard to identifying a stable context for measuring, so that measuring is reliable, and the discussion has a definite content about a stable pattern.

That is, it is only in the context of stable shapes, where measuring has validity, and where the set-containment of an actual pattern (which has measurable properties) can be defined, so that definitive properties can be identified and discussed (ie precisely described), and then used in a "practical" context, where "practical" use in higher-dimensions might seem unusual, or difficult to conceive.

Math patterns can exist in an abstract context of high-dimensions, related to both properties (or functions, eg solution functions to a physical system's (partial) differential equation) and/

or a language organized around stable shapes in a discrete, discontinuous, multi-dimensional context of set-containment. Furthermore, in this new context, the idea of measuring of a system's properties, in a "place" (or containment set of stable shapes) where large-scale, discontinuous changes in scale can occur between different adjacent dimensional-levels, so that:

new material, and
new geometries of interaction, and
new shape-spectral relations exist, and

can occur and (thus) form, in a new multi-dimensional context for "material" existence, in a context of discrete discontinuous changes, which are a natural part of the new descriptive construct.

In this new context, new materials are trapped in their own open-closed metric-space topology, ie where (partial) differential equations of a two-body system can be defined by (within) a (global or two-patch) differential structure, which possesses new geometric relations for interactions, and these two-body (partial) differential equations can quantitatively adjust (or perturb) the orbital-spectral properties of either the material objects trapped in the envelopes of orbit-defining metric-space shapes (where the orbiting material is contained in the metric-space) or stable metric-space shapes associated with finite-dimensional fiber Lie groups.

It should be noted that, set-containment, and size-of-spectra-orbits depends on a particular dimension-and-subspace partition (and an associated containment-tree, of a dimensional-and-subspace dependent partition) of an 11-dimensional hyperbolic metric-space, where the partition is based on a finite set of stable circle-space shapes.

So that it is within such a containment space for existence, within which life may be best defined as an abstract math pattern existing in high-dimensions, and, furthermore, life's purpose is about extending this abstract math construct of stable shapes in a many-dimensional context.

Quantum physics

The math-physics history after classical physics led to quantum physics, since local point-particle-spectral events were seen to be random events in space, so that the math of quantum physics was based on randomness (or probability) of random point-particle-spectral events in space and time, but quantum physics could never perform at the level of the high-math-ideals in which it was formulated

The descriptive context, of quantum physics, is defined by the randomness of locally identified particle-spectral events, but which were quite often related to spectral properties of systems, which had great stability. But this set of stable spectra for general quantum systems cannot be identified by using the math methods associated to quantum physics.

The wave-particle model . . . inter-related to a system's observed particle-spectral events and its set of operators, but the observed particle-spectral events were interpreted as probability-waves associated to a system's defining energy, so that this descriptive structure was about algebras of (unitary, or Hermitian) operators acting on a quantum-system modeled as a function-space, where geometric and system spectral cut-offs (or limits, or constraints) and (local) algebraic properties . . . (were used) to find the system's stable and fixed spectral set.

But this was (has not been) not possible in general.

Thus, a structure of randomness was imposed for the descriptions of quantum systems, but which could not be used to find the spectral sets of general quantum systems, whereas identifying the properties for general systems was a main attribute of science, under the viewpoint of classical physics.

Furthermore, many of the fundamental general quantum spectral-systems, composed of many-(but-few)-charged components, are very stable, and this implies a linear, solvable context.

So basing a descriptive structure on randomness, but not being able to identify the spectral structures of stable general systems is a (new) type of indefinable randomness, wherein such (fake) quantitative models have no validity.

The only real question in quantum physics is "why does 'this type' of model for the H-atom work so well?"

The H-atom is a two-body problem while, also, Newton's law of gravity only works for the two-body problem, though Newton's non-linear equation, in a many component system, have critical points, which can relate these gravitational equations of satellites to feedback satellite-guidance systems.

Furthermore, when particle-collisions were considered in particle-accelerators . . . , [where this is done primarily to determine the cross-sections of particles (mostly considered in the context of a nuclear reaction in a nuclear weapon)] . . . , it was found that the collision patterns of the elementary-particles (and particle-families) were also unitary, as was the function-space's algebraic structure, but, for particles, it was finite dimensional.

However, the events considered (or observed) for elementary-particles are unstable events, since most elementary-particles are either unstable, or not observable.

This is again a form of indefinable randomness where quantitative properties have no validity.

The arbitrary authorities want to always hide their abstract, unverifiable relations (hidden in abstract [and invalid] math structures) so that their authority will not be questioned. The main authorities of academic institutions are psychotically dominant types, as are all the key individuals which make-up both the ruling-few and those few who are used and manipulated by our spy-based social-control of social-power.

Apparently, this relation of truth to hidden (unverifiable) abstractions, is also true for the investment interests, since a set of arbitrary authorities who are supposed to espouse a truth which is unintelligible, fits into the structure of the propaganda system, so that such a social relation of absolute and arbitrary authority has more social value than does actually finding a valid truth about the very stable spectral-orbital systems which exist at all size-scales.

Furthermore, the abstraction which the authorities espouse fits into the rather specialized use of particle-physics information, namely its identification of particle-collision cross-sections, used in the design of nuclear weapons, so that new contexts for creativity which result from the development of truly new knowledge do not create new types of business competitions for the monopolistic businesses, and so that this new knowledge would place their investments at greater risks.

Uncertainty in quantum physics

Since the math of quantum physics is based on probability, this means that there is an uncertainty principle defined between dual (or Fourier transformable) differential (or local) variable-types, where in quantum physics this uncertainty principle can be represented (or interpreted) as a statement that a system's "fixed" geometric properties cannot be quantitatively related (at a point) to its dynamical geometric changes.

But this means that a particle-collision of point-particles cannot take place, and still be contained in a math structure based on probability.

But elementary-particles are modeled as point-particles.

Furthermore, the H-atom uses a fixed geometric potential-energy term in a dynamical system of a randomly moving point-particle (electron) component, and so that the positive charged point-particle component is placed at the singularity point of the $1/r$ potential-energy term.

Furthermore, there is the issue of dynamically moving point-particles which are charges, so that these charged-point-particles compose the (dynamical, charged) components of many of the fundamental quantum systems, eg nuclei, atoms etc, yet it is not considered that these dynamically-changing charged-point-particle components are radiating electromagnetic radiation, which they would need to be doing in a classical system, and which needs to be considered in particle-accelerators when charged point-particles are being accelerated.

[It should be noted that if charged-components are trapped in a closed metric-space shape (as they would be in the new theory) then these electromagnetic energies associated to radiation caused by accelerations would also be trapped by the closed metric-space shape.]

That is, in the current math model (of quantum physics), a somewhat arbitrary quantitative structure, based on probability, is imposed on a set of physically measurable properties, but the dynamical properties associated to the point-particle model of the quantum system's charged components are ignored, where, apparently, this is based on a wave-of-the-mathematical-hand so that some properties of the quantitative structure are ignored, because of the hand-wave, and the imposition of an arbitrary probability-based quantitative structure, wherein the dynamics of charged point-particle components are ignored.

That is, the axiomatic formalization is used in regard to identifying a descriptive context where quantitative structures are identified, so that both measurable and non-measurable quantitative properties can be identified in this context, so that only the measurable properties are associated to math structures by means of abstractly imposing this quantitative structure through formalized axioms, which discard the so called non-measurable properties, which, in turn, can also be associated to the descriptive context, ie a probability-base associated to quantitative (and theoretical measurable) properties allows radiation "due to charged-particle accelerations" to be ignored (as a quantitative structure, even though these same properties of radiating accelerating-charges are applied to charged components in a particle-accelerator).

Again the stable solvable properties of general "quantum systems" (ie material components with discrete spectral properties) suggests that the linear, solvable properties of circle-space shapes should be considered as a new basis for physical description, especially, since A Sommerfeld's elliptical-orbital adjustments to Bohr's circular model of the H-atom provide a very precise set of theoretical relations to observed measurable properties.

Furthermore, Darwin's model of evolution now associated to mutation and natural selection is an indefinably random quantitative descriptive context, about which the numbers seem to not support the notion of moving toward greater complexity, but rather the numbers (the statistics) support the idea that the original life-form is the most DNA-complex, and further life-forms have less DNA-complexity.

Thus, Lamarck's viewpoint about free-will affecting the development of life-forms seems to be just as valid as an indefinably random basis.

That is, the math community has failed at providing guidance in regard to using math patterns to identify the basis for the observed stable patterns of the physical world, but it is also a problem for the notion of materialism, since the idea of materialism is causing (providing) an unnecessary constraint on the descriptive quantitative context.

Furthermore, it should (might) be noted that the data from particle-collisions might be best interpreted to mean that existence is many-dimensional and it has a strong relation to the finite-dimensional unitary group.

In higher-dimensions there are (can be) many new models for life.

It is the origins of life which is central to understanding "what life is," and "how it develops."

54.

How to define value?

How to define value?
(Should investment bankers and their political puppets determine "what has value within society"?)

To compare capitalism with socialism, is to compare one collectivist and hierarchical social system (ie ruled by a few), which is based on the idea of both materialism and violence . . . . , where violence is needed to maintain the narrow way in which a hierarchical society is to organize material (and material use) in society . . . , with another such arbitrarily hierarchical social system, with its need (or both of their needs) for great violence in order to maintain an artificial social hierarchy.

This social power structure is dependent on science, which is organized in such a manner so that the beliefs of science support the bankers interests. Thus, the dogmatic authority of science and math needs to be challenged, unless one believes in the inequality of a totalitarian society, and its associated propaganda-education system.

The teachings about the spirit (by the religions of the world) are, in turn, about "everyone being equal," so there are not to be artificially, and thus falsely and/or arbitrarily, defined narrow categories of absolute-value, within which a person's (absolute) worth is to be determined (in a hierarchical society).

Yet it is this concept which is attributed by the public to the propaganda system.

This is a result of the fact that those who are defined (by the propaganda-education system) as intellectuals (or the experts to whom the propaganda-education system turns so as to establish, what are represented in this propaganda system as, objective facts) accept a viewpoint that a dogmatic authority be attributable to science and math, even though the "quantitative structures" . . . used by a great many (if not all) of these so called scientific disciplines . . . are not reliable.

It is also implicitly assumed that making expressions about information can be done in an "objective manner" (however, information has a context and a set of assumptions associated to itself).

Furthermore, measuring does not make a description "objective," (ie identifying only well-established facts) where the ideas (or the model) of Ptolemy, about the motions of planets in the heavens, were (was) measurably verified, but the Ptolemaic system is now considered to be wrong.

That is, in regard to both spirit and understanding existence;

It is not a: materialism vs. spiritualism, dichotomy; rather both the material-world and the spirit are beckoning us to explore beyond the material world.

The left identifies itself as "being a failure" when (since) they do not realize that their intellectual viewpoint is a narrow absolutist (dogmatically authoritarian) viewpoint, which is both derived from and supports the interests of the ruling class. This means that, their viewpoint is an arbitrary and narrow dogmatic intellectual structure, which is upheld by the same violence which upholds the unequal (or hierarchical) society run by the bankers (the owners of society).

It is the bankers-oilmen-military big business concerns who determine how material is organized in society, where the interests of the ruling-class are about controlling material and controlling the knowledge of society.

It needs to be noted that a narrowly defined absolutist (or totalitarian) intellectualism, is specifically related to creating a narrow range of material products, which, in turn (through the economy), these products (which, by the way, are related to the knowledge of 19th century science) support the power of the ruling class. This system of material-based creativity, which is dependent on 19th century science, is, in turn, based on, or upheld by, the rich and (violently) powerful few, where these very-rich few are allowed (by the justice system) steal from the rest of society, so that society violently upholds the right "of these very rich few" to steal.

This "license to steal" is provided by the institutions of: law, the government, and the propaganda-education system.

However, this means that . . . "what the ruling class believes" . . . . is (or has become) the basis for the left's overly-valued absolutist viewpoint of knowledge.

That is, the knowledge which the intellectual-left represent (or "in which they believe") is far too narrow and absolutist, and thus the left is taking-up sides with a failed intellectual construct. That is, monopolies, based on narrowly defined knowledge, end-up failing in very destructive ways.

The left represent the narrow vision of knowledge which best supports the ruling class, ie the high-intellectual-value which the ruling class defines.

The right is "all about" serving and cheer-leading for the high-value of the rich, where this service is, essentially, about a "spiritual idea of creativity," which is based on both (1) the belief that the very rich are (assumed to be) "favored by God," and (2) a belief in a materialistic context of (for) creativity, so as to relegate the idea of non-materialistic creativity "not to new visions of an abstract existence, (of a very real, but abstract existence, which can be accessed by anyone as an expression of knowledge), but rather to the context of either creating material objects (eg pictures and books) ie creating an illusion of the existence of an arbitrarily determined belief in a high-valued set of (disciplined) artistic or spiritual creations, which are advertised and distributed over communications systems, and then by using violence so as to force the lower classes to accept and buy the narrow set of ideas expressed by high-art or, so called, high-spirituality.

Again high-value is defined by means of the violence involved in defining a narrow context of an arbitrary high-socially-valued vision, and again forcing the lower classes to buy and cherish these created objects or artistic creations (which are narrowly defined (usually with a narrowly defined theme) in an arbitrary vision of high-value). This results in a viewpoint that, within society, it is only material based creativity, or communication-system based creativity, through which a narrow set of ideas are expressed, are the types of creativity and the range of the expression of ideas which will be "the closest that the society can get 'to the spiritual existence' defined by creativity."

That is, creativity is defined in a material context, or as a dogmatic belief in knowledge (associated to either social organization or product creation), so that art is also to be defined in a narrow context of an artistic discipline, which is provided a market (so as to define its value), by the communication systems, or is institutionally displayed, so as to possess an arbitrary high-social-value, eg architecture.

It needs to be noted that the Christian religion's central (but abstract) personality, ie Christ, said to stay away from "Caesar," but nonetheless "the church of Paul," where Paul was the church's main administrator and founder (or his subsequent high-raking followers), made a deal with "Caesar," so as to effectively define the spirit as creativity in a material context, and so that ideas must (only) serve the interests of the Emperor, and such high-valued creativity is defined in a hierarchical society sustained by extreme violence and sustained by the propaganda of a hierarchical religion, ie the Bible was made by the Roman Emperor Constantine so as to be the basis for the propaganda of the Roman-Empire.

Thus, one can claim a different viewpoint about organizing society, based on the equality of individual creators, which is an idea which comes from the right . . . , but where in the European west one separates "the institutionalized religion of Paul" from the material world, so as to define the material world vs. the spiritual-world, . . . , but instead one defines value (in an abstract context) as being related to both (1) knowledge (identified in many contexts) and its associated (2) creativity.

In the new society, value is to be defined by the many new creative contexts associated to the different contexts in which knowledge (and its associated creativity) can be expressed.

That is, one divorces value from (being defined by Caesar as) being both material and the coercive violence of Caesar . . . , where Caesar needs coercive violence so as to enforce this narrow definition of value, . . . , and instead one should consider a wider range of knowledge and the "creative value" to which such knowledge is (or can be) related.

In Caesar's society, the left represents (is) the knowledge upon which both (1) material creativity, and (2) the limited range of ideas which are used to justify and maintain a hierarchical society, is based (which apparently the left sees as being a knowledge which is separate from Caesar, but the left is very wrong about this), where material creativity is, essentially, the brick-laying of Rome's armies, but which "is now" the engine, circuit, and bomb building of the western culture.

In turn, the right is defined by those who possess a reverence for the violence "needed" to maintain an arbitrary high-social-value, ie violence is needed to enforce the arbitrary high-creative value of Caesar, but the "need for violence" is to be expressed as a principle of freedom, but, in reality, it is the freedom of the emperors to do "whatever they want."

It is in the context of narrowness of: ideas, knowledge, creativity, and value; whose (narrow) range is defined by the coercive influence of the emperor (or of the ruling class, and their intellectual elites, ie the propaganda-education system), where the sequence of . . . ,

(1) "created object," eg brick-laying of the Roman legions or the electrical circuits associated to communications channels today, or the narrowly confined "disciplined (or skillful)" artist, or mathematician etc, and

(2) an association with beauty or high-value of the object created by the high-valued experts of the empire, and

(3) the accepted belief in the desirability, or demonstration of genius in the discovery of knowledge, of these material-centered creations . . . , which defines the market-place.

For example, the narrow dogmas of the intellectual community . . . , where these highly-valued dogmas are not capable of solving the observed stable patterns of spectral-orbital properties for many-(but-few)-body systems . . . . , are claimed to define a valid intellectual market-place of ideas, and the, so called, knowledge which they claim to be uncovering, in this so called market place of ideas, is always associated to the ideas which are claimed to possess the property of "highly-valued beauty of the intellect."

However, this narrow dogmatic and subjective construct of the intellect (ie subjectively seeking intellectual beauty) is not a valid place (or market-context) within which to express ideas, since the dogmatic authority of these ideas has led to a context of an intellectual-community whose members are a part of a highly disciplined "absolute truth," so as to define a context of intellectual dominance by a few. Those few who are considered to be the intellectually-dominant few, are also the people whom follow the intellectual traditions and the authority of personality (defined in an academic context which has always supported the bankers interests), so as to define an "absolute truth," but it is an "absolute truth" which is based on the same type of belief in an intellectual superiority which the puritans used to justify exterminating the native people, in a similar manner as the current, failed, intellectual-aristocrats use their dogmas to exclude (or exterminate) other ideas, which, the experts claim, do not possess the same beauty of intellect which the intellectual-authority's belief structure possesses.

That is, the politics of the left and right is the same politics of the violent exclusion of those who are not within the top-spots of the social and intellectual hierarchy. And this politics of violence is also the politics which defines the market-place, as well as "what it is" that the authoritative experts believe, as well as the disciplines which the skillful artists master.

This is all a part of the propaganda-education system which defines the markets which support the ruling-class, and underlying "it all" is violence and arbitrary distinctions (or arbitrary choices) [namely, the choice to be on the winning side, in a collectivist social hierarchy].

This is the natural barbarism of an elitist society.

Note: The west is really defined by the religions of Judeo-Christian-Islam, as well as the political oligarchies based on inequality, and its associated violence, but the religions, or social codes, of Hinduism and Confucianism also seem to also be hierarchically-western in their nature.

Why has the western civilization which has emerged from Rome been so dominating throughout the world? Its activities are organized around violence, it is hierarchical, it is autistic in its narrowness, and it is delusional.

[Not: Guns, Germs, and Steel, though Germs were decisive in the Americas, but Guns and Steel are simply expressions of an overly narrow focus on violence. The world-view of the Native-American people was much more sophisticated than the European world-view.]

History of the west:

Western civilization grew out of the Roman-empire.

Rome began as a few aristocrats, so that each aristocrat had their own army. These aristocrats used their armies to attack their neighboring cities to steal the wealth of these other cities.

This structure of Rome was surprisingly stable, perhaps because it was consistently violent, but eventually "one of the aristocrats" became emperor. The Roman-Empire expanded in this same way, but now the Empire built roads, water-systems, sewer-systems, and grand public buildings in its conquered territories.

Centrally controlled, narrowly defined and regimented communities were built around these public works, and laws proclaimed from the steps of grand buildings, throughout the Empire.

This was a relatively stable social structure, and the roads allowed more trade, and the narrow regimentation of society allowed banking and mercantilism to form, ie uniformly narrow.

Constantine enhanced the basis for propaganda of a hierarchical society by accepting Christianity, and by writing the Bible, which was used as the basis for propaganda in a delusional, hierarchical society.

Thus, the banks, the military, public works (ie brick-laying by the Roman legions), and religion became the nucleus from which European culture developed, though the military and religion were the most apparent leaders. The leadership in regard to social activity later became banking, and the military depended on banking, whereas the military activity, science, and religion were integrated into a propaganda-education system).

The banks became the new-Caesars, though this was usually a hidden fact. Nonetheless, by 1400 the Medici banking-family was selecting the pope, and this was also about the time in which the eastern part of the Roman-Empire finally fell (the European crusades, to the middle-east, had been organized through Constantinople).

However, the "new US" (1780) did not need either banking, or the military.

Yet, the US proceeded to create a military so as to attack and steal from the Native-American peoples, and much of (the large institutional) development in the US was controlled by banking.

The original American alternative to the Roman model for civilization

However, in the new context for society, high-value is not defined by the ruling class, but rather high-value is defined by individually developed knowledge (where everyone can engage in equal free-inquiry, so as to not be violently brow-beat by the left and violently coerced by the right) in a manner so as to relate this set of individual knowledge to the equal "creative capacities" of each individual.

Liberation from materialism can be realized by using science and math (and liberation from delusion and autistic narrowness)

(note: these (following ideas) are ideas which will not be published in the professional science and math journals, where such journals are really about identifying the science and math interests of the ruling class)

This is given a much wider context for creativity, which can now be clearly identified, because there is now (2013) a math model of existence which encompasses the material world as a proper subset, but which is not based on the idea of materialism (ie it transcends materialism).

That is, there is now expressed, new math ideas, which are:

(1) many-dimensional, and
(2) based on stable math patterns, and
(3) defined in a context where measuring is reliable,

so it provides a quantitative platform for measurable descriptive knowledge so that creativity is: "not defined only in the context of the material-world," but rather creativity is now defined in a broader concept of existence, whose description is reliably measurable, and the description is based on stable math patterns, which are used to, finally be able to describe the observed stable properties of both material and living systems.

Note: It provides a math model of the many-(but-few)-body systems so that these models are stable.

Social structure needed in order for new knowledge to be considered within society

The (so called) new context for society, based on equality, is really "the original US idea" of the purpose of the US Revolutionary War, where it was claimed that society should "base law" on equality, and not on property rights, (as declared by the Declaration of Independence), and so that, according to "the preamble of the US Constitution," the government is to serve the individually-equal people, and promote the common welfare amongst an equal public . . . , where each individual is an equal creator, but where value is defined by both knowledge and creativity (broadly defined), where knowledge and creativity are related to both freedom of expression and freedom of belief, as defined in the Bill of Rights, ie knowledge is not dogmatically narrow, and creativity is not to be defined only in a material context by the ruling few.

If the law is based on equality, in a context of "selfless creativity," (eg the magic tortoise lives on air and freely provides creative gifts to all existence and to all living-beings, I Ching), then the law would be opposed to selfish actions.

Yet, it is only in such a context of "equality of individuals" that a truly free market can exist, where in such a free-market "advertising" consists of providing the measurable specifications of the creation, and identifying how to use the creation, but the creative context may transcend the material-world.

That is, there is no fixed context through which the value of a creation can be judged.

Yet, it (creativity) should still be thought of as a form of practical creativity.

Artistic creations (or expressions) would still be there, but they would not be hyped for profit, nor confined to elitist-determined narrow disciplines, whose themes are also very narrow (the artist and their expressions are often now (2013) expressions of inequality [defined by a narrow discipline] and personality-cult, they are (now, 2013) more or less autistic expressions placed in an artistic context, eg musical-jingles, movies, etc, used in the propaganda system to re-enforce a form of a narrow autistic viewpoint).

The issue is not about an equal share of material, rather the issue is about an equal (social) share in regard to being a creative and knowledgeable person.

A hierarchical society demands distinctions in value, which are upheld by violence, eg the value of people and the value of knowledge which is narrowly defined, and a value for image.

At (or within) the individual level, value is arbitrary, and often, in an immediate time-frame, only depends on one's attention.

It is only in a collective, which is defined by a hierarchy, ie a fixed (or narrow) idea about "what has value," where value comes into existence as being a part of the propaganda-governing system, where "the value" ends-up being defined for society (through the legal system) as property and as scarce resources, where, in such a context, a hierarchy ends-up being defined by stealing and the violence needed to enforce the theft, so as to also define the few at the top of the hierarchy (so law was based on property rights and the implicit enforcement structure which must be based on violence which is needed to maintain a social hierarchy based on both material [and knowledge] ownership).

Furthermore, it is this same hierarchy which also defines (and identifies) the knowledge, which is applied to "how" both the property and the scarce "resources are used," so that this, so called, useful knowledge (which uses resources in certain narrow ways) is identified as being superior knowledge.

That is, high-minded ideas (existing within a narrow context) depend for their high-value on the violence which upholds the social hierarchy which these ideas serve.

In a social hierarchy, the ideas which are allowed to be expressed are those which are consistent with the social hierarchy, or consistent with the interests of the ruling-class, but also consistent with the motivation for the violence which is associated to the society defining and maintaining a (the) social hierarchy all based on the ownership and use of resources.

For example, a belief in the importance of a social hierarchy is an allowed expression.

That is, both racism and the idea of (or belief concerning) the intellectual superiority of one idea over another different idea, are the results of an expression of the violence, which is deemed necessary to maintain the social hierarchy, and how "socially acceptable" violence aligns itself with (or is hired by) the social hierarchy, ie or the needed violence which defines and maintains a social hierarchy.

This goes back to the way in which Rome organized its society based on the violence, defined by the privately owned armies associated with the beginning Roman aristocracy (ie 500-400 BC), and their subsequent hierarchical society defined by ownership and control of resources, which were used in very narrowly defined contexts, ie to serve the interests of the Roman aristocracy.

Thus, the types of language (or propaganda) used "in a society which is hierarchical," are associated with the illusion of intellectual high-value, which is knowledge used in specific narrow ways which define a social hierarchy.

Thus, if one is consistent with what is considered to the intellectually superior viewpoint of the times, then one can write from a viewpoint of possessing authority. Note: As a wage-slave, one must pay careful attention to these types of issues concerning knowledge in order to have a successful academic career.

If one is an authority (ie one's beliefs are consistent with the current authoritative viewpoint) then one's sentences can be complicated, with many pronouns, whose reference is unclear, and one can reference the knowledge (which the authoritative presenter possesses) as having an absolute high-value, and an embedded condescending method of "testing the reader," . . . . , as to whether the reader has yet to possess the authoritative language . . . , is used.

This is the structure of how the education system is managed, so that an "absolute truth" (ie the narrow ranges of knowledge used by the ruling class and applied to their resources) is indoctrinated into the students, so as to be consistent with the interests of the ruling-class.

These are administrative decisions.

European culture is totalitarian, and thus dogmatic, and based on violence.

The US was formed because distance allowed independence for the colonies, which, in turn, allowed for a new upper-class (within the US) to form.

However, the US had some fundamental expressions about equality and freedom in its founding documents, and in its history.

For example, breaking from the tyranny of fixed dogmas, eg R Williams and religious freedom, and interpreting the Bible in new ways, eg the evangelicals, and the most important large society in the US colonies was the egalitarian Quaker colony, and their high-energy society, which dealt fairly and equally with the native people, . . . , but unfortunately, by being European they (the Quaker society) got swallowed-up by the totalitarian culture, about which they were surrounded, by the other colonies.

For example the totalitarian collective puritans and the more Roman-like aristocracy of the big European planters in the south.

The Hamilton-types (the Federalists) stood for the totalitarian social structure ruled by the few.

On the other hand the main aristocrat, who was an egalitarian . . . , and who was allowed in the discussion, mostly since it was the theme of the revolution . . . , was the egalitarian T Jefferson.

But Jefferson's life was engulfed in aristocratic totalitarianism of the south, and he did not see that he could not personally sustain his expressed ideals for equality.

Note: This does not mean that his vision of an equal society was wrong, rather it is the much better viewpoint about how society should be organized so as to support the development of knowledge and creativity in an equal society, the true goal of religions, ie the goal of equality and creativity within society for each individual.

So in the US society, there is the "dogmatic authoritarian left" who represent an arbitrary high-intellectual-value (the meritorious academics), which is also related to "good works," and all the, so called, science which is both limited in its vision (with its narrow focus on military interests), and so that what has been chosen as its logical-basis supports the idea of the natural-ness of inequality, but whose intellectual platform of high-value is based on indefinable baloney (see below, why the intellectual-left should be so ashamed).

Nonetheless, it is the left (those who seek good-works) who define the dogma of high-social-value (or high-cultural-value) for society, which is used as the main justification for violence.

Yet, it is the right who self-righteously express the need for violence based on a proclaimed set of personal values which the elite few owners of society personify (the virtuous rich).

Just as the Romans followed blueprints so as to build "with bricks:" water systems, and roads, and huge public buildings "from which totalitarian rule was expressed," so today (2013), engineers follow the blueprints of electrical circuits to build communication systems (used to spy on the enemy, ie the public), and they follow the blueprints of particle-collision probabilities to build bombs (so as to express the need for extreme violence to be used by a totalitarian society); but (in a hierarchical society) one needs the activity of building within society, and what is built must "further the totalitarian cause."

It is this aspect (of building) within the Roman propaganda system, along with defining the Bible, which were the two main properties of the Roman propaganda, which was the heritage of the western culture, based on a propaganda-education system, and which has resulted in a collective hierarchical society within the (European) west.

However, there is the chance to express ideas with language in a clear manner, but without the cloak of authority.

According to the propaganda-education system, it is only the obvious failures of the experts, which requires that new ideas be expressed. [In reality, new ideas and new contexts for creativity should always be expressed, by everyone.]

Of course such failures (which leave the propaganda-education system open to criticism) are marginalized and pushed aside, and other false claims . . . :

eg controlling fusion-energy and
developing a quantum computer etc, and
complicated irrelevant issues:
black-holes,
big bangs, and
finding more cross-sections of elementary particles, etc,

. . . , fill the "written space" (in authoritative science and math journals) where the fundamental failures of physical science and mathematics should be expressed.

However, the monopolistic power of the social hierarchy is based on the structure of knowledge which already exists. This means that the authority of the expert science and math people cannot be challenged.

It is about addressing the failing of knowledge for which the authoritarian, and supposedly highly meritorious, intellectual-left are being used (so as to hide these intellectual failures).

The elitist intellectual-left is being used to hide these intellectual failures, so that the intellectual-left must be seen (in the propaganda system) as being very important, so as to have a very destructive affect on education, it is for these actions as to why the intellectual-left should be so ashamed.

The obvious failure is that:

There is a wide range of very stable, spectral-orbital, material systems, ie systems which possess a stability which suggests that they are linear and solvable systems, yet these very many stable material-systems go without valid (or without any) measurable descriptions (neither accurate nor useful), where the descriptions are, supposed to be, based on the, so called, laws of physics. These are all very stable physical systems composed of many-(but-few)-components. [However, these systems are solvable (have now been solved) by using new math techniques, ie see scribd.com put m concoyle in the web-site's search-bar.]

In philosophy, or in religion

There is the notion of descriptive knowledge
vs.
the perceptive knowledge (for a person who "perceives the world as it really is.")

This can be represented as the:

"western scientist"
vs.
"the spirit defined as something which exists beyond the material world, but upon which science is based"

Thus, the precise verifiable description is to be used as the basis (or is viewed as the source) for scientific authority, but unfortunately the descriptive epicycle system of Ptolemy was also measurably verified, so a measurably verifiable descriptive language, which is based on a lot-of complication . . . , as are: quantum physics, particle-physics and string-theory, . . . , could (all) easily be, yet, more examples of epicycle language-constructs, so as to be another form of an incorrect context for the descriptive language which is being verified.

That is, this same, overly, authoritative language is not providing either a valid, precise-enough descriptive structure for the observed patterns of physical systems, or descriptive context, in regard to either a wide range of descriptive successes, or in regard to the practically-creative usefulness,

It is failures in this context of both accuracy and usefulness which identify the failings, over a wide range, which can be associated to this overly, authoritative language, in regard to its: categorizing and organizing of the "information constructs," about which the language is supposed to deal.

The weakest link in this communication-propaganda system of the totalitarian society is related to the math and sciences.

It is claimed that the circuit building, and bomb building are based on the highest level of scientific knowledge.

But instead (or rather) it is based on 19th century science.

But (nonetheless) it is the knowledge which serves the investment-oil-military interests quite well, as brick-laying "served Rome well."

The imperative of the propaganda-education system is to not interfere with monopolistic profits, ie do not interfere with the emperor's interests, and 19th century serves these monopolistic interests quite well.

Oil is still king, even though intellectual rationality is in opposition to the reign of oil (in regard to oil destroying the earth's eco-systems), but the opposition is wallowing in its own ineptitude, ie its opposition are the authorities which oil supports, ie the intellectual beliefs of the opposing authorities, in fact, support oil's interests.

That is, the main problem facing science and math is concerning the stability of fundamental physical systems composed of only a few components. The, so called, "laws of physics" are neither accurate, in regard to describing a wide-enough range of general systems which are stable, and these. So called, "laws of physics" are also not related to any range of practical creative developments.

This is (also) why the intellectual-left should be so ashamed.

That is, the failure of the materialism viewpoint is demonstrated by the fact that there are many very stable fundamental systems with observed stable spectral-orbital properties which exist at all size-scales so as to be composed of many-(but-few)-bodies, and the properties of these systems cannot be described within the assumption of materialism.

These failures were well known by the 1950's (see Dirac), but at that time Teller took-over the military-physics connection, which was (is) associated with government funding of university research, so as to push the funding of physics toward the bomb-engineering properties of particle cross-sections, ie probabilities of particle-collisions, which are associated to rates of reactions.

The assumption of materialism . . . , either geometric and locally measured (classical physics), or random and reduced to small components, but which (though random and with local particle-spectral-event properties, they) also possess global (or non-local) properties, whose measurable properties are, supposed, to be found by means of the algebraic properties of sets of operators . . . . , (so that such a viewpoint) leads to both non-linearity and indefinable randomness, where the random descriptive language is based on both unstable events and an assumption of containment and solution, but, essentially, no solutions exist, ie incalculable randomness (hence, the use of the expression, indefinable randomness).

That is, the many-(but-few)-body problem cannot be solved in this descriptive context, yet the stability of these fundamental systems implies linearity and solvability.

On the other hand, Particle-physics leads to complicated absurdity, and string-theory follows this complicated absurdity.

Since the stable spectral properties of nuclei cannot be described by these methods, thus, it is clear that they are useless methods.

What can be calculated by these methods of particle-physics, etc, are the cross-sections of unstable elementary particles, a property which might (or might not, since these particles are unstable) be useful in bomb engineering, ie it is a knowledge which is being followed based on monopolistic military interests.

That is science and math of the high-valued culture, which "the left champions," is based on non-linearity and indefinable randomness, where non-linearity is a quantitatively inconsistent descriptive context which leads to chaos, which can be related (but in an uncertain context) to feedback systems, and the uselessness of indefinable randomness was mentioned above. And all of these highly-valued ideas are about concerns of the military businesses. [If the left supports these intellectual constructs then the left cannot be opposed to war and still remain intellectually consistent, their intellectual responsibility (if they are opposed to war) is to provide an alternative set of scientific and mathematical ideas.]

Why all this indefinable baloney is attributed with such high-intellectual-value, yet to be neither accurate nor possessing any practical usefulness, especially, in regard to solving the many-(but-few)-body problem, is a great achievement of the propaganda system and a result of the inertia (or ineptitude) of the intellectual left.

That is, the dossiers of the people in high-cultural positions are used, so that those in the "top-spots" in academia are domineering psychopaths. They will support their own dominant positions and not question their intellectual assumptions.

We are a society built upon layers of meddling (by the spies who work for the owners of society), where the main issue in regard to the apparent stability associated to this meddling, is attributable to the control that the propaganda system has over the population, so as to rigidly control the language and the thoughts of the population.

Thus, the irrelevant concerns, espoused by the intellectual authorities, are not seen as being irrelevant because the propaganda system is saying that these types of (irrelevant) issues are important.

Thus, the propaganda system is seen as the sole voice of an absolute and completely objective truth, which is a truth which is superior to any other such expressions of truth, where other expressions of truth are coming from those who have no merit.

This is intellectual totalitarianism with exactly the same irrational structure as is the belief in racism.

But the many-(but-few)-body problem is solvable using Thurston-Perelman's geometrization, and basing measurable description on stable geometric structures in a many-dimensional context for existence, where both the material components and the containing metric-spaces (as well as the structure-process of material interactions) are given (or related to) stable shapes in "different dimensional-level subspaces" which are contained in an 11-dimensional (hyperbolic) metric-space, which has been partitioned by these (same type of) stable shapes, so as to identify a stable finite spectral set, for the full containment set, and so that existence, either material or metric-space, depends on these existing stable structures being in resonance with the finite spectral set.

This construct is capable of encompassing both inert material properties and the properties of living systems.

The new context provides the correct knowledge needed in order to understand material existence, and existence in general, and life's relation to existence, and life's purpose in regard to creating the properties of existence.

That is, it identifies the form of both existence and life and the properties and the context through which creative expansion can occur directed by life and based on correct knowledge.

The stable spectral-orbital properties of a many-(but-few)-body system are defined as material contained on these stable shapes, which are allowed by the resonance-containment construct, which exist in the correct dimensional level (and in the correct subspace).

This context for existence is the type of context which people who believe that the feelings of living are "of such a nature" that one's life is connected to a spirit, might find very interesting.

However (or unfortunately), just-like the elite intellectual left these people will feel too important (again due to the propaganda-education system) to believe a math description which has not been given the approval of the propaganda-education system, ie intellectual-left, eg accepting the dogmatic authority of the propaganda system (which expresses the beliefs and interests of the banking-oil-military ruling class) is similar to accepting the puritan's desire to exterminate the native peoples so as to get their land because of the belief on the part of the puritans that they possess an intellectual superiority over the natives, ie they will exterminate ideas due to following the dogmas of the superior upper-classes.

However, our immediate living connection to higher-dimensions seems to require that we experience ourselves, delve into our life's connection to the spirit, as being an energy-structure which is spatially very-large.

That is, this set of measurable ideas which brings the context of human life into the context of the origins of living systems, and this simultaneously brings the context of the origins of human life into the mythology (or religious histories) of being an energy-structure which is spatially very-large so as to still be associated to the purpose and context of human-life.

But this true mythological existence is not the true context of creativity. In fact, turning, what is materially an unreasonable context of being very large, into the focus of human (or life's) existence so as to be motivated by staying in such a particular dimensional level also leads to misconceptions about life's purpose. Thus, the mythology of the selfish actions of the "gods," in this higher-dimensional context of (possibly) being very large, is a mis-representation of life's relation to creativity and correct knowledge.

However, it is in the 4-dimensional context of being very large where the properties of material can be examined in the 4-dimensional context, but the 3-space structure of the solar-system is only discontinuously related to the 4-space properties, ie in 3-space there is a natural barrier to experiencing a 4-space material structure, and one aspect of this dimensional barrier is an open-closed topology in 3-space.

Material-man confined to the earth vs. a 4-dimensional living-man confined to the galaxy or confined to a 6-dimensional universe, or more-likely, human-life is (even) more dimensionally diverse than this.

55.

How the media and education are organized

Both intellectual efforts and politics (determining what is done, and how it is done) are controlled by banking and investment interests. Thus, it will be "physics" as the banking and military businesses expect "physics" to be expressed, ie to serve their interests. Thus, very fixed and authoritarian ideas about science are expressed in propaganda publications.

Organization of media

The media, concerning politics (ie politics is, supposedly, about determining what the society will do), is set-up in a context of a religious morality-play. This might well have been the way in which the Roman emperor Constantine developed public communication around the Christian church, where this same church eventually took-over running the Western-Empire, but the banks, eventually won-out, in regard to the top-spot in the western social hierarchy.

The lefts is:

1. elitist (the intellectual elite) they have "won" their social positions within dogmatically based educational contests, so they believe both the intellectual dogmas and their rightful superior social positions,
2. collectivist, ie the belief that the public needs to be guided by the intellectually superior people (this is mistakenly taken to be the position of science, but the main part of science is about opposing dogmas which have lost their practical useful relation to existence),
3. The left are the traditional-dogmatic-authoritarians, they believed in the dogmas upon which educational contests are based, and they have won their high-spots in the social

hierarchy, (ie they side with the meritocracy, and the idea of equal opportunity in [a rigged and manipulated social context, associated with] an intellectual-contest which is defined within the education institutions, but it is a contest, which is set-up to represent the intellectual interests of the banks and its relation to investment and business), and

4.   those on the left seem to believe completely in the cult-of-personality, eg they are prone to revere Einstein's, so called, great intellect, etc.

5.   The left accepts the symbolism of the idea that "we are all sinners," and in the context of being selfish those on the left support good-works, where the left believes in the principle of helping others, but their elitist beliefs make their sentiment for helping others a hypocrisy, (the bankrupt claim of "the greatest good for the most people," is often errantly expressed, where the errors stem from a delusional model of society, the exploitative extraction, and the subsequent product development, and distribution systems compose a destructive "wealth accumulation" process which defines social class, "the greatest good" is defined by advertising and an illusionary world of "product demand" made in the context of a highly controlled market)

(note: by contrast the right accepts, and follows, the high-virtue of the owners-of-society, and thus "the right is good" since it worships "that which has high-value"),

6.   Thus, one also sees that the elitist left seems to believe that, "sin belongs to the social condition of being poor," yet the left believes that the superior people need to be filtered from the inferior people, where intellectual superiority is measured based on the narrow constriction of intellect associated to bank investment schemes, intelligence is a measure of a person's rush, by a person, toward adhering to the culture which most favors banking interests, and

7.   Thus, the left implicitly acknowledges (as does the right) that the rich and their high-works are pure (and to these pure-works of high-value, which the, intellectual left greatly contributes, and the left competes in this narrow, non-intellectual, context with great energy).

8.   The left are the, so called, "intellectual superiors of society," yet they are servile to the rich, since their beliefs fit into the intellectual construct which the rich have identified, and it is through the high-value defined by the rich within which the distinctions between the superior and inferior are seen.

The ideas of the left are always those ideas which both come from . . . , and serve . . . . , the owners-of-society. Though the high-rungs of academia are filled with plenty of those on the right, those on the left who are given voice come from the academic institutions, while the right owns the media.

In the contexts which define both political domains, the right and the left, both sides believe that they can "say just-about anything, as long as they either are the authority, or they represent authority."

The right is:

1. self-righteously, and piously moral, and violently abusive in this role (whereas the left are intellectually violent in their judgments of others, in regard to high-social-value)
2. "The right" stands for a vague notion of high-social-value of the rich, where this worship of the rich apparently results from "the right" acknowledging the piety of the rich, which is, apparently, based on the Puritan-viewpoint "that the rich are favored by God, and they (the rich) are virtuous."
3. That is, the right are a bunch of flattering worshipers of the ruling-elites, ie brown-nosers, and
4. In their servile-state as cheer-leaders for the rich, the right represents the interests of the owners-of-society.
5. since the media allows the righteous-right (the helpers of the rich) to "set the context" of most media discussions, where this is apparently, due to the, assumed, high-morality which they (the right) represent, and
6. what they (the right) say, instigates an, essentially, religious dialog between the
   Right (the virtuous (few)) vs. elitists (the few of real merit, who serve
   the virtuous rich, yet they still seek to do good-works))

. . . , and because "what is proclaimed by the media" are, usually, the expressions given by the right, thus, the left must react to what the media says.

7. The right is all-for the violence, which is framed as the violence needed to maintain and protect the virtuous, where the necessary violence is often framed in regard to protecting the property-rights of the virtuous-rich (ie the virtuous, who, must sometimes exterminate some of the sinners of the world, ie those whose selfish interests are interfering with the interests of the world's most virtuous people, ie the rich few).
8. The right can be summed-up as supporting the actions which say that:

"Life (of the rich) is sacred, but to protect the virtuous (sometimes) there is a need by the rich to exterminate the poor and servile classes, and the intellectual structure of the culture must serve the rich's interests, ie another form of extermination."

9. The obvious dilemma of the right is: that they are collectivists, they actively encourage a "society based on collectivism" whose collective actions are to support the owners of society, ie "the right" is no different from the communists (who they claim to revile so much) but their collectivism is a result of their belief in property-rights, the society of "the right" is a collective society which supports property-rights and supports the extreme violence which is needed to maintain property rights,

10. The right is the most hypocritical, they act as supporting individualism (and their bogus claim that they believe in small government), but the only individuals who they support [and need for their careers in the media, ie in politics or the media] are the very rich of the ruling-class, and a very-big government is needed to violently protect the property-rights of the very rich few within society and around the world,
(ie the right supports the central-planning committee (ie the rich few), who plan things for society based on the selfish interests of the very rich few),

"The virtuous," vs. "The sinner"

In regard to the dialog, about sex, within society is designed to increase "sex actions" within the population, so that the population will be increased, ie stir-up a lot of consideration, and thus a lot of interest, about sex.

Thus, there is the moralistic division:

"the virtuous whom control their sexuality" vs. "the sinners who want birth control"

In regard to equality, the (real) religious stand concerning sexual actions of people would be,

"try to have sex (only) when a person (a pair of people) want to have children."

However,

If one wants: individuality, truly free markets, small government, and where violence is decreased, and in a highly knowledgeable and practically creative society, then law needs to be based on equality, and the context of equality is in regard to acquiring knowledge and using knowledge for a wide range of practical creativity.

That is, Law needs to be based on equality, not based on property-rights and minority-rule,

That is, the great American experiment is about breaking from the traditions of western-society, it is opposed to collectives built around banking interests, and instead supports the free, creative individual, who is, supposed to be, allowed to express their (knowledge-creative) interests through equal free-inquiry, and knowledge's subsequent relation to practical creativity.

The context of social activity is knowledge and practical creativity,

This is, the collective endeavor (of developing knowledge and its relation to practical creativity) to which equal individual-acts are to contribute.

But this requires a government which serves the people, and does not serve an elite-few who are very-rich, and this government (of the people) is (should be) trying its best to provide for the common welfare, so that a state of equality in society can be achieved, in regard to developing new knowledge and relating knowledge to a wide range of practical creativity, and thus forming truly free-markets.

(Why are we now (2013) paying private companies for public services? Garbage pick-up, electricity, connection to a communication system, health service, education-systems ie this is simply a system of unfair taxation, so that in this model of "taxes given to monopolies" one finds that the monopolies only exploit and dominate the market for their selfish interest.)

Only when people are considered to be "equal creators" can there be a (truly) free-market for trade.

Advertising is about providing the context for an instrument's use, and its specs (and that is all, period).

Language always needs to be re-organized (this is about education); where the re-organization is about the: assumptions, contexts, interpretations, containment sets, and if a measurable description then the measuring context needs to be measurably reliable, and the patterns used need to be stable-patterns, where the point of this, is that, language actually needs to both possess and express content, where one is trying to quickly get practically useful information about the observed properties of (physical) systems.

It is in this context . . . , where there exists equal free-inquiry, as expressed by Socrates, and correct knowledge's relation to "practical" creativity, but where the creative context does not "have to" be based on the material world.

. . . , that we are all equal creators, so that this is the correct way in which to interpret the US creed of "basing law on equality" as was stated in the Declaration of Independence (now we need to live, so as to realize our creed) (and not basing the US society on property rights [and minority rule], and subsequently on the narrow monopolistic interests of the oil-banking-military businesses)

. . . , so that the government, which serves the people, must address the common welfare of a country, so as to serve a community (society) of equal people, who are motivated to gain knowledge and to be (practically) creative, and not motivated to be selfish, nor dominant, nor violent, where these behavior traits could be continually thwarted by the government, . . . , just as the government now (2013) continually thwarts the interests (of the general welfare) of the people, in order to serve the owners of society, ie a "state-and-owner-of-society" partnership.

Consider free-inquiry, and its relation to the media

The media, ie the propaganda-education system, creates a demand for particular intellectual viewpoints, by means of its advertising for particular narrow authoritative contexts and interpretations, in regard to the relation that the observed patterns have to a precisely determined language, ie language based on measurable properties. This is not about "objective constructs," rather it is about selecting viewpoints which are consistent with the interests which control the propaganda-education system.

Note: (Commentary about science and propaganda)
Reply to the article in the 8-13 monthly science journal, "The Scientific American," titled "What is real?" wherein the discussion is in regard to the reality of quantum physics and its particle-physics extensions, where the "extensions" are made by introducing an "internal particle-state interpretation and construct" which is placed onto a quantum system's wave-function, in a similar way in which the property of "spin" was added to the wave-function, where these internal particle-states are being related to the data on particles obtained from particle-accelerators. However, the emphasis placed on particle-accelerator data is to focus on the cross-sections (or probabilities of collision) of elementary-particles, where the point of focusing on cross-sections concerning the probabilities of elementary-particle collisions is related to the rates of reactions in nuclear weapons engineering.

That is, the military-industry and the-banks are controlling: language, knowledge, and thought within the society for their narrow interests.

Then in the process of getting the public to agree to these particular constructs and interpretations which are to be embedded in the technical language of the education system, both the wage-slave social status of the public, as well as the close relation that human thinking has to both (1) ranking the value of things and ideas, and "in this context of arbitrary high-value" using (2) autism and its subsequent relation to obsessions, where the property of a person's autism is found by means of testing in the education system, where these obsessions can be defined by means of dogmatically authoritative intellectual viewpoints which, in turn, are used, and manipulated, and their high-value identified. Thus, causing in an autistic person a rush to adhere to these promoted ideas (promoted by the propaganda system of the virtuous few), so as to define a difficult to achieve high-value, which is further-complicated because of both (1) its relation to obsessive people and (2) the poor fit which exists between its language structures and the observed patterns (which are to be described), so that narrow business interests are satisfied within society.

This results in such highly controlled set of business "markets," which are so narrowly defined that valid development of knowledge is destroyed, and the type of creativity needed for the expansion of the creative context needed for the expansion of true markets does not exist.

It is speed of both agreement and use (or assimilation) of a language which arbitrarily define (or are called) intelligence, ie intelligence tests, ie tests for obsessive behaviors related to language, where intelligence is a property which has a high-rank in society, as identified by the propaganda system, and its high-valued properties which are circularly identified by journalists and their reliance on the opinions of indoctrinated experts.

The ideas which are given high-ranks are the high-valued aspects of culture, which are the high-values defined by the banking and investment interests.

That is the value identified by the, so called, highly virtuous rich, is the value of society existing in a regimented order (note: these were also the main results of the actions of Rome and its armies), but wherein a society the main behavior is sexual behavior, and this behavior plays a central part of a propagandistic morality story, and a story of marriage where a female selects a man of high-value.

Furthermore, this regimented context for society is also about opposing the high-value of being human, ie humans are knowledgeable and creative, nor is it a social system which best fits into the life-sustaining capacities of the earth.

The story of quantum physics and particle-physics

Quantum physics and particle-physics are both about taking all the attributes of materialism, ie (1) containment-space and (2) material and (3) a way to identify physical properties, then (4) reducing material to be modeled as unstable random particle-events, and then (5) in a probability based descriptive context, to try to extend these traditional physical ideas beyond the ideas of both material containment and material existence, so as to describe a "quantum field theory" associated to the attributes of unstable particles, whose properties are those properties seen and determined in particle-accelerators (in the context of the particle-paths in bubble-chambers, which occur after particle-collisions).

This is an intellectual process which is simply about moving from a classical particle-model of material, ie material which has properties in the space which the material properties define, into an incomprehensible and undefined language whose only purpose is to measure the cross-sections of the collisions of elementary-particles (ie the probabilities of particle-collisions, which are associated to rates of nuclear reactions).

The descriptive process associated to (5) is about attaching an internal particle-state structure to a quantum system's (solution) wave-function, but this construct cannot identify any new (or calculable) measurable properties, which are either observed or predicated to be observed, in any general context of its application, with the exception of the two or three of the most elementary quantum systems, eg a free-electron and an electron in an H-atom, and (?), and in these cases "infinity is subtracted from infinity" (ie an indefinable math operation) to get a number, which is supposedly, precise to 16-significant digits.

Thus, with this great precision, one would expect to be able to easily apply these precise math techniques to the large set of general quantum systems (but the answer is, NO). The point of a valid physical law is that it can be applied to a wide range of very general (quantum) systems, but the laws of quantum physics and particle-physics cannot be applied to a wide range of general quantum systems.

The principle of quantum physics, that . . . "a random particle-spectral event structure forms around a quantum system's energy-geometry properties, so as to be based on the observed stable discrete spectral properties of the quantum system" . . . , is an idea which cannot be formulated.

The reason is that such a random model, whose spectra is not calculable, moves outside the capability of a math descriptive context of reliable measuring, which, in turn, is based on both set-containment of the system and a valid (stable) method of identifying a (the) system's properties, ie the needed geometric properties associated to a quantum system's energy potential term, where the geometric potential term in the energy operator for a quantum system is incompatible with the random basis for the description.

That is, the descriptive language is unrelated to calculable or useful descriptive properties . . . , it at-best fits an almost empty set of data in regard to the properties of a wide array of physical systems . . . . , it has no relation to the set of observed system properties of any significant applicable (descriptive) range, and has no relation to practical usefulness.

The authority of such a practically useless descriptive language should be challenged in all of its aspects, and at its most fundamental levels.

It is a (failed) descriptive language, which is only relatable to nuclear weapons engineering.

Despite its narrowness, and essentially, useless measurable descriptive value, "it, along with string-theory" have been defining the focus of the attention of professional physic community for nearly 100 years. This is all about the power of propaganda and its relation to the education systems, where this relation between propaganda and the education system has been strengthen by the militarization of the US society, ie the managerial class, since WW II.

There are ideas in the monthly science journal, "The Scientific American," in the article "What is real?"

which are the pronouncements of an arrogant intellectually authoritative community, whose authority is given to them by social forces, not by the validity of the ideas which they represent, whose practical achievements should have been a basis for such an intellectual viewpoint to have been discarded 60 years ago (by 1955), but the military-industry has maintained and protected such a useless viewpoint, because it is all about the data needed for the physical models of nuclear weapons engineering.

The apparent message of the article, in "The Scientific American," is that, further scientific development needs to be based on developing the traditional authority of university science,

even though particle-physics depends on attributing classical physical properties to a set of apparent-particles (they are apparent-particles since they are unstable and thus their "existence" needs to be highly qualified), which are observed after (or resulting from) creating particle-collisions of an assumed set of elementary-particles (ie the particle-paths in bubble-chambers are observed and measured) . . . ,

Where the data are the observed, and interpreted in relation to the bubble-induced-paths of an assumed set of elementary-particles, which are seen after their collisions in particle-accelerators (or particle-colliders), (and) so that cross-sections can be determined for these particle-collisions, where this can be done by counting the different particle-events in the cloud-chamber (even though these elementary-particles are unstable, and thus cannot be counted in a reliable context of counting), [nothing about particle-physics has ever made any sense. In this "The Scientific American," article particle-physics is not being (and never has been) described in a context of a coherent, well-defined, and precise language.]

. . . . , nonetheless, (and what follows in "The Scientific American," article is the further protection of these useless ideas, where this protection is being done by the media) these useless ideas of particle-physics can be re-cycled and developed in their own failed context, but it is claimed to be based on a new viewpoint, but it is a viewpoint for "rational" absurdity, namely, the article claims that:

1. "particles, whose cross-sections are determined, are not really particles" and
2. "their paths of bubbles are not really paths"
   and an assumed underlying structure of
3. "the math constructs which are needed to identify (or define) the containment set of the observed physical properties of the particle-collisions (and their paths) cannot be a well defined containment set (because it does not contain valid sets of physical properties),"

   That is, we will deny the properties through which particle-physics is related to determining particle-collision cross-sections, but still keep the useless context of particle-physics, ie no new models through which new interpretations can emerge.

   [That is, we, the intelligencia, (and our association with the propaganda-education system) transcend mathematics, but we do not define the context which does transcend mathematics, you simply have to trust us, and that our intellects are superior to all of the publics, so that whatever we say is true, a truth which the public cannot know, because of their inferior intellects. (Does this sound familiar? It should. This has been the claim of the national security state for the past 65 years. That is, the irrational crap which they get away-with in science, is also the same type of irrational viewpoint in which the national political dialog has been copied, and framed)] and

4. "the numbers of particles in a quantum system can change based on moving in an accelerated coordinate frame" even though a very precise set of spectral properties associated to atoms are fairly precisely determined by knowing both an atom's atomic number and that same atom's atomic weight (mostly the stable spectral properties of an atomic system are based on atomic number, and these spectral properties stay stable in general motions, with maybe slight variations in spectra, where these measured variations are dependent on the measuring context).

{The claim of #4, is a claim, which sounds like the usual type of a measurably unverifiable theoretical claim, which is made in modern physics all the time (or which have measured properties which are interpreted in the context of the accepted theory, eg this is the same as the Ptolemaic system being verified by data interpreted within the theory of the Ptolemaic system), but which possesses no (valid) content, especially, since the description of a vacuum is that of unstable turmoil of elementary-particles, so the claim "seems to be" that an accelerated frame provides a stable attribute (a measurable change) of elementary-particle properties associated to the (accelerated) vacuum. Nonetheless, general relativity is based on the belief in an inability to detect any changes of properties of a uniformly accelerated frame, within the frame, ie this is a claim by "The Scientific American," article that general relativity is wrong, but the basis for this claim is an invalid basis, since an atomic number is central to an atomic quantum system's stable spectral properties, and atoms do not fall-apart in accelerated frames.}

That is, the point of this article in "The Scientific American," is that quantum physics and particle-physics are not placed into a valid or coherently defined descriptive, and measurable, language, and the authoritative personalities (in charge of this dogmatically authoritative knowledge) can say whatever they like about such issues as,

"what the language (of either quantum physics or particle-physics) means," and
"how the language and data is to be interpreted," for whatever context is being considered, so as to
"use different languages and different interpretations in the different contexts,"
ie the precise language is not precise, "its meaning changes as the authority wants its meaning to change."

Subsequently, the public must accept these "new authoritarian truths," so as to remain "in awe" of the superior intellects of the meritorious authorities, who provide these changes in meaning and interpretations for the different contexts of measuring (and these authoritative proclamations, subsequently, becomes the curriculum of university physics departments).

This is similar to how in the politics-media context "the right" is allowed to speak in arbitrary ways on the media, and this arbitrary way of speaking is "that to which the public must relate (or adhere)" if anyone of the public wants to be within the functioning communication constructs of society, as defined by those who control the (valid) communication channels (ie the media) within the US society. Whereas the authority of the expert particle-physics people is the same authority as the authoritarian left. There are no places for other ideas, and this narrow context of communication is defined through the propaganda-education system.

That is, the words of particle-physics or quantum physics mean "what-ever the authorities say that they mean" and in whatever context they are using (or measuring) the instrument related (dependent) language of particle-physics.

Where, in particle-physics, the context is point-particle-collisions, in regard to the observed particle-paths, as observed in bubble-chambers, so as to use this set of information to try to calculate the cross-sections of the elementary-particles (to determine rates of particle-collisions during a chaotic transition process, which exists between two relatively stable states of material), yet the elementary-particles are either unstable (or undetectable, as in the case of quarks) and thus the probabilities cannot be accurately (or reliably) determined (ie counting is not defined in a valid manner), yet (for the military industry) any estimate of such a set of particle cross-sections can (may, or may not) be related to the energy generated from such particle-collisions between unstable entities, and thus may or may not be relatable to nuclear weapons engineering.

while

In the context of quantum physics the entire-system is modeled as a global wave-function, so that in this context, it is now assumed that random particle-events are associated to a quantum-system's global wave-function, and this wave-function is (now) endowed with a both meaningless and useless attached-set of internal particle-states (except the property of spin), which are attached to the (a) global wave-function, where (for wave-functions) the context is random, ie not well-defined point-particle collisions (where such a collision geometry is not allowed by the uncertainty principle) and it is assumed that any physical properties associated to a global wave-function are related to sets of operators, which act on the wave-function's function-space, but still the context "to which the math structures of quantum physics are related" is about the random particle-spectral-events in space . . . , but in this mathematical context the spectra of relatively stable general quantum systems cannot be determined, so the construct is not valid.

Note: It should be noted that "the claim is being made that" since the particle-corrections (due to a quantum system's hidden particle-states), which are attached to the wave-function, and which affect the wave-function's relation to a quantum system's particle-structure, are global constructs, thus these types of corrections should not be dependent on "the point in space" where they are determined for the quantum system, thus these corrections should be the same for each point in space, and thus either such an affect should be the same everywhere, so "how such a uniform structure of particle-collisions between the vacuum and the system could perturb the system in any way, which is different from an average null affect, is difficult to identify?"

or

It needs to be noted that there is a local relation which exists between particle-types and the particle-structure of a quantum system, which exists at a particular point in space,
this would be because, in the laws of particle-physics, the particles which compose the system are supposed to affect the nature of the particle-collision interactions, eg certain families of hadrons interact with those same families of hadrons etc, but this would indicate that there is a local structure to the particle-collision perturbing structure, ie it is not a global structure.

In either case, in the adjustments of a quantum system's descriptive properties which are due to a quantum system's relation to local particle-collisions between "the particle which compose the system" and "the particles coming from the vacuum" there is not a coherent (or comprehensible) idea about such a quantum system's properties, ie (1) set-containment is not defined (2) the physical properties cannot be defined which determine a general quantum system's spectra (3) thus the construct is indefinably random and (4) thus outside of math descriptive capacities.

What is a valid math context for a non-local relation between separate elementary-particles? (This has been explored mathematically by using the stable shapes of discrete hyperbolic shapes placed into a many-dimensional context for containment (cut-off after 11-dimensions), but such a novel use of language is not allowed within the professionally published communication channels.)

The invalid and meaningless models of quantum physics and particle-physics, and the way in which these descriptive language structures have virtually no relation to any form of practical creativity means that these language should be criticized endlessly as being useless.

Particle-physics and quantum physics are examples of a (fake) math construct based on indefinably random properties, ie properties which cannot fit into a valid math structure.

That is, the math descriptions of a general many-body quantum system cannot even be formulated within the math context of quantum physics, let alone calculated. So, again, the type of descriptive language used in quantum physics is based on the idea of indefinable randomness, and is not mathematically valid, ie the descriptive language has no content.

That is, these types of quantitative descriptions do not identify a set of stable spectral properties for a general quantum systems, and they do not relate to a reliable context for measuring.

Yet, the stable spectral properties of quantum systems, eg atomic spectra stable and definite based on the atomic number, indicate a: linear, metric-invariant, and continuously commutative almost everywhere system, and thus a solvable and controllable system should be associated to a many-body stable quantum system.

Such a property of a stable shape being associated to quantum systems and their stable spectral properties is realized in a new alternative descriptive context.

But this would require new investments for banks, and the monopolistic business structure, which rules society, is functioning very well without new alternative ideas . . . , so this type of valid knowledge needs to be ignored.

Consider alternative ideas

Consider an 11-dimensional hyperbolic metric-space, which is used to represent the containing space of existence, can be thought of as having a partition into various subspaces of various dimensions, and when these subspaces are less than 6-dimensions, then the partition would also be about "cubes" of various sizes partitioning the subspaces of various dimensions

While the energy associated to a "physical existence" may not manifest as a 10-dimensional cubical subspace divisions of the different dimensional subspaces, where some of these shapes would (could) have faces of odd-genus 9-dimensional faces, or can contain 9-dimensional shapes whose genus is odd, so as to be 9-dimensional oscillating, energy-generating, shapes, and thus, these shapes would be creating an energetic (or energy-filled) partition structure of existence, related to "cubical partitions" which is filled with sets of relatively stable energy-filled structures of spatial shapes.

Thus, there is a partition of the 11-dimensional space so that at the level of the 5-dimensional subspaces each subspace can be related to bounded "cubes" so as to be filled with various sized 5-dimensional cubical partitions, which in turn contain various sized 4-dimensional cubical partitions, etc, so that each dimensional level will have a largest bounded-sized cubical partition, and a set of various other smaller-sized cubical partitions, which fit into "containment trees associated to a dimensional sequence of bounded cubical containment."

For a world modeled as the assumed structure of particle-physics, and its associated string-theory construct, in the model of the above paragraph this would be a set of 3-cubes whose three and lower-dimension cubical partitions would be many very, very small cubical partitions, but such a construct would not allow for a 3-dimensional solar-system to have a stable set of planetary orbits associated to itself and this would not allow for the way in which the low dimension partition for a material existence as we experience it.

An 11-dimensional containing space is partitioned into "cubes" of various sizes, and associated to various dimensional levels from 5- to 1-dimensional levels, which, all together, identify a dimensional-tree of subspace containment of a cube-sized containment tree-complex, up to dimension-5.

The 3-shape of earth is a "cube" the size of the solar-system. Thus our 3-subspace can contain 1-, 2-, and 3-dimensional "cubical partitions" (of cube sizes) which are smaller than the solar-system, and which resonant with the spectrum of the 11-dimensional space, where this spectral set (of the 11-dimensional hyperbolic metric-space containment space) is defined by its partition into shapes, wherein for each subspace of each dimensional level there is a largest cubical size, and it is that largest size which defines a member of the partition.

The stable spectral systems in the 3-space of the solar-system are composed of positive- and negative-charged shapes of both compact (ie hadrons) where the compact property allows spatial position to be identified for (or within) a complex system and a combination of a bounded and an unbounded (or semi-bounded) lepton shape of "electron and neutrino shapes" so the neutrino-shape either may be bounded by either its higher-dimension or larger shapes (than the electron) containment metric-space (eg 3-dimension solar-system shape) or it might be unbounded.

However, due to Dehmelt's "isolating an electron," it has been shown that the electron shape can be bounded.

Thus, the relation of an electron to a neutrino is a relation between bounded and unbounded or semi-bounded or semi-unbounded discrete hyperbolic shapes.

The (relatively) unbounded neutrino-shape relates the lower dimensions (5-dimensions and less) to the unbounded higher-dimensional containment properties. This is allowed since the 1- to 5-dimensional discrete hyperbolic shapes can be both bounded and unbounded stable shapes.

How do the bounded stable spectral shapes of 5-dimensions and less relate to higher-dimensional subspace containment structures in the 11-dimensional containment space?

Do the unbounded higher-dimensional shapes have a semi-bounded facial structure, through which a definite finite spectral set can be easily (naturally) carried by unbounded (infinite-extent) shape, or is a finite spectral set based on subspace set-containment within the 6- and higher-dimension subspaces, which are independent of the 11-dimensional space's finite spectral partition-set structure?

But an independent 5- or less-dimensional shape, in a 6- or higher-dimensional subspace, can exist in regard to containment in a higher-dimensional subspace, based on its resonance with the 11-space's finite (bounded) spectra.

Does the set of an unbounded-shape structures (which can also be associated with the lower-dimensions) allow for discontinuous discrete changes to exist (between a) in a dimensional subspace which also changes the spectral containment structure of (one's) experience, ie a change in subspace could also change the spectra due to size restrictions which might exist in regard to any particular subspace, ie due to the cubical partition structure of subspaces?

The neutrino ie unbounded subspace organization, would be affected by changes in subspaces and an associated set of cubical-sizes associated to a subspace's partition structure, ie dependent on the sets of largest-sized cubes in a dimensional and subspace partition.

The different dimensional levels 4- to 10-dimensional, ie seven different dimensional-levels, within which the subspace (and possibly spectral organization) can change, so that clearly changes within a context of 4- and 5-dimensional discrete changes in subspace shapes, can be spectrally organized differently (after the discrete changes), than in regard to how the subspaces associated to our 3-space solar-system subspace structures are organized.

(however)

If human life (or all life) is 7- or 9-dimensional, ie unbounded shapes, then there is an all encompassing capacity for both life's awareness and actions, but what about its relation to bounded stable structures in a 1- through 5-dimensional set of dimensional levels of bounded stable structures, in particular 5-dimensions. Does the 5-dimensional stable shape associated to a human-life contain (within its bounds) the solar-system, since the solar-system (upon which the human material form exists as a 3-dimensional structure) is also a 3-dimensional structure?

Note; A 5-dimensional discrete hyperbolic shape is naturally contained in a 6-dimensional hyperbolic metric-space.

Thus, the question is "to what range?" can the bounded spectral-orbital structure related to a human life be extended in space?

That is, what is the range of our material and spatial experience, where our life-form structure is still in a bounded context?

Such a bounded life-form structure would possess a stable set of spectral-orbital properties, so that the spatial extent of these properties would provide a context within which to travel in a context where material still identifies spatial positions so that material forms can be identified, so that travel between different points in 3-space, by means of moving in a "fast manner" through stable orbital geometries, can also be modeled.

56.

Differential structures

Geometry and (partial) differential equations, stable math patterns and measurable descriptions which possess practically useful information, ie the descriptive language (which is associated to stable math patterns) possesses content, ie it is not simply irrelevant babble.

Differential structures are, essentially, the fiber groups which define the range of allowed local coordinate changes, which in the new descriptive context, and which identify a context of a measurable geometric interactions associated with discretely defined differential operators, where discreteness is defined by both discrete time-intervals and discontinuously discrete relations between adjacent dimensional levels of (metric-invariant) metric-spaces (where the metric-spaces contain interacting material-components).

Note: The differential structure is related to the sets of transformations which exist as being defined between local coordinates of a manifold's "coordinate cover," where two patches of the manifold's "cover of local coordinates" over-lap. Often these sets of local coordinate transformations define a fiber group.

"Material-interactions in space" do not (in general) align with the stable shapes (defined by the stable structures of the metric-space and its fiber group), where stable shapes, in turn, define the stable geometric-spectral-orbital material-components and metric-space configurations, within a high-dimensional containment-set, in turn, the containment-set is defined by a hyperbolic metric-space.

The most general fiber group for differential structures is

1. the diffeomorphism group, ie invertible at a point, while

2. the isometry groups can be associated with metric-invariant metric-functions, as well as (under certain conditions) to linear shapes which have metric-invariant metric-functions, while

3. the discrete isometry subgroups can also be associated to linear shapes which have metric-invariant metric-functions, and which are continuously commutative [almost] everywhere (within themselves) [except for the shape's distinguished-point, ie the vertices of the fundamental domain], thus they are stable shapes and [when modeled as (partial) differential equations] they are both solvable and controllable, whereas when opposite metric-space states are included in a descriptive structure, then these are

4. the associated unitary fiber groups, which exist within classical Lie group theory and are associated with complex-coordinates, ie the real and pure-imaginary subsets of the complex-coordinates contain the two separate metric-space states (but which mix in complex-coordinates).

Parallel and/or tangent material-interaction structures

There are two types of geometric shapes associated with inertial material-interactions, which exist at the different dimensional levels (or more, intuitively, at the dimensional levels of 2- and 3-dimensions), and these two types of interaction shapes are:

(1) the "tangent" types, where a toral component of a tangent-type of interaction shape fits between the material shapes (with each changing time-interval) so as to be tangent to both material shapes which have the same dimension, so as to tangently touch both material shapes, so that the normal to the tangent to the point of touch of either of the tangent points, in turn, points to the other interacting material-component, and

(2) the "parallel" type, where a material component is touching in a parallel relation with the material component's containing metric-space (there is only one point of touch in the parallel types of interacting shapes), ie the dimensions of the parallel shapes may be the same dimension or they may exist at adjacent (different) dimensional levels. Note: The context of the two interacting shapes (the material-component and the metric-space) touching at their boundaries, depends on (the observer) being in the context of the adjacent one-higher-dimension space which contains the metric-space.

In the "parallel interaction geometry," it is both the metric-space's and the material's associated toral components, which are defining the parallel-geometries of the interaction, [ie inertial properties are defined by discrete Euclidean shapes, which are in resonance with the toral components of the stable material-defining discrete hyperbolic shapes].

The tangent-types of material interactions are, in general, non-linear, since they depend on a derivative-connection math pattern for their definition, while the parallel-types of material interactions depend on a stable containment shape, (for the material-containing metric-space).

However, the parallel-type of interaction identifies the idea which is central to general relativity, wherein the shape-of-space affects a material component's properties of motion, but now it is being described in a stable, linear, metric-invariant context, which is (can be) associated to various dimensional levels of containment.

However, within a stable shape, which is acting on a material-component in a parallel fashion, the material-components can also (still) possess a tangent-type of interaction, defined between the material-components which are in relatively stable orbits, within the metric-space shape (where the shape of the metric-space defines an envelop of orbital stability, which guides each orbiting material component). The envelop of orbital stability can, effectively, isolate material components from one another, so as to, effectively, define a 2-body tangent-type of interaction, which exists between the orbiting components of the orbiting-system, eg the planet and the sun.

The effective isolation of the two-bodies from a many-body system, is caused because the properties of the parallel-interaction of a material-component with the shape of the metric-space is stronger (bigger) than the other tangent-types of interactions with the other distant material components, where these other distant material components are not the sun (or not the massive [or highly charged] center component of the orbital structure). Thus, the tangent-type 2-body interactions, eg between a planet and the sun, perturbs the effects of the parallel-type of interactions, which exists between the planet and its containing metric-space shape.

The discrete time-intervals relate discrete Euclidean shapes [which form an action-at-a-distance interaction shape, whose tangent geometric structures are related to the derivative-connection-form, which is associated with the fiber group, and whereas these discrete time-intervals also have a relation to opposite metric-space states, which play an active part of the dynamic process], while the discontinuous discrete dimensional relations (of either to the derivative-connection, or to parallel interactions) allow for:

1. discrete changes of shapes, and it allows for other different contexts within the same shape, eg
2. changing to a new containment space (this is due to the properties of the infinite-extent shapes of many of the discrete hyperbolic shapes which are part of the description of material systems, eg neutrinos and light (but these unbounded shapes have different spin-properties), and these infinite-extent shapes exist at all dimensional-levels, so that this allows a coupling mechanism (or coupling context) at-infinity, so that at-infinity the containment structure can discretely change),

3. discrete changes in shapes (to re-iterate), or
4. affecting the way in which energy-flows within a shape's (toral, or angular-momentum) components of a system-defining discrete hyperbolic shape.

Discrete changes in shape

The discrete changes in shape . . . , most often are associated with a collision-interaction between (a pair of) material components . . . . , and are properties which can occur due to the relation which the existing (or newly emerging) shapes (of material components) have to a finite spectral set of an over-all high-dimension containing metric-space, where resonances (of either material-components or metric-spaces) with this finite spectral set defines both material and metric-space existence within a given 11-dimensional hyperbolic metric-space containing space, ie the containing space of all material components and all metric-spaces (and their shapes) in this 11-dimensional metric-space.

This finite spectral set (which determines both stable material components and stable metric-space shapes) implies that both the continuum is unnecessary, and that the entire range of the rational-numbers is also unnecessary, ie there are limits to both the biggest and the smallest spectral values and geometric sizes when within a physically descriptive context of (1) material-components, (2) metric-space containment, and (3) properties of material-interactions or material-metric-space interactions.

However, (partial) differential equations can be related to a continuum of rational numbers associated to the discrete Euclidean shapes, in turn, associated to inertial properties of both types of material interactions.

Issues associated to either the continuum or the rational numbers exist in physical description due to a non-linear derivative-connection as a part of a material interaction structure, in regard to tangent-types of interactions.

Note: The diffeomorphism type of differential structure can be related to a system's non-linear defining (partial) differential equation, which classical physics defines so often, so that the critical points of this (partial) differential equation can define a context for limit-cycles and subsequently, feedback, but the stability of such a descriptive context can evaporate for no reason, since it is not based on a stable shape.

However, the stable shapes into which material-components and containing metric-spaces form (or evolve, or to which they are confined, or bounded within) are based on a finitely generated quantitative structure of geometric-spectral values.

The differential structures, ie the fiber groups, associated to the different dimensional levels (where the different dimensional levels are discontinuously related to one another in a many-dimensional containing space) can affect the allowed finite set of possible discrete, spectral-orbital structures which are defined at each dimensional level.

The fiber group $SO(4) = SO(3) \times SO(3)$, allows for many differential structures as possibilities (eg see M Freedman, S Donaldson). However, the fact that a new oscillating energy generating shape exists "as an odd-genus 3-dimensional discrete hyperbolic shape (or hyperbolic space-form)" implies that a new material comes into existence, as a shape, so as the be contained in 4-space. This means that the natural differential structure (ie $SO(3) \times SO(3)$) would be the simple division of material types, ie one $SO(3)$ for each different type of material, or each material-type is associated to a different $SO(3)$ factor group of the $SO(4)$ fiber group, ie each $SO(3)$ is associated to a 3-metric-space structure, so that the two 3-spaces have a common 2-plane associated to each 3-space.

Could the 2-plane nature of the solar-system be a result of this pair of 3-spaces division of $SO(4)$, with their shared 2-plane?

however

The 2-plane structure of the solar-system could also have a relation, in regard to inertial material being defined as a 1-shape (ie a circle) and its interaction structure would be defined on a 2-plane.

We (our living systems) are related to the types of 3-shapes which naturally oscillate, while there are other 3-shapes, which are "inert" material shapes. But in this spatial division of 4-space, into two 3-spaces, we choose to see our (pair of both oscillating 3-shape selves and inert material selves) in relation to our inert material parts, which possesses an inert form of stability (not our oscillating part), which is a part of the condensed material which defines our planet, and we do not see our energy-selves, whose shape might be very large (ie does our life-form include the 3-subspace defined by the solar-system, which is defined by the common 2-plane of the solar-system, in regard to the pair of 3-subspace division of 4-space).

This differential structure for the $SO(4)$ inertial-material interactions in 4-space, causes our attention, which is associated with 4-space, to re-focus back on the 3-space which contains ourselves as inert-material, thus, causing us to be "so assured" that materialism is an absolute truth.

However, though $SO(4)$ has a natural choice for differential structures, ie a division between the two material types which are 3-dimensional shapes, the higher-dimensional levels and their fiber groups can be related to more types of variety, in regard to the types of geometric relations, which exist between material interaction geometry and the structure, properties, and geometry of the fiber groups.

Consider, the 7-sphere is contained in 8-space, and it can have many different differential structures.

However, the differential structures may be more limited when the fiber group is being related to a base metric-space whose shape is a discrete hyperbolic shape and or Euclidean tori.

How can these considerations about the algebraic structures of number-fields be related to a metric-space's differential structure?

The line (or line segment) and the circle can be used to define stable shapes due to the fact that the real line and the complex plane, ie a circle and a ray, are quantitatively consistent with one another.

Furthermore, the complex-coordinates can be used to separate the two opposite metric-space states which occur when metric-spaces are associated with properties (math or physical).

and

The differential structure of SO(4) seems to be a natural way of separating inert-shapes from oscillating-shapes which can exist as 3-shapes in 4-space.

The interesting thing about using stable shapes as a basis for a measurable descriptive language is that one can actually talking about something. That is, the discussion has content.

The genius of the education system is to develop language so that one can talk about quantities which are not associated to stable forms so the discussion has no content. Its set of deduced properties, obtained from formalized axiomatic (whose patterns are either algebraic or about convergence within a continuum), are only properties of a language which has no content, they are not patterns which can be associated to experience in a reliable manner.

Conjectures which can be made in a context where patterns are stable

It is clear that inertial shapes, ie discrete Euclidean shapes (or tori), can be many-dimensional, so these higher-dimensional discrete Euclidean shapes can be used to model inertia in any dimensional set-up, but inertia consistently identifies a 2-plane, so finding reasons for this can be set-up in the context of discrete isometry shapes so as to be given an actual content, since the discrete hyperbolic shapes are stable patterns and they can be associated, by resonance, to (the continuous) toral shapes associated to the inertial properties of material systems.

Thus, one can ask, that since a galaxy (or a spiral galaxy) has somewhat of a 2-plane shape, and if a galaxy can be related to an 11-dimensional containment metric-space, which has the property of low-dimension bounded-ness, so that, nonetheless, these galaxies can also be related to 2-plane inertial properties, then can these same galaxies be thought of as condensed material-components of a larger scale "material and metric-space system construct" so that the galaxy components are modeled as having inertial structures in a 2-plane (the 2-plane which the spirals seem to define) in another 11-dimensional hyperbolic metric-space containing space, which, in turn, is built-up with the galaxies modeled as pairs of opposite metric-space state spectral flows on a 2-dimensional discrete hyperbolic shape which represents charged material contained in 3-space (3-hyperbolic-space), where only the galaxies associated to real-flows are observed by us, ie the galaxies are the "real-number" charged structures (namely, the real aspects of a 2-dimensional discrete hyperbolic shape which models charged components), and upon which a very large-scale material system is based, a structure similar to the "material and metric-space structures defined over dimensional levels" which exist within each of the galaxies?

In such a context, where a galaxy represents the bounded low-dimension part of an 11-dimensional material containing hyperbolic metric-space, the drift between galaxies could be the natural expansion identified by Perelman for discrete hyperbolic shapes (and described in wikipedia in its geometrization entry [8-30-13]), or it could be a result of a natural group conjugation process associated to the fiber group and its attached principle fiber bundle, ie the fiber group and its base-space. Such a conjugation would be associated with a drifting motion of the principle fiber bundle.

Note: This could be the cause of the expansion of discrete hyperbolic shapes which Perelman described, ie the Ricci flow manifolds (as mentioned in wikipedia) are (ie are they) defined by fiber group actions (?).

Note: The discrete discontinuous properties which exist between dimensional levels allow (or are a result of) attaching a constant factor between discrete changes, ie defined by discrete operators, ie discrete changes in dimension, so there is (can be) a new size-scale defined between dimensional levels.

SO(8) acts on an 8-dimensional local coordinate space, SU(9) has rank 8, SU(3) has dimension 8, How can this be related to octions? and

SO(4) acts on an 4-dimensional local coordinate space, SU(5) has rank 4, SU(2) has dimension 3, How can this be related to quaternions?

57.

Einstein was wrong

All of science needs to be challenged and alternative ideas provided.

One needs to distinguish the type of intellectual development which Einstein represented vs. the intellectual development which Copernicus represented. Einstein was consistent with the rationality of his age, while Copernicus was in opposition to the rationality of his historic era. That is, Copernicus was a revolutionary enemy of what was considered to be rational in his era while Einstein was embraced as a great genius, yet it is Einstein who took the weaker intellectual position when compared with Copernicus in the horse race of winners and losers of the "game of truth," since there is direct evidence that he was wrong and a review of the intellectual position of his era; where axiomatic formalism and indefinable randomness and non-linearity were being embraced as a basis for a valid descriptive language but now it is seen that such a language has no substantial content, rather it is descriptions of an illusionary world (which can be fairly easily adjusted to fit data), whose information is unrelated-able to practical creative development. In this context of embracing the illusion which the professional rational community (of around 1900) was embracing, eg fitting data as the Ptolemaic system was designed to fit data (thus the belief in there being a scientific verification), Einstein was touted as an intellectual revolutionary similar in importance as was Copernicus, but he really was the opposite, rather he was consistent with the authority of his age. Though Einstein is one of the better examples of one who followed a descriptive heritage of paying attention to assumption and interpretation, his failures, in regard to general relativity and in not identifying in strong enough terms the data following structure of quantum physics and its inability to identify within its descriptive constructs a valid relation to the stability of the many-(but-few)-body systems (also something which general relativity fails to deal).

Though Einstein writes at the level of intuition and assumptions, these assumptions are consistent with the rational community of that era, which at around 1900 was the rationality of materialism (ie those rational people who learn to use descriptive language in order to get jobs), yet the wave properties of light did not fit into the framework of Newton, (Newton was also consistent with the rationality built-up by Copernicus, Kepler, and Galileo), and how to fit the idea of inertial material into the context of charged material systems and how to understand the stability of the atoms. Einstein's beliefs seem to center around the idea that material would have to be understood in terms of the force-fields which are to be modeled in the context of partial differential equations.

[But open-closed topologies of metric-spaces and a mis-understanding concerning containment and the flawed idea of materialism leading to a mis-representation concerning the nature of physical properties led to failure of the partial differential equation based force-field models]

Unfortunately, we do not live in a scientific age, rather we live in an age of domination and celebrity, ie inequality (similar to the Roman-Empire), where such social issues about who can dominate result in fixed authority, and subsequently, an exclusion the unworthy (who cannot cope with this fixed authority), not unlike the church in the age of Copernicus, but now it is done in the name of science (a fixed authoritative contest to demonstrate worth in being a professional science person) (ie not done in the name of religion, yet that is still what organized religion is all about). While professional scientists are wage-slaves, and thus, their thoughts can be controlled by the owners of society. Thus the embrace of materialism, as well as the "survival of the fittest and indefinably random" models of life's development provided by Darwin, and the particle-collision model of physics which fit into the models of nuclear weapons. That is, social domination (by the owners of society) requires that certain ideas be allowed expression . . . . , while the criticisms of these ideas and their alternatives be suppressed . . . , just as the church acted, concerning ideas, in its role as the head of a collective empire.

In this era (around 1900), there was discussion about "the reality of motion." Einstein believed, as did Mach, that motion was only physically distinguishable in relation to two bodies. Thus, one body could be put into a still frame while the other body (in the still coordinate frame) would be considered moving, but what Einstein did, was to point-out that the metric-function needed to be changed from a Euclidean to a space-time metric-function, the metric-function to which the electromagnetic wave-equation was invariant.

But does everything fit into space-time?

In the affirmative, is what Einstein believed, and it was consistent with the rationality of that era, where the professional intellects of that era also believed that there is materialism and its proper coordinate frame must be space-time (the point which Einstein was able to state so simply, and along with its implications), and that "all the physical properties of material systems will fit into space-time," where these properties can be represented as functions whose domain space is space-time.

Thus, these properties can be investigated quantitatively by their (are) being related to local linear measuring models of a material system and its measurable properties.

But this did not agree with the simple ideas of being both measuring reliable and related to stable patterns, rather it was assumed that all locally defined (or apparently distinguishable) quantitative constructs could be consistently placed into a fixed math context . . . , (of formalized axiomatic's, where fixed axioms are stated and their relations derived, and whenever there is something like a quantitative structure (ie something can be measured) then it is assumed that then all of these formalized axiomatic relations are applicable to the context of what is being measured, but this has led onto a path where "illusion is being followed") . . . , and that the discussion would possess meaningful quantitative content, but this is done without these quantitative properties being related to stable patterns, thus the measured context has no "reliable or practically useful" content [ie whereas stable patterns actually do possess quantitative consistency, and thus, are related to a context in which measuring is reliable, and the information is relatable to practical uses].

The question considered in this context of formally-quantitative viewpoint of materialism (and frames of a material system's space-time coordinate system) was whether acceleration has a valid definition "as to whether the second-order spatial derivative of a body's spatial position in absolute-space can be a valid a measurable construct" where it was assumed that acceleration only has a reality in regard to relative motions, ie this would be relative acceleration properties between two bodies [Note: The linear context of translational motion requires two-bodies (unless the other bodies [ie more than two] were also on the line)]. That is, special relativity was formulated based on the relative velocity-motions, while general relativity was formulated based on the idea of relative accelerated-motions.

Newton defined acceleration based on a belief in space having measurable properties associated to itself, so that this was called an absolute viewpoint of space.

However, physics uses two contexts in which to describe the properties of material in a material system's containing domain space, either by means of motions caused by force-fields acting on point-material-objects in space or by identifying spectra associated to function-space models of the physical-system, where the spectra are observed to be (random) point-spectral events in space.

The starting-point for Einstein's claim for general relativity (which is analogous to the starting-point for special relativity) was that that "acceleration was also a relative motion (as were velocities)," and thus, any accelerating-frame is equivalent in regard to describing the (partial) differential equations which are related to the laws of physics ie locally measuring the properties of a material system contained in a space-time metric-space, ie geometry or randomness.

The partial differential equation related laws of electromagnetic-fields were invariant to the isometry group of the space-time metric-space [Is the Lorentz force, also invariant? Apparently, yes, if one follows the rules of special relativity in order to transform the mechanical formulas for force of Lorentz to be consistent with the space-time metric-function.]

However, the Newtonian laws of gravity are not consistent with space-time metric-invariance.

The laws of electromagnetism possess covariant invariance, ie they relate to measurable constructs of physical properties which depend on (locally linear) geometric measures which are consistent with the space-time metric-function. That is, the vector-field is consistent with a second-order metric-invariant partial differential operator which is applied to a vector-field can be organized so that the force-field is a 2-form so as to be related to a space-time metric-invariant measure of locally linear 2-dimensional (planar) subspace regions of the containing metric-space.

Thus, in regard to general relativity, a second-order metric-invariant partial differential operator applied to the metric-function might be constructed so as to satisfy the property of metric-invariance or general covariance.

The motivation for this type of a "quantitatively inconsistent" math construct, was to be philosophically consistent with the thought process in regard to either absolute space (of Newton) or relatively determinable motions of two bodies (of Einstein).

If one believes materialism then there is no reason to expect space to have any properties of its own, since all of space's properties are derived from the properties of material in space, ie measuring length depends on using a rigid material measuring-rod.

Thus, all physically measurable properties of motion must be those measurable properties which can be defined relative to the two bodies, where the model for measuring these properties is the local linear derivative operator, and can be formulated in regard to coordinate frames which are moving as either first order or second order motions (ie orders of derivatives of time).

The assumption is that . . . since the metric-function's matrix is symmetric then it can always be made diagonal at a point in space, by the proper choice of local coordinate directions, this means that . . . . the general group for such general coordinate transformations is the locally invertible group of diffeomorphisms (and not the space-time isometry group of special relativity). However, the locally diagonal coordinate-frame, which is defined only at a single-point in the containing space, is interpreted to be a local space-time frame for the system's properties defined only at that point.

But this construct is almost always non-linear. Thus, it is also quantitatively inconsistent, eg chaos is due to the bifurcation of quantities . . . . at random points in the coordinate space (or domain space) . . . . of a solution function to a non-linear (partial) differential equation, and this does not lead to the idea that such solution functions (to non-linear (partial) differential equations) can represent stable patterns.

Thus, the idea that general frames form a basis for physical description cannot be used to describe the most fundamental set of observed properties of the material world, namely, that there exist many of the most fundamental very stable material systems, and which exist at all size-scales, including the stable solar-system, and all of which go without valid descriptions when the basis for description is solutions to (partial) differential equations, which form the, so called, laws of physics, ie these systems are either not formulated or non-linear and thus its solution function is chaotic and unstable, but these systems are stable.

In both Newton's model of a physical description of inertia, and in Faraday's model of charged material interactions, either only two point-bodies or one-body and a (electrical) current-component . . . types of configured systems (which exist in the natural coordinates of the system's shape, which are commutative everywhere) can be solved, while the two-body opposite-charge system is unstable, since its natural orbital shape causes the accelerated charges to radiate electromagnetic field-energy, apparently always descending into being a 1-body neutron, unless the radiation, which the charges would be radiating in their forcefully constrained orbits, happened to also cause (by field-interactions with the charges) the two charges to be pushed-apart (but again resulting in an unstable system of two-charges).

Note: Charged systems, which show fixed separations of charges in crystalline type geometries, seem to exist, but the descriptions of the material interactions which can cause such systems to exist, in turn, does not exist.

So getting the language consistent with philosophical viewpoints is pleasing, but only in a narrow context of language, and such pleasing viewpoints can be consistent with the rationality of the age, thus, it can get a person a job, and it can also be adopted as the authoritative structures of the professional communities, but it seems to not lead to being able to describe the very stable material systems which are observed.

On the other hand, the methods of attributing the properties of waves to physical properties, by using the Planck constant, and then modeling material components as waves, so $e^{\wedge}i(kx-wt)$ is changed (by using Planck's constant) to $e^{\wedge}i(px-Et)$, so as to allows partial differential wave equations to be defined for quantum wave-systems associated to a set of many-(but-few)-body discrete charged-material components. That is, sets of operators applied to sets of harmonic functions can be used . . . , where the operators are attributed to physically measurable properties . . . , so as to form energy wave-equations . . . (and possibly other types of wave-equations based on a set of operators which act on the physical system's function-space, where a function-space is the model of the quantum physical system), . . . whose solution functions would (are supposed to) represent the stable spectral properties of the quantum system (or wave-system).

However, in general, this set-up cannot be either formulated and/or solved for the many-(but-few)-body system. Again the notion that a solution to a partial differential equation, which represents the laws of physics, can describe the observed order seen within quantum physical systems is shown to be not true.

This wave-equation set-up is supposed to allow a quantum system to be associated to a set of spectral values (and associated spectral wave-functions) which are observed to be the local measures of the (locally determined) spectra of the individual components of the quantum system. Waves interact with material at localized positions in space, and this happens in a random fashion, so the wave-functions are attributed to probabilities of the distributions of spectral properties (or events) in space, but where the spectral events are localized in space.

Essentially, the only types of physical systems made of material components which can be described accurately and to sufficient precision based on the laws of physics being represented as (partial) differential equations are the two-body systems contained in Euclidean space (or in space-time by using the Dirac operator for the H-atom, but the math methods for solution of the Dirac operator result in math constructs whose properties are quite similar to the Schrodinger-equation of the H-atom defined in Euclidean space), while general relativity only has one solution, and that is to the one-body system which possesses spherical symmetry.

There is also the concern about whether the descriptions of quantum physics ie wave-functions of quantum systems, are non-local (ie non-causal in regard to a space-time continuum) or local (ie what can be considered to be causal in a space-time continuum). One would have this concern, since a wave-function is global, and their domain spaces are Euclidean metric-spaces (eg the Schrodinger wave-equation), which possess the property of action-at-a-distance.

The experiment of A Aspect showed that quantum properties are non-local.

That is, Newton's formulation concerning absolute space, and action-at-a-distance is correct.

However, one can go further, in that, one can understand, "from what material and its stable properties depend for their existence," namely, the stability of material system properties depends on the properties of space (of several metric-spaces, which each contribute to the observable properties of the physical system) and on the natural stable shapes of some metric-spaces.

In fact, the derivative operator needs to be re-defined as a discrete operator, which is defined between:

1. Discretely (or discontinuously) different dimensional levels,
2. between discrete time intervals, and
3. between discrete Weyl-angles, and
4. Between (or with the help of) different metric-spaces, ie metric-space shapes, whose properties can be mixed together so as to allow for the many physical properties to exist within a physical system, ie not sets of operators which represent physical properties but rather sets of different metric-space which represent the different physical properties that are observed.

That is, space is "more than real," it is fundamental to understanding both the properties of reliable measuring contexts and the existence of stable (physical) patterns.

That is, space plays the most substantial role of a descriptive language which can be used to describe stable physical properties, while differential equations, unless the derivative is re-defined as a discrete operator, cannot be used to identify either a context of measuring reliability or so that the description provides the constructs (and/or solution functions) which are based on stable patterns.

That is, one needs to be able to actually use a descriptive language which possesses content in regard to descriptions about some actual thing which has stable properties, and not a fleeting and unstable quantitatively inconsistent, vaguely discernable, set of patterns, where these fleeting, unstable descriptive patterns are associated with the use of a viewpoint of materialism and axiomatic formalization.

The idea of materialism implies that all material systems are contained in space-time, and all the properties of a material system can be derived either from a local measuring process, or can be provided by wave-function function-spaces and their associated spectral-values (determined by sets of operators) which relate to locally measurable component-spectral properties. In both of these two cases (geometric based motions or the randomness of a local component related spectra) the function's domain space is either the space-time space, or Euclidean space, defined by materialism, and the implied containment of all properties of material systems within either space-time (ie local space-time) or Euclidean space.

Instead

It is space which is the only thing which exists.
Space which exists in:

1. several types of different spaces [each associated to its own physical (or mathematical) property],
2. Space is defined for several different and separate dimensional levels, so that
3. it is the natural stable shapes of each of these different metric-spaces . . .
   {different dimensions and with different metric-function signatures, where a metric-function signature is associated to the spatial-dimension, s, and temporal-dimension, t, and which is a division of the metric-space's subspace structure, eg $R(s,t)$, where $s + t = n$, but the key-dimension, in regard to material properties and their descriptions, is the spatial-subspace-dimension, s}
   . . . where the stable shapes of each of these different metric-spaces determine:

   (a) the stable material components and such material-system's set of stable spectral properties,

(b) the shapes of metric-spaces determine the stable orbital structures of the condensed material contained in the metric-space, and

(c) these shapes also determine the structures of material interactions, where the interactions of n-dimensional material component-shapes are defined in an (n+1)-dimensional metric-space [where the material-containing metric-space also has a shape of its own, a shape which determines the stable orbits, into which (condensed) material components fit into stable orbits, about some stable central shape, where the structure of interaction caused by the envelope of orbital stability (upon the condensed material which is within the orbital envelope) has a different geometric relation of interaction between the shape of the condensed material and the shape of the envelope of orbital stability than does the interaction between separated shapes of material-components].

4. It is within a metric-space, wherein a (partial) differential equation is defined . . . , in regard to either force-fields and material interactions, or in regard to the spectral structures of material systems [where this type of a descriptive process needs to be about shapes of metric-spaces which model spectral structures, ie the functions which compose the function space are stable metric-space shapes, and thus the functions are a part of the descriptions of the material system of interests], . . . , but these (partial) differential equations can only adjust, or perturb, the stable properties (which include the stable spectral properties of material systems, or be made consistent with the stable shapes of metric-spaces), where the stable material system properties are, essentially, all defined by stable metric-space shapes which define the stable orbits by the shape, of which geodesics are a part, where the adjustments to the orbits (determined by the shape of the material-containing metric-space), are a result of material interactions with its containing metric-space shape, where (we state again that) the interaction of the condensed material with the metric-space shape (which is the envelope of orbital stability) has a different geometric context than does the shapes of space which define direct material interactions.

That is, the shape of space determines stable orbits by a different geometric inter-relationship which exists between the shape of the material and the shape of the material-containing metric-space, than (does) the shape of material interactions (where a shape connects the two interacting material-shapes).

It is only in the context of a material containing metric-space shape which provides an envelope of orbital stability where the two-body (partial) differential equation makes sense as a two-body problem, since the envelope of stability shields, or overwhelms, the effects of the bodies which are different from the two bodies of interest. The (partial) differential equation is a substructure with which to the complete geometric context of a physical system's containment and material-interaction structure.

This idea allows both the limited structures of either (partial) differential equations of fields associated to motions or the or the function-space and its associated set of (physically motivated) differential operators to be transcended, so that the shape of space in a diverse containment set-structure allows for the descriptions of the stable properties of the many-(but-few)-body systems to be determined to sufficient precision.

The shape of a metric-invariant metric-space . . . , whose metric-function has only constant coefficients, within which a local measuring property is defined . . . , are the shapes which most determine the properties of a stable material component, and which determine the approximate (stable) shape which a solution function can possess, where the solution function (of a system's partial differential equation) would be defined, in regard to either local geometry or a locally determinable set of spectral properties of the system.

That is, a system's organization is not a result of finding a solution function to a (partial) differential equation, rather its observed stable properties are a result of the stable shapes of space, either the apparent shape of a material component, or the shape of the metric-space within which the condensed material components are contained, so that the shape of the containing space determines the envelopes of orbital stability for the contained material which make-up the many-(but-few)-bodies of condensed material which possess stable orbits.

Where do the stable shapes come from, and why are they so naturally a part of physical description?

The isometry group [of a metric-space which possesses a particular metric-function (so that the metric-function has constant coefficients and the metric-space has non-positive constant curvature)] has discrete isometry subgroups (which are associated to stable shapes) as well as possessing maximal tori subgroups, where both of these different subgroups are circle-spaces, and represent the math context of commutative types of number-relations that can exist as stable patterns in a many-dimensional coordinate metric-space) these represent the stable patterns which material systems can possess, and which can exist. Where one of these stable patterns (either discrete subgroups or maximal tori) is a shape (ie the discrete isometry subgroups), while the other can be thought of more "as a set of diagonal matrices" which can transform the local coordinates in the metric-space, so the results of these transformations will be both linear and metric-invariant.

The spectra of the over-all containing space can be defined by means of a partition of the different subspaces of all the different dimensional levels by these stable discrete isometry shapes so that in each subspace of each dimensional level there is a largest-scale-size discrete isometric shape so that all the other shapes which exist must be in resonance with one of these largest shapes of the partition.

The maximal tori are spaces associated to each point in the metric-space, and are spaces which possess the capability to resonate with many spectral sets.

Thus, awareness can be defined as the set of resonances which relate the spectral set defined on a maximal toral set to the partition set of an over-all high-dimensional (ie an 11-dimensional hyperbolic metric-space) and subsequently to the material-spectral set in the dimension of the metric-space within which our attention is placed (focused).

Then after we perceive by paying attention to an external spectral structure, where the local spectral-material structures around us resonate the strongest, and then we use our knowledge of the resonances which compose our own living system to direct energy into various spectral subsets within ourselves, and we can do this in relation to the outside world which we are also perceiving by means of detecting resonances.

These math structures of stable shapes defined in metric-spaces are those which allow both reliable measuring and provide for the existence of stable patterns (which are measurable). These are the discrete shapes of the metric-invariant metric-spaces whose fiber groups are the isometry groups. These different metric-space, ie both different dimensions and different metric-function signatures, are associated to physical properties so that a mixture of these metric-space are needed to describe the composition of a material system and its various physical properties, ie differential operators do not define physical properties rather different metric-space define physical properties.

These properties of metric-spaces exist as pairs of opposite metric-space states, which need a complex coordinate system to both contain and to describe . . . (note: the rotation between opposite metric-space states defines the spin-rotation of states, and its period of rotation can be used to define discrete time intervals) . . . and this leads to the change from isometry groups to an associated set of unitary fiber groups.

Within a metric-space a set of second-order, metric-invariant, partial differential equations can be defined, so that these equations are now defined to be discrete operators, but these new definitions result in math structure which is quite similar to the laws of classical physics.

These (partial) differential equations do describe material interactions, but such interactions can only perturb (or adjust) the properties of a material system whose properties are mostly determined by the shape of space, ie the material system is a stable metric-space shape whose properties can be slightly perturbed because of inertial properties related to a material interaction whose math structure is similar to classical physics.

Furthermore, the function space techniques of quantum physics can be re-organized so that the functions which compose a quantum system's function-space are sets of stable discrete isometry shapes, ie the shapes of the stable material systems called quantum systems.

What is a (partial) differential equation ultimately trying to describe? Answer: The stable shapes and the stable spectral sets associated to physical systems, so that these stable systems result from material interactions modeled as (partial) differential equations. This has not been accomplished, after over 100 years of trying.

However, for both shape and spectral sets, such stable properties can (only) be related to very stable patterns through the discrete isometry subgroup shapes (discrete Euclidean shapes and the discrete hyperbolic shapes), ie related to the properties of space and the natural stable shapes associated to some metric-spaces.

That is, Einstein was exactly wrong, because the notion of needing a context of general frames of reference . . . . , {since space cannot have any properties of its own, and thus, that the laws of physics are about (partial) differential equations whose forms are covariantly invariant to a diffeomorphism fiber group (so that inertia is defined on the geodesics of the general shape, which in the case wherein Einstein is referencing the shape of space, this general shape is an unstable non-linear space)}, . . . , is not the way in which to organize mathematical constructs so as to be able to describe the stable properties, which are observed for the many fundamental material systems which exist in very stable states.

However, the idea that the shape of space can cause "an effective (inertial) force" on a material component, can be true, but the cause is (again) a result of the shape of the stable shapes of metric-spaces, and the effect of these shapes upon the shapes of the material components, which are contained in the metric-space-shape {where metric-space shapes model both material, and material interactions, as well as the metric-spaces which contain material} are the motions which stay within their orbital envelopes within which the condensed material is contained.

58.

Equality 5

The triumph of Copernicus, is not a triumph of science over religion, rather the brave act of Copernicus was to challenge dogmatic authority, the same type of dogmatic authoritative truth which is now being associated to science. This is clearly true, since the most fundamental of physical problems, understanding "the stability of the many-body system" cannot even be formulated, let-alone solved, but all that modern (2013) physics provides as patterns of knowledge, to be used to address this problem, are hopelessly complicated descriptive language based on, indefinably random, and non-linear descriptive contexts, with their definitive statement (about the properties of these observed stable systems) being that "these systems are too complicated, so as to not be able to describe their properties based on the laws of physics." Yet, that this stability exists implies a: linear, metric-invariant, and continuously commutative everywhere, type of a physical system.

Why will people (who serve banking interests, the professional science and math communities) not allow these types of challenges to the dogmatic authority of modern science?

If the new model of existence . . . , based on a finite set of discrete isometry shapes (of non-positive constant curvature metric-spaces whose metric-functions have constant coefficients), which define the spectra for all such discrete hyperbolic shapes, which are contained in a many-dimensional context, but the dimension is limited by an 11-dimensional hyperbolic metric-space, so that these shapes are the stable shapes of which (almost) all stable material systems are composed, . . . , is true, then one can describe to sufficient precision the stable spectral-orbital properties of a many-body system at any size-scale. Furthermore, life and mind are a natural property of such an abstract math context, and the material world is a proper subset.

Furthermore, this is the proper context within which to understand the fundamental attributes of all existence, and thus one can consider what types of the properties of existence can be changed, or need to be changed in order for new attributes to be brought into existence, so that one can understand that the purpose of life is to create existence, ie to extend the descriptive bounds (or to extend the bounds of containment) of existence, where this extension is based on true knowledge (or of one's true 'and complete' awareness) of existence's properties and the relation which our life-form has to this existence.

In regard to the properties of the material world being a proper subset of the new descriptive context:

Consider a property associated to particle-physics, one can model the SU(3) part of the fiber group of the internal particle-states (of hadrons, ie in the nucleus, where a nucleus identifies an atom's position in space) as being related to a cubical partition of a complex-number spatial region, where this cubical partition exists around where a particle-collision happens, so that these cubical partitions of space contain a sequence of opposite metric-space states which are associated to the dynamical path of a particle-collision, and where one of the cubes of the partition is at the collision site, so that the complex-number cube is broken apart (by the particle-collision) into its three-directions of the complex-number cube, and where the broken cube is also composed of three sets of opposite metric-space states.

[In fact there is extensive discussion (in the authoritative physical literature) about such sets of mixtures of opposite time-states as being a necessary part of the particle-physics' model of a particle-collision.]

The so called, Higgs (scalar) field-particle would be the remnants (after the collision) of the break-up of the inertial "discrete Euclidean shapes," which are associated to both the 2-shapes and the 3-shapes of the interacting charges which are involved in such a collision.

This same construct could be applicable to a similar type of cubical energy-partition of a thermal system, also statistically characterized by component-collisions, where in thermal systems such a partition is also related to the entropy properties of the thermal system.

If one possesses a true knowledge of existence, so that in this context (of true knowledge about existence) we are all equal creators, and each person is equally likely to extend the reach of existence by the intent of their life, guided by correct knowledge, ie life-forms are the perfect instruments through which the structures of existence can be extended and changed.

Thus, we need to be considered equal, so as to seek the basis for knowledge which relates to our true creative capacities. Thus, we all might become the mythical creature . . . , as mentioned by R Wilhelm's translation of the book, I Ching . . . , of the magic tortoise who breathes (and feeds on) only air, and provides gifts of creativity to existence (or to the earth) and to all life-forms (as an expression of love).

However, the propaganda-education system, the basis for what we believe-in as a meritocracy, has a limited, and very tailored, descriptive knowledge so as to be (1) deceptively complicated, and (2) used in very narrow ways, so as to mostly serve the interests of the owners (or rulers) of society.

If law is based on property rights, and an associated control over all aspects of society, then the society will be constrained to narrowness by extremely violent, suppressive, and coercive methods, by both physical and intellectual coercion.

It is just as violent to:

1. claim to possess an absolute truth, so that one self-righteously proclaims that all other ideas about the subject are inferior, ie one can assert one's own "believed truth" without needing to suppress other's beliefs, as it is to
2. establish an arbitrary claim of ownership of property, and to "make that claim stand" by means of violence, as it is to
3. extract resources after military victory, or through violent control of the property, by violent and corrupt means.

Thus, in our education we rejoice that, historically, Copernicus challenged the knowledge of the ruling church in his age, but then mis-interpret his act, so as to proclaim that he established the truth of a "material based science" as opposed to an authoritative religious language, where religious language mostly provides stories about history, not about measurable properties (other than being measured in regard to past-time).

The triumph of Copernicus, is not a triumph of science over religion, rather the brave act of Copernicus was to challenge dogmatic authority, the same type of dogmatic authoritative truth which is now being associated to science.

This is clearly true, since the most fundamental of physical problems, understanding "the stability of the many-body system" cannot even be formulated, let-alone solved. Yet, that this stability exists implies a: linear, metric-invariant, and continuously commutative everywhere, type of a system, but all that modern (2013) physics provides as patterns of knowledge, to be used to address this problem, are hopelessly complicated descriptive language based on, indefinably random, and non-linear descriptive contexts, with their definitive statement (about the properties of these observed stable systems) being that "these systems are too complicated, so as to not be able to describe their properties based on the laws of physics."

That is, at all levels of intellectual or propagandistic expression, which fit into the interests and beliefs of the ruling class, one is confronted with meaningless babble, but of course with the exception of those who maintain and adjust the complicated instruments upon which much of the social power of the ruling class depends, ie but it needs to be noted that these are instruments which are based on the concepts of 19th century physics. That is, the meaningless babble is very ineffectively directed at the development of new technologies and the development of new contexts within which to be practically creative.

Despite such selective usefulness and complete failure at solving some difficult problems which face the endeavor of physical description, nonetheless, on the inter-net, one finds that, on most of the, so called, free-speech websites, there is a concerted effort by editors (or by a clamoring public) to hold language fixed, and there is a demand that one is to conform to all proper categories of authority, one needs to be consistent with the objective form of rationality (as expressed by the media-education system) and one's expressions of ideas need to remain consistent with the proper hierarchical social institutions (so as to protect the public from hearing an untruth), so that any words, which don't do this, are to be excluded, as either being the expressions of an incompetent voice of truth, which is not consistent with the social hierarchy of rationality, yet these institutional truths are the truths which most serve the interests of monopolistic business interests, ie they are the truths which serve the ruling-class, so to challenge the ruling-class one needs to challenge their overly authoritative, so called, truths, or if one is expressing new ideas, but then the propaganda-education system judges that the new ideas are being expressed by a person of low social-standing, a person who does not have the social standing to challenge the absolute truth of the experts, ie the experts who serve and compete for the positions of the narrow intellectual interests which support the narrow interests of the owners of society (or both).

Apparently, these editors (or the public) are being directed (or led) to use censorship (or self-censorship) so as to serve the narrow interests of the owners of society, in regard to the knowledge which the society is allowed to use. Unfortunately, the delusions upon which propaganda is so successfully based have entered into and taken-over the institutional discourse concerning truth.

The truth which is being protected is the truth expressed either by dogmatic institutions, or by those represented to be in the top set of personalities in a particular category of intellectual interests. That is, more than anything else it seems to be about protecting personality-cult where truth is now what personalities proclaim, it is not a function of rationality.

That is, in our society, "truth" is that which is expressed by an authoritative personality.

Thus, the knowledge of our society stays fixed (since authority can only be judged within a fixed dogma), and the language used to express ideas in "an acceptable rational way" stays fixed, so that the ruling-class can do the things it wants to do, so that no new ideas interfere with their investments.

That is, the ruling-class does things, since

It is only the will of the ruling-class which determines what society is allowed to create or do, and then the ruling-class frame their actions through the media, so that the public responds to "how the media frames these actions of the ruling class."

Thus, in this way, the public always remains ineffective, ie always following the assumed merit which the Plutocracy defines, ie always following (or reacting to) the propaganda-education system, so that in the social institutions there are some who occasionally express truth about social power in our western society, eg that the US society is a Plutocracy (not a big surprise), but these meritorious intellects, who are always allowed to express ideas, ie the journalists and their professor leaders, refuse to clarify sufficiently . . . , (more likely they do not know, or they do not see, or they do not deduce that it is) . . . . , the extreme violence which places them in their social positions as experts, where there rational constructs are deteriorating, and, instead, they supply red-herrings concerning the main issues facing society, and that main issue (really) is that; "the ruling-class has created a system which is completely rotten and corrupt, and it is now failing due to its ignorance (in regard to rational development, and in determining the consequences of acting on partial knowledge as they have done in regard to DNA and radioactivity, etc, where their actions based on partial knowledge are leading to wide spread destruction).

So let "me" explain:

"why this is"
and
"what to do":

Property rights results in violence, and in an arbitrary way in which language comes to be used, since merit is judged by the plutocrats, and experts depend on someone else (ie the plutocrats) to judge the merit of them the experts who have successfully competed in a contest based on dogma, ie then the merit of the experts is supposed to determine what the knowledge of society is capable or doing in a successful manner, but their dogmas are too narrow and, mostly, can only be used as they are being used (and so the circle goes) [though electronics and its relation to controlling spectra and fast-switches has been applied over a fairly wide-range].

This results in the fact that language, in regard to developing both new ideas and new contexts within which to be practically creative, has virtually no content associated to its use . . . , except in regard to the original use of the language, ie technical development is all based on 19th century science (ie this is how the technical language of the experts was originally used) and it remains the only practical value which the old fixed language possesses, yet the (limited but useful) language has gotten a lot more complicated.

. . . . , The expert physicists, mathematicians, economists, biologists, psychologists, etc are using precise languages which are based on math constructs which do not allow their, so called, measurable descriptive contexts to be in a state of their application where there is reliable measuring, and the math patterns used within these languages . . . as well as the resulting patterns (in their descriptive structures) . . . are not stable.

It is a language which is quantitatively inconsistent, it is non-linear, (virtually) never "commutative almost everywhere," and it is most often based on indefinable randomness (ie the events whose probabilities being determined are not stable events [this means counting such events is not reliable]) so statistics can be manipulated in arbitrary ways, since it is not being related to a stable pattern (ie vaguely distinguishable states are (most often) not stable events).

It is a descriptive construct which is without the capacity to contain any content, and thus, it is easily manipulated so as to appear to be a valid measurable confirmation.

For example, the stable many-body systems are too complicated to describe, etc. (yet validity of actions is verified through bogus sets of numbers (or measures) which, in reality, have no meaning, ie arbitrary actions can be validated within an arbitrary context [by counting events which are unstable])

So the actions of the plutocrats are framed in a context which does not have any meaning, ie it is arbitrary, and the public is required to react to how the media-education system arbitrarily frames things.

Calvinism and the western culture

Calvinism (the belief that those who possess money are favored by God) extended Constantine's use of religion as the propaganda of the western Empire . . . (where institutional religion is an arbitrary worship of hierarchy) . . . . to a religious based propaganda system for a (western) banking empire.

[note: the Judeo-Christian-Islam religions, as well as capitalism and collectivist-socialism (often called communism) possess the central beliefs of the western culture, ie where in all western societies the collective efforts of the people of a society are focused-on supporting an arbitrary social hierarchy ruled by a few people]

{note: by 1400 AD the Medici banking-family were choosing the Popes for the Holy-Roman-Empire, where the Popes, essentially, ruled western Europe}

Education, in western culture, came to be about the indoctrination of the absolute truths associated to the society's economic interests.

To show that this "indoctrination of the absolute truths associated to the society's economic interests" is intrinsic to the education system of the west, consider that, because of the nuclear weapons development during WWII, E Teller was able to change the physics departments of all (US) universities into nuclear-weapons engineering departments, by (about) 1952.

Teller's nuclear weapons viewpoint of physics is about describing the chaotic transition state characterized by a model of random collisions of the components which were broken apart from the original "stable state of material" whose break-up was caused by the onset of an unstable transitioning context (eg gathering and then putting certain types of material together), where the (now) chaotic system transition is occurring between the pairs of relatively stable-states of material systems, ie the initial and final states of stable material systems. This is a chaotically transitioning model of material between stable states, which totally ignores the issues concerning the stability of material systems, where determining the reason for "the stability of material systems" is the much more fundamental problem, but physics departments seem oblivious to this fact.

However, axiomatic formalization of mathematics, expressed through the Bourbaki publications (around 1920), where the "axiomatic formalization of mathematics" took over the professional math community between 1890 and 1910, and it was a prominent intellectual action, apparently led by D Hilbert, which pulled precisely formulated technical issues, away from expressions of ideas at an elementary level of discourse, so as to instead place the discourse as being based on an arbitrary and vague set of quantitative patterns, which (it is now assumed) can be placed onto any context where either measuring or numbers can emerge.

Thus, quantitative descriptions became ever more arbitrary and manipulate-able, and the obvious failures of this precise descriptive language, eg its inability to describe observed stable patterns, could be "papered over" with a vast array of complicated number relations, and the claim that when these abstract quantizations of vaguely general (and unstable) patterns become fully integrated with one another, then the failings of the descriptions would be corrected. Thus, such a focus on a fixed set of axioms ignores the content of Godel's incompleteness theorem, which states that a fixed set of axioms has great limitations as to what patterns it is capable of describing.

Axiomatic formalization takes existing math patterns, which are correlated with (successful) descriptive constructs, and pushes them into more general contexts with greater complications, where the types of number-relations are not distinguished (non-linear, non-commutative, etc), and instead a vision of extreme generality is considered, upon which sets of vague number patterns of descriptive properties are imposed, so that mathematics lost its relation to stable sets of math patterns and subsequently, the vague generality has no descriptive content, it floats on a general context of complicated-ness.

This leads to delusional rationality, and the precise descriptive context loses any valid form of content.

But nonetheless such practically useless quantitative constructs are determined, and then used in arbitrary ways, ie can be used to deceive and manipulate others based on a claim to the pattern being quantitative (so it must be true). Apparently, the idea that one can statistically (using arbitrary statistical constructs) determine when a person will commit a crime, so that the authorities can pre-emotively exterminate the person. One can see that this would be an efficient way in which to exterminate unwanted people.

For example,

1. the solutions to non-linear partial differential equations may be computed, but the validity of the numbers may not be considered, ie it may be over-looked, yet people might believe that "number never lie," and
2. the idea of differentiation is pushed into a structure in which all orders of differentiation are constructed, and since high-orders of differentiation (of a function) can be associated to polynomials, then ranges in the domain space where convergence can be defined, and where, outside these ranges, rates of convergence for small values can be determined, so that "whatever the pattern," nonetheless, a quantitative relation with a polynomial can be maintained, and thus manipulated numerically.
3. The derivative can also be related to a general metric-function whose invertability range (ie distance away from a given point in the domain space so as to remain invertible) may be limited to a single point, ie it is non-linear, thus, it is quantitatively inconsistent, and it is a math context which is unrelated to a reliable measuring context (so if a pattern is identified, nonetheless, this pattern cannot be used in a context where measuring is reliable), so such a pattern has a very limited range of practical uses. Etc etc etc.

But again, the assumed goal is that, "the complications will (eventually) be finally integrated into a unifying system of ideas" and everything resolved.

Thus one might ask "What will the unifying math property be?"

Will it be: the unit of measuring, the arithmetic operation of addition, or multiplication, or perhaps it will be the point, or the ray, or the angle? Perhaps the unifying principle is continuously commutative.

Could unifying principle for quantitative consistency be the circle and the line?

(Perhaps, but in a much more limiting math context, where generality is restrained by a need for using stable math patterns)

The way in which the media is organized and how it acts, is that the intellectual-class, ie the university professors and journalists and published artists, express a limited (and limiting) dogma (the dogma which best supports big business interests as well as their own interests) and proclaim this dogma as (almost) being "the absolute truth," and then this media-education complex acts to exclude any other expressions of ideas, because those who express other ideas must not have any merit, ie they have no intellectual worth.

This is also how the Puritans characterized the Indians.

So the actions of the plutocrats are framed in a context which does not have any meaning, ie it is arbitrary, and the public is required to react to how the media-education system arbitrarily frames things.

This irrational context for society is leading to its failures.

First let me point-out that:

Marx and socialism (as well as capitalism) are about extending the idea of the "collective" which is the basis for rule by plutocracy, ie the main idea of western culture is inequality, instead of the Roman-Emperor it is now the banker-oilman-military business interests, [and the interests of this elite group can be represented by as few as three people, ie one from each category]. That is, the western culture is a "collective" which upholds (by both violence and failed expressions of rationality) an arbitrary social hierarchy. The viewpoint of western culture is that of a collective upholding an arbitrary and vague idea about high-value, ie the high-value of the ruling-class and the intellectual constructs which support the ruling-class.

Though, the main point of religion is that people are equal, eg love your enemy, ie create things for everyone, but since Constantine, religion has been the main propaganda for empires, ie expressing an arbitrarily hierarchical context (but, we are all equal creators). Apparently, Islam grew due to its propaganda-governing model of society.

However,

Law needs to be based on equality

That is, the great American experiment is about breaking from the traditions of western society, it is opposed to collectives, and instead supports the free, creative individual, who is allowed to express their interests through equal free-inquiry, and knowledge's subsequent relation to practical creativity.

The context of social activity is knowledge and practical creativity,

This is, the collective endeavor to which equal individual acts are to contribute.

But this requires a government which serves the people, and this government (of the people) is (should be) trying its best to provide for the common welfare, (why are we paying private companies for public services? Garbage pick-up, electricity, connection to a communication system, health service, education-systems ie this is simply a system of unfair taxation, so that in this model of "taxation given to monopolies" one finds that the monopolies only exploit and dominate the market for their selfish interest)

572

Only when people are considered to be "equal creators" can there be a (truly) free market for trade.

Advertising is about providing the context for an instrument's use, and its specs (and that is all, period).

Language always needs to be re-organized (this is about education); where the re-organization is about the: assumptions, contexts, interpretations, containment sets, and if a measurable description then the measuring context needs to be measurably reliable, and the patterns used need to be stable-patterns, where the point of this, is that, language actually needs to both possess and express content, where one is trying to quickly get practically useful information about the observed properties of (physical) systems.

It is in this context . . . , where equal free-inquiry, as expressed by Socrates, and correct knowledge's relation to "practical" creativity, but where the creative context does not "have to" be based on the material world . . . , that we are all equal creators, so that this is the correct way in which to interpret the US creed of "basing law on equality" as was stated in the Declaration of Independence (now we need to live, so as to realize our creed) (and not basing the US society on property rights [and minority rule]) so that the government, which serves the people, must address the common welfare of a country, so as to serve a community (society) of equal people, who are motivated to gain knowledge and to be (practically) creative, and not motivated to be selfish, nor dominant, nor violent, where these behavior traits could be continually thwarted by the government, . . . , just as the government now (2013) continually thwarts the interests (of the general welfare) of the people, in order to serve the owners of society, ie a "state-and-owner-of-society" partnership.

As it is now

Since the propaganda system continually supports the notion of people being unequal, and an authoritative structure of fairly useless (and very narrow viewpoint concerning) knowledge (always directly related to business interests), these above assumptions concerning rationality, and the belief expressed (at the beginning) about the position of living entities within this (true) context of existence, most will neither believe nor identify with such an abstract vision about existence.

Nonetheless they instead believe in delusional knowledge which leads to idiotic and destructive behaviors.

Thus, to proclaim that "Einstein was wrong," where the experiments of A Aspect, concerning non-localness, have already has shown that "Einstein was wrong," and his model of inertial systems, essentially, can be interpreted to mean that all inertial systems are too complicated to describe, and all orbital systems are all chaotic (yet the solar-system possesses a set of stable orbital patterns);

And then to provide a set of, well thought-out, alternatives, {where it needs to be noted that the non-linear descriptive context of general relativity is useless, and without content, unless it is placed into a descriptive context which is based on stable patterns, ie and this is where the alternative viewpoint about existence enters the discussion}, . . . , but all of this (rational discussion) will (automatically) be censored, since in the land of intellectual delusions, it is the authority of personality, which is the basis for truth and these social hierarchies of authority are to be protected by the propaganda-education system.

If one does not challenge authority at its highest levels then the violence of these authorities on other's psyches (eg classifying valid rational discussion as being inept) is quite similar to, and just as damaging, as the violence which the justice system applied to the occupy movement and to the low-income communities, the violence which the justice system applied to society so as to turn people into wage-slaves who serve the ruling-class.

The delusions caused by the propaganda system, and subsequently upheld by people, is obvious

People oppose nuclear weapons, yet they revere the physics taught at universities, but the principle focus . . . , of the "knowledge about physics," which is being developed at university physics departments, . . . , is on making nuclear weapons, ie particle physics, general relativity, and string theory, are all forms of knowledge whose primary relation to practical creative efforts is their relation to nuclear weapons, and to nothing else, etc

People oppose "genetically modified foods" yet they revere the mostly ineffective complications associated to the chemistry of DNA (it is ineffective knowledge since it is based on an extremely limited understanding of chemistry), furthermore, the public is not sufficiently critical about the (clearly) insufficient evidence which supports the authoritative claim that the DNA, alone, is sufficient to make a copy of the life-form which is associated to any particular DNA-type, when it is clear that the epi-genome determines the majority of a life-form's embryonic development.

That is, the knowledge upon which GMO's are based is not adequate to be able to judge the affects of creating and then using GMO's,

ie the effects which "DNA modification" can have on
the life-form and on
the environment and on
using the GMO's as food,

. . . , cannot be known because of the inadequate model of life which our culture possesses.

By discovering and using radioactivity and DNA our culture, by its actions, has greatly exceeded the knowledge which we actually possess, and through which the uses of these properties and systems are to be be evaluated.

In these cases where there is a lack of knowledge, the "precautionary principle" should be used to determine "what is safe."

On the other hand, to error on the side of safety ie the protection of the public (the precautionary principle) is not even considered, when it comes to global warming, and all of the science of global warming is questioned by the ruling class of oilmen. That is, the owners can criticize science, which is based on statistics, {where it should be noted that using statistics as a basis for science should be criticized}, but the public is not allowed to criticize any form of science.

It may be that the public does not challenge authority because "if they express the idea that they do not believe the authorities" then they will never be able to get jobs, ie they are terrorized by their wage-slave social status (this is possible due to the extreme violence of the justice system).

Yet, the public should be critical of a science, which is not capable of describing the observed patterns, and is either not being related to practical development, or the practical development being done, ie in regard to manipulating DNA, is being done in a context of insufficient knowledge about the nature of living systems, and the relation of DNA to both the living system and the environment.

59.

Calvinism

Calvinism and the western culture

Calvinism (the belief that those who possess money are favored by God) extended Constantine's use of religion as the propaganda of the western Empire . . . (where institutional religion is an arbitrary worship of hierarchy) . . . . to a religious based propaganda system for a (western) banking empire.

[note: the Judeo-Christian-Islam religions, as well as capitalism and collectivist-socialism (often called communism) possess the central beliefs of the western culture, ie where in western societies the collective efforts of the people of a society are focused-on supporting an arbitrary social hierarchy ruled by a few people]

{note: by 1400 AD the Medici banking-family were choosing the Popes for the Holy-Roman-Empire, where the Popes, essentially, ruled western Europe}

Education in western culture came to be about the indoctrination of the absolute truths associated to the society's economic interests.

To show that this "indoctrination of the absolute truths associated to the society's economic interests" is intrinsic to the education system of the west, consider that, because of the nuclear weapons development during WWII, E Teller was able to change the physics departments of all (US) universities into nuclear-weapons engineering departments, by (about) 1952.

Teller's nuclear weapons viewpoint of physics is about describing the chaotic transition state characterized by a model of random collisions of the components which were broken apart form the original stable state of material whose break-up was caused by the onset of an unstable transitioning context (eg gathering certain types of material together), where the (now) chaotic system transition is occurring between pairs of relatively stable-states of material systems, ie the initial and final states of stable material systems. This is a chaotically transitioning model of material between stable states, which totally ignores the issues concerning the stability of material systems, where determining the reason for "the stability of material systems" is the much more fundamental problem.

However, axiomatic formalization of mathematics, expressed through the Bourbaki publications (around 1920), where the "axiomatic formalization of mathematics" took over the professional math community between 1890 and 1910, and it was a prominent intellectual action which pulled precisely formulated technical issues, away from expressions of ideas at an elementary level of discourse, so as to instead place the discourse as being based on an arbitrary and vague set of quantitative patterns which can be placed onto any context where either measuring or numbers can emerge. Thus, quantitative descriptions became ever more arbitrary and manipulability, and the obvious failures of a precise descriptive language, eg its inability to describe observed stable patterns, could be "papered over" with a vast array of complicated number relations, and the claim that when these abstract quantizations of vaguely general (and unstable) patterns become fully integrated with one another, then the failings of the descriptions would be corrected. Thus, such a focus on a fixed set of axioms ignores the content of Godel's incompleteness theorem, which states that a fixed set of axioms has great limitations as to what patterns it is capable of describing.

Axiomatic formalization takes existing math patterns, which are correlated with (successful) descriptive constructs, and pushes them into greater complications, where the types of number-relations are not distinguished, and instead a vision of extreme generality is considered, upon which sets of vague number patterns of descriptive properties are imposed, so that mathematics lost its relation to stable sets of math patterns and subsequently, the vague generality has no descriptive content.

This leads to delusional rationality, and the precise descriptive context loses any valid form of content.

But nonetheless such practically useless quantitative constructs are determined, and then used in arbitrary ways, ie can be used to deceive and manipulate others based on a claim to the pattern being quantitative (so it must be true).

For example,

1. the solutions to non-linear partial differential equations may be computed, but the validity of the numbers may not be considered, ie it may be over-looked, and
2. the idea of differentiation is pushed into a structure in which all orders of differentiation are constructed, since high-orders of differentiation can be associated to polynomials then ranges in the domain space and outside these ranges rates of convergence for small values

can be determined so that "whatever the pattern," nonetheless, a quantitative relation with a polynomial can be maintained.

3. The derivative can also be related to a general metric-function whose invertability range (ie distance away from a given point in the domain space so as to remain invertible) may be limited to a single point, ie it is non-linear, thus, it is quantitatively inconsistent, and it is a math context which is unrelated to a reliable measuring context (so if a pattern is identified, nonetheless, this pattern cannot be used in a context where measuring is reliable, so such a pattern has a very limited range of practical uses. Etc etc etc.

But again, the assumed goal is that, "the complications will (eventually) be finally integrated into a unifying system of ideas" and everything resolved.

Thus one might ask "What will the unifying math property be?"

Will it be: the unit of measuring, the arithmetic operation of addition, or multiplication, or perhaps it will be the point, or the ray, or the angle?

Could it be the circle and the line? (Perhaps, but in a much more limiting math context, where generality is restrained by a need for using stable math patterns)

The way in which the media is organized and how it acts is that the intellectual-class, ie the university professors and journalists and published artists express a limited (and limiting) dogma (the dogma which best supports big business interests) and proclaim this dogma as (almost) being "the absolute truth," and then this media-education complex acts to exclude any other expressions of ideas, because those who express other ideas must not have any merit, ie they have no intellectual worth.

This is also how the Puritans characterized the Indians.

But consider the types of ideas which are being expressed on the inter-net today (8-18-13)

There is show #484 of M Keiser on (rt.com), where Keiser describes how the spy-state is (violently) invading every human's existence (so as to support the interests of monopolistic business interests), so as to either sell something or to behavior-modify people's behavioral-economic actions, as well as infiltrate and intimidate any group whose dialog (or whose expression of ideas) opposes the "state-owners-of-society" interests.

So that we have a society where there are those who can watch all the other lesser people, but the society cannot, in turn, watch those who are "doing the watching" (of all public behavior which electronic information systems can scoop-up and interpret in any context in which the owners of society want that data to be interpreted, ie to gain insider information, to steal, to engage in fraud, to gain sexual favors from people, to exploit and manipulate people, etc . . . etc . . . . etc.

This is effectively the social set-up which supports the one supreme god, the arbitrary god of monopolistic business interests.

M Keiser seems to, occasionally, provide some of the more truthful social commentary in the media (he does not try to uphold the hierarchical rotten-ness which is so obvious to see).

Then

There is the #5 Chomsky interview by D Gee on wildwildleft,com, where Chomsky, as usual, demonstrates his "elitist stand with traditional authority."

D Gee concludes the interview, asking Chomsky about the trend toward ever more extreme violence perpetrated by the "state-and-owner-of-society" partnership, which is arrayed against the public . . . ,

though Chomsky acknowledges the war between "the state" and "the public," [where the state supports inequality, thus, making a collective intent within the public, which is in opposition to equality]

 . . . , whereas Chomsky's reply (concerning Gee's question about increased violence of the state towards the public) is that the "state-and-owner-of-society" partnership has always been extremely violent, but (Chomsky goes-on) the US is the least violent society, and getting less violent.

This is both baloney, and it is double-talk.

Though it is true western civilization has always been extremely violent (apparently it deep prevalence makes it invisible to Chomsky), he is excluding the information that the violence can now be directed systematically at whole categories of people, who are now being classified by single words, the justice system is being Googlized in the name of efficiency.

He seems to be echoing the expression of M Thatcher, "this is the only way,"

(perhaps J Caesar could have expressed the same viewpoint).

Where the same attitude exists within the media, in regard to our society's, so called, superior rational systems (that there is only one way), due to propaganda strategies, that science possesses an absolute authoritative truth.

That is, Chomsky speaks out of both sides of his mouth, similar to Obama and all the other politicians and media-hounds. Though, if one pays attention, one sees that he suggests he is for equality, but the over-all structure of his discussion points-out that he is an elitist, and stands for authoritative dogmas. He stands with those who believe-in delusional idiotic behaviors, ie those who have fit into the arbitrary structures of social privilege. He is not critical of the experts as he should be, but rather he seems to be in total support of the narrowly defined and delusional authorities.

Furthermore, he (Chomsky) seems to be unaware that private economic interests are in the detention business, so more numbers are needed in the prisons for both validation which they would claim from the practical numbers of detentions, which measure the number of bad people in society, and with more prisoners there will be more profits, perhaps Chomsky sees this as a non-violent way in which to deal with the population, as well as being unaware of the context of society, which M Keiser has recently described (see above), a violent invasion of privacy, which allows "anyone" to be targeted and fake statistics presented, based information gotten by spying, "which is only to be interpreted by the authorities" who then show that that "the anyone who was targeted" should be incarcerated (where the judge then instructs the jury to only interpret the experts viewpoints to be true), or exterminated. It may be that the case against the former Alabama governor was created by means of this targeted spying, perhaps at the request of K Rove (during Bush II).

Chomsky approaches the debate on "inequality vs. equality," based on distinctions he makes between two elitist thinkers, Aristotle vs. Madison, where he claims that Aristotle stood for "elitism but also for a welfare state" and Madison stood for "domination over society by the aristocracy." Thus, Chomsky ignores the egalitarian ideas about knowledge and creativity expressed by Socrates, as well as the importance of independent thinkers, eg such as Tesla. That is, the structure of his argument is based on a red-herring.

Furthermore, Chomsky ignores the capacity of the "state-owner-of-society" partnership to be able to assassinate people by means of controlling (or handling) people who are capable of . . . or are prone to . . . assassinate others, ie people who are either professionals or who possess hatreds based on their self-righteous beliefs in an absolute context, ie based on delusions. That is, these violent hateful people act on beliefs which are unrelated to practical creativity, rather they want to use their knowledge so as to join a side in a battle over dogmas and absolutes, which when examined, turn-out to be arbitrary expressions of superiority and domination. That is, after over 3000 years of an arbitrary hierarchical social structure based on violence, the population is full of these types of sociopaths.

Note: Many of the (native) cultures of the Americas were very egalitarian, and their base in knowledge was expressed by C Castaneda, and it is more sophisticated than the European knowledge.

Chomsky also believes in a self-righteously expressed set of absolute knowledge, developed out of (or based on) authoritative intellectual traditions, ie intellectual authority is not to be

580

challenged, because then there does exists an absolute knowledge upon which a person's worth can be measured.

Chomsky is a high-standing member of the propaganda-education system, so ultimately, as a wage-slave, and a "contender for merit," he will support the system which, in turn, supports himself, but this is true for all journalism and meritorious intellectuals, they manipulate and twist their words to support their arbitrary positions of social privilege.

They are all a bunch of gas-bags.

[Perhaps these gas-bags should get together with R Limbaugh and sing a gas-bag chorus.]

The activity of these gas-bags is to express the rationality (the dogmatic authority) upon which the social hierarchy is based, and to protect this social hierarchy and its associated very limited rationality.

60.

Declaration

It needs to be noted: That the truth of the new math construct (as presented by m concoyle) cannot be judged by today's experts, since that is like having the authorities of the Ptolemaic system judge the truth of Copernicus's system in that age. They are different ideas expressed in languages whose assumptions, contexts, containment sets and interpretations, etc are quite different and quite incompatible with one another.

That is, the truth of the new math-physics construct will be found based on both (1) its descriptions of new properties, and descriptions of new contexts within a new interpretive format, and (2) the practically useful instruments, and inventions, as well as new creative contexts, which emerge from the new math-physics descriptive structure.

The propagandists will claim that "its truth" (the truth espoused by the experts and journalists who serve the propaganda system) must be determined by "the intellectually superior people," those superior people, the experts whom the propagandists uphold, and which the media has cataloged in their push for the public to worship the personality-cult defined by the people at the top of institutions which the propaganda system supports, where personality-cult is the religion which the media develops and maintains (and propagates), and the personal history (the CIA records, the records which serve the interests of (and are available to) the owners of society) concerning "m concoyle," does not show him to be either a genius, or at all that capable, so this description "he gives" cannot be true, based on the risk assessment, which is used to determine investment risks. That is, this "truth" (which the propaganda system serves) is based on the type of personality which best serves the interests of the owners of society.

Note: Now the intelligence of an individual is defined to be "a person's testing behavior," which can be learned (or improved) through a scheme of behavior modification, ie as stated on a PBS

8-9-13 TV-show about the brain. Intelligence is (was) measured as achievement, in regard to "an assumed set of educational development levels," ie it was a measure of high-cultural advancement, where "high-culture" is, effectively, determined by the owners of society. That is, intelligence (in a manner similar to defining an economy) has no meaning, except as an emotional attribute, which expresses a value of high and low, and which is used by the propaganda system, which, in turn, is vaguely supported by experts whom gather statistics about people's behavior patterns.

But this is the way the propaganda system operates.

The journalists are essentially mindless brown-nosers, and the experts are pathetically indoctrinated obsessive types, who are also mindless, ie without a capacity to judge the value of a new idea by themselves, the experts have always been guided (and motivated) by the propaganda-education system.

So one sees the circular-ness of the meaningless logic, which is the basis for modern rationality, a rationality which is based on both propaganda which guides the experts, and experts, who simply compete within the narrow rules which are provided to them (provided for them by the propaganda-education system), but these competitive people have not judged "truth" for themselves, except in terms of irrelevant details about their data collecting processes, where data is always interpreted inside a theory, which the data really does not uphold, ie the data is filtered by the theory, ie the theory directs the attention as to what is counted as data. The expressions of "ideas" of both the media and the wage-slave experts, are based on this circular-rationality, because these efforts (of the media and the experts) are all about serving the interests of the owners of society.

Neither the math nor physicist asks:

"Why all this pointless complication (in a process which supports the authoritative theory)?"

"Why are the very stable properties of a wide range of general and very fundamental physical systems not being described based on the laws of physics and mathematics?"

They seem to accept the ridiculous claim that "these systems are too complicated to describe," even though they are stable systems, and can be distinguished from one another based on these stable and precise properties. But the existence of these properties, of stability and precise measurable properties, imply a linear, stable, metric-invariant, and continuously commutative (almost) everywhere, set-structure, of both physical properties as well as these math properties, being related to such consistently measurable stable physical systems.

"Who is there who can judge fundamental truths?"

"Who is there to judge the value of someone's creative efforts?"

The truth of an idea or of a construct is related to the range of understanding, that the viewpoint provides, and the range of practically creative things, and the new range of "the creative context," which determine such value to be associated to new ideas and new constructs.

The bankers and the Roman-Emperors were both incorrectly attributed to possess such judgment about both value and to be able to judge truth, ie investment risks. But investments and directing the development of already "known technologies," are not about developing new ideas, rather they are about exploiting traditions and exploiting animosities and competitions within a narrow context.

We need to re-kindle the idea that the law of society is to be based on equality, in relation to knowledge and creativity, as Socrates saw it.

The main issue confronting the US public has been framed by the Occupy Movement, it is "the 1% vs. the 99%." Because US law follows the law of the Roman-empire, and is based on property rights and minority rule. This means that the public institutions are supporting, in a collective fashion, the right of a few to steal from everyone else, so that this "right to steal" is supported by both public institutions and by extreme violence. It is claimed that, this social structure is supported, so that this ruling few can supply to the masses the, so called, great works of technology. But the fixed structure of authoritative knowledge, where its fixedness has caused "knowledge based technology" to become stunted in its expansion, due to control of information and control of what is proclaimed to possess high-value.

But high-value in human terms is knowledge and creativity, not material possession and highly controlled authoritative dogmas. Science has been re-instituted as an authoritative dogmatic institution, no different from the authoritative church in the age of Copernicus. One of the ways in which this has been done, is through formalized axiomatic, where math is "not" based on intuition and context and interpretation and new containment structures, rather mathematics was given "fixed rules," which were assumed could be applicable to all contexts, where quantitative structures might apply.

Thus, one has the carefully determined set of, so called, truths, which are true only within the rules of language, and these patterns of language have no relation to precise measurable descriptions of existence, nor to practical useful ideas associated to existence. Axiomatic formalization creates only a literary context for creativity, it is unrelated to existence as we can perceive it.

This is possible, since the ruling class is allowed to identify a narrow range of knowledge (ie the knowledge upon which the power of the ruling class depends) to which an abnormally high amount of (arbitrary) value is associated. But it only has high-value to the ruling class, since their great products the ruling class provides to the public, in a similar way as the Romans gave to its conquered people, the great works based on the narrow set of ideas associated to brick-technology. Similarly, today, the narrow set of ideas associated to "19th century science" of thermal and electrical science, are the basis for the so called great works, which it is claimed that the ruling class provides to the public.

But the issue is that the "1%" are at war with the "99%" and this is because the "1%" want to make sure that there are no new ideas which emerge from the "99%" whose technical value is greater than 19th century science, so that this would allow other ideas and other inventions to compete for the propagandistic title of the greatest "brick layer," or the greatest "technology." That is, the Roman-emperors provided brick-laying to the public, while the ruling-class today (2013) provides the technology of 19th century science. Furthermore, this is the proof that the high-value defined by the ruling class is superior to the value of the public.

Thus, this knowledge of the relation of 19th century science to the complicated instruments of today, is guarded by: making it very complicated, and highly competitive, in a game which is unfair.

As S Jobs showed "the children of the electrical engineers" can be used to base a high-tech company. That is, this is the same structure as trade-guilds and the monopolistic relation that trade-guilds had to bankers in the 11th century, where this was the transition time between, power being held by a solitary ruler, to a similar social power being possessed by bankers, in association with the church, where the church is the center of authority and the basis for propaganda.

So that this social-domination by bankers was well realized by the 15th century, where the Medici would be a good example of the power of bankers in that age.

Science was supposed to have changed this formula for social control ie the bankers and the church controlling the idea of high-social-value within society, but science quickly became professional, and the autistic Newton disdained relating his ideas to society, so as to result in social knowledge, ie an elitist protected knowledge of science, serving the moneyed-interests.

This was allowed since the social-institution of wage-slavery was developed in Rome (or perhaps Sumeria), and based on extreme violence, and the possession of special knowledge, ie brick-laying.

Note: The best social position for the middle-class was to be in the Roman-army, which required an entry-investment, which only someone in the middle-class could afford.

Furthermore, this is a house of cards, up-held mostly by propaganda and traditions. The rulers have no more value than anyone else, but the language of the propaganda-education system is based on this fraud of high-value.

This is exactly the point at which the ideas of Socrates, about equal free-inquiry, enter. The ideas of Socrates are ideas which take-down the illusion of high-value which the ruling class has cleverly instituted within society, but the idea that there be a ruling-class is also based on extreme violence, which is needed to support law being based on property rights and minority rule, ie the basis for Roman-law, ie the right to steal and be supported in a collective manner (by society) for that theft.

The Declaration of Independence stated that US law was to be based on equality, so as to break from the western traditions of violence which were set in place by Rome where law has been based on property rights and minority rule and by the violence through which this social structure could be maintained, but the US society and the administrators of the law could not maintain this ideal, mostly because of the conditioning of the European culture of which the aristocracy of the US who led the revolution were a part. But nonetheless the reason the US colonies thrived as much as it did was about the egalitarian Quakers. It is the Quaker colonies which was the most thriving and it was the most energetic of all the colonies. This is the rule for western civilization, where very equal societies developed the thought-of-the-culture and the energy for a culture, over a very brief time duration, and then the context of violence and dominant thought patterns within which these egalitarian communities were surrounded and where these dominant thought patterns were prevalent within the equal community so that some of the individuals in the equal societies started developing advantages over the other individuals in the community and they used these advantages to gain greater social power so to subsequently dominate the community.

This is the story of the community from which Socrates existed, where Socrates expressed the main property of an individual who exists in an equal society, namely, that an individual has the right to equal free-inquiry. That is, any person has the right to gain knowledge (through equal free-inquiry) and to use that knowledge to create new things, so that this creativity is motivated by a selfless desire to give gifts back to the earth, where the earth is the main nourisher of one's own life.

The human being is an entity who is:

curious,
gentle, and a
creative entity;

while a person within unequal societies learns to be authoritative, violent, and narrow, and to live with an "us vs. them" world-viewpoint, an "inside of the group vs. someone outside the group" awareness, eg an outsider questions the validity and authority of the narrow viewpoint of the inside group, but where the inside group represents (an arbitrary) high-social-value.

But in science this has been re-interpreted to also be an "inside vs. outside" the group where value is identified in a context defined by a mythological existence of an absolute truth which is associated with possessing the correct viewpoint (ie the insider's viewpoint).

Thus, there is an elitist viewpoint that after the earth was "put into the correct frame of reference" by Copernicus, and "materialism was associated to measuring" [but dependent on the measuring existing at a certain size-scale], and a continuum was developed, so as to define the construct of local measuring of properties defined on the sets within which the system (whose properties are contained) is contained (where the containment is based on the idea of materialism) and the material identifies a particular size-scale, then if one is outside this narrow dogma then one is again considered to be outside the group of the superior intellects who accept this dogma of materialism and its absolute relation to differential equations. But is a differential equation a model of local measuring as defined and used by Newton, or is it about sets of operators defined on function spaces, without clear physical restrictions placed onto the properties of the function space (then), the operator viewpoint has not been able to identify the observed order of physical systems, so that modeling a physical system as a function space has not been valid, and modeling descriptive contexts which depend on the idea of "indefinable randomness" (randomness based on unstable [or unidentifiable] patterns) have not shown themselves to be valid. That is, formalized axiomatic has led to delusional viewpoints concerning modeling by "formalized axiomatic patterns" properties observed within existence.

Should societies be based on collectivism which is in support of the stealing which is successfully undertaken by a special elite-few, a collective support for a society based on stealing, and fixed dogmas, and this highly stratified society is maintained by violence and terror, but the high-value of such a stratified and unequal society is arbitrary and leads to failure,

Or

Should societies be based on individualism and equality, and ideas developed based on equal free-inquiry?

The issue of a described truth can now be placed onto computers as written descriptions, where the computer can identify letter-strings and word-strings, which, in turn, can be related to selecting the expression's (within a written paper) set of assumptions, contexts, interpretations . . . , etc, which are being expressed within the descriptive language as belonging to different categories "associated to these properties of language" (or of a descriptive structure) in regard to the descriptive language one is trying to build (or simply by requiring that a person who is expressing ideas, also must identify the assumptions contexts etc, by which they are basing their descriptions) and this should be true, especially, when one is in the context of equal free-inquiry, and trying to determine a truth within the very limiting context of precise descriptive language, where the limitations are expressed by both (1) the idea that one's precise description be based on the reliability of measuring and based on the existence of stable patterns, as well a (2) by Godel's incompleteness theorem (ie that there are many patterns which a precise descriptive language cannot describe), in regard to precise descriptions, based on developing the idea of quantity based descriptions in turn (in regard to Godel's ideas) based-on identifying a uniform unit of measuring and then counting to form the basis for measuring comparisons (these are the simple ideas about quantity upon which Gödel based the proof of his incompleteness theorem).

To promote the idea of law based on equality, as stated by the Declaration of Independence, so as to break from the western tradition of violence, as defined by the Roman civilization, equality is to be interpreted to mean that we are equal creators, ie each individual is an independent creator, where each person is to discern truth develop the knowledge which they need to create a new thing and then create that thing. Then to sell this idea or thing within a truly free-market, where the point of government is to support each person's individual equal creative capacity and to not allow markets to become dominated by the so called good ideas but, perhaps, more importantly not to allow advertising other than presenting the specifications and uses of a created material or mental product, in a categorized list of products. Where the more demanded products need to be collectivized and controlled in regard to their over-all affects (or even possible affects) on the earth. The baloney about individuals need to profit from their creations is a bunch of baloney it is creativeness in a context where judgment of the value of a creation is not allowed, since such judgment determines an inside and outside to the idea of value and these insides vs. outsides are maintained by violence, where mental violence of this type of value-judgment is the more dangerous type (or way) of promoting violence. The main ideas on which the creative attributes of our society are built were developed by Copernicus, Galileo, Newton, Boyle, a few thermal scientists, Faraday, and Tesla, so the argument that those with the good ideas should profit, should have also resulted in these (just listed) families being the richest families in the world. But clearly they are not. This can only mean that those families, which are so rich, must have gained their riches from stealing and an associated set of behaviors by those in society which result in institutional violence and direct violence, where this institutional violence supports the thieves and it opposes the public. These behaviors are preformed by: the justice system, the military, the professionals who maintain the complicated instruments for the rich, and most notably by the propaganda-education system which is only allowed to express the ideas of a certain type of a high-valued truth, whose truth must be protected, for fear that the public be mis-informed.

Yeah . . . , right . . . , like the media expresses truth.

The media is all about irrelevant baloney, and digging-into the details concerning all the red-herrings (ie information which mis-directs and mis-informs), which are planted in the media, and framed in language which supports inequality and a mis-guided viewpoint about truth.

All the issues about injustice needs to be re-framed as re-formulating law in the US so that US law is based on equality and not based on property rights and minority rule, where property rights and minority rule was the basis for the law of the Roman-Empire, as well as monarchies, as well as the basis for rule by the bankers-oilmen-military big business people, ie the basis for the 1% to rule.

All issues about violence and the ineptitude of the justice system (ie injustice), and the military violence, need to be framed in terms of "basing law" on equality.

Instead of social collectivism based on violence (or imposed by violence), which is the distinguishing feature of both capitalism, and Marxism (except Marx was arguing for an elite set of leaders who were not necessarily rich) consider the idea of equality where the main point of society is the creativity of each individual.

The individual must:

1. determine what they want to create
2. They must develop a precise language within which whose context the creation can take place
3. Within this endeavor they must discern truth for themselves or
4. Supply creative alternatives to the problems which they cannot surmount.

Particle-physics is an endless discussion about irrelevancies, where one can know that this is true, since there is not anything that one can create within the descriptive context of particle-physics, particle-physics is only about determining the cross-sections of elementary-particles, and this only relates to "rates of nuclear-reactions."

Basing the organization of society on the equal creative powers of each individual, in an individual-based society, ie not a collective society, is incorrectly related to Marxism . . . , (planning society based on committees of elites (or Marxism) vs. planning society based on committees of selfishly motivated elites of capitalism; is essentially, the same form of government.) . . . , with the claim that "a society of equal individuals is a form of Marxism." This statement is completely incorrect, and is a smoke-screen to mis-direct the public into not paying attention to the fact that capitalism is the collective and it is about governing based on a committee who plan society based on their selfish interests and the collective society does the bidding of these selfish few people.

The reality is that there is virtually no difference between Marxism and capitalism, as the example of China shows (or does it show that Marxism is a better collective society for capitalism than capitalism can form a collective society based on stealing and violence).

These types of mis-leading statements are a result of one of the main tactics of propaganda. Namely, that of supplying society with an endless supply of irrelevant ideas.

This is the main condition of quantum physics and particle-physics (ie expressing non-useful ideas), especially particle-physics, while quantum physics is a result of a highly protected dogma associated to an authoritative intellectual tradition, ie it is a result of requiring intellectual elitism, ie a manufactured elitism needed to make the control of language much easier. That is, the idea of the existence of an expert who knows the absolute truth about a subject. The thermal and electrical engineers associated to 19th century science are the only experts who possess knowledge related to making products. But given the mysteries which have been observed for the structure of our existence, and the failure of the idea of materialism to provide answers concerning the structure of the observed properties of material and the failure to provide new ideas about practical creative development, means that the highly-valued truth of the experts is a "failed-truth."

The "1%" have failed to provide truth which can result in expansion of creativity, based on practical creativity (as opposed to the somewhat sterile literary creativity, though it is the western tradition to based creativity on a described truth, but in regard to practicality one wants a measurable description, where measuring is reliable and the patterns being described are stable). Furthermore, the fact that the re-newable energy sources should have been developed back in 1900, but instead the justice-system gave to the oil-barons a free-reign to dominate society and destroy the environment. This was done by the justice system, and this alone should mean that the law of the land should be changed so that law is based on equality to know and create.

That the "1%" have failed to provide truth which can result in expansion of creativity, means that the high-value (provided to the public by the owners of society) is nothing but an illusion, and that the social institutions are corrupt, and they failed in showing any type of valid capacity for judgment.

The US needs a new continental congress, so as to establish that "US law be based on equality," so that the government serves so as to address "the common welfare of the public," and so that each person is seen as an equal creator, who can contribute their creations to a truly free-market, where the market is to be protected from domination by a few, or by a few products.

This is not the collectivism of capitalism, which is run by a few (very very rich) for their selfish interests.

Start by creating a central bank controlled by the continental congress. Protect some institutional systems, which can be changed later, but pinch-off all the debt and the trillions in paper-money bets.

The problem is that we are born into the world, and the world supports and nourishes our lives.

but

On the other hand, capitalism (as well as any type of collectivism) is based on stealing the valuable parts of the earth, upon which we are nourished and upon which we base our material creations, and then stealing the creative ideas of others, so as that this theft is maintained by the social institutions which are based on violence, and which support the thieves, so as to force people to buy these stolen constructs from the main thieves of society, the thieves who create an image of possessing high-value.

This is maintained primarily with propaganda, where the best example of propaganda and its close relation with the maintenance of the social position of the thieves, is Constantine (ie propaganda is not new) and how he crafted a manual (or book) of an Empire's propaganda, and he called it (or it was already called) the Bible, where Constantine followed Paul (the true creator of the Christian church) where Paul was a master of propaganda.

Using language to influence people so as to support an external structure of which the public are but a little part (this is the structure of both the church and of the Roman-Empire). Thus, the people are supporting ideas which are in opposition to their own interests.

The trick, in regard to Paul's propaganda, is that the main point of religion, is the statement that people are equal, eg love your enemy, etc, but the institutional church is about expressing institutional inequality, and this is done in the name of the person who loved his enemies.

Constantine saw the way the church's propaganda, in fact, could be used to support an unequal empire and empire almost exclusively based on violence, (the early Christians were themselves quite violent, in regard to their different, competing, sects, apparently, this violence was a result of the intrinsic violence of the society of which they were a part), although the Roman army built many substantial things with bricks.

Again this is the blueprint for capitalism, where there are some products which the society values, thermal and mechanical constructs and an electrical technology which almost all is derived from the ideas of Faraday and Tesla.

Where Faraday and Tesla show that people primarily like to create, and they do so mostly in a selfless manner, as both Faraday and Tesla demanded very little from those who used their ideas.

The main idea of TV was invented in 1910, as a mechanical device, and this idea was realized by fast switches on a cathode-ray tube, and then the fast switches led to the computer. The transistor and the circuit-board allowed full development and miniaturization, where the transistor is not so much about quantum description but rather about thermal techniques which allowed for impurities to be introduced into the crystals lattice, and then incorporated locally on a crystal (or isolated, with chemical layers, so as to be local), so as to build circuit boards on crystals. Though they use quantum properties of the semi-conductor, they are built based on classical thermal-chemical and classical optical properties so the quantum properties are incorporated into, and used within, a classical electric circuit.

One of the strongest holds over society is the belief by the public (as well as the investor class) in the absolute-truth which is espoused by the authorities through the propaganda system. Truths as identified by the experts and the investors, where the investors have a very strong influence over the education system and research in very strong ways.

This is the belief structures of the collective, which is gathered together through (by means of): direct violence, the violence of the social institutions which uphold the right of the minority to steal and the propaganda-education system, and the intrinsic violence of the economic system which maintains this collective support for the (very selfish) ruling-few.

This is controlled by the propaganda and violent cultural traditions of a very violent western culture, where one of the results of this institutional violence is the condition of wage-slavery, a condition imposed on the public by the justice system where law is based on property rights, ie the right to steal, and minority rule, ie the assumption of government by the few that people are not equal. This imposition of wage-slavery was done in a manner similar to the way in which the justice system re-instituted, essential, slavery in the southern US in the 1870's and the 1880's.

Wage-slavery was imposed in the US by both property rights and controlling immigration ie the politicians were always controlled by their "being a part of the propaganda system" . . . (through which many individual politicians could make great fortunes), . . . , and by squeezing the poor, who did not own property.

It also requires a system of domestic terrorism and spying, where the reason for the spying is to be able to precisely control and target the, so called, "enemies" of the ruling-class, the few who stray outside the confines of the propaganda system.

But most enemies have already made deals with the ruling-class, so they are caught-up in the propaganda system, while thinking on one's own "is (can be) outside the propaganda system," but it is not done by an enemy, but rather it is done by a true competitor, someone with better ideas.

The new context for realistic possibilities, ie the new ideas about set-containment of a measured existence, is expansive.

The ultimate context which it identifies, is the realm where thoughtful living systems consider "How to extend the scope and properties of existence itself."

"How to extend to higher-dimensions."

"How to fashion new shapes with new properties."

Or the more mundane, yet, which might seem even more exotic:

"How to use finite spectral-sets so as to have both stability of pattern (of existence) and a "fluid" use of existing patterns and shapes?"

The practical use of existence is about: harnessing energy, using stable existing angular momentum structures (associated to the stable circle-spaces defined over many-dimensions), using the relations of many-dimensional existence and its relation to both size-scale and to infinite extent relations to the (an) over-all containment set, especially, as applied to some living-systems, which are subsets to this many-dimensional structure.

61.

Stable

Introduce the need for such a book, as the book "Describing physical stability"

Issues in regard to "What is real?" and "How to describe observed physical properties?" should have a deep affect on science and physics and mathematics.

The point is that modern physics (and its math constructs) cannot describe, to sufficient precision, the stable spectral properties of the general many-(but-few)-component physical systems so that these descriptions are based on the, so called, "accepted laws-of-physics." Thus, one needs to re-consider the more difficult question as to "What is real?" and, "Is the idea of materialism valid?"

Consider that:

1. Materialism identifies an open-closed topology, upon which the (partial)-differential-equation-models for the "laws of physics" are defined. But these laws cannot be used to identify (to sufficient precision) either "the cause" or "the structure" of the observed spectral-orbital properties of fundamental physical systems which exist at all size-scales, eg nuclei, solar-system, etc. Neither . . . the laws based on models of local measuring of physical properties (as originated by Newton), Nor . . . laws based on the combination of "function-space and sets of operators," which are models of both randomness and (supposedly, the stable) spectra of quantum physics . . . , can provide these needed descriptions.

    That is, in regard to both quantum techniques and particle-collision techniques; the stable spectra of general fundamental physical systems do not emerge as descriptive properties which possess sufficient precision, in regard to either the algebra-operator-structures applied to function-spaces or when non-linear operators are applied to internal particle-states which are attached to the functions of the function-space, where, supposedly, the particle-states are supposed to perturb the wave-functions of the quantum techniques so as to make the calculations more precise.

2. The claim to be able to describe reality, based on the idea of materialism and sets of differential equations, seems to be "doubtful at best," especially, since, after nearly one-hundred years of trying to use function-spaces as the basis for these types of descriptions (of the observed stable spectral properties), and, essentially, coming-up empty, in regard to being able to "sufficiently precisely" describe the spectral properties of general quantum-systems.

    Furthermore, in regard to macroscopic physical systems, the non-linearity of general relativity means that it mostly describes chaotic patterns (ie unstable patterns), and this means that general relativity is not capable of describing the observed stability of the solar-system. This has resulted in applying . . . general relativity . . . "only to cosmology," where the assumptions about the uniform properties of space and that physical law, as defined by differential equations, are applicable everywhere, cannot be verified (within the context of materialism), by all the measurements which are, in fact, needed "in order for these assumptions to be considered valid" so that these speculative descriptions can be considered to be science. Rather, it is speculation, and the data is interpreted through physical theories, which might not be applicable. Thus, cosmology is a mostly irrelevant discussion.

    Quantum physics cannot solve the difficult problems and general relativity is irrelevant.

    Thus, one can conclude that, it is . . . worse to follow the ideas of the modern (overly dogmatic) physics . . . than it is to follow new ideas.

3. One way, in which to extend beyond the construct of materialism, is for one to find a math model which possesses "more dimensions" than "the idea of materialism allows," and yet "the properties of materialism" form a subset within that new descriptive context.

    The new definition of a many-dimensional model of existence requires that the different dimensional-levels be stable shapes, and the models of material components are also stable shapes, but the stable shapes which are material-components are one-dimension less than the dimension of the stable metric-space shapes within which the material components are contained, so that all of these shapes (both material-components and metric-space shapes) have an open-closed topology, so that the dimensional levels are discretely discontinuous (or boundaries are defined) when one changes dimensional-levels, and there are many different possible size-relations which can exist between different dimensional levels (yet the set of different sizes, for these shapes, is a finite set). This is a new model of existence, within which materialism is a subset, and within which the stable spectral-orbital properties of the general many-(but-few)-component physical systems can be described to sufficient precision.

4. The new context is a math model which has surprising implications, but these, apparently, strange implications, are "strange," because of the limitations of language . . . {where our language is a language which has been built more on "faith in materialism," than it was built, so as, to determine a hidden physical reality}. A hidden physical reality, which exists beyond the idea of materialism, can be described when that "hidden structure of reality" is mathematically modeled.

So, why not follow math constructs which actually solve the most difficult problems . . . , ie being able to describe (or determine from a math model) the stable spectral-orbital properties of the general (many-(but-few)-component) physical systems . . . , facing physical description, rather than following those (math-physics) ideas which do not solve these difficult problems?

62.

Outline of stable shapes

Stable shapes and descriptions of existence

Math is about quantity and shape, where a math description can only possess meaningful content (ie accurate and practically useful, within the context of the existence which we actually experience), if it stays in a context where measuring is reliable, and the patterns which are either used to make a description, or which are identified (in a deductive manner) by the description, are stable patterns.

The main properties, which must be maintained, in order for a math context to be stable is if: (both) (1) a unit of measuring is uniformly stable and reliable, and (2) the shapes and patterns, which are being described by math (in a reliably measurable context), are very stable.

The point of Godel's incompleteness theorem is that the set of (true, and/or stable) patterns which a given precise language . . . (which is associated to some fixed set of axioms and contexts) . . . . is capable of describing are very limited.

Thus, in order to re-generate a useful language, so as to develop a new practically useful and measurable descriptive [possibility (or context for) the] language for math, then the language must be continually re-configured in the contexts of: assumption, interpretation, set-containment, determination of stability and uniformity, and the re-organization of (stable) the patterns, (all) used in a precise language, etc, etc.

Experts cannot be used as "the people who could most efficiently do such activity," such as, re-organizing and re-interpreting language and data, since experts are those "who relish fixed dogmas within which competitions can be defined," and in such a context the experts have competed with complete faith in the truth of the language which they use.

That is, the context of new assumptions and new interpretations and new viewpoints, is the place where the ideas of Socrates about equal free-inquiry is the only relevant viewpoint in regard to developing a new language (or new descriptive) basis for both knowledge and (practical) creativity.

The context of equal free-inquiry is the true context of science, and it should define the social relations between humans as a society of equal creators. That is, according to Godel, the only way to develop (new) knowledge which is both accurate and practically useful, is if everyone in all of soceity is considered to be an equal in such an endeavor.

It is assumed that describing the properties of a physical system depends on formulating, and then solving, a system's (partial) differential equation, where these (partial) differential equation (models of physical systems) embody physical law, and are, supposedly, invariant to arbitrary changes in the coordinate frames, ie the coordinate frames within which these equations are formulated, ie an assumption about physical containment, where one of the main assumptions about containment is the idea of materialism.

However, there are a great number of very stable physical systems which at exist at all size scales (nuclei to the solar system) and whose properties are both stable and precisely observed. This precise stability of many such physical systems suggests that they are linear, solvable, and controllable systems.

However, today (2013) the equations of these many-but-few-component systems cannot be formulated, let alone solved, where the equations used to describe the properties of many of these systems are (is) based on an assumption of (either) randomness (or non-linearity, or both) so that the observed random events are either not defined . . . , ie neither can the physical system's equations be formulated nor are their properties calculable, or . . . , the fundamental events of the description are not stable . . . . , ie the patterns of "unstable elementary particle event transitions" form the basis for physical law (within particle-physics), ie at the base of the equations describing material-interactions (where material has been reduced to a small set of elementary-particle events) is a non-linear derivative (which is related to the internal particle-states of elementary-particles) . . . . , and the subsequent math context (even in regard to macroscopic geometry) is:

1. non-linear, and
    (as just described above)
2. indefinably random (ie random events are either not identifiable [ie calculable] or they are not stable), and
3. non-commutative,
      so that the descriptions of physical system's properties cannot be formulated or solved in a such a math context,

so that only particle cross-sections and a property of spherical symmetry is being described in such a dys-functional descriptive (or mathematical) context (ie the context of the description is about rates of reactions for a spherically symmetric explosion),

wherein no other properties of observed physical systems are capable of being described within this descriptive context, ie the descriptive context of particle-physics and general relativity and their derived string-theory, etc.

These contexts of math descriptions do not qualify as valid math models, they are not quantitatively consistent, and they are illogical (eg probability based on a set of unstable events), and there is no a stable pattern about which the descriptive content can have meaning, because the observed stable patterns cannot be calculated.

That is the observed patterns of physical systems are well-beyond the capability of such (or this type of) a dys-functional precise descriptive language to describe.

Thus, the language of physics needs to be re-formulated in new types of assumptions and contexts etc.

One cannot describe either the stability or the charge-neutrality of the atomic and molecular systems without the use of models for these systems in which these systems are both "stable and topologically-closed" discrete metric-space shapes.

Assume that shape is the only part of a math context which allows for stable math patterns to be defined

Math is about quantity and shape, but the stable quantitative patterns depend on describing stable shapes. For example, the quantitatively consistent properties of real and complex-numbers are about describing the line and the plane, where the plane is about measuring rays and circles.

It should also be noted that the continuum and the "axiom of choice" result in defining a quantitative set, such as the real-numbers, which is 'too big' of a set. This overly-big "set property" leads to many logical problems, such as filling a descriptive context with too many contradictory math constructs, eg particle-physics based on point-particle-collisions defined in a context of randomness, where, in turn, the assumption of randomness requires an uncertainty relation between dual (Fourier transformed) quantities, ie position and motion. Thus, the uncertainty principle does not allow for point-particle-collisions.

Note: However, point-particle collisions are an indispensable part of the models of nuclear reactions. This suggests that the nuclear industry is determining the descriptive context of physics, for professional physics communities.

This is how commercial interests categorize and pigeon-hole the ways in which the language of math is used so as to make "descriptive knowledge" serve commercial interests. But this means that science is not about an indifferent development of knowledge, rather its knowledge is identified and categorized by commercial interests and knowledge becomes fixed, so as to serve these commercial interests.

Assume that stable patterns in math . . . , which possess relations with (to) accurate and precise math descriptions, and so that the descriptive structures of these stable math patterns provide practically useful information (associated to such a math language) . . . , are the stable patterns of geometric shape.

Stable shapes are:

1. linear,
2. metric-invariant (ie the unit of length measurement stays uniformly stable),
3. where the metric-spaces must possess metric-functions with constant coefficients,
4. The metric-spaces are metric-space-shapes, which possess non-positive constant curvature, ie they are both discrete Euclidean shapes and discrete hyperbolic shapes.
5. these stable shapes are associated to "rectangular"—related polyhedral fundamental domains [of (checkerboard) lattices (of metric-spaces)], so that the moded-out shapes (so as to mostly result in circle-space-shapes) are continuously commutative (almost) everywhere (perhaps with the exception of a single-point). The atlas of these shapes would be either one-patch or at most two-patches so that these covering sets relate to the differential structure of the isometry fiber groups or the finite dimensional classical Lie groups, where the transformations of local coordinates (by the Lie algebra elements of the Lie fiber group) would always be done with diagonal matrices, ie continuously commutative coordinate-related shapes.

The properties of the discrete Euclidean shapes and discrete hyperbolic shapes

These shapes, when viewed within their own space possess an open-closed topology. Thus, the boundaries as seen in the next higher-dimension metric-space within which these shapes are contained would not be apparent (from within), and no holes would be seen on the space's fundamental domain (polyhedron), when within the shape.

Thus, in the Euclidean shapes, what appear to be rectangular coordinates within, in turn, appear to possess properties which seem like latitude when viewing their properties from the next higher-dimension and thus would appear to possess curvature and the metric-function would not possess coefficients which are constants (as similar to the case of the sphere). The cylindrical circles would be geodesics but only the most far reaching and the most inner circles or the toral shape when viewed from a higher-dimension would be geodesics, but within the discrete Euclidean shape there would still be the "cylindrical circles," which are geodesics, so there would be no "push" due to quantitatively inconsistent shapes, ie there is flatness within the shape.

For the discrete hyperbolic shapes, when viewed in the shape's (next-higher-dimension) containing metric-space, one would see a coordinate system with a few circles [which define the geodesics of the shape's rigid and very stable shape and stable set of spectral properties where the spectra are determined by the lengths of the geodesics of the shape] while the rest of the coordinate system would be sets of orthogonal hyperbolic shapes (where in 2-dimensions it would be hyperbolic-curves). In such a shape "the hyperbolic shapes would not be geodesics," and thus, would push any material-component, which might be contained in this metric-space shape . . . , (and where the material-component does not already possess the shape of one of the geodesic circles, ie if the material occupies [or fills] the geodesic spectral-flow) . . . . , towards the few geodesic circles of the discrete hyperbolic shape.

This push, by the discrete hyperbolic shape's natural orthogonal hyperbolic-coordinates (except for the few circles, which intersect the hyperbolic coordinate shapes), would also exist within the shape. This is an example of the claim made by general relativity that "the shape of space determines inertial motions," but where the shape is a linear, solvable, stable shape of a discrete hyperbolic shape.

The conditions of bounded-ness and unbounded properties of discrete hyperbolic shapes, and a set of properties which restrict these shapes from being continuously deformable into one another (as discrete Euclidean shapes can be continuously deformable into different sized discrete Euclidean shapes of the same dimension), where these restrictive properties are associated to the existence of a discrete hyperbolic shape's fundamental domain, ie an allowed "discrete hyperbolic shape" depends on both the lengths of circle-segments of the rectangular polyhedron fundamental domain and the reflected angles defined around the various vertices of this polyhedron (so that the fundamental domain defines a lattice on the hyperbolic metric-space). Many of the properties for discrete hyperbolic shapes were determined by D Coxeter, by considering the symmetry groups defined on the vertices and faces of the fundamental domains of the discrete hyperbolic shapes, while the properties of rigidness and stability (of the geometric facial measures of a fundamental domain of a discrete hyperbolic shape) were determined by various others eg M Gromov.

Some properties of discrete hyperbolic shapes "of note" (which have not been mentioned) are: the 5-dimensional shapes are the last bounded discrete hyperbolic shapes, and the unbounded 10-dimensional discrete hyperbolic shapes are the last dimension within which discrete hyperbolic shapes can be defined.

Note: The fundamental domains of discrete Euclidean shapes are simply the rectangular shapes, which are similar to the square-shapes which fill a checkerboard lattice.

That the spaces of constant curvature (in particular the non-positive constant curvature), and the set of discrete shapes defined on these metric-spaces, determine the stable shapes of geometry, so as to be the geometric shapes towards which other non-linear shapes evolve (or decay), is the essence of the geometrization idea of Thurston-Perelman.

Coordinate base-spaces and their fiber groups

Such sets of fundamental domains (of a lattice partitioned metric-space) exist within a real-number context of discrete isometry (ie metric-invariant) subgroups, or in a context of either isometry or unitary (with complex-number coordinates also associated to Hermitian-forms, like metric-functions, and the property of Hermitian-invariance) classical finite dimensional Lie fiber groups, whose base spaces, in the real-case, are the metric-spaces associated to isometric (ie metric-invariant) changes of (real) coordinates (although these fiber Lie groups would also be such a stable-shape's differential structure, but these stable shapes usually only have one or two patches in the shape's atlas, ie the exceptional single-point of an otherwise continuously commutative shape). Where if there is one-patch in an "atlas cover" of a shape (or manifold) then this means that at each point well-defined orthogonal coordinates are determined by the directions of the natural coordinates of the shape, where, for circle-space-shapes, these natural coordinates are a bunch of circles, whose local directions are always locally orthogonal to one another, so that the natural coordinates of the shape define the shape in a global manner, ie there is no need for local coordinate patches so as to have well-defined (local) rectangular coordinates.

That is, these stable shapes (defined in the real-numbers), ie when considered in their real context, are related to the metric-spaces of:

1. R(n,0) (Euclidean space, and associated to inertia), and
2. R(n,1) (general space-time spaces, which are equivalent to hyperbolic spaces, and associated to charge and energy) and

3.   R(s,t), where s is the dimension of the spatial subspace, and t is the dimension of the temporal subspace, (new [Euclidean] metric-spaces which are associated to new material types, where such new material is defined (or contained in) in metric-spaces whose spatial subspace has a dimension higher than 3-dimensions, ie beyond the dimensional-limiting idea of materialism). Note: Two-dimensional discrete isometry shapes are contained in 3-space.

Metric-space states, and the (physical) properties which a particular type of metric-space can describe

Opposite metric-space states create a unitary relationship associated to fundamental symmetries of the (physical) properties which are (or need to be) associated to physical descriptions. This (new) relationship depends on new coordinates which are based on the complex-numbers, and so as to have local measures of length be dependent on a Hermitian-form, [ie not the real-case of metric-functions], where the complex-coordinates relate to the two opposite states as each different state being contained in either the real or the pure-imaginary subsets in the complex-coordinates, where in each subset there is contained one of the pair of the separate and independent opposite metric-space states.

These opposite metric-space states are associated to the symmetries of spatial displacements which are related to Euclidean space, which is related to the physical property of inertia, and temporal displacements whose symmetries are associated to conservation of energy, eg stable patterns, and which are related to hyperbolic metric-spaces.

The opposite state of a spatial position in Euclidean space is a spatial subspace, where the frame of a spatial position is the distant fixed-stars, and where the frame of rotating distant-stars is the opposite frame, where the idea of "rotation of the fixed-stars at infinity" is an idea associated to a fixed and well-defined subspace (although higher-dimensional properties may be what we see at a great distance) which extends to infinity.

The stable energetic properties of physical systems, ie a hyperbolic material-system-component's stable shape, which possess dynamic motions, which are modeled in a local context, so that in this context there are a pair of metric-space states of opposite time-directions of a motion caused by a material-interaction.

The existence of such opposite time-states (also) manifests in regard to a charged system's wave-equation (of a current-charge caused [or related] "potential" 1-form) in R(n,1), ie the advanced and retarded potentials, and its subsequent dynamical dimensional-shape relation (of the potential's field structure) to differential-forms in R(n+1,0), (as described in the new context of dynamical material-interaction descriptions).

Material-component dynamical interactions are determined by the tangent properties of touching (intersecting) shapes, which are a part of the material-interaction process . . . , ie material-shapes and the Euclidean-shapes of interaction (which is related to the separated material-components which are interacting) . . . , which are being related to:

1. the local geometric properties of the material-interaction process, and
2. the local geometric properties of the fiber group, ie a derivative-connection type operator, which determines (along with the opposite metric-space states, see #3) the discrete spatial displacement which occurs in a dynamic process whose description depends on the definition of discrete, discontinuous time-intervals, and
3. opposite metric-space states of time, which are defined locally
   (the changes of a local spatial displacements are locally an involution in the two opposite metric-space states, where a transformation [between metric-space states] is an involution, when it is its own inverse, that is the local spatial displacements in the two opposite metric-space states are equal and opposite,

Furthermore, the two opposite metric-space states (if the real-state is multiplied by i) define vectors whose commutators of (or) the Lie bracket stays in one or the other of the two vector spaces (which exist in either the real or pure-imaginary subsets).)

Properties of (stable) metric-space shapes

The discrete Euclidean shapes can be related to different size-shapes in a continuous manner, and they are compact, and form during a material-interaction (so as to form within each discrete discontinuous time-interval) in a (the) spatial context of action-at-a-distance,
while the discrete hyperbolic shapes are stable, and discretely and discontinuously related to different sized shapes, and they are either compact or unbounded.

The compact-unbounded properties of discrete hyperbolic shapes are related to the position-subspace opposite properties of the Euclidean-inertial properties of material shapes.

Both a material-component and a metric-space are modeled as stable metric-space shapes of various dimensions (up-to an 11-dimensional hyperbolic metric-space).

That is, the discrete hyperbolic shapes are stable, while the discrete Euclidean shapes form within a condition of resonance with the toral components of the stable discrete hyperbolic shapes.

Note: The smallest or highest-energy toral component of a discrete hyperbolic shape becomes the dominant discrete Euclidean shape and thus the dominant inertial property of the material system.

The model of a material component depends for its existence (in the over-all high-dimension containing space) on its being in resonance with a finite spectra, which is defined by a partition of an 11-dimensional hyperbolic metric-space by a finite set of discrete hyperbolic shapes (which represent hyperbolic metric-spaces), where the 11-dimensional hyperbolic metric-space is the over-all high-dimension containing space.

Since metric-spaces are also stable shapes (as are material-components, but the material-components are contained in metric-spaces, which are 1-dimension greater than the dimension of the material-component, thus, size of a metric-space's shape is an important in regard to its containment of material-components).

This construct, which is made in a higher-dimensional context (than is allowed by the idea of materialism), implies that there can be a dimensional-subspace partition of a high-dimensional hyperbolic metric-space (namely an 11-dimensional hyperbolic metric-space), where this partition is defined by the set of largest compact (as long as they exist; note: the last dimension in which discrete hyperbolic shapes can be compact is hyperbolic-dimension-5), discrete hyperbolic shape which exists in each of the different dimensional-subspace sets, in the 11-dimensional hyperbolic metric-space.

The property of being in resonance with some element of the partition must be true for any of the smaller material-components which is contained in a metric-space shape (subspace) to exist within a metric-space (of the correct dimension and the correct size).

The partition defines a finite spectral-orbital set.

This means that our measurable existence is based on a finite spectral-orbital set of quantities, ie this is better than basing number-systems on the idea of a continuum, as in the case of the real-numbers.

A central construct which is associated to material and metric-space properties, is a tree of containment ie a subspace containment which is dependent on properties of: dimension, subspace, and size (of the subspace shape).

Material interactions

A set of both stable components and stable metric-space structures, or set-containment configurations, together determine a stable "scaffolding" which is related to the dynamic changes of material-components, (which, in turn, are contained in metric-spaces which possess shapes), where the material changes can either be

1. Dynamic

Or

2. Changes in the properties of material-systems, which are collision-related, and during the collision the changes in the shapes and spectra of the colliding material-components is dependent on the properties of resonance between the colliding, interacting material-components and (or with) the finite spectral-set of the 11-dimensional hyperbolic metric-space (which is the over-all high-dimension containing space), and the result also depends on the energy of the collision.

There are two contexts within which material is contained within (or occupies) a metric-space shape:

(1) the material component is about the same size as its containing metric-space shape, and the material occupies the shape of the metric-space's spectral-flows (or main geodesic shapes) which exist within the metric-space, and

(2) where the metric-space is very large and contains many material components, so that much of the material contained within the metric-space shape is a set of many small components, which, in turn, form into condensed material, so that the condensed material interacts with both

A. the metric-spaces geodesic structures and

B. the material defines second-order metric-invariant (partial) differential equations, which perturb the dynamics of the condensed material, which is also interacting with the metric-space coordinate-shapes which, in turn, are related to the set of the shape's geodesic coordinate-shapes.

Then there are two-types of the dynamic interactions of "free-material components," (ie most often not colliding material-components) where these interactions depend on geometry, and the dynamics are usually defined in a context of second order (metric-invariant) (partial) differential equations, which are defined on material-components in a particular-dimension metric-spaces, namely, the two types of interactions are:

1. Independent material-interactions (classical material interactions)
2. Material-components interacting with the shape of the material-containing metric-space (the interaction of material with the shape of the metric-space, within which the material is contained (ie an idea associated to general relativity, but now defined in a context of linear-shapes and possibly linear solvability of the dynamic system).

In regard to #2, the partial differential equations (which will be a model of a 2-body system, since the envelops of orbital stability are much greater in influence within the interaction than are the interactions with other-bodies in the system) are used to perturb the orbital-spectral properties of the stable orbiting material-component.

It should be noted that this type of geometric-model for material-interactions when applied to many very small material-components would define a set of Brownian motion dynamic paths for the material components, and subsequently, (due to E Nelson, Princeton, (1967)), this type of Brownian motion defines the same type of random properties as observed for quantum systems, ie quantum randomness is derivable for systems composed of many, small material-components when their dynamics are based on the new geometric model of both material-components and material-interactions which are defined in a discretely discontinuous way.

Important properties of the descriptive context

The relation between dimensional levels is discrete and discontinuous.

There can be defined a (new) scalar factor between discrete discontinuous changes of dimensional levels.

This discontinuous relation between shapes in different dimensional levels can also be a part of material-interactions, where the material-interactions (of material contained in a particular metric-space of a particular dimension) is defined in a metric-space which is 1-dimension higher than the dimension of the material-containing metric-space.

The relative sizes of material-components which exist in two adjacent metric-space containing-spaces can either determine or hide the detect-ability of inter-dimensional material properties.

Hypothesis:

Both "van der Waal's forces result from higher-dimensional interaction structure," and Dark-matter manifests at large size-scales, due to a large scaling-difference which exists between two adjacent dimensional levels.

There are also discrete and discontinuous Weyl-angle folds, which can be defined between the toral-components of stable discrete hyperbolic shapes.

A scalar factor can also be defined in the context of these discretely discontinuous Weyl-angle folds (ie folds between toral components of discrete hyperbolic shapes).

Quantitative consistency

The set of line-segments and circles determine the general context of quantitative consistency, as indicated by the same algebraic structures of the real numbers, ie the line, and the complex numbers, ie defined on the complex-2-plane. That is, the circle and the ray shapes which are used as measures of length and position for the 2-plane, where there is quantitative consistent relationship which exist between circles and line-segments in the complex-number system, and this allows for both the existence and the construction of stable shapes out of sets of circles and line-segments, where these stable shapes can be both material-components and metric-spaces, where, in turn, metric-spaces can contain (small enough) lower-dimension material-components within themselves.

Can stable shapes exist beyond the context of discrete hyperbolic shapes, which was described by Coxeter?

Are stable hyperbolic shapes only relatable to discrete hyperbolic shapes (subgroups) and certain types of fundamental domains or can there be many types of stable shapes which can be constructed from discrete, discontinuous changes related to . . . :

dimensional,
angular,
scalar, and
cylindrical component,
. . . , changes,

for these discontinuously discrete contexts, which are of great importance within the new descriptive construct, ie where these constructs are different from circle-space shapes ie shapes only related to toral-components, where these (new) constructs are (can be) related to either fundamental domains or to quasi-stable (or naturally oscillating) hyperbolic shapes?

In regard to determining the general context within which the line-segments and the circles can be put-together so as to form stable shapes, then consider:

A pool of hot-lava which cools very slowly. The high-temperature implies that charged material composes such a pool of lava during some of the important stages of its cooling.

The pool of lava, when slowly cooled, can partition itself into discrete hyperbolic polygon (or polyhedral) shapes so as to solidify into a crystal partition. A crystalline structure can be imposed on basalt-rock.

This, apparently, happened at "Devil's Postpile" in CA, where there exist 5-, 6-, and 7-sided polygons which also define smooth walls along cylindrical columns, so that the polygons were occasionally curved as is a hyperbolic polygon in a 2-dimensional hyperbolic metric-space.

These polygons, which formed upon the slowly cooling lava, filled a lattice which was defined on both the top and the bottom of (what was) the lava pool.

This suggests that the discrete hyperbolic shapes do not have to be genus-related, ie holes in shapes, they do not have to be related to the types of polygons whose 1-face (or edge) structures are (usually defined) in multiples of 4 (four), ie one-hole in a discrete hyperbolic shape for each multiple of four 1-faces in the polygon of the shape's fundamental domain.

This implies that there could be a relation of discrete hyperbolic shapes to rectangular polyhedrons which have various numbers of faces (1-sides) associated to their own shapes, which can be related to a set of cylindrical (or column-like) components of stable shapes, ie an un-moded-out 3-shape.

Furthermore, such shapes could also be folded by Weyl-angles, where scaling-factors could also exist between discrete Weyl-angle folds.

Appendix:

The mysteries about existence are there "plain to see," eg why are the fundamental many-but-few-component systems so very stable?
And
in order to try to find an answer to this question;
anyone can try to consider these mysteries, and
anyone's considerations are as valid as anyone else's.

Thus, if the propagandists do not themselves consider extending knowledge in regard to knowledge's obvious failures, eg not being able to describe the stable spectral-orbital properties of many fundamental physical systems, then "what the propaganda that these propagandists are espousing" is not worth anything, ie they are being paid to espouse the knowledge which is of interest to the monopolistic investment forces within society. In fact, it is an expression of arbitrary high-value, in the same way in which racism is also such an expression of arbitrary high-value.

Knowledge with a wide-scope, concerning existence and about both life and "material," flounders inside narrowly defined categorical-boxes. The most damaging "narrow viewpoint" is materialism.

This is because materialism is the pillar of:

1. property rights,
2. Darwin's survival of the fittest in a context of material scarcity, ie competition,
3. corporate productivity,
4. control over (material) instruments (because of property rights), and
5. militarism and violent coercion.

Further causes of the confining narrowness, into which knowledge is being forced, are due to a particular knowledge's relation to the investment of product and/or process of monopolistic businesses:

1. Physics has become bomb engineering, (this has a debilitating affect on the other sciences, eg chemistry),
2. biology is DNA (as the best-bet for getting control over medical knowledge),
3. a very low-level viewpoint about chemistry, (where the knowledge about the chemistry of enzymes is quite limited, due the very low-level knowledge about atoms and molecules)

    Furthermore, chemistry has been dominated by oil interests and its branching into farming (control of the world's food supplies), but the knowledge upon which these viewpoints of:

    (a) oil,
    (b) DNA, and
    (c) enzymes,

    are based is a very low-level form of knowledge, and this is, in part, due to the confinement of physics to bomb engineering,
4. Communication channels (computers) are about military and propagandistic control of society, a very materialistic viewpoint.

The unification of physics should be about the categories of:

1. space,
2. material (with a desire to transcend the idea of materialism),
2 (b) and a need to transcend the idea of materialism, so as to model complicated systems which seem to have non-understood properties of unification, Eg
3. chemistry, and

4. life,

where part of the problem in biology is an invalid model of life, and this interferes in chemistry, but chemistry is also hindered by an invalid model of atoms and molecules.

Unification should be about existence's

1. containing space,
2. its processes,
3. properties, and
4. its set-containment structures.

The new ideas provide a viewpoint which transcends the idea of materialism and also provides a set of stable geometric models of charged systems which are stable and can be neutrally-charged, and it provides a model of both life and mind.

How is this related to the "laws of physics" and the idea of "materialism"?

The unification of the laws of physics, placed in a context of materialism, is considered in regard to finding the "correct" set of arbitrary frame-invariant (or (local) diffeomorphism invariant) non-linear partial differential equations (for a system and its material-interactions) placed in a context of both materialism and randomness, ie the solution function is assumed to be a part of a harmonic wave-function, which provides probabilities of random point-particle-events (ie both physical property [eigenvalue] and particle-type) in space (or space-time), defined in a context of particle-collisions, a model which only relates to a process which is a part of bomb-engineering models, and particle-collisions have virtually no relation to understanding the properties of:

atoms,
molecules, and
living systems,

. . . , since the, so called, unifying equations are all about particle-type, and only most-marginally related to a system's physical properties, ie it is a probability model of a system composed of many-but-few-components, and since these types of systems are so very stable, then this means that such a random model is ludicrous.

63.

Quantum physics and propaganda

Consider quantum physics, and how it is situated in the propaganda-education system. One finds a very limited range in regard to valid descriptions (based on the, so called, laws of quantum physics) of the (actual) wide range of existing quantum systems,

But also consider how one can re-interpret the observed properties of quantum systems within the context of stable geometric shapes.

But also consider how one can re-interpret the observed properties of quantum systems within the context of stable geometric shapes, instead of relying on a descriptive language based on randomness and instead, change the descriptive context to a much more realistic set of models of very stable discrete hyperbolic shapes, ie circle-space shapes, which very easily fit the context of the observed properties of quantum systems, and with a capacity for both much greater accuracy and precision of the physical system's properties.

Take a text on quantum physics, for example, Quantum Physics for Dummies, by S Holzner (2013), which is fairly well organized, and quite easy to read, but one needs to consider its central purpose as a "cog in the propaganda-education system" ie it has propaganda-education value.

That is, one (most often) finds, in quantum-physics textbooks, that claims are made about the validity of quantum-physics which are more about propaganda, than these textbooks are about an accurate and practically useful truth.

Where this contention (about the propaganda-education system) can be proved, by the fact that, no valid descriptions of quantum systems exist for the wide range of very fundamental quantum systems.

That is, the observed properties of some of the most prevalent and fundamental quantum systems and which possess very stable spectral-orbital properties cannot be described using the laws of quantum physics (where the laws of quantum physics are based on both the idea of both randomness and that physical properties (of random quantum systems) are determined by (often differential) operators, which act on function spaces, which model a quantum system's randomness, so that solving this system identifies the quantum system's discrete, measurable, spectral properties).

This book makes most of the "usual over-reaching claims" about the validity of quantum physics.

Descriptive contexts (within technical texts) can pivot on terse explanations, which are invalid

That is, one finds that in physics (and often-times math) textbooks (especially in texts on quantum physics) there exist solitary (or isolated) phrases, and/or solitary sentences, which are central to the flow of information of a text, so that these short phrases are used to re-define the content of a description, so that these solitary pieces of information can (and often do) change the entire logical sequence of the discourse.

And,

One also finds that the content of these crucial sentences are not critically examined, and often they are related to a math-process which is being described, so that the process, and the content and meaning of the description (or the content of the text) is completely changed by these terse solitary phrases or sentences (which possess so much importance).

For example, it is often stated, that the periodic table of the elements can be understood from both the angular-momentum, and energy, properties found for the H-atom, so as to be able to identify the periodic structure of "the periodic table of the elements," where it is claimed that the periodic-structure can be determined from the degeneracy of the energy-levels identified (caused) by the angular-momentum spectral properties of the eigen-wave-functions of the spherically symmetric energy-operator of the H-atom.

The energy of the H-atom depends on both the solution functions to the radial equation and the discrete angular-momentum properties, L, with eigenvalues, L, where the energy term (in the energy-operator) which is contributed by the angular-momentum properties (of the H-atom) is proportional to "L^2 divided by the diagonal-values of the diagonalized moment-of-inertia tensor,

I," (or multiplied by the inverse of the symmetric, 2-tensor, I) (or most simply considered in its diagonalized form, ie the diagonalized . . . , symmetric, 2-tensor, I . . . , matrix)).

There is the energy-level degeneracy due to the projection of L to the z-axis, and the several such (discrete) L(z)'s (ie related to a set of the latitude angle-measures of a 2-sphere) are associated to a given energy state (except the ground state), so as to cause $(2L+1)$-degeneracies for each energy-level (associated to energy-angular-momentum eigenvalue, *L).*

(apparently) There is a correlation between "the periodic table of the elements" and the degeneracies of energies of the H-atom, where these energy degeneracies are due to the angular-momentum properties of the H-atom, [and the energy-degeneracy of different latitude measures on the sphere (associated to each energy-level), which result from its spherically symmetric potential energy term].

But this correlation between an element's position in the periodic table and the energy degeneracies of the H-atom cannot be made definite (it seems to be an unexplainable relation between H-atom energy-degeneracies and the length of the rows of the periodic table):

(1) the other elements (which are different from the 2-body H-atom) in the "periodic table" cannot be assumed to possess a potential-energy term which is spherically symmetric (this is because there is "now" more than one electron orbiting the nucleus, ie the system is more than a 2-body system reduced to the single center-of-mass point, about which spherical symmetry can be defined). Thus, the energy degeneracies cannot be identified by the angular-momentum properties of the H-atom.

and

(2) the radial equation cannot be properly identified, let alone, solved, for a general atom in "the periodic table of the elements."

This means that the "principle quantum numbers" as well as the energy resulting from the angular-momentum of the system of a general atomic-element are not known from the calculations based on the laws of quantum physics applied within the context of the energy-operator (or the quantum system's set of physical-property-defining operators) for a general atomic-element in the Periodic Table.

Thus, claiming that such a general element shares the H-atom's energy degeneracies cannot be (proven to be) true.

The failure to show that "the H-atom's energy structure is related to a general element's energy structure" is proven by a new interpretation in regard to the case of the Stern-Gerlach experiment, in which spin was detected.

The Stern-Gerlach experiment

Ag-atoms, whose atomic numbers are 47, and whose spectral energy properties, as determined by assuming them to be related to the angular-momentum properties of the H-atom, implies that the 47th electron would have principle quantum number 5, and its angular momentum would either by 0 or 1, so that the three states of different angular-momentum quantum-numbers,-1, 0, and 1, (associated to current-flows of the Ag-atoms) should be separated into these three-states by a magnetic field, if the Ag-atoms are passed by a magnet, as was done in the original Stern-Gerlach experiment.

But instead only two-energy-states appeared, or only two-energy-states were separated by the magnetic-field.

This means that the 47th electron was not in the energy structure which was related to the H-atom, but rather its (the 47th electron's) energy-degeneracy was only two, ie the up-spin-state and the down-spin-state. Note: Why is this true? (these two spin-states should possess an energy spectral-value which is close to the same atomic-energy level)

The spin-structure of the Ag-atom and its energy properties dominate the Ag-atom, in this Stern-Gerlach experiment. That is, the Ag-atom acts more like a Fermion-component than an Ag-atomic spectral-orbital energy structure.

Consider the details

How do the energies of a Fermion compare with the, assumed to be true, energies of the Ag-atom, if one assumes that the energy-degeneracies of the Ag-atom can be correctly related to the energy-degeneracy properties of the H-atom?

The results of the original Stern-Gerlach experiment says (implies) that the assumption . . . of the energy-orbital structures of the Ag-atom being the same as (or closely related to) the energy-orbital structures of the H-atom . . . is all wrong.

Note: an atom's energy is determined by three quantum numbers the principle energy number, n, the energy of the angular-momentum number, L, and the energy of the spin-number, S.

[for:
E   (5) the energy of the 5th principle quantum number,
L   is the H-atom's angular-momentum, and
S   is the spin-angular-momentum of (either) an un-paired electron within an atom, ie of an atom with an odd-atomic-number, (or) (Is S only the spin-angular-momentum of the Ag-atom?)]

The energies of the Ag-atom would be, E(5) + (L^2/I + S^2), where E(5) is negative, compared with the energy of a spin-½ Fermion, which would be S^2, ie the spin energy of the 47th electron is S^2 (and the spin-energy of a spin-½ Fermion is also S^2), but the total energy, E(5), of the Ag-atom is negative,

Thus, a solitary spin-½ Fermion would be a higher-energy system, than would the approximate energy of the Ag-atom's energy would be, and thus it is not the lower-energy state Furthermore, (that is) the 47th electron belongs to the Ag-atom, and is not free, so the S^2 energy (term) only makes sense in the context of this 47th electron being bound to the Ag-atom, in which case it (the electron bound to the Ag-atom) would have a lower-energy than the energy which a free spin-½ Fermion would possess.

So, "How should the results of the Ag-atom Stern-Gerlach experiment be interpreted?" and "How should the Ag-atom be modeled, for this experiment?"

Alternative

Perhaps a quantum system is not about defining a set of operators associated to a quantum system's physical properties so that the quantum system is to be modeled as a function space, rather a quantum system is a stable circle-space shape, whose energy is determined by the circle-space shape's resonance with a finite spectral set defined by the structure of an over-all high-dimension "material"-containing hyperbolic metric-space (for a description of the structure of the finite spectral-set for the containing space, see other papers). Thus, an Ag-atom may be a "folded circle-space shape," so that the 47th electron has a different "fold-structure" than do the other 46-electrons in an Ag-atom, so the Ag-atom's shape requires that it interact with an external magnetic-field as a solitary spin-½ Fermion.

Nonetheless, the usual claim that . . . , "the periodic table of the elements can be understood, based on the energy-degeneracies of the wave-functions of the H-atom" . . . , is made in this book, without the additional statement that "the energy-states of the electron components of the different elements of the Periodic table" do not actually fit into H-atom wave-functions, which were "computed" as the (radial) energy-states along with additional angular-momentum energy-states and its (the H-atom's) angular-momentum-related energy degeneracies, where the word, "computed," is in quotes, since the arbitrary truncation of the radial series of the divergent radial-equation solution-function for the H-atom is not actually a computational technique (see below).

The author, at another point in the book, states that "the actual Hamiltonians (or energy-operators) for a many-body atom are not solvable." Thus, how this statement (about many-body atoms not being solvable) can be logically consistent with the belief that the energy degeneracies of the atoms in the Periodic-Table are consistent with the energy degeneracies of the H-atom, is not known. Nonetheless, the statement, is made in the book, that "the Pauli-exclusion principle allows the energy degeneracies of the atoms in the Periodic-Table to be exactly consistent with the energy degeneracies of the H-atom" seems to be a necessary extra-added propaganda statement required of the intellectual class, similar to the statement that "the US and its ruling 'private corporate entities' are not spying on everyone in the US," is required to be stated within the propaganda system.

That is, if a system's wave-function and energy-states cannot be found (calculated from the laws of quantum physics), then "how can energy-degeneracy be determined?"

Note: I have not actually identified all the energy-degeneracies of the H-atom so as to see if these degeneracies, in fact, do coincide with the periods of the Periodic Table.

Finding the solution to the radial equation, or more to the point, "adjusting math to fit the data"

Now consider the issue of (or the issue about) solving the radial equation for the H-atom.

The problem, in regard to solving the energy-operator for the H-atom, is that the solution to the radial equation diverges for large radial-values . . . , (when the spherically symmetric, 2-body H-atom is placed in center-of-mass coordinates, and in spherical coordinates of the radial coordinate and the two-spherical (longitude and latitude) angular coordinates) . . . . , where the divergence occurs in a manner similar to how the exponential function diverges for large radial values, eg e^r as r goes to infinity, where the exponential function is, e^x (or $\exp(x)$).

That the radial solution diverges, is made very clear in this book.

Note: This type of direct statement is slightly rare for elementary quantum physics textbooks.

This divergence is then called "non-physical."

So "How to fix this divergence problem?"

This problem is "fixed" by arbitrarily truncating the series representation of the exponential function.

But this has the same affect as simply assuming that the radial function is defined on a circle, ie e^x is changed to e^{ix} (where e^{ix} is a periodic function defined on a circle).

This truncation is exactly the same, as what Bohr did, when he assumed that the angular-momentum of the H-atom was discrete and it was defined on circles.

The proof of this (ie that the truncation is equivalent to re-producing [or following] the ideas of Bohr) is that the two methods (both Bohr's and solving the radial equation by arbitrary truncation) come-up with exactly the same energy formula for the principle quantum-number.

That is, based on a simple phrase, "this is non-physical," all of the carefully described math context was totally ignored and discarded, and a completely different math context was created (namely, that circles were associated to different radial distances), and then the math problem is reconsidered in the new mathematical context, so as to "solve the problem of non-physicality."

Thus, the idea that atoms possess discrete energies "is an assumed property," and it is not derived from the laws of randomness, ie the property of randomness is the property upon which the laws of quantum physics are based.

The assumption of the discreteness, in regard to the energy-levels of the H-atom, is equivalent to defining the H-atom on circles, and thus it is "not as principled" as placing the H-atom onto a stable circle-space shape (or modeling an H-atom as a discrete hyperbolic shape), ie a discrete hyperbolic shape which has been folded in a particular way (but folded in a way which is consistent with the allowed coordinate transformations which are defined by the fiber group (either in the context of energy or in the context of inertia), and the energy properties of this circle-space model of the system are determined by the resonance properties the circle-space shape has to the finite spectral-set defined on the over-all high-dimension containing space (for all of existence).

The "principle" involved in the new viewpoint is that a precise physical description must be done in a context of reliable measuring and the (geometric) math patterns used need to be stable.

It is the assumption of stability of (geometric) math patterns, which implies that these quantum systems be defined on the very stable discrete-hyperbolic-shapes, and it is in this stable geometric context . . . , (which is significantly more diverse, in its mathematical content, than are somewhat arbitrary circles (of Bohr), which are implicitly assumed to represent the orbits of the electrons [in the solution by "arbitrary truncation of a solution function"]) . . . , in which the property of discreteness for quantum-systems is identified.

While randomness can be derived from the new descriptive context:
It is the interaction structure in the new context which implies the, apparent, random properties for small components.

My outline of (regular) quantum physics

Quantum physics is based on reducing all material systems to small (quantum) components which compose a quantum system, whose behaviors are assumed to be random, which is modeled as a (harmonic) function-space, and there is an associated set of Hermitian (local differential) operators which both act on the function-space, and represent physical properties associated to the quantum system, where when the entire set of operators, which represent all of the physical properties of the quantum system (which can be measured as spectral values), is found to commute, then the eigenvalue structure (or spectral structure, ie the full set of the system's spectra) of the quantum system is said to have been found, where, in this context, the quantum-system identifies a discrete set of (very stable) energy-states (and associated physical properties which are associated to spectral values which can be measured), which are occupied by the quantum-components, where the random event-probabilities of the system's components are found in space and time, where this probability can be determined from the wave-function of the quantum system.

That is, quantum physics is based on the idea of randomness (associated to observed point-particle-spectral events in space and time), where the wave-function for a quantum system is, supposed to be, related to the probability of finding a quantum component of a quantum system in space and time, where the random-event point-particle-component will be (usually) associated to a particular spectral-value when the random component-event is observed.

(regular) Quantum physics is both a "linear," and it is contained in a "Galilean" structure.

But, in general, the spectra of very stable (and general, but very fundamental) quantum systems . . . , which can be observed . . . , cannot be computed by finding a set of commuting operators which represent the physically measurable properties of the quantum system and which apply to the quantum system's function-space, where the spectra of a quantum system can only be found if a complete set of commuting Hermitian operators can also be found, and such a set of operators can very seldom be found (if ever).

That is, the spectra for large sets of general quantum systems cannot be computed using the methods and physical laws of quantum physics.

This is the failing of the viewpoint.

But this is never discussed in the propaganda-education system.

Brief review of a quantum text

The text begins with the features which distinguish quantum physics from classical physics.

Namely, the wave-particle duality, where "what are considered to be (small) material point-components" can possess wave-properties, so that the wave-functions are related to probabilities of point-particle spectral-events in space and time,

And, vise-versa

what were considered waves, eg light, can possess particle-properties.

Even (or especially) statistical, or thermal, systems have quantum properties, eg black-body radiation and the free-electron gas in (often metallic) crystals.

Alternative

The new geometric model of quantum systems can derive the probabilistic features of (small) material systems composed of small components.

This is about the new structure of material interactions and its relation to Brownian motions, and the subsequent realization (identified by E Nelson, Princeton 1967, 1957) that such Brownian motions of the set of (system) components is equivalent to quantum wave-properties.

The new geometric model is based on the discrete structures of stable spectral-orbital systems, ie a discrete model of spectra, which is somewhat similar to Bohr's model of the H-atom, but it can also apply to (closed) bounded systems of macroscopic size, but discretely related, in a geometric manner, to many small components.

The new model for light is based on identifying a subspace by means of an unbounded discrete hyperbolic shape, in 3-space, so as to have an odd-genus and then applying (usually) an outward moving spherical-shell of toral shapes associated to the wave-length and frequency of the light wave, where the shell structure of the light-signal depends on the oscillating properties of the odd-genus and odd-dimensional unbounded discrete hyperbolic shape which identifies a particular subspace within which the light is confined to propagate.

These two models (Brownian motions of small components associated to, what appears to be, a fundamental randomness for these small components, and a geometric model of light) can account for the elementary quantum properties upon which quantum physics was based, but they are instead modeled by stable geometric patterns, from which both randomness and the particulate nature of light waves can both be modeled.

Note: the non-oscillating even-dimensional unbounded discrete hyperbolic shapes do define an outward moving shell, but, the inside this shell, is filled with light-particle components, which reflect back from the shell's shape with each discrete time-interval of the new model of physical structures, (this is about Huygens's principle, concerning the properties of light waves contained in either even- or odd-dimension spaces).

Back to book-review

The apparent main statement (in the quantum text) about the applicability of quantum physics to the descriptions of physical systems is contained in the statement that . . . , the wave-functions of:

1. free-particles (eg wave-packets),
2. particles trapped by potential energy wells, including a more formal treatment of the harmonic-oscillator, (most of this discussion about the shape of a quantum system's potential energy term, is done based on a box-like form to the quantum system's potential energy term, so as to determine boundary conditions, for the wave, and how waves, with particular boundary properties, harmonically fit into such a box)

   The potential energies of very many quantum systems are provided with a box-like geometric structure, eg radioactivity of a nucleus is described based on a potential-energy shape of a wall, with a finite height, so as to associate nuclear-decay rates with tunneling probabilities, ie a very contrived way in which to fit data. That is, this viewpoint of potential energies which possess a box-like shape, is used to fit data "within the probability based description of quantum physics," where the point of a probability based description is, exactly, to fit data.

   The box-potential energy is the basic model for a crystal's energy states, (ie however these are geometric potential energy structures, but for material components contained in a random quantum system, and this is a logical inconsistency), and
3. Models of point-particle-collision properties of scattering, (highly dependent on the Born approximation, of plane-waves going-out, after the scattering event) and
4. the distinction between the statistical properties of Bosons (which are represented by symmetric wave-functions), and Fermions, (which are represented by anti-symmetric wave-functions), . . . , are the main types of systems which this particular text on quantum physics "describes."

This is almost entirely a propagandistic statement, which except for (3) and (4), ** has very little relation to science, except in a heuristic context (the context of the story being told [about quantum systems] has some value, but it is not to be considered to be a valid "precise description" of the physical world).

The assumption is; that waves determine the probabilities of the quantum components, so that the frequency and the wave-number of a wave-function, are associated to energy and momentum of the quantum-components related (by probability and by spectral-values) to the quantum-waves which are being used to model a physical quantum-system.

Essentially, only a handful of physical systems can be modeled in this unrealistic manner and still fit data, and the main example, the H-atom, has already been shown to be a non-mathematical model.

That is, "what the equations are trying to show," ie the discreteness of energies for components in quantum systems, are actually being assumed within a "calculation charade."

Furthermore, the viewpoint that physical law defined for quantum physics; is a "fairly accurate description of a wide-range of general quantum systems;" is not being satisfied by the relation that the quantum descriptions have to the observed properties of quantum systems.

The statement given is that:

> "Using this kind of physics one can predict the behavior of all kinds of physical quantum systems."

This is not true.

Rather, there is only a vague, heuristic, relation between quantum systems and quantum descriptions.

That is, only a handful of "real quantum systems" can be described (or can have the behaviors of their components described) within the structure of physical law, which is associated to quantum physics, where the laws of quantum physics are based on the property of randomness, and that sets of operators can be found which are related to the set of measurable properties which (for a stable quantum system) [uniquely] determine the quantum system.

Consider the harmonic oscillator:

The quantum model of the harmonic oscillator distorts the very idea of a harmonic oscillator, since a harmonic oscillator is a classical system whose solution function are the trig functions of the sine and cosine. But, this is what the quantum harmonic oscillator also wants to be, but it (such a descriptive structure) is best defined on a circle-space shape, a discrete-hyperbolic-shape whose toral components define envelopes of spectral-orbital oscillation stability, so that these orbits could even be "distorted sine-waves," ie orbit-shapes deformed from the expected domain of circle-shapes (which is the expected domain for a wave-function).

Alternative model of the harmonic oscillator

Assume that, the discrete hyperbolic shape of a harmonic oscillator has the same-size toral components, which are folded repeatedly onto themselves, so as to allow for many-charged components to have the same orbit structure.

Propagandistic motivation within ordinary quantum physics

The reason for thinking about quantum harmonic oscillators in the context of an x^2 potential-energy term has to-do with the vague notion that, if one has a classical Hamiltonian then this can be translated into the context of quantum physics.

But the real point of a harmonic-oscillator is about how physical systems would be related to the shapes of circles, or in a more sophisticated viewpoint, (physical systems being) related to orbital envelopes defined on very stable circle-space shapes, most notably the discrete-hyperbolic-shapes which have been folded.

Physical motivations for considering harmonic oscillators

The physical motivation for considering harmonic oscillators is that there are very many micro-contexts wherein the harmonic oscillator is a good model of the small material component's oscillatory-behavior, quite often this could be a charge in an electromagnetic field, and there is also the atom "rigidly held" in a crystal lattice.

These examples are often small charged components, whose properties of motion are being determined by an electromagnetic field.

Except for an atom held in a lattice (and oscillating), these are more-or-less about wave-motions, which are thought of as being the back-and-forth motions of an object where these motions identify a cylindrical shape (where the object identifies the shape of the cylinder's base), but the motions of the object follow the motions of a uniformly rotating circle (note: waves can be defined on circles).

That is, one might ask,

"Are these motions cylindrical motions?"
or
"Are they toral motions?"

When Dehmelt isolated an electron (see below), the electron oscillated (or orbited) on the geometry of a circle-space, toral, shape, and not in an oscillatory motion which defined a cylinder.

The back-and-forth cylindrical motions are modeled, classically, as a spring with an x^2 potential-energy term. But it is difficult to determine what micro-physical property would result in an x^2 geometric potential-energy shape, whereas it is natural to consider circle-space shapes defined by their property of resonating with "the over-all containing space's" finite spectral properties.

The energy expression for harmonic systems made of small components is hw(n+½), where when n=0 there is still a ground state.

But this (non-zero ground state) implies, not a force modeled as a spring, but rather a circular-motion with a lowest rotation rate.

Oscillations which are deformed by (say) an (2-body elliptical) orbital property (of condensed material confined to a circle-space shape) would (could) be consistent with oscillations defined by an object confined to circle-space motions, which are either defined on the circle (so as to defined circular function-domains), or (better yet) are orbits defined within an orbit-confining toral envelope.

Thus, the "oscillatory solution function" (of the cylindrical harmonic oscillator) is very closely related to (perturbed) orbits on a circle-space shape.

In fact, a circle-space shape may-well be a much better model of a (quantum) harmonic-oscillator, especially, if the harmonic oscillators are based on (or built from) charges and "the electromagnetic-fields" of light-waves, since electromagnetism naturally fits within the context of a hyperbolic metric-space.

The actual (alternative) geometric model of the harmonic-oscillator

If this is to be (or it can be better) modeled by a circle-space geometry, then this would be a discrete-hyperbolic-shape composed of uniform toral component-shapes, and then folded on-top of one another, with as many toral component-layers as the number, n, in regard to the energy expression of hw(n+½).

This could also be the model, upon which light manifests within a metric-space, in relation to a shape associated to light-propagation, but where (now) the discrete-hyperbolic-shape is unbounded, but (nonetheless) with each unbounded toral component of the unbounded discrete hyperbolic shape being uniformly shaped.

This model applied to the propagation of light in a metric-space

Thus, a 3-space discrete hyperbolic shape (which is contained in a 4-dimensional hyperbolic metric-space) would essentially be a toral component with many layers. Thus, it is natural for the odd-dimensional models of light (propagation shapes) in metric-spaces to be oscillating shapes, so as to propagate light as expanding spherical shells, due to destructive interference (caused by the oscillating unbounded shape) acting on any reflective light signals defined for each discrete time-interval of all dynamic processes in metric-spaces (the time-interval is the "period of rotation" of the spin-rotation of [hyperbolic] metric-space states).

This shape, of light-propagation, would look-like a somewhat bounded-looking shape (centered around the light source) with spikes sticking-out, or perhaps more like a pin-cushion (in shape), but the pins would be unbounded, and thus would appear to be more-like lines stretching-out to infinity from some bounded-looking shape (which was centered around the light source), or vise-versa the source could be out at infinity propagating back to the "somewhat bounded appearing shape" with an approximate position in space, ie time moving in its opposite direction.

The harmonic oscillator, contrasted to an atom, where both systems are discrete hyperbolic shapes

Whereas an atom would be folded, so as to have concentric toral-components, so as to make-up the atom's shape, ie the toral components would not be uniform within the atom's geometric shape, where the atom's geometric structure of concentric toral-components is also a result of it (the shape of the atom) being a folded discrete-hyperbolic-shape.

Consider the property of spin-½

The spin-½ math-pattern is a (complex-number) 2-dimensional representation of the SO(3) isometry Lie group, and it is related to the SU(2) unitary Lie group.

The value associated to the spin-½ property has two-states, the up-state, (h/2), and the down-state, (-h/2), associated to slightly different energies.

These two spin-½ energy-states and their rotational context, deal with the spin-rotation between opposite metric-space states. This property of opposite metric-space states can be seen in regard to the (infinite extent shape) of the spin-states of the two-opposite-states of the neutrino, which possesses a (h/2) spin-state, and the anti-neutrino, which possesses a (-h/2) spin-state, where these two opposite metric-space-state neutrinos are unbounded shapes, and define the metric-space subspaces for each of the two opposite spin-states of the metric-space (and the metric-space's two opposite metric-space states).

The spin-½ properties are defined on (lower-dimensional) material components which are contained in the metric-space, within which the observations are being framed, while the integer-spin properties are associated to the metric-space, and to properties which are observed to exist within a single metric-space state, eg motions caused by material-interactions.

Note about perception

A discrete hyperbolic shape is not "the inertial object" which we sense,
Where "what we sense," is:
either
Being sensed through an inertial model of light ie of inertial packets (or tori) moving along the lines of a light-field's unbounded shape out to infinity (where there are toral packets [or inertial packets] which move along the lines, of an oscillating electromagnetic-field shape, so as to model the electromagnetic field of propagating-light)
or
Being sensed by means of touching an inertial model of a rigid-body, that is, the inertial properties associated to "established geometric-shapes, ie the inertia associated with the rigidness felt during touch (of a material object), and this inertia is partly about the toral shapes, which resonate with the discrete-hyperbolic-shaped atomic components, and which are both

(1) associated to the atomic components' rigid positions in the crystal, as well as
(2) there existing another such toral shape being related (by resonance) to the entire crystal's inertia, as a whole, which, in turn, is associated to a crystal's independent discrete-hyperbolic-shape, where this shape is sort-of superposed on the crystal, which is modeled as many very-long strings of (folded) toral components, which compose condensed material, so as to form a complex of many discrete-hyperbolic-shapes.

We do not perceive the discrete-hyperbolic-shapes directly

When Dehmelt isolated an electron by a process of electromagnetic-discharge, in regard to a cooling process, and then trapped the relatively-cool and relatively-still electron, in a containment bottle . . . , in "a particular shaped electromagnetic field," in the "still electron's" containing chamber (or containment bottle), . . . , the electron "came to rest" so as to identify a circular-motion, or in an orbital structure, which is similar to that of an atom (or similar to the orbital structure of a nucleus), that is the (still) electron's orbital motion [also] defined a "skinny" toral-shape (or a flow on a toral-component of a discrete hyperbolic shape composed of a set of folded concentric toral components).

However, the cooling and trapping "bottle" required that there exist certain electromagnetic properties which needed to be maintained, in order to keep the electron trapped.

Thus, the discrete-hyperbolic-shape into which the electron "came to rest (in its orbital motion)," may be defined by the trapping device, itself, and not defined by a natural discrete-hyperbolic-shape, which is in resonance with the finite spectra of the entire high-dimension (over-all) containing hyperbolic metric-space.

But nonetheless, in the context of the solitary electron's semi-stable resting place, of a discrete hyperbolic shape, it needs to be noted that this discrete-hyperbolic-shape is not seen, and [perhaps] the discrete Euclidean shape . . . . of its inertia (which is in resonance with the toral components of the discrete hyperbolic shape) . . . . is relatively large (and thus, with a small property of inertia) since the discrete-hyperbolic-shape of an electron's orbit might be very large, in relation to the sizes of atoms, whose masses would be greater (and whose resonating inertial shapes would be smaller, thus, identifying greater mass).

Appendix (details about the properties of the solution functions)

If one considers the Laguerre polynomials, one sees that, as a sequence of polynomials, they are alternating polynomials, the coefficients are positive, decreasing, and the limit of the sequence of coefficients is zero, so in the sequence of Laguerre polynomials . . . , in the limit as the sequence identifier (of these polynomials) increases . . . . , they are convergent (for every variable in the domain), and thus they are (tend to be) closely related to functions whose values are bounded, such as the series representations of the sine and cosine functions, (and the various representations of these sine and cosine trigonometric functions, eg the double- or half-angle representations of the sine and cosine functions), [will (could) be closely related to the Laguerre polynomials].

Furthermore, the harmonic oscillator solution function is "the Hermite polynomials multiplied by the normal function (of statistics)," where the normal function is a distribution which has a center (defined by its mean), which is (in its natural form) zero, and it has a width (defined by its variance, or its standard deviations) while the Hermite polynomials are alternating polynomials, and when multiplied by a normal distribution then the product acts, so as to be somewhat similar (very closely related) to the sine and cosine trigonometric functions, but whose oscillations are to be defined on a finite interval, eg related to the size of the cylindrical shape of the oscillating object, and so as to be a standing-wave within the harmonic oscillator's, assumed to be, bounding parabolic-shaped potential-energy function.

This harmonic-oscillator model is a bit contrived, since it is taking into account the "finite cylindrical length" of a cylindrical-model of an oscillating system (this finite cylindrical-length causes deformations to the oscillating-pattern). It is not clear that this is ever a valid model of the micro-structure of material systems composed of atoms and molecules, etc.

64.

Peer-review

In the US society one can:

buy politicians, buy the law, buy "whatever 'justice' which one might seek," and
one can buy "what our society considers to be the scientific-truth."

(so that this arbitrary "scientific-truth," in turn, identifies high-intellectual value within our propaganda-education institutions, and it also identifies the set of assumptions, to which any "media-recognized-intellectual" must agree.)

It should be noted that in the US society one can buy politicians (in the context of the propaganda system), and thus one can get any law passed that one wants passed, one can "only use the justice system" if "one has a lot of money," and that money can often buy the desired results within the justice system, and for monopolistic-businesses knowledge can be "bent" to serve narrow commercial interests, at the expense of losing an ability to "discern a truth," where identifying inconsistencies concerning an authoritative truth can change the context of creativity.

(this control that investment has over truth, depends on the fact that investment has come to determine what is created within society [both practically and in regard to art-propaganda] and the people are forced into a subservient condition of wage-slavery which funnels their interests into the narrow commercial creative context of the US society).

That is, money is determining what our society considers to be true.

{News is about the behaviors which result from society's "creative efforts" but conditioned on the fact that the control of money over society results from a condition of identifying a few people who are entitled-to-steal, and they also determine "what is true" within society, and the associated condition of the extreme violence required to protect and maintain these entitlements.}

Peer-review is an expression of a state-of-knowledge which has been forced to "have a relation with" a set of narrowly defined commercial interests. That is, it is also, intellectually, an expression of violent actions, ie identifying and maintaining an arbitrary (but, nonetheless an) authoritative-truth.

These violent intellectual-expressions are quite similar to the violent actions taken against those who followed Copernicus (in that age), but the investor-class has now (2013) supplanted the church.

It should be noted that though peer-review is interpreted by the public to be "the badge" of one of our culture's absolute-truths, ie an established-truth whose truth can be relied-upon, wherein the idea of an absolute truth (or an officially established reliable truth) is a failing of our society, but it is a failing instilled into the public by the society's propaganda-education system, which, in turn, is interpreted (by the public) to be the sole voice of a reliable truth, while all other ideas not expressed in the published (professional) literature are to be marginalized and extinguished.

It should be noted that peer-review is about categories of knowledge which are being classified by their relation to supporting commercial interests, nonetheless, even professionally ie in the context of a controlled and highly-edited expression of the "allowed ideas" within our society, peer-review is seen as the expression of an (arbitrary) authoritative intellectual high-value.

But this arbitrary high-value is of exactly the same character as the arbitrary high-value which is expressed by racism, or any other arbitrary-value used to express the idea that people are not equal.

Peer-review science and math identifies the narrow intellectual rules of an intellectual contest, where the form of knowledge being expressed . . . , in such a contest . . . , is based on the interests of the banker-oil-military business monopolies, and its narrow precise language does not have a valid relation to either . . . , wide-ranging and accurate (descriptive) knowledge (the properties of fundamental stable physical systems are not being described) . . . , where wide-ranging applicability is supposed to be one of the main interests of the idea of unification of physical law . . . , and/but these professional dogmas about particle-physics (or derived from particle-physics and general relativity) only claim precision in regard to a very narrow context, ie measuring the cross-sections of elementary-particles, but this narrow context is irrelevant to a unified and accurate knowledge concerning describing the observed properties of the material-world (nonetheless it is expressed in the language of unification) or . . . , it is it a knowledge which has virtually no relation to practical useful value, other than its usually very narrow relation to commercial interests, eg turning physics into bomb-engineering so as to please the military-monopolies.

That is, peer-review expresses authoritative dogmas which are given high-value, and this is used in a similar way as the Puritans used their belief in arbitrary dogmas to justify exterminating, the (then [1600's]) assumed to be, intellectually and morally inferior native-peoples.

Peer-review is equivalent to this same type of racism, and other forms of arbitrary high-value.

Peer-review is used by the propaganda system as a central component to express both an intellectually superior class of people, and to express an authoritative truth, which (it is understood by everyone) the public is not capable of attaining, or understanding.

It is the perfect way to express inequality, and to express, what is believed to be (by the public), a carefully acquired truth, which the propaganda system controls . . . , but which is not true.

The propaganda system is so construed so as to create the image that the public is believed to be too inferior, for anyone of the public, to be able to determine and then express the error of such a peer-reviewed dogmatic truth, ie the professionals are those who accept the dogma (the set of assumptions, which are required to be believed, if one is to be able to discuss the truth), and it is assumed that these professionals guard the truth for all of society (yeah, right!).

The errors of the dogma (ie a dogma which is expressed in peer-review) cannot be expressed by a member of the public, and then be believed by the public.

Both inequality and (a false) high-truth are inextricably intertwined.

The professional experts are weak-minded people, who have been tricked by the propaganda-education system into believing that their personal worth, as well as truth, is determined by how well they can impress the bankers (ie the investment-class), in their pseudo-high-value uppity-ness, so as to serve the interests of traditional, conservative, and relatively un-creative investment structures within society. The mask of uppity-ness is used to justify the violent oppression (of ideas and creative efforts), which is expressed in society, ie the extreme violence of an unequal society is claimed to serve high-value and truth.

Thus, what is occurring is, there is an expression of weak-minded people, who become the authoritative experts, because they are willing to acquiesce to a competitive, mindless, narrow set of pigeon-hole dogmas, pigeon-holes (or language categories) which serve monopolistic investment interests.

A set of uppity-people (ie the lap-dog experts of the bankers) who, in their competitive-zeal are trying to impress the banker-oilmen-military's monopolies, of their intellectual superiority, by winning the educational competitions which are based on failed dogmas. Furthermore, they are exactly the authoritative dogmas which are related to the creative efforts of the commercial investments of the bankers-oilmen-military monopolies. Where these dogmas fail when they are related to contexts which are different from the "narrow context of use" within the banker-oilmen-military's monopolies.

Though the educational competitions are a charade, ie for those in the education system who are seeking to confer on themselves high-social-value, this "seeking of high-value" (within the educational institutions) is picked-up (or reported) by the propaganda-education system, so as to represent a march towards an absolute truth, as conceived by selfish monopolistic business interests, while it is really a set of descriptive structures, in regard to the development of knowledge, which are mostly invalid, except their being narrowly related to trade-secrets, ie monopolistic investment which (along with the propaganda-education system) is the biggest enemy of knowledge.

The mysteries about existence "are there" "plain to see," eg why are the fundamental many-but-few-component systems so very stable, and anyone can try to consider these mysteries, and anyone's considerations are as valid as anyone else's.

Thus, if the propagandists do not "themselves" consider extending knowledge in regard to knowledge's obvious failures . . . , eg not being able to describe the stable spectral-orbital properties of many of the most fundamental physical systems . . . , then "what the propaganda is claiming (and which these propagandists are espousing)" is not worth anything, ie the propagandists are being paid to espouse "the knowledge" which is only of great interest to the monopolistic investment forces within society.

In fact, it is an expression of arbitrary high-value, in the same way in which racism is also such an expression of arbitrary high-value.

Knowledge with a wide scope, concerning existence and about both life and "material," flounders inside narrowly defined categorical-boxes.

The most damaging "narrow viewpoint" is materialism.

This is because materialism is the pillar of:

1. property rights,
2. Darwin's survival of the fittest in a context of material scarcity, ie competition,
3. corporate productivity,
4. control over (material) instruments (because of property rights, or the rights of investors), and
5. militarism and violent coercion.

Further causes of the confining narrowness, into which knowledge is being forced, are due to "a particular knowledge's" relation to the investment of product and/or process of monopolistic businesses:

1. Physics has become bomb engineering, (this has a debilitating affect on the other sciences, eg chemistry),
2. biology is DNA ([considered by the monopolists] as the best-bet for getting control over microscopic medical knowledge),
3. a very low-level viewpoint about chemistry, (where the knowledge about the chemistry of enzymes is quite limited, due the very low-level knowledge about atoms and molecules, as well as living-systems [in regard to bio-chemistry])

   Furthermore, chemistry has been dominated by oil interests, and its branching into farming (control of the world's food supplies), but the knowledge upon which these viewpoints of:

   (a) oil,
   (b) DNA, and
   (c) enzymes,
      are based is a very low-level form of knowledge, and this is, in part, due to the confinement of physics to bomb engineering,

4. Communication channels (computers) are about military and propagandistic control of society, a very materialistic viewpoint. Communication systems were developed from the descriptions of electromagnetism provided by Faraday, and turned into modern "electrical circuitry and antennas" by Tesla, ie it is 19th century science.

Alternative viewpoints

The unification of physics should be about the categories of:

1. space,
2. material (with a desire to transcend the idea of materialism),
   2 (b) and a need to transcend the idea of materialism, so as to model complicated systems which seem to have non-understood properties of unification, eg atoms, molecules, crystals and living-systems, Eg
3. chemistry, and
4. life,

where part of the problem in biology is an invalid model of life, and this interferes in chemistry, but chemistry is also hindered by an invalid model of atoms and molecules.

Unification should be about existence's

1. containing space,
    1(b) "what it is" that is contained,
2. the processes (interactions, and inter-relations) of the contained-components,
3. Properties (most often related to observations) of material- and living-systems, and
4. its set-containment structure (of all the parts).

There are new ideas which provide a viewpoint which transcends the idea of materialism, and also provides a set of stable geometric models of charged systems, which are spectrally stable and can be neutrally-charged in their stable-state, and it provides a model of both life and mind.

How is this related to the "laws of physics" and the idea of "materialism"?

The unification of the laws of physics, placed in a context of materialism, is considered in regard to finding the "correct" set of arbitrary frame-invariant (or diffeomorphism invariant) non-linear partial differential equations (for a system and its material-interactions) placed in a context of both materialism and randomness, ie (in regard to randomness) the solution function is assumed to be a part of a harmonic wave-function, which provides probabilities of random point-particle-events (ie both physical property [eigenvalue] and particle-type) in space (or space-time), defined in a context of particle-collisions, a model which only relates to a process which is a part of bomb-engineering models, and particle-collisions have virtually no relation to understanding the properties of:

atoms,
molecules, and
living systems,

. . . , since the, so called, unifying equations are all about particle-type, and only most-marginally related to a system's physical properties, thus, it is a "probability model" of a stable system (composed of many-but-few-components), and since these types of (quantum) systems are so very stable, then this means that models of such stable systems which are based on randomness is a ludicrous idea.

Cultural considerations about propaganda

Note: upon reading C Hedges "Sparks of Rebellion" Truth-dig 9-30-13, one sees a propagandist aligned against the destructive forces of domineering violence. He explains how the state controls the society's communication channels, and thus can get information needed to stop the people who oppose the ruling-elite, where those who control social violence, as well as spying, represent the ruling-class.

These forces of violent social dominance, are the same as the aristocracy of the Roman-Empire, but Hedges seems to miss this aspect of the European (or western) colonization of the world.

The only "way out of Rome" is to escape from the idea of materialism, and the notion of inequality, where a delusional vision of a "natural inequality" is easy to prove in a material world upon which are placed arbitrary rules.

The side "for violence" is also a side which supports arbitrary ideas about high-value, but it does so with a vengeance, and "the intellectuals" side with violence, where they show their support for violence by being intellectuals, but for these intellectuals they only consider (as true) thoughts which support the aristocracy (they are trapped, mentally, by the propaganda-education system), as was also the case in the Roman-Empire (ie "creative belief" and "related actions" [for the Romans] were only about those actions which supported the Empire),

for otherwise the free-thinking intellectuals would be aware that the greatest revolution has already occurred within science, namely, that materialism has been transcended, where the material world is a proper-subset of the new mathematical containment structures.

The world needs to align itself with the efforts of the native people, who C Castaneda followed, and that (the difficult to understand how C Castaneda's works got published, but) the expressions of C Castaneda are works which are based on the practical experiences of those native-people, who are aware that "the material world is not a real world," ie creativity and knowledge are not confined to a material world, and that religion and science are the same attempt to seek a "life's relation" . . . . (ie to try to exist), so as . . . . to build a relation between knowledge and "the creativity of existence, itself."

This is the context of the new many-dimensional mathematical containment structures for existence, which is partitioned by stable geometric shapes.

For propagandists, such as Hedges, this is ignored and/or excluded from consideration by the "intellectualness" such as C Hedges (where intellectualness is based on a narrow set of common assumptions, and these assumptions are the assumptions which define the narrow commercial interests of the investment-class), as well as (any new ideas which are not supported by the bankers, are) being arbitrarily excluded by the intellectuals he (Hedges) follows.

If one expresses ideas which are outside of a narrow set of common assumptions, then they are not considered within the propaganda-education system.

This "circular mental structure" is one of the main forces which leads to the failures of the system (of the society).

Apparently, these intellectuals need the propaganda system to tell them "the truth."

Apparently, they do not have the capacity to decide "what possesses truth" for themselves.

The intellectuals of western societies, if a revolution occurred, will only re-build the same institutions of extreme violence as they oppose, eg Lenin, because they follow the same elitist social structrues, which, today, are being determined by the bankers, or the investment-class

This is because of their intellectualism, which is based on being a member of an elitist institution (whose interests are being determined by the investment-class), where their intellectualism does not follow the requirement of the need to build knowledge from the most elementary aspects of language: ie assumption, interpretation, set-containment, organization of patterns etc.

If the building of a descriptive language, to be used to describe the truth, is not built in such elementary terms (of new assumptions etc), then the established ideas (which are continuously brought into the, so called, new social structure) must be arbitrarily authoritative, and depend upon a narrowly defined language, and this narrowness will lead to arbitrary dogmatic authority, which quickly becomes irrelevant when these overly authoritative ideas (established, and continuously used to establish new social contexts) are pulled-away from their narrow context.

One needs to think in terms of Socrates' idea of "equal free-inquiry," and do not think in terms of the elitist Plato and his Republic.

However, in the context of elementary language, everyone needs to be granted the social position necessary for equal free-inquiry, so the thirst for knowledge is to be related to a desire to create, based on knowledge, so that the creation is to be given as a gift to others.

This was the way of Tesla, especially in his younger days (later in life he seemed to be forced into considering violent destructive instruments).

That is, the most extreme violence of the Roman-western ruling aristocrats is the act of excluding ideas, and the subsequent destruction of ideas, and the intellectual class is all about exactly this type of social action.

Furthermore, art is all about the person whose capabilities identify them as being of such "unequalled capability" that they are worth money from markets.

Western religion is all about trying to be recognized by the ruling aristocrats.

Science serves production and ideology and propaganda.

Thus, in the west, its intellectuals cannot describe an existence which transcends materialism.

The intellectual aristocrats of the west do not consider new math-science ideas which transcend the idea of materialism.

So all that the intellectual aristocrats (of the west) know, are arbitrary ideas about materialism, which are related to extreme and arbitrary violence.

But the new ideas, which transcend materialism, do emerge from western science and math, but they are not pigeon-holed into (or contained within) invalid intellectual pursuits (associated to social domination).

65.

Categories

Life, human-life, is not primarily related to a material world, though material is a natural structural outcome of a creative impulse which creates both life and material, where creativity and material are fundamental properties of living-systems, knowledge and an intention for a creative extension, when guided by (in) the context of a correct knowledge structure. However, the investor-ruling class requires that technical precise language be based on the idea of materialism, and the investor-class always "gets its way."

The economy and its associated material organization of society, and of the material world, where this organization is done in a context to expand and increase human-life, but the expansion of human-life in a material-world has limits (or boundaries) based on material resources, especially when resources are used in narrow ways. When they are used in narrow categories associated to resource organization (then there exist resource limits), so that society supports a collective hierarchy, based on a particular way of organizing resources, and it does so in a particularly narrow organizational context, so as to maintain stability of the top of the arbitrary social hierarchy, but it is a social-hierarchy which is defined and maintained by extreme violence.

This viewpoint of narrowly defined "exploitative" expansion, defines material scarcity, which is associated to the narrow categories, in turn, associated to a form of monopolistic trade domination over society, which is exercised by a few.

This narrow viewpoint is the basis for scarcity, in turn, attributed to Darwinian evolution and to (the dis-claimed notion of) social-Darwinism [despite the rhetoric, society is based on social-Darwinism, and wars in the middle-east are about oil, and elsewhere they are about stealing material-resources], so that the theory of society is; that there are changes which either increase-value in society or they do not survive (in the free-market) where these results are base on the "law" which "demands excellence," the law of survival of the fittest. All of this is dis-ingenuous, there is no free-market, and what is called excellence, in fact, possesses arbitrary-value (upheld by the propaganda-education system).

That is, social-Darwinism is a material based prescription for social hierarchy, whose root-cause can be "rightfully" based on the violence of "the, so called, strong," (where the idea of the moral-virtue of the rich, the idea upon which Puritanism is based, provides for a "divine right" of the wealthy to rule) and it also justifies the, subsequent, violence needed to expand and maintain the social-economic expansion of a narrowly defined and arbitrary social-value, and hierarchical social-order.

The main error is the idea of materialism
(but this is the "de-fault" position [or belief] of everyone)

But humans are not primarily related to the material-world, rather they are knowledgeable creators (equal creators) in a context of existence which transcends the idea of materialism.

This world (or existence) is now (2013) correctly described (in many-dimensions, and based on stable shapes) and can be experienced and/or perceived.

This greater existence is now provided with a math model which identifies humans' "proper creative context" within that same existence, which is erroneously identified as an absolutely material-world.

From this new description, both the properties of the material world are derived, as well as the, apparently, confusing property of fundamental randomness, a property which is derivable (and of secondary importance), in regard to understanding and using "the fundamental properties of existence" to further the extension of existence.

The more fundamental pattern (in a precise descriptive language) is about reliable measuring and the stability of patterns, where the stability of patterns derives from sets of very stable shapes, in turn, whose geometric properties are rather simple (ie mostly circle-spaces, ie toral-shapes and toral-shapes attached to each other).

The observed properties of stability in fundamental material systems of all size-scales imply that these material systems . . . , with their observed properties of stability . . . , be related to stable simple shapes.

Descriptive knowledge is categorized, and pigeon-holed, into knowledge about material categories, which are associated to trade and trade-monopolies, which, in turn, allow for monopolistic power in an economic social structure built on banking, oil, and the military. This ability to control descriptive knowledge to support the social hierarchy, was a result of the social structure of the Roman-empire, which facilitated trade (the Roman-empire both stabilized money and a set of uniformly narrow set of standard products) so that when the Roman-empire fell-apart, it was patched back-together by religion and commerce, where the commerce was organized around banking, which, in turn, was organized around the narrow social-product context, instituted by the Romans.

The monopolistic power structure being centered around trade, resulted in the canonization of trade-knowledge, ie the knowledge of production, and its relation to craft-guild monopolies.

Science developed along the lines of practical trade (or craft) knowledge, but the stories of: Newton and mechanics, Boyle and thermodynamics, and Faraday and the development of knowledge-of and the use of electricity were remarkable.

Science after Newton and Boyle was: gravity, mechanical systems related to simple patterns of angular-momentum, and thermodynamics, where eventually the atomic-hypothesis was put-forth and the idea of reducing thermal systems to many material (atomic) components, where a thermal-system's open-ness or closed-ness were fundamental to the developed invariance-properties of:

1. conserved energy,
2. conserved mass, and
3. conserved momentum,

involved in component-centered processes which were used to model the chemical (eg chemical-reactions) and statistical-thermal descriptions, along with the thermal concept of increased randomness, or entropy, which was "energy which is (was) unavailable to be used for work."

Newton's gravity could describe and solve the two-body gravitational system, and then the model of the randomness of point-particle-collisions (so prevalent in statistical-thermal models) to which it was assumed that all material systems could be reduced, were modeled as harmonic-wave-related probabilities, which was applied to the two-body model of the charged H-atom in the context of the (parabolic, or heat) wave-equation and it was solved, where the math context could be framed in a context of invariant properties (and invariant equations) defined on "harmonic function spaces," eg the spherical symmetry of the 1/r potential energy term in the energy-operator, and a new idea of discrete conservation of energy.

The component collision-rate model applied to systems which were assumed to be reducible to material-components (at least for systems transitioning to new states, due to high-energy component-collisions) was applied to nuclear-reactions with success, and subsequently all material systems were reduced to a set of elementary-particle components, but which defined unstable events, so that the elementary-particles were classified into families, in turn, related to one another by a property of internal-symmetries, which is defined by finite dimensional unitary Lie groups, so as to be (sort-of) consistent with the unitary invariance's (or unitary properties) of the harmonic function-spaces, which are used to model quantum systems.

But all of this math structure being based on both a set of unstable events, and a very limited set of solvable systems, has resulted in this descriptive context to be without any content.

This is shown to be true, since it is a descriptive structure which can neither describe observed properties of fundamental quantum systems, which are supposed to be composed of these elementary-particle components, nor has it been knowledge (of the reduction of all material systems to elementary-particles) which is useful in any other realm, which is different from the very narrow and very specialized realm of nuclear weapons engineering.

Finite unitary invariance's which were applied in the context of infinite unitary invariance's were canonized into fundamental physical-laws, but the stable properties of the many-but-few-component quantum-systems have never been described in this context, either in the infinite context or with the finite context of (internal) elementary-point-particle states which was added-on. Thus, the indescribable states, of the observed properties, for randomly modeled systems implies a context of indefinable randomness, and it is a descriptive context which also has no content.

The entire notion of arbitrary-frame invariance's for the types of (partial) differential equations which make-up the context of physical-law, to be canonized as the fundamental property of physical law, has led to non-linearity and quantitative inconsistency, and subsequently, it has become a descriptive context without any content. For example, the solar system is stable, but general relativity cannot be used to model such a many-but-few-component system, let alone, determine the cause of its stability.

The canonization of limited viewpoint concerning descriptive knowledge has led to all of the fundamental descriptive structures of theoretical physics to have no content.

On the other hand partitioning a high-dimensional model of existence by stable shapes (discrete hyperbolic shapes) and using this to define both material components and metric-spaces, and it results in also defining a finite spectral set to which all components (material or metric-spaces) . . . which exist in such a high-dimensional model . . . , must resonate.

This defines existence, in a way, in which the very stable properties of the many-but-few-component material systems can be modeled (or described in a precise and accurate manner).

It allows the previous constructs concerning material systems, and it also expands and clarifies the actual context of existence, [and in this context it is like the Copernican model, which clarified and changed the canonized data-fitting techniques of Ptolemy's model], and it provides the context of a many-dimensional subsystem structure with which one may begin to be able to understand some of the many complex, but controllable, systems which we observe, systems such as life.

The trouble with the propaganda-education system is that

(1) it serves the narrow pigeon-hole vision of knowledge which serves the interests of the ruling-class, and

(2) there is within it an intellect-morality play which is the main instrument of the information about governing, we are being governed by the most virtuous people in society, where the left is composed of people who are must try to do good-works (to atone for their immoral behavior), but who act as though are the intellectual superiors of the nation, and

The self-righteous supporters of the ruling-class, where the ruling-class represents moral-superiority who dys-ingenuously espouse populists concerns.

The government is a propaganda-system side-show and the actors are picked by the business community to fit into the propaganda system, where the propaganda about government is expressed as a morality-play.

But since the intellectual structure of the society is based on the pigeon-hole categories of knowledge and language, which support the interests of the ruling-class, the equally dis-ingenuous striving for good-works of the so called intellectually-superior left, their, so called, good-works are about doing the things which support the interests of the ruling-class.

Namely, the intellectual left does not question the authority of those who serve the ruling-class in their intellectual endeavors, eg they do not challenge theoretical physics and its basis on logical and/or language structures which possess no content.

The other big division in the propaganda system is also a disingenuous distinction between "what are two collective viewpoints about civilization"

whose only difference is how the propaganda-education should be structured within society, concerning how the collective-societies are to serve the elite few.

This is, the capitalism vs. Marxism distinction (a different viewpoint about how to organize propaganda for a collective hierarchical society, should propaganda be based on promoting social welfare or based on morality).

An alternative is a belief in individualism and a belief in equality and a belief in knowledge and creativity

This was the idea upon which the American Revolution was based

That is, the Declaration of Independence stated a truly different way in which to organize society, namely, organize society around a public which are equal creators, where equality is defined in terms of knowledge and creativity and equal free-inquiry, where the government is to take-care of the common-welfare of its citizens.

That is, the emphasis is on the individual and their relation to knowledge and creativity, but it is clear that the collective-hierarchical notion of society is going to lead to destruction.

It is a sign of the weakness of the intellectual-class when one sees how they analyze the problem within the propaganda-education system.

The main issue is not that the society is a plutocracy which steals from other nations, rather

It is that in this society, the violence directed at the public, is applied by the justice system, and in doing this the society is exactly the same as the roman-empire, where the law was (is) based on property rights and minority rule, and that the actions of the society are based on the propaganda-education system.

Furthermore,

The thing about the propaganda-education system, which is so overwhelmingly destructive, is that, it categorizes and pigeon-holes the language so as to be aligned with commerce, while commerce is not an economic system based on a, so called, free-market, but rather it is a way to narrowly divide the society into organizational units, which have fixed social-value, and are easily linked to a banking and investment way of organizing power within society.

The propaganda system is owned by the investors, and thus, the politics of the society is about the owners of society selecting the politicians who properly fit into their propaganda system.

Thus, since the justice system has violently forced the public to become wage-slaves, in a pigeon-holed society, where the context of discourse is based on there already being a given set of assumptions, which all parties in a communication channel share.

This results in a uniformity of thought.

Namely, that the point of being human is to be a consumer in a social structure which upholds the banker-oilmen-military monopolistic industries, so that this is done within a set of fixed categories (with fixed ways of discussing these structures, and a fixed way in which the organization of material, society, and knowledge is done) so that one can only compete within a structure, in which it is required to believe that it (the given social and language organization) is the only way society can be structured, and the only way in which knowledge can be considered in the society, so that one's actions will remain actions, which will both "win one a wage" and which will support the interests of the ruling class.

"Manufacturing consent" has been an art since the beginning of the Roman-empire.

It is about the illusion of value and deception.

Knowledge fits into the arbitrary category of high-value, when it, in turn, properly fits into the organization of commerce, ie the social organization imposed by investment, within the collective social structure which is organized to support the ruling investor class.

If one considers new ideas then one must use language in ways which are in opposition to how language is used, where it is assumed that the communicating parties must possess the same set of assumptions, in order to communicate, namely, using the categories of language which best into the social system of commercial interests.

In regard to expressing new ideas, the main division within society is the relation of a collective society to a society which is based on the individual (each individual with individual new ideas), and the individual freedom to engage in equal free-inquiry, where the assumptions are not already in place.

The propaganda system erroneously identifies the ruling class as examples of those people who believe in the individual freedoms, and having used the rules of society, and the desire to build, and work-hard, so as to build their commercial empires (to become the winners of the game).

But

This is clearly not true, since "those at the top of society" are all about the type of people who are interested-in enslaving people (society) into a social condition of supporting the ruling-few.

That is, the rules are put in place by the ruling class, so as to create an unequal game, so that those who are willing to compete in an unfair way, will become the winners of the game.

The issue is about knowledge based on language, and its relation to: assumption, context, interpretations, organization of patterns, the ways in which existence can be modeled within a containment-set, etc.

The idea of materialism has made the idea of containment-set a fixed part of the set of assumptions which the scientists are supposed to hold-to when conversing with other scientists.

Newton intertwined the idea of material with the (partial) differential equation, and the laws of physics were defined as properties of local measures of material constructs.

The (partial) differential equation has not been understood by the mathematicians, but it extended math patterns outside of rigid, fixed geometric (or algebraic) patterns (of old).

The problems with (partial) differential equation are the problems about the prevalence of non-linearity and non-commutative math patterns, when considering (partial) differential equations.

It has come to be about the algebraic properties of sets of operators (which are, in general, not commutative) acting on infinite-dimensional function-spaces, where numerical values (or the capacity for a math patterns to be reduced to something simpler) are being determined by convergences, and/or types of divergences ie asymptotic properties, and singularities of a function's form, eg the Dirac delta function.

This has not led to much useful information, unless the system is a highly controlled system in which limits and cut-offs can be imposed on the information being processed by the system, eg computer switching-systems.

Or

In the case in which the critical-points and the limit cycles of non-linear (partial) differential equations can be used, so that feedback may be introduced to the system, in relation to the identified limit-cycles

(If the (partial) differential equation of the non-linear system remains relevant, a big If), ie the system's containing-space is changed and then an exact replica of the original containing space is re-introduced after the feedback is provided to the system.

The mathematics becomes questionable, if not illogical, or straight-out wrong, and it becomes very complicated, but as a category, it is likely to have some relation to a commercial processes, ie it fits into the narrow set of commercial categories about which the society is organized to serve the interests of the ruling-few.

Or

The uselessness of these strangely developed math languages means that new creative contexts cannot be considered, and thus there is not the further competition of new products.

Thus, in mathematics it becomes the set of assumptions which define math, and introducing other patterns into the professional literature is not allowed, ie the new ideas represent a context too different from the assumed context of what are to be "the math structures," within which a mathematician must compete so as to identify themselves as being "worthy of working on the commercial patterns of interest" to the investment class.

There is to be no "equal free-inquiry."

The information needed to understand complex systems cannot emerge from the inquiry which is focused only on (partial) differential equations. Rather, the new information concerning complex systems must (might) come from the context which identifies the set of stable patterns to which (partial) differential equations can, in turn, be related. Namely, the stable discrete shapes of the non-positive constant curvature discrete shapes whose context is many-dimensional. These stable shapes would form the stable framework within which the other aspects (or properties) of the system are (can be) described (or defined), some of which are (might be) defined (or can be identified) by (partial) differential equations, and possibly used.

But the primary use of the new containment structure would come from the awareness about these stable patterns, when considered in their (or which form the) proper containment-set structure, so that how this containment structure would (could) affect the results of more general systems, which can be determined, which could (might) be defined by (partial) differential equations.

One needs to consider fundamental questions:

Is existence all about materialism and the set of non-linear (partial) differential equations which are used to define material-systems, and which almost always define a non-commutative (algebraic) math structure?

or

Is existence many-dimensional and based on stable geometric patterns which are both linear and continuously commutative almost everywhere, (so that the relevance of the (partial) differential equation . . . , in regard to defining a physical system's properties . . . , has bounds)?

The math structures which are the dominant (or primary) structures in regard to describing the properties of existence are the many-dimensional and based on stable geometric patterns.

What has value?

Is human-value about knowledge and creativity, or is it about property?

Does one want to live in a creative and equal society?

or

Does one want to live in a hierarchical society which is based on arbitrary-value and which depends on extreme violence to maintain its own charade of arbitrary-value?

Note: Both capitalism and Marxist socialism are ideas about society which assume that the society is a collective, so that this collective upholds the social positions of the "special and arbitrary set of" the ruling-few.

It is clear that a belief in a, so called, superior basis in knowledge, upon which the social hierarchy depends, has come to be the cause of all the social failures of our age.

Science and math are failing, the economic-propaganda-education system has failed, the media has been turned into a spy-system, instead of using the many channels of communication as a means to develop knowledge, and express a wide range of ideas, whose idea structure can be classified and the set of assumptions upon which the ideas are based identified and put into categories about assumptions not categories about commerce.

That is, a person may claim to have new ideas but when the assumptions are determined, it might be found that the assumptions used are in an "already identified set of assumptions," thus, the discussion would proceed in the context of the assumptions used.

Does one want to live in a society which respects individual creative-value, a society which bases law on equality?

or

Does one want to live a in collective society which necessarily is a hierarchy, based on arbitrariness, so that the collective society supports a small set of people, which composes the ruling-class, who, in turn, destroy human knowledge by their selfish narrowness, and who protect their social positions with extreme institutional violence?

To compete for high-intellectual-creative-social-positions within a collective (and hierarchical) society . . . , (a society run by a propaganda-education system, where knowledge becomes fixed, and discussions become based on there existing a common set of assumptions, where all of this social (relation to fixed use of language) structure is based on its relation to commercial interests) . . . , means that one is functioning within a discussion ruled by fixed assumptions, ie it is an exploratory discussion (often about the level of the fixed form of knowledge) rather than analytic, where analytic means considering a new language, new assumptions, and new types of discussions.

When a small group of people, who are given great social power, have their discussions, which are all based on a fixed set of assumptions, then such a group can be quite tyrannical and quite narrow in their outlook.

For example, this could be small groups defined by spies, politicians, swat-teams, violent people (or groups) who are controlled (or managed) by agents from spy institutions, journalists, academics, etc.

We are supposed to consume, we seek products which fit into the large-ticket products we have already bought, and we buy, . . . from the, supposedly, wide-variety of products . . . , which are available, then

We are supposed to sell ourselves.

This was the idea of high-school students back in the 1960's. This is not a new idea about thinking of ourselves as commodities, which are to be defined on a very narrowly defined market place.

Though there may be many things for sale, but it is similar to the 500-channel TV stations, all the channels are saying the same thing, and similarly, all the products available are all tied together in the same way of organizing the power structures of society, and they all support that same social structure.

But the development of knowledge requires many new ways to consider contexts and the organization of both language and society ie the society needs to be open to new contexts of creativity, and "the truth" needs to be analyzed in truthful manner.

Thus, when we are asked "what do you want to be?" The set of careers seems to be quite broad.

Yet, all the best paying jobs are narrowly defined, and very competitive, so as to re-enforce the notion that people are not-equal when one lives in a hierarchical society, while the highest-educational-related jobs are the most narrowly defined. Yet these narrow categories of language claim to be related to the widest scope concerning the descriptions of existence, so that, it is erroneously claimed that our culture can describe the properties of all physical systems to their, most essential, true context.

Yet, if one simply analyzes this idea for a moment, then consider that . . . , nuclei, atoms with more than 4-charged-components, molecules and molecular shape, crystals, the stability of the solar system, the basic idea about "what life is," understanding the complex nature of the molecular processes of a living form, etc etc, . . . , are all physical systems which do not have descriptions of their observed properties, which can be derived (ie deduced) from what is considered to be "physical law."

Thus, one can plainly see that the vast majority of the physical systems which exist, have no valid descriptions (based on what is considered to be physical-law).

Thus, the narrowing of the categories of knowledge is not about knowledge, ie it is not about "discerning truth," rather it is about fitting into the categories of knowledge which investment defines, and fitting into the vocational structures of the wage-slave status, which the justice system has forced (by using extreme violence) onto the US public.

Such an expression of arbitrary high-value in relation to knowledge (a knowledge which is essentially dys-functional), is no different from the arbitrary propaganda structures of racism.

The only way (or the best way) to resolve this situation concerning the failing of knowledge is to institute equal free-inquiry and the only way to do this is to base the laws of the society on equality and not on property rights and minority rule.

That is, the experts and the investor class are the same as ignoramuses in regard to discerning truth.

"Being able to discern truth," would be a good definition for intelligence and not defining intelligence in the context of "how fast a person learns," which is how it is defined in a function way by educational institutions, but this only makes sense if only the ruling class can really assess truth (but that is the underlying assumption of our society).

But no one knows "the truth," so "being able to discern truth" is an unknown, which no one is capable of determining, so it can not be measured.

Yet the propaganda-education system is busy convincing the public that the authorities (who support the interests of the ruling-class) "do know the truth," and our society has access to the very highest of absolute-truths, which only the intellectually superior people can discern.

This of course is a lot of baloney.

However, the wage-slave servants of the ruling-class only analyze truth, in the context of the interests of the ruling-class. That is, the set of assumptions contexts interpretations and organizing structures are set-up by the interests of the ruling-class, thus within such a social norm for such a narrow and relative truth, then quickness of learning is the property that the investment-class would bet-on as identifying intelligence.

That is, in the age of Copernicus it was the ignorance of the church's authority which undermined the attempt to discern truth, but today (2013) it is the dogmatic authority which is associated to the investors interests, associated to the categories and processes which are associated to their narrow concerns about the commerce (into which they have invested), which most interferes with a person's true ability to discern truth.

Since, again, as in the age of Copernicus, the truth is overwhelmed with the arbitrary authority of the ruling-class, in an arbitrary hierarchical society, which is the same type of arbitrariness which is also the basis for the belief in racism, or any other set of barbarous belief-constructs.

That is, the hierarchical society, upon which capitalism depends, is a collective society, which is all about the public supporting the ruling-few, where it is claimed "to be a scientific fact" that biological systems develop based on survival of the fittest, in a world with scarce resources, in other words "the world of biology is competitive and based on inequality," but this is far from being true.

That is, instead of the Church dictating an authoritative truth to society, today (2013) it is the investor-class determining—and then dictating—an authoritative truth to society, through their propaganda-education system, though this seems to only be indirectly related to the activities of the (so called, upper) intellectual-class of society's so called top institutions, but this is not true, where the subtle relation is being determined by the commercial-categorization of language, and the social-power of funding, where politicians play more of a role of a propagandist serving investment interests than as their "stated role" (as stated in the education system) as public law-makers.

That is, instead of the church determining an arbitrary but authoritative truth for all of society, now, it is the bankers who determine this same type of arbitrary authoritative truth for all of society.

And

Essentially, this is still being based on "the virtuousness of the ruling-class," which is their role within a propaganda system, and (being based on) the authority of what are considered to be a separate class of intellectual expert-class, but the intellectual-class is controlled by the propaganda-education system, which is serving the interests of the ruling-class.

Nonetheless, today (2013), the propaganda system is still similar to the propaganda-system of the age when the church was all-authoritative because of its great virtue.

Consider the expert models of life

(ultimately, models of life-forms depend on the knowledge of physics and chemistry, it is these material-based systems of knowledge (or systems of precise language) which are failing to provide a "correct" context for comprehensive, accurate, and useful discussions about life (or models of life).

Consider that,

Life began almost immediately (a few million years) after the earth was believed to have cooled (where it is believed that the earth cooled about 4-billion years ago), and the forms of life which came forth, eg bacteria and one-celled plants etc, seem to have possessed the most complicated forms of DNA which still exist, while the, so called, more advanced species (seem to) have less complicated DNA.

There is no model for this in a theory which is based on the indefinable random context of mutation and survival of the fittest.
That is,
Can small changes define an extreme advantage for survival?
Can destructive processes lead evolution towards superior attributes for survival?

Clearly we do not possess a valid model of life.

The way in which our society is organized is called "free-market capitalism within a democracy."

But none of these words are even close to meaning what they are defined to mean, where the definition of capitalism might best be "organizing society's creativity around bank investments (where the bankers form the ruling class)," since the social structure is really a Plutocracy, which is organized around commerce (which, in turn, is used to pigeon-hole and categorize knowledge) and It (commercially-motivated categorization) is used to organize society in the context of material distribution, and more fundamentally, it is organized around the propaganda-education system in regard to categories such as: law, violence, military, energy, politics, and economy (which is a system about taxing people for the material-sustenance needed for life and the instruments which are used to control information and language and "artistic culture"), where law is based on property-rights, ie a way to organize society so as for society to support a certain type of stealing done by the ruling-class, ie society is based on the idea of materialism, but the justification for a material-based-society is (based on) an arbitrary form of morality and virtue, derived from the church.

It is not enough to provide information, though truly correct information could have a force of its own, but a force, due to information, is not to be expected to be true. Instead, one must create, and the context of the creativity must be related to some form of community, as least when creativity is within the so called real-world of politics.

But community does not (really) form the basis for knowledge and creativity.

However, community can easily by organized to represent a concerted effort to oppose the validity of "true creativity" and its correct creative context.

The "true" alternative to capitalism, is a society of individuals, where each individual has the same value as an individual-creator, and that the main value of a human life is about their individual creative efforts, and that the collective is not based on "the truth of materialism and its proclaimed relation to a Darwinian social structure of survival of the fittest, ie organized violence prevails over the truly-strong and those who are very creative (which is everyone)"

That is, the Marxist brand of a collective society is also based on materialism, and its propaganda disingenuously based on equality.

Thus, within either the Marxist or the capitalist ways of organizing society, there is concern with, or they are based on, the idea of each person possessing equal amounts of material (or money), ie the distribution of material and material instruments.

But material is not the highest value of being human (and the value of money is arbitrary, mostly upheld by extreme violence) rather

The highest value which can be associated to people is each person's knowledge and creative capabilities, and these capabilities cannot be measured and ranked, since descriptive knowledge is based on always changing language at the level of assumption, interpretation, and a set-containment context, etc and one wants a description which can be used in a practical context . . . , (though the material realm does not constrain the notion of practicality to the realm of materialism, there are higher-dimensional material realms, but the geometry of "material-interactions" in higher-dimensions are quite different than the geometry of material interaction in 3-space) . . . , so that one wants measuring to be reliable, and the patterns (which one considers) are to be stable patterns, ie the context of quantity and shape (where shape is what mostly determines "the idea of 'what a stable pattern is'").

That is, the description of a many-dimensional context for existence, in which the present idea about materialism is a proper subset, is a context which can be highly related to practical creative efforts.

Addenda:

All the propagandist based "high-value baloney" used to entice the professional-expert, jackasses, to prance-around (in institutions associated with high-social-value) with their banker (paymaster) friends, exalting in their own belief about their superiority, based on the propaganda-education reward system, about their high-caliber intellectual talents, but their illusions about their own abilities rest on the "spare-no-expense to-deceive the public" propaganda system, and its necessary relation to the needed existence of extreme and arbitrary violence.

This way of organizing institutional structures is all based on using, or manipulating, psychopathic personalities, ie people with obsessive behavior which they apply to their quest for domination and/or interfering-with and/or hurting others, but These, so called, great-intellects have no clue as to how to solve the: "many-but-few-component systems which display both very stable spectral-orbital properties, as well as often being, in their stable forms, neutral material-components, though often composed of charges, these are commonly observed properties associated to very many fundamental physical systems, including life-forms."

Yet the, so called, great intellects, who are being manufactured by the propaganda-education system, have nothing of any relevance to say about these systems.

The intellectual authorities are those who possess the empty statistics, good on paper (ie the paper stat-sheets, provided to society by the propaganda-education system) but in relation to victory (or in relation to discerning truth), the acclaimed intellectual authorities are to be relegated to irrelevance.

The propaganda-education system is a paper-tiger, which nonetheless controls; how language is "used" in society, and this determines both education and politics (as well as, what are called, markets).

What passes for high-valued intellectualism in society, are those "intellects" who score-high when intellectualism is defined in regard to the investors, who are in the exalted social position to act as "those who determine truth" and "those who judge social-value," (in the propaganda-education system), but when intellectualism is defined in regard to "discerning the 'real' truth," these (posturing) intellects are measured to be quite low on the scale of intellectual-worth, where the highly-promoted intellects "find rigorous precise word-relations," but these 'word-relations" are unrelated to the actual patterns which exist within our experiences. That is, they are not solving the fundamental problems, they do not even consider these fundamental problems, the answers they provide (eg business risks) should not be trusted, and they wield their oppressive authority against any new ideas.

That is, despite the high personal capacities conferred on them by the propaganda-meisters, in the real-world . . . . of: existence, knowledge, and creativity . . . . , these highly touted intellects fail, in a similar manner as the propaganda-based (controlled)-market-ruling structure is failing. (Maestros)

Note that;
The propaganda-based "(controlled)-market-ruling social structure," ie central-planning based on selfish interests, is:

1. destroying the planet,
2. leading an expansion which leads to population explosion,
    but (nonetheless)
3. is engaged in techniques of population extermination.

It is a system which supports the illusions of a few, and it does this by its necessary exploitation, destruction, and exclusion of everyone else, and everything else.

That is, the bogus top-intellects fit perfectly with the delusional ruling-class, and all of this (baloney) is (mostly) about the control of language, ie the propaganda-education system, along with the needed violence which upholds all this social structure of arbitrary value.

It is not that:

> "many (or some) of the math patterns the professional mathematicians and physicists consider are not interesting,"

rather

The issue is, "who are they to close the door" an discussions about other math patterns which are presented in new contexts based on new sets of assumptions,
so that their judgment to close and constrain the discussion is based on their personality-cult style of religion of arbitrary authority.

Now they are claiming that they are presenting, so called, "rigorous truths" about math, but the math patterns which they are presenting do not apply to the patterns of experience,

Consider that, the calculations of risks, though they may be rigorous (within the word-agreements accepted in professional math communities), nonetheless, they do not hold-water (and their failure is related to the economic collapse of 9-15-08) and [this is because math patterns, in both math and physics, are now (2013) based on: non-linearity, patterns of indefinable randomness, and patterns which are not commutative],

But, the main failing of the professional intellects is: about the existence of very stable and neutrally charged systems, which are composed of charges, and which possess (stable) spectral-orbital properties, a condition which implies both solvability and controllability of the properties of these systems, yet these systems are not being described using the math techniques associated to the laws of physics, eg solving generally frame-invariant partial differential equations which are supposed to model the measurable properties of physically measurable systems.

If these fundamental descriptions are not being accomplished by these, so called, great intellects, who prance around with powerful bankers, swaggering about their great intellectual capabilities, in austere intellectual institutions, eg universities and often institutions of commerce, then it must be clear that these highly-touted intellectuals are not so-bright (they appear to either be quite stupid, or they are psychopaths who are most interested in positions of domination), and they must be required to humble themselves (or house cleanings of the high-faulting, so called, public institutions, would also be a good idea) to accept new ideas, ideas not associated to great prize-winnings, but rather directed at providing accurate and practically useful knowledge, an issue which apparently the great intellects of society's, so called, great institutions, have forgot to consider, when they have become so highly revered within the propaganda-education system, for no obvious reason, except that, they do what the bankers want them to do and say unintelligible things which clearly have no content and no relevance.

This is a community of intellects who are familiar with Godel's incompleteness theorem, so either they do not understand it, or their teachers mis-led them about the content of Godel's incompleteness theorem, but they should understand that Godel's incompleteness theorem, which implies that mathematical knowledge (as well as knowledge about physics) is limited when held fixed, because a precise language has limitations as to what patterns the language is capable of describing,

That is, rigid language structures (especially math languages) must always be challenged at the level of its: assumptions, contexts, interpretations, pattern organization, and the containment-set structure being employed, so that new types of precise (or mathematical) languages based on ideas at this elementary level, upon which a language can be formed, identify the main point of language at which math and physics can be creative and be able to discover new ways to consider math problems.

That is, authority only can be defined in a fixed context, so the only strategy about precise language is to consider all the new ways of organizing and building a new precise language.

A computer might (could easily) be able to distinguish different descriptions based on different sets of assumptions etc.

66.

Religion is still dominant

It is an outrage that the expression of technical ideas are so carefully controlled by the ruling-class.

The current technical knowledge is mired in inaccuracy, irrelevancy, and a deficiency in regard to providing direction for practical creative efforts.

Thus, any new context for measurable description which is also based on stable patterns, which are geometric, so that the new patterns which are identified (in the new descriptive language) also define an inter-related pattern which is stable, and thus such a stable pattern can be used, and clearly considered in a (linear and continuously commutative) geometric context, and (thus) is a new language which is vastly superior to what is now available to the technical expert-world.

But since now (2013) knowledge is structured to fit into commerce, the ruling investor class still can greatly influence the minds of their independently-ineffectual experts. (eg the global-warming experts must break-away from their institutions to protest the mis-use of science)

That is, "the experts do not influence the ideas which are considered true" rather "the investor class determines what is considered to be true," and the experts adhere to the dogmas determined by commerce, ie dogmas upheld by peer-review, and which have a narrow relation to commerce.

Thus, the statement that there is a new form of "superior knowledge" . . . ,
Though (because of its superiority) it fits into the scenario within which the violent cult of stealing fits, ie the legal institutions claim to support stealing if it is related to superior knowledge, . . . , is a challenge to society, about the nature of value and the nature of truth,

But it is a statement which the public will interpret as "the claims of one who does not possess the proper authority" to make such claims, whereas that type of authority and value is the type of social-authority which can only be granted by the banker-, or investment-class.

Though the public is oppressed by the bankers, and belittled by the bankers, nonetheless the public acts as if it worships the superior bankers.

But the social-class which the public believes they are worshipping, is the social-class invented by the propaganda-system, namely, intellectually-superior social-class of experts, but the public apparently is unaware that the expert-class become experts because of their compliance with . . . , and their willingness . . . , to serve and follow the investment-class and their narrow interests.

Can an independent individual discern truth, or is society dependent on the institutions which support banking interests to tell the collective society . . . . , where the main purpose of "the collective" is to support the bankers . . . "what is true?" ???

If the new ideas are superior, then the bankers and oilmen must fall from their ruling-positions within society, since it is them who would be in greatest opposition to such new ideas about existence, ie science is now being controlled by them, but the type of science which the bankers-oilmen control has failed.

The truth in which the professional physics community believes is just a bunch of baloney.

The free expression of ideas

One can not even see why the expression of different ideas about "how society should be governed" should be so carefully guarded, since the main ideas which the, so called, intellectual class come-up with are also oligarchical and collective social-systems, as is the one in which we now live, the intellectual-class wants new leaders imposed on the same type of a social system. This is because the intellectuals are also elitists, whose elite intellectual status was defined in an elitist social system, which is being defined by the investor-class.

Furthermore, now the nature of the ideas can be so easily identified in regard to the construct of the assumptions and interpretations (by a computer search) so could be easily placed into various categories of ideas, where the different ideas could be compared and considered in an attempt at rational-ness, ie the age of science (or the age of enlightenment), but it is only in a society governed by arbitrary-values, which, in turn, are used to define the ruling-class (in a society governed by the law of property-rights and minority rule, ie business investment or revenue defines the ruling-class), eg the use of oil, and thus "the big business of oil" being the principle basis for identifying the ruling-class, where then the ruling-class needs to be so careful in its control over the communication channels, or over ideas which are expressed, in order to protect their business.

This all implies that the so called free-market has been an expression which has lost its meaning long-ago, ie by controlling information one can gain a lot of control over markets.

Control over "technical ideas expressed" is about controlling the relation of knowledge to creativity, again the ruling-class wants to determine what is created, so as to lessen the risks of their investments. Thus, they also must control knowledge. The control of knowledge is based on "the control of funding," and on the way the academic community is organized so as to fit into the needs of the commercial interests, which have been built and organized by the investors.

Furthermore, most of the experts come from the social-class of wage-slaves so they are the types of people who are desperate to follow that which is identified as possessing high-social-value (ie the categories of experts are defined in the propaganda system).

The point of modern human-life is:

1. to be part of a narrowly regimented collective hierarchical and extremely violent empire,
2. to believe in materialism,
3. to claim to believe-in a "lyrical relation" to a spirit, which is essentially (or translates into) a belief in inequality, and then this collective-empire's intent is to expand this narrow construct ie to noticeably expand the population.

There are two watch-dog types which jealously guard the banker's interests, but the watch-dogs believe they are autonomous

(1) The violent types who are given the social position of people who are in the social-position of being the morally-superior social-judges of the public (they have a close relation to the top administrators of the justice system, ie the judges) but they are the types of people who are given to believe-in and act-on stereotypes such as racist viewpoints, where the violent actions of these social-judges against the public, say, the 99%, are well publicized, where these acts of violence are taken so as to support the justice-(morality-play)-system, whose function is to terrorize the public.
    They are groups of people which are similar to the "old" southern-vigilante groups (of the right). And
(2) The dogma-bullies (the true-believers in arbitrary abstractions, which have, despite their arbitrariness, been identified as defining an absolute authority)
    These are the intellectuals, who demand intellectual conformity to the commercially-motivated set of categories of technical "intellectual-thought,"

ie peer-reviewed science and math, where social function of peer-review is to establish the idea in the public's mind of the existence of a "natural" intellectual inequality, an idea promoted in the propaganda-education system.
(of the left).

Note: The idea of materialism both destroys technical knowledge and it destroys the spiritual experience.

New mathematical descriptive structures exist which transcend the idea of materialism, and contain the material-world as a proper subset.

In a society, which functions based on a hierarchy of social-value, there are new ideas which define a new superior intellectual context for creativity, but the lack of curiosity in such new, and revolutionary, ideas (especially, compared to the dysfunctional ideas about "math science and religion" which are now (2013) accepted within society) proves the narrow delusional mental-state into which the empire has forced humanity, as well as forcing society into an arbitrary basis, in regard to society's measures of value and truth being related to a commercial context, upon which the empire's hierarchy is built.

If one wants to understand: . . . .

1. the chemistry of enzymes
   (eg inter-system messages related to action-at-a-distance, within a high-dimensional system, while DNA allows molecules to be built, locally, to comply with [distant] messages),
2. the greater (higher-dimensional) context of existence (to which particle-physics points) . . . , (in physics there is no notice, by the, so called, experts, of the dysfunctional-ness of the physics-story, apparently this is because the, so called, experts are absorbed in the complications of the description, and trying to make sense out of words which should possess meaning, but do not),
3. the higher-dimensional structures of life,
4. a new context within which "the world of the spirit" can be explored, ie the real world which transcends materialism, and
5. new ways in which to consider connections between distant places, by means of action-at-a-distance, and by the existence and structure of very-large unified systems . . . ,
   eg the solar-system has a unity and the galaxy has a unity which is an actual shape, ... ,
   can be made,
   ie travel to distant galaxies in an instant (outside the confines of materialism), and

6. a place of new invention, and a place of new types of energy-sources can be considered and/or imagined, ... then one should be quite interested in new mathematical ways by which existence can be modeled, in a new math structure, where measuring is reliable, and the patterns are stable, and it is a new idea in which "the idea of materialism" is a proper subset.

Why wait for the "OK" of the current science authorities, since they are bogged-down in the idea of materialism, and its subsequent incomprehensible complexities, and the fact (seemingly unbeknownst to them) that their, so called, rigorous knowledge is so highly dependent on its relation to the commercial categories, which are defined by the investment class, eg bomb-engineering, information-systems, etc, all of whose practical contexts are defined by 19th century science, "Is this all that different from the technology of the Romans being based on brick-laying?"

That is, one might try to make a decision about a fundamental truth on one's own.

Be a mental-revolutionary instead of a slave to monetary-needs.

The current model of physics cannot describe the observed properties of the (assumed to be material) world, either accurately to sufficient precision and with wide enough applicability, or in a practically useful manner.

One needs a new math context to describe the properties of the measurable existence which is being observed.

The propaganda system has, since the Roman-empire, been organized around proclamations about products which are made by the empire, and "superior" technology, but instead of the emperor controlling the knowledge and organization which is needed to build large-scale civic-minded projects using bricks, now (certainly since the 1400's) the bankers invest in the monopolies organized around products and services, eg the monopolistic guild-associations of Europe, whose market is guaranteed by the very stable life-styles of the public, ie products and services which depend on 19th century science for their conception and production.

Note that violence and the regimentation of society also allows the value of money to remain stable, within a collective society which supports the banker's investments. This social-regimentation is implemented by both the propaganda-system and the extremely violent justice system.

Within the narrow context of a particular life-style, eg using oil, it is claimed that one can purchase any material-good which one needs, ie the bankers now control our vision of life-styles and an associated set of "needed" products.

The context of use (or in regard to framing) for the propaganda system, in regard to both law and governing, the frame (of communication) is placed in a context of religion, so that law and politics are represented within a (religious) morality-story, and based on arbitrary identification of high-value (high-social-value), where value is interpreted in the context of Calvinism.

Namely, "being very rich" means "having great favor with God," ie one's value is determined by how rich one is, according to the Calvinists (ie (historically) the rich do not need to buy indulgences).

[Note: The propaganda system equals the political system, the ideas about "civics" mean nothing when the communication-channels for the society are so highly controlled].

This arbitrariness "by which both value and truth are defined" inevitably leads to failure.

Technology reaches the limits of the society's limited-knowledge, but the limitation is self imposed, due to adhering to authoritative dogmas which relate to both materialism and commerce and to "educational" competitions which are designed to show-case the talents of the academics, so as to prove that they (the academics) can work-on (or are qualified to work-on) the narrow set of technical problems which are of interest to the investment-class, but just as importantly, the professionals must "use a special complicated language," which hides the knowledge from the public, but it really hides knowledge itself.

The established dogmas cannot be questioned, they can only be worked-on (by wage-slaves), in a rigorous manner, and in ever more complicated and incomprehensible representations.

(they are incomprehensible, since they are descriptions which either do not lead to any further practical creativity, beyond a narrowly defined application, or they only lead to one particular category of invention).

It is the requirement that people make judgments about value in relation to established value (or established truth), where this truth is ruled-over by the expert-class, within the structure of the social-hierarchy, where it is the requirement that "people adhere to an externally determined truth," which has made people within society befuddled and helpless . . . , (they become dependent on experts, yet in the context of competition (used to determine who "the experts are") the winners become arrogant, yet they remain too dependent on the beliefs of others, namely, the other experts, and subsequently people (the experts themselves) cannot make their own judgments about either value or truth), . . . , though they often adopt an assuredness associated to the ruling-class, or the assuredness of the experts who work for the ruling-class, ie they believe in a particular set of dogmas associated to categories of commerce and social domination, which is really being presided-over by the "virtuous" very-rich people.

In poor societies this means either obey, ie "be slaughtered or conform," while in rich countries this means arbitrary institutional judgments are to be followed, eg racism or peer-review etc, or fail as a wage-slave. That is, in rich countries it is the same dictum, but it is couched in a more developed set of institutions and covered with greater amounts of propaganda, so that it is "obey or 'be without resources.'"

The arbitrary identification of high-value is institutionally based on "the value of material" and "property rights," where property (including material) is mainly obtained by means of (from a process of) stealing, and then being able to hold-onto the property by means of violence, where the validity of the theft (stealing) is supported by the justice system, and this is often justified through the pompous-claim about the "productive" use of the property (the claim being that there is better use of the property [by the superior people] because they possess both superior knowledge and access to efficient large-scale productive capacities).

The stolen property is then used in narrow ways, associated to both life-styles and commerce.

And the rich have been granted the right (by the legal-political system) to control and organize "how property is used" in society, namely, to support the monopolistic businesses of the rich.

In turn, "intellectual property" is also associated with theft ie the Faraday and Tesla families should be the main families which are getting rich from information-system electronics, since it is they who possess the natural copyrights to these inventions, but in fact, "intellectual property" is associated to either slight adjustments to *either* an engineering-science creed, wherein categories of descriptive knowledge which are related to commerce, *or* to an arbitrary traditional intellectual authority (ie to the professional academics) . . . , [which is (now, since the 1910's) related to an axiomatic formalization associated to either quantity and pattern (or shape) [ie math], or to "physical law" (frame-invariant partial differential equations) [ie science]] . . . , where the "main use" of this narrow authoritative intellectualism . . . (beyond its, usually, narrow commercial use) . . . is about "the place of high-intellectual-value" in the propaganda system, where it (incorrectly) represents the "superior knowledge of the culture," (it is really knowledge associated to commercial categories)

Or (in regard to further considerations about "intellectual property")

Consider the "creative arts," which service both

(1) the market of arbitrary-value (associated to traditions and to disciplines), and
(2) the propaganda-system,
    which usually, only allows art which expresses the idea of:

    1. inequality and
    2. arbitrary hierarchy (ie the main context of (western) religion), and
    3. "superior intellectual ability."

Commercial art is all about propaganda, and its principle practitioners (the main producers of content) are the, so called, journalists.

And the rich have been granted the right (by the legal-political system) to control and organize "how intellectual-property is used" in society, namely, to support the monopolies of the rich.

The journalists are, supposedly, those who objectively point-to and interpret societal occurrences in a manner consistent with the investments of bankers, who invest in narrowly defined creative endeavors, essentially, built around life-styles whose organization (either) has a strong dependence on (or was strongly influenced by) investments.

To a journalist the idea of being objective, and finding truth, is to seek-out, and interview the authorities, those who reign over a small range of a narrowly defined "truth," which has fundamental connections to either a commercial category of investment, or a relation to the main-point of the propaganda-system, ie that of expressing an "unequal representation of arbitrary high-value, in particular, high-intellectual-value" which is consistent with the interests of the ruling-class.

It is all a model of a arbitrary hierarchy, which is also the main-idea of institutional religions, ie "the leader is the spiritual-adept who everyone else must follow."
(but in this case "spiritual" means being able to acquire money and social-power, since the material world has not been transcended)

Note that, the organization of the empire is balanced on the idea of materialism as an absolute objective truth, and the spirit being about "Lyrical creativity," such as using propaganda to support both the ruling-class's right to be socially dominating, and the idea of inequality (eg survival of the fittest in a world where material is scarce, or the triumph of virtuous and God-favored rich-people).

The (righteous) right represent the "superior people," so considered to be superior by the propaganda system.
The "superior people," are those who compose the ruling-class.
They represent an arbitrary measure of high-value, ie the superior people of the society, the investment class, whose traditional role has been to fit into the traditional narrow structure, and narrowly defined set of categories through which the life-style of a community (or society) is defined, and it is also a "collective community," whose purpose is to support the ruling-class.
Material has come to be controlled by monetary-value, where some people wish that the value "attributed to money" should be associate to "the value of precious metals."

(this distinction is irrelevant, since value is about knowledge and creativity)

Thus, the, so called, "scientific viewpoint of trading-value (where it is assumed that science is based on the idea of materialism)," of having material-value (or barter-value) being associated to the scarcity of the material.

Thus, the arbitrary value of culture, of people, etc, is linked to an arbitrary value of material.

Knowledge and creativity are tied-to the value of material, and the use of property (or material) in regard to a narrowly defined context of creativity, namely, the creativity into which the investor-class has invested.

The pragmatism is that material defines existence.

But this idea destroys both knowledge and creativity.

That is, all of this "arbitrary judgments of value," in turn, determines the structure of the "math and science professional communities," who follow arbitrary traditional authority, narrowly defined and adhering to the categories of commercial use of knowledge.

This "blind following of authority" (and the idea of being led by authorities) has led these professional communities into a descriptive language, which though, it appears to be, upheld by axiomatic formalization, but it is "knowledge" which has no content, ie it is not comprehend-able,

though convergences (which are defined on a containment set which is far too-big) create the appearance of, apparently, valid quantitative constructs, eg diverging-sums are given an arbitrary structure, a structure which fits the answers, which are expected, from calculations which are based on diverging quantitative patterns, etc, etc.

These arbitrary hierarchical social or community structures are all quite similar to the institutional structures of (western) religions.

The left represent high-intellectual value (the professional-class competes, in the propaganda-education system, for their high-salaried social positions) but the intellectuals are sinful (they do not always support the ruling-class directly, however, the ruling-class supports the intellectual-class, who, in turn, fit-into such a narrowly defined viewpoint of knowledge, in which their knowledge is only (practically) related to those ideas which support the investments of the ruling-class, but this professional intellectual-community occasionally express ideas which are in opposition to the ruling-class, eg global-warming) and must atone for this sin, by doing "good works," and the intellectual-class is also "soft," they too often express the golden-rule (which most simply-put is, "love your enemies"), whereas the ruling-class must "smite the inferior-class," as it is so-stated in "Constantine's Bible," the propaganda-manual of the (Holy-)Roman-Empire.

It is the much-populated context of "the inferior," which, in turn, defines "the superior" (in specific narrowly defined contexts): . . . rich and poor, (ie strong thieves vs. weak thieves), smart and stupid, (ie intellectualism defined in the narrow context of the ideas which support the ruling-class's interests, not in the general context of "the natural-failure of fixed precise languages," ie Godel's incompleteness theorem, about the natural limitations of precise languages, namely, an inability of any precise language to identify many true patterns) hard-working and lazy (hard-work which, by luck, provided a big pay-off), . . . , all these narrowly defined contexts are defined in a context of scarcity . . . , (except for all the material-value and intellectual [copyrights] which the ruling-class has stolen, it is the ideas of inequality, materialism, and material-scarcity which are all forced onto the public) . . . , and according to Darwin, where the fittest survive, ie those endowed with talent and intelligence, where the associated ideas of worth and value are defined in the narrow monopolistic context of investment and traditional academic authority. (all these "measures of value" are defined on an arbitrary context for which there is a material manifestation)

Those who possess talent and intelligence get rich, while those without either talent or intelligence must be relegated to the inferior poor, who must be punished for their inabilities. A very Calvinistic social scenario, where (fearless-leader) "Calvin" takes the role of the worst of any possible tyrant.

In a nutshell

That is, "the right" is simply uppity and violent, while "the left" is intellectually-uppity in the delusional context within which they frame their belief in intellectual-value.

The "working assumption" of this society based on material-product and invest-and-consume . . . . (monopolies are primarily defined on narrowly conceived expectations about life-styles, ie the material necessities of life) . . . . , is that:

"one can buy whatever one might need"

That is, all needs (within a narrowly defined vision of a life-style) can be satisfied by the invention and investments which are directed by the investment-class.

The uppity-left believes that anything, which is related to technical inventiveness . . . , either can be invented or must be invented . . . , in the context which is currently defined by the authoritative dogmas of the professional intellectuals.

But the ability to build a "clean cheap energy source" is not possible, the ability to "travel to many-places very-quickly" is not possible, the capacity to "understand life," . . . within the materialistic-based authority of the experts . . . , is not possible.

Thus, the left's vision of intellectualism, ie knowledge serving investments, is a very flawed idea.

The context of life is a "wage-slave existence" defined by products, which one is required to purchase. Work and competitive domination gives life its, so called, proper meaning, in an assumed context of social Darwinism, and such an idea's dependence on the notion of materialism.

This context for life, within society, is held together by lying, stealing, and violence, where the collective experience is all about "being forced into narrow competitions" within which "one is to identify their own worth or social-value" and being forced into a life-style, but it is a social context based on an intellectual structure which is failing.

It is a social experience which is all about both an arbitrary self-righteous uppity-ness (both morally and intellectually) and the extreme violence needed to maintain such a hierarchical power structure, and identifying those who can express this (so called, needed) violence, these people need to be identified since it such violent people who can protect a (the) particular uppity viewpoint (of the ruling-class).

It was decided by the federalists, that the many indebted revolutionary-war veterans could not ignore the foreclosing bankers (eg Shay's rebellion), these veterans were not to be allowed to interfere with the interests of the ruling-class, where the social interests of the (then, 1786) upper-social-class were organized around banking ands commerce.

Thus, the constitutional convention was convened, even though it was opposed to "the point of the revolutionary-war," where the point was

1. to establish the law of equality, and
2. establish a common-welfare in the context of equality,
   ie law was to be based on the principle of equality,
   the main principle about which the Quakers organized their social behaviors.
   but

   The ruling-class social-structure of the south was allowed to dominate "how the new US society was going to be organized."

   Though the constitution was only made valid, after the Bill of Rights was amended to the constitution, but the ruling-class has never enforced the Bill of Rights. (eg J Adams signed the Sedition laws)

The technological creations of the intellectual elite have run their course, and they are not capable of providing any types of new inventions. That is, the technology of today is all based on 19th century science.

Instead of basing physics on partial differential equations, ie the laws of physics represented as frame-invariant partial differential equations associated to non-linear geometry and randomness, or more to the point indefinable randomness;

One can (instead) consider the context of stable geometry in a new many-dimensional containment space where measuring is reliable, and materialism is no longer true.

That is, partial differential equations are but substructures, which can be used to identify perturbations upon the stable geometric structures, into which material systems are organized, within a many-dimensional existence.

That is, for ten-years (1776-1787) the rationalists saw the cure for "European social tyranny" as, having law based on equality, in turn, based on the egalitarian Quaker-society.

But the European commerce from-which the upper-class of the new America depended was diminished, and this resulted in the upper-class re-instituting the law of the Roman-Empire, which in Europe was then expressed as "rule by the ruling-class," ie the bankers.

So the constitution, whose Bill of Rights was never enforced, was the re-establishment of "rule-by-the-emperors."

But a society of individuals, where law is based on equality, and where the context of equality is creativity and knowledge, and where the material side of existence is about the general welfare, which knowledge and creativity can provide for society, is a very good model of a society, which is very different from the idea of "rule by the ruling-class," ie the continuation of the idea of a collective society supporting the ruler, ie models of the rule of the Roman-empire, which are the common ideas expressed by the western intellectuals such as A Smith and K Marx.

But it is clear that one wants a free-market for a wide variety of creations, about which many different types of life-styles can be envisioned and realized, but if the trading basis is money then the market-players who possess "too much money" will find-ways in which to destroy the free-market, ie they will steal the market and make it one of their own possessions to control as they wish, as is the problem today (2013).

That is very popular items within a free-market need to be made collective, since they are defining a collective (viewpoint).

Thus, in order to preserve the main value of a society . . . , ie that being "its knowledge and creativity," so everyone is an equal creator, and can equally be a part of the free-market, . . . , (then) the very successful parts (products) of the free-market will need to become collective.

That is, market success is not a valid measure of high-value, rather creations are gifts which are given selflessly.

The world is not about material-scarcity and the confinement of behavior, or the confinement of thought, where such confinement causes people to identify an unequal state of existence for human-life, within the confines of a material-based context.

This is because the material-world is a small (and relatively unimportant) subset of existence.

Alternative ideas about math

There are two categories of orthodoxed mathematics, (1) geometrization and (2) the identified set of properties of discrete hyperbolic shape (as well as the rigid and very stable properties of discrete hyperbolic shapes), as well as (3) the new experimental finding about non-locality in physics, which can be considered together, so as to lead to new interpretations concerning the organization and structure of the language of mathematics, so as to form new contexts for descriptions of both math and physical patterns.

Geometrization and the relation of circle-spaces to stable and measurably-reliable contexts of:

1. linear,
2. metric-invariant, and
3. continuously commutative everywhere (except a one-point)

which lead to solvable and controllable descriptive contexts.

Thus these properties (linear, metric-invariant, commutative) identify the main attributes of stable systems, where these are properties which all point-to the properties and the context of discrete hyperbolic shapes, whereas their properties have been described by D Coxeter, as well as their very stable geometric-and-spectral properties, where, in turn, the properties of these shapes identify a many-dimensional context where stable patterns of circle-space shapes can exist and thus define a stable measurable context for a many-dimensional existence.

Physical description

In regard to frame-invariant partial differential equations representing the laws of physics, one finds a:

1. quantitatively inconsistent context, as well as
2. logically inconsistent context, and
3. an indefinable context,

as being the general attributes of the descriptive structures and capacities of such a language.

That is, in the descriptive context which focuses on partial differential equations as the basis for the descriptive language, one finds that

1. Descriptions of physical systems cannot be formulated,
2. non-linear systems cannot be solved,
   the structure of systems are not known yet
3. the orthodoxy of material-based laws of physics described in the context of partial differential equations and defined in
4. containment sets which are too-big, where
5. singular indefinable math processes are turned into orthodoxy
   which describes essentially nothing
   since it is a context of:
   1. non-linearity,
   2. non-commutativity,
   3. measuring is not reliable, and
   4. randomness is not definable,
   5. there are no stable patterns, and
   6. the descriptions are incoherent.

Apparently, the experts, themselves are unaware of this functionally-useless context for precise descriptions.

The experts of science and math toil in this meaningless descriptive context,

However, there are categories of math descriptions (or constructs), ie math patterns, which can be used to organize math and science descriptive language in a new context where these categories of math descriptions (or constructs) possess no obvious relation to a new global math language.

The current descriptive contexts is a general structure in which "the language of math has lost its capacity to be comprehend-able" (at the level of accurate descriptions, and at the level of practical applications of this general structure of the math language)

It is abstract word-agreement used by the professional experts, but it is a dys-functional language which has been fixed to an axiomatic formalization (this formalized-axiomatics began about 1910), where the words bring the description into a (dys-functional) "professionally real" context, but where measuring is unreliable, and there are no stable patterns in the description

There may exist distinguishable states, which are either un-relatable to quantities, or are only relatable to unstable patterns, but the type of language describing such a non-quantitative context is, nonetheless, considered to identify a valid set of topics for professional discussions, but rigorous word relations, which are used in this context, have no bearing on a measurable or a practically useful context.

If the professional math community is unaware of these apparently subtle or hidden failings of their descriptive language axiomatic formalization is "a difficult deception to uncover" then, nonetheless, they should be aware of the limitations of precise language and a need for alternative ways to organize new precise languages.

But such alternative languages are not in relation to either commerce or arbitrary traditional authority, so it is not possible for these new alternative language to be peer-reviewed.

The math community (reluctantly) granted a presentation of these new ideas, whose basis is centered around the idea of geometrization, at their annual get-together (in 2013, San Diego), but excluded it from the get-together's central focus, which is a major failing, in its own right, by the math community.

67.

Preface (Preface 66)

This paper begins (in part I) with a review of the current context for the physical sciences, and its fundamental failures.

Then it goes over the new descriptive language, where in the new language-constructs, the properties of stable material systems can be described (both stability as well as "randomness and chaos").

Then it gives a historic over-view, to point-out where the wrong path (of language) was taken, ie an over-view of a precise descriptive language and its, supposed, purpose.

Part II is about how the US (or western) society controls language and thought. The western viewpoint is still the same viewpoint as that of the (old) Roman-Empire (The US revolution, and the Declaration of Independence tried to change this viewpoint, but it has, so far, failed).

Part I

According to professional peer-reviewed physicists, ie according to their authoritative dogmas, the dogmas upon which peer-review is based, physical systems are composed of material, ie mass or charge, contained in space-time, within which their "behaviors" are determined by the (partial) differential equations (the laws of physics), which define (or inter-relate, by means of material interactions) the system's measurable properties whose math-structures determine that within which the material-system is construed, where the equations are generally-frame-invariant and the containing coordinates possess a general metric-function so as to usually define curved metric-spaces.

However, the notion of materialism has led scientists to believe that all material systems can be reduced to "elementary" point-particles, which define sets of random events in space-time, where these events are random spectral-point-particle material events (but the elementary-particle events, to which material is believed to be reducible, are unstable events), so that in this context (of reduction and its associated randomness) mass and charge, which are contained in a system, are defined by function-spaces with internal-symmetry vector structures associated to the wave-function, where the wave-function's spectra can, supposedly, be determined by finding (complete) sets of commuting (Hermitian) operators, and where the (differential) operators are connections (or square-root 1st-order operators), which act on both the wave-function and its internal-vector-particle-state structure, where the connection is only relevant during the collisions of elementary-point-particles, where these collisions model (quantum) material-interactions (within the quantum system), where, supposedly, these point-particle-collisions perturb the quantum system's wave-function (about the spectra) of its (main-defining) energy-operator, which is supposed to define a quantum system's measurable spectral-energy properties. But this is a model which is based on randomness, so a point-particle collision geometry does not make logical-sense, and this also causes irreconcilable diverges, as also does the 1/r term in a charged system's energy operator (where 1/r diverges for point-particles), furthermore, the elementary-particle events are unstable (so it is a random structure [whose elementary-events are not stable] thus the randomness is indefinable). Furthermore, for most stable quantum systems, the energy-operator is not definable (ie the set of operators, which are supposed to identify the quantum system's measurable properties, are not identifiable), thus the spectral structures of these indefinable systems cannot be calculated, thus, such systems are based on the idea of indefinable-randomness, and there is no set of spectral-values to perturb, ie they are systems whose description is based on non-sense.

Furthermore, such an internal-particle-state operator structure is never commutative, and never linear, thus, its metric-structure is also unknown.

Thus, this defines a math descriptive context where measuring is un-reliable (unknown) and the (math) patterns are not stable, and the systems cannot be solved (or they cannot be formulated).

Yet the vast majority of fundamental quantum systems are characterized by their great stability, a condition which implies that the quantum system is: formulate-able (ie so in quantum physics this would mean that the system's set of commuting operators should be findable, but, in general, they are not findable), and linear, and solvable, and thus, controllable.

Note: This is also true for the stable solar-system, but general relativity is non-linear, thus the partial differential equation for the stable solar-system is not determinable, within the descriptive context of general relativity.

Note: Elementary-particle collisions are only of practical importance to bomb-engineering.

Physical systems, composed on many-(but-few)-components, and characterized by stable spectral-orbital properties, exist at all size-scales, and none of the stable properties of these systems are describable by applying the, so called, laws of physics, and then solving the equations, where, mostly, the equations for these systems are indefinable, ie the equations for these stable systems should be formulate-able, but they cannot be formulated, let-alone, solved (ie they are stable systems so they should be both solvable as well as being controllable).

Whereas

One wants a descriptive context where measuring is reliable and the (math) patterns are stable.

In the new descriptive context; existence is many-dimensional, metric-invariant, and it is composed of sets of stable shapes, where the shapes can possess properties of mass, charge, or other energy-generating properties, so that a physical system exists . . . , if it is a stable shape, or it is condensed material, which is:

either
. . . , as a material component with a definite shape, it is in resonance with some part of the finite spectral set, where the spectral-resonances are dimension-dependent, or
. . . , as a condensed material component, and it is contained in a metric-space shape, where the metric-space shape is in resonance with some part of the finite spectral set (or the metric-space shape is a part of the partition which defines the finite spectral set), and the metric-space's shape is affecting (or can affect) the motion of the condensed material.

In the new context there are natural geometric structures within metric-spaces of constant curvature (where the metric-functions can only possess constant coefficients) which determine both material-interactions, and the (partial) differential equations associated to these interactions, but these (partial) differential equations either determine the energy-structure of a (usually) collision interaction, wherein a new stable system-component might form due to resonance, where the system's resonant-set can abruptly change during a collision process, or they perturb the orbital-spectral structure of the material, which either composes the system, in either an "object" "metric-space-shape" inter-relationship (as defined in the context of general relativity, but now the "shape of space" is defined in a linear context), or (the material property) occupies the material-component's spectral-flows in the material component's stable shape.

That is, a stable system is not defined by "a (partial) differential equation which is generally frame invariant," yet this condition can be satisfied within a system, but (if the system is stable then) the system will be: linear, metric-invariant, and (if a stable shape) the shape of the system will be continuously commutative everywhere, except possibly at a single-point, ie it is solvable, and thus it is both stable and controllable.

Material's relation to stable shapes is about stable shape being related to either Euclidean or hyperbolic metric-spaces.

Note that, hyperbolic space is space-time with time divided-out (or moded-out, that is, 4-dimensional space-time becomes 3-dimensional hyperbolic space).

The Euclidean or hyperbolic metric-spaces are spaces with non-positive constant curvature, where the coefficients of the metric-functions are constants.

D Coxeter identified the properties of discrete hyperbolic shapes, ie the most (prevalent of the) stable shapes in geometry, so that the last discrete hyperbolic shape is of hyperbolic-dimension-10. Thus, the set of different types of discrete hyperbolic shapes would be shapes contained in a hyperbolic metric-space of dimension-11.

The last bounded discrete hyperbolic shape is of hyperbolic dimension-5.

Thus to determine the finite spectral set associated to existence, the existence is defined to be an 11-dimensional hyperbolic metric-space, which is partitioned into a set of discrete hyperbolic shapes for each subspace, and for each dimensional level, but all such bounded shapes would be 5-hyperbolic-dimensions or less, so that the "partition of shapes" is identified as the set of the largest shapes for each subspace at each dimensional level for all dimensional levels, note: the largest discrete hyperbolic shapes of hyperbolic-dimension-6 through hyperbolic-dimension-10 are all unbounded shapes.

The spectral properties of discrete hyperbolic shapes are very stable, and the number of spectra on a bounded discrete hyperbolic shape depends on the genus of the shape.

The shape of a bounded discrete hyperbolic shape is a set of toral-components attached to each other, where a torus, or a doughnut-shape, is a discrete Euclidean shape, whose geometric-spectral properties are not so stable as those of a discrete hyperbolic shape, and where the discrete Euclidean shapes exist by being resonant with the toral-components of a discrete hyperbolic shape.

The number of toral components of (a bounded or unbounded) discrete hyperbolic shape also defines the genus number of the discrete hyperbolic shape.

One can count the number of subspaces of each dimensional level, so when this number is associated to (multiplied-by) the genus numbers of the bounded discrete hyperbolic shapes, which compose the partition set, which, in turn, determines "the finite spectral set" for the 11-dimension hyperbolic metric-space, within which all of existence may well be (is) contained.

The partition set defines a set of metric-spaces within which other lower-dimensional material-components may be contained, if they are both (1) of a small enough size and (2) (based on whether) if they are in resonance with some part of the (bounded) finite spectral set of the over-all 11-dimensional containing space.

Note: There are, 11, 1-dimensional subspaces in the 11-dimensional over-all containment set, while there are, 55, 2-dimensional subspaces, 165, 3-dimensional subspaces, 330, 4-dimensional subspaces, 462, 5-dimensional subspaces, 462, 6-dimensional subspaces, etc . . . ie the number of subspaces is a set of symmetric numbers after dimension-5 . . . ,

The shapes contained in this 11-dimensional hyperbolic metric-space containment-set, relate to one another, especially between dimensions, in a discontinuous and discretely-changing manner, but they also (relate to each other in a discontinuous manner) between discrete folds defined between toral components, and between discrete time-intervals (defined by the [time] period of the spin-rotation between opposite metric-space states).

Within a shape (ie when one is contained within a metric-space of a particular dimension and in a particular subspace of that given dimension; then), one "sees" (ie one experiences) an open-closed topology, and thus one experiences the internal curvature of a hyperbolic space, but since we are on (or within) the orbit of the planet-earth, thus we are within a folded toral-component of a hyperbolic metric-space shape, thus, we may mostly experience the toral shape geometric structures, which are flat.

If one is in a metric-space, which contains within itself a lower-dimensional shape, then the boundary of the contained-shape is identifiable (within the higher-dimensional shape-containing metric-space).

The shape contained in a metric-space will have properties of inertia and possibly charge, ie lower-dimensional metric-space shapes contained within a higher-dimensional metric-space represent material components, though the shape may possess a neutral charge.

Thus, the geometry noticed by the boundary of the shape of the material-component, contained within a higher-dimensional space, may be the curved shape and the geodesic structure of the metric-space within which the material-component is contained.

That is, a lower-dimensional shape, which is contained in a metric-space shape, can also experience (feel) the shape of the metric-space, so that the geometric relation of the material component with the shape of its containing metric-space shape can define a material interaction (ie the interaction of the material component with the shape of the metric-space [within which the material component is contained], ie the same type of interaction structure envisioned by general relativity, ie the material component trying to follow the geodesics of the metric-space, within which the material component is contained).

There is a scale-factor which is defined, as a discretely identified scalar factor, which exists between dimensional levels. Thus, the size-scales of material shapes can greatly change between dimensional levels (the cause of the, supposedly, unseen dark-matter).

The geometric structure of material-interactions, and the scale-factors defined between dimensional levels, and the discontinuous nature of the relation between dimensional levels, can cause the existence of higher-dimensions to be very difficult to experience when one is (measuring) within a given dimensional level, ie within the metric-space within which (we believe) our experience is defined.

The newly defined process of material-interactions, result in (partial) differential equations which result from the new geometry of material-interactions, where this geometry is intrinsic to the properties of stable shapes, ie discrete-Euclidean-shapes and discrete-hyperbolic-shapes, where the properties of these stable shapes most characterize the new descriptive context, for all of existence.

Apparently this dependence for both existence (metric-spaces and material-components) and of the (partial) differential equations, which perturb the behavior of material-components within metric-spaces, are defined based on stable shapes, where such a shape-dependent context, is required in a mathematical (or a quantitative) descriptive language because of the fact that stability is so fundamental to our experience of existence.

The finite spectral set, which now (ie in the new descriptive context) is used to define material existence, but it also allows for a quantitative structure which is built from a finite set of quantities. Thus, "the continuum" is an unnecessary math construct for (or in regard to) the description of the measurable properties of existence. However, issues of "infinity" (or unbounded-ness) are still a fundamental part of the new description, but they are related to stable discrete hyperbolic shapes, ie they are related to geometries.

The unbounded discrete hyperbolic shapes, which determine the higher-dimensional geometric shapes, can have profound relations to the confinement to a particular dimension subspace, eg the properties of both neutrinos and light, and about defining relations to other 11-dimensional hyperbolic metric-space containment-sets, as well as to the properties . . . , which the odd-dimensional and odd-genus models of . . . , life-forms (can) possess.

Review

This is a completely new context within which to view both math and physics it is a completely new re-arrangement of mathematical language which liberates these descriptions from (Newton's) (partial) differential equations, which seem to have never been placed into their correct mathematical context and thus the idea of a (partial) differential equation have never been understood, they are really perturbing operators which emanate from the natural geometric shapes of material interactions or within material interactions within a metric-space which possesses a shape which acts as an envelope of orbital stability for the perturbations of the perturbing differential operator. That is, the main context of quantitative and physical descriptions deal with the stable shapes associated to charges, nuclei, atoms, molecules and the more general (or transitional) macroscopic-microscopic shape-structures of crystals, but it is also about macroscopic shapes, such as the solar-system, as well as macroscopic structures within galaxies, where the stable contexts of angular-momentum could be similar in their relation as a complex and controllable system structure which might be analogous to linear (solvable) electric circuits.

That is, Galileo's and Kepler's contexts of physical descriptions are more fundamental than are Newton's viewpoint about (partial) differential equations, which describe local measurable properties, which are associated to perturbing stable geometric structures, which, in turn are associated to the relation of these geometries to similar relatively-simple-stable-shapes. That is, Kepler's idea about stable shapes is more fundamental than has been Newton's ideas about the (partial) differential equation being the main math structure which allows for a wide range of relatively precise physical descriptions.

The atomic hypothesis of "material being composed of small stable systems" which are best described as being stable shapes (which have some perturbed properties), ie the atomic hypothesis allows material to be understood as, mostly, microscopic stable shapes. Whereas the chaos of both the macroscopic, mostly, non-linear systems when modeled as (partial) differential equations, as well as the apparent randomness of the microscopic, or very small, atomic material-components, which is due to interactions between micro-components which result in Brownian motions, (these random and chaotic behaviors) emerge, exactly, from the descriptive context of (partial) differential equations.

Kepler tried to explain the stable solar-system with his model of the solar-system based on regular geometric shapes, whereas no model of the gravitational-force, which is based on partial differential equations, has been able to describe the structure (or interaction process) which allows the solar-system to be stable.

However, Kepler's model of "regular" geometries has some significant relationships to the discrete stable geometric shapes of both Euclidean space and hyperbolic space, though it is the stable (or consistent) quantitative relation that exists between the line and the circle which is the core geometric structure of the new context of descriptive containment of all of existence (however, many regular-geometries can be related to circle-space shapes).

Further notes

Another mystery is: "Why is the solar-system confined to a 2-plane?" There are a couple of possible answers, which might be considered (in regard to this mystery of the 2-plane) when one considers the stable shapes as being the fundamental basis for existence (as opposed to the idea that the (partial) differential equation is the fundamental way in which to understand measurable properties of material systems), eg either inertia is (simply) defined in a 2-dimensional context in Euclidean space; and/or the structure of SO(4) = SO(3) x SO(3), implies a common 2-plane, in regard to an inertial interaction in 3-space whose interaction geometry is naturally contained in 4-space (note: this is the "4-space structure" which defines the inertial properties of the electromagnetic material-interaction).

Note: The interaction structure can also account for "the apparent randomness" of the small material-components which compose material, ie the small material-components of the atomic hypothesis. Furthermore, the interaction structure can also account for the fact that "spherical symmetry of material interactions" is only a property of material-interactions which exist in 3-space, but the geometry of the interaction is still based on the stable shapes, but the force-fields identified in "3-space interactions" possess spherical symmetry, where spherical shapes are non-linear and are only "stable" when they stay-in exactly the shape of spheres, ie spheres are not stable shapes, since in reality (unless they are modeled as rigid-bodies) they will always be perturbed, since the sphere is the (assumed) shape of the field (gravitational or the electric-field) of a solitary point-object in 3-space.

Further Review

Newton's derivative was a local (instantaneous) (linear) model of measuring for the physically measurable properties (eg solution functions to differential equations) which are changing (in time and space), where these changes are caused by a distant geometry of other material.

The operator (or quantum physics) viewpoint is about function-space averages defined by both (1) sets of operators, and (or including) (2) geometric material relationships (or material geometries).

But this descriptive context cannot resolve the (this) relation between shape and averages of a random math-structure (which are associated to operators used to describe the properties of physical systems).

But included in this (current) descriptive context, is the idea that material can be reduced to small elementary-point-particles (which are defined by unstable events), and subsequently it is assumed that material-interactions depend on point-particle-collisions of field-point-particles and spin-½ point-particles (associated, mostly, to energy operators).

This descriptive context is about trying to reconcile limit properties on the functions' domain space, which is modeled as a continuum, with limit properties defined on an even larger-set of values, which are defined within the context of the function-space.

The continuum is too-big of a set, which results (or allows for) logical inconsistencies.

Problems with containment-sets which are too-big

It is this type of indefinableness, and logically inconsistent descriptive context, which seems to vaguely work as a descriptive format but only when either the answer is already known or the constraints on the system being described are very easily controlled.

The continuum is too big of a set, and is used to define point-particle-collisions in a descriptive context which is based on randomness, where position and motion are dual properties (or dual operators) for the uncertainty principle, ie it can be used to describe logically inconsistent (descriptive) contexts.

The new context

The new viewpoint assumes that material systems cannot be reduced without losing physical structure, and field-particles are actually shapes, which are often associated to instantaneously changing geometric shapes, such as discrete Euclidean shapes, {where this metric-space structure is to be, further, placed in the complex-coordinate space, where the two opposite metric-space states can be identified on the real and pure-imaginary subsets, respectively, and thus these instantaneous discrete Euclidean shapes would also have a relation to the unitary fiber groups, which are so much a part of particle-physic's differential operators}.

[note: the neutrino and the anti-neutrino . . . , which in the new descriptive context would be unbounded discrete hyperbolic shapes, so as to identify the two opposite metric-space state subsets . . . , have opposite spin-½ properties.

Thus, proving that the opposite metric-space state model is true.]
And
The new viewpoint claims that stable shapes are the primary attributes of existence, and that these primary stable shapes exist when they are in resonance with the finite spectral set of the over-all high-dimension containing space of all existence, while material-interactions are secondary descriptive structures, and either create chaos, and/or uncertainty, or they perturb material properties for the material which is contained within a stable shape.

Though the property of unbounded-ness still exists in the new descriptive context, nonetheless, in the new context, the property of being unbounded is (can be) given a stable shape, and placed in a new set-containment context, where an unbounded property may be relative to a higher-dimensional "bounded" metric-space, wherein the apparent unbounded property might be contained.

Furthermore, unbounded-ness might (also) be related to either distant properties, or to completely different over-all containment-sets etc.

Part II

The right-left media-dialog and the diabolical and anti-intellectualism of the left

If one is living in a meritocracy then this means that there exists within society an arbitrary judgment of value, and such a society will be both violent, and it will be doomed to failure, due to an arbitrary dogmatic authority which will be associated to judging value (and to judging "what is true").

The diabolical role of the intellectual-left in the media, which, apparently, most of the intellectual-left, themselves, are unaware of their role, though it is essentially the role of an undercover agent, a double-agent whom expresses dissatisfaction with the corporate-state, but, in fact, their actions, beliefs, and expressions support the corporate-state, so that revolutionary ideas ie new ideas and new contexts of creativity cannot be realized, and any such new ideas are opposed by the intellectual-class, whom suppress the type of (new) ideas (new ideas, which the left believe do not possess sufficient authority, and) which the left believes can not expressed by the intellectually-inferior public (ideas which question the arbitrary intellectual-values of the ruling-class), and instead new ideas are stifled and excluded by the elitist intellectual-left, ie the reporter-scientist-propagandist intellectual-complex.

The current extremely violent and repressive social-system fundamentally depends on: (1) the justice system, so as to legitimize its (such a social-system's) need for extremely-violent protection, for its basis in theft, and (2) the (rulers of) society express their social-domination in what seems to be a quantitative structure of an "economy of trade," based on a uniform stable currency, again upheld by the violence of the justice-political system, though (3) the political system is really an arm of the propaganda-education system, which is used to maintain the narrow focus of the collective-society which supports the ruling-class, where (4) the ruling-class identify, for "all of society," an arbitrary idea about high-social-and-intellectual-value, which, in turn, identifies the (narrow) context of the social-domination by the ruling-class.

It is a social-structure which depends on: materialism, the control of "language and thought," and extreme violence.

It is a society which will be opposed to any form of (new) science and math language-structure, which over-turns "the idea of materialism," and which "supports the independent capabilities of the public."

The public must remain wage-slaves, and they are taught to "clamor about their need for the ruling-class to provide them with jobs," and guide the public to greater social-economic expansion based on an intellectual structure which is so narrow (1) it leads to both resource-depletion and resource-destruction, and which (2) has become irrelevant in regard to greater creativity.

Spying and stealing (who are these virtuous spies?)

Perhaps the spy-system should be maintained, so as to stay connected to business actions, where the economy (controlled by big-business monopolies) has become the social-vehicle through which selfishness and social-domination are expressed in the current social context,

where in the new context one is assuming the justice-system (and its deep connection to violent methods) has been rectified.

A new political system which protects the individual's right to be an "equal creator" can be run as a democracy, where the abundance of the earth is the basis for the people's survival. These are the ideas of the Declaration of Independence and the Continental Congress and the original colony of Quakers.

That is, stealing and then saying others must "pay the thieves" for the stolen material in order to survive, is not a valid model of a "moral" society. Moral is about being human upon an earth where the earth nurtures everyone's survival, and being human is about acquiring knowledge and using knowledge to be creative.

To elaborate on the nature of the superior-intellectual-left's mind

Note: Watching a liberal TV show, which exposed both the destructive and manipulative nature of the ad-campaigns of corporations to instill brand-names into infant's minds, in regard to (or which is allowed by) the prevalence and absolute control over the broadcasting (and publishing) media by the few monopolistic businesses (ie banking-oil-military),

whereas the liberal commentators point-out that babies (or children), up to adolescence, need to explore the world with their hands and their own experience in order to develop mentally,

yet the liberal commentators ignore the fact that the college student (and the adolescent) also need to explore ideas with their own language and using their own ideas, in order to gain knowledge, which they can claim for themselves.

An educational competition is a sham about posturing to be seen by the ruling-class.

Yet the college curriculum is very narrow, and overly authoritative, so as to be made into an intellectual-contest, of which only a few can win, but the point of the narrowness is to force the population into the narrow context of the language which is associated to the monopolistically dominated markets, so as to not have available to the public (both) any other ideas and/or any new creative contexts, which might challenge the interests of the monopolies.

This forcing of the mind (of a college student) into a narrow dogma is exactly the point of Godel's incompleteness theorem which uncovers the limitations of fixed (precise) languages, especially, in regard to developing new creative contexts.

The conclusion of the incompleteness theorem is that the public (or the college student) needs freedom of thought, in the same way as the child also needs to be free of the (language) onslaughts of a narrowly defined language of the media, if one wants new ideas and greater creativity to be developed.

Yet the left is willing to call the narrowness of the authoritative dogma . . . , which is categorized by monopolistic business interests . . . , of a college curriculum; the expression of "equal opportunity to compete" within society.

Such a statement is imbecilic, and is a false characterization of the idea of equality.

Rather

It is a statement about a belief in inequality which exists amongst the people

Equality is about both (1) truly free-markets (no product or person is allowed to dominate either a market or trading structure) and [more importantly] (2)the equal free-inquiry of an individual in regard to their own quest for knowledge and creativity.

Equality is about being equal-creators, it is not about the material world, and it is not about who "gets to control" material and knowledge and creativity for all of society.

Such an idea about centralized control is an expression of a collective society, whereas equality is about equal-individuals and the creative gifts which all the individuals can bestow on their fellow-man, within a loose-knit society of equal people.

"The political right" is characterized as being superior, and in charge, (they represent both the highly virtuous "owners of society," ie the owners of society whom possess the religious property of supreme-virtuousness, at least based on the Protestant-Calvinist moral-standards of the US justice system, as well as being the few people who own and control the media) while "The political intellectual-left" is characterized as the superior intellectual (those who, supposedly, have won the intellectual-contests in college, and they get the well-paying jobs of the monopolistic based business society) so that the left states things in great detail whereas the right states things in an arbitrary manner but the left always defers to the right, ie the right guides the dialog (and has the upper-hand in the dialog).

That is, the left-leaning intellectual believes the baloney about their being proven superior-intellects for all of society (by, apparently, winning the academic contests), whereas in reality (in the academic setting) they were indoctrinated to a high-level of language-competency (eg being able to relate math to social structures) and they have come to believe the dogmas as representing an absolute truth (since, after-all, they are proven to be superior intellectuals [since they (or those whom they follow) have high-wage jobs]).

The intellectual-left does not see that science and math have been bent, so as to fit into language categories, which, in turn, serve the monopolistic business interests.

They believe that the academic math-and-sciences are actually saying something of substance, but this is clearly not true, ie the stable physical systems go without valid descriptions within the professional math-science dialog (and this problem has been known since the 1930's, but clouded-over (or hidden) by the propaganda-system, and the, supposed, set of illuminati (ie the high-valued intellectuals) about which the propaganda system reports).

That is, the left (has been tricked so as to) espouse the intellectual-dogmas which support the interests of the (monopolistic) banker's investments.

The left represent those who have been intellectually trapped by the monopolistic business interests of the US society.

The corporations (or equivalently, the owners of society) define the elite context of the right-left media dialog, which is called a debate of the stakeholders. It is an incessant expression of lies, propaganda, mis-representations, and total domination of the language of society by the people of the ruling-class who control the propaganda-education system.

Consider

The global warming "debate"

The intellectual-left is indignant, and claim "the right" (ie the owners of society) must acquiesce to the intellectual framework upon which businesses are based (and deal with global-warming), but the (hypocritical) left never indignantly claims that the right must acquiesce to the more encompassing intellectual demand associated to Godel's incompleteness theorem which implies that technical development ie the development of precise technical languages, requires free-speech, ie not allowing only the (so called) experts to have a voice in technical matters.

Nonetheless, "the right" never obliges the left, since "the right" controls the media, and thus controls the conversation, ie it controls language and in so doing it controls thought.

The intellectual-left are hypocrites and elitists.

For the left, "intelligent articulations" are not about ideas and the relation of ideas to creativity, rather it is about gaining a social-position within the intellectual structure of the authoritative dogmas, which support the monopolistic businesses within society, where the sinful-left must demonstrate their complicit-ness with "the right," by expressing the virtue of "loving one's enemy," namely, by expressing their love the amoral and anti-intellectual right and being-joyful about the high-wages they (the intellectual-left) earn from the corporate state.

But
The narrowness of monopoly causes its own demise.

Yet, the totalitarian US corporate-government and the thoughtlessness of the liberal intellectual (concerning intellectual development) . . . , in the phony left-right language construct, which is the model of the media political dialog, and which also encompasses the intellectual efforts of the institutions of intellectualism, which are all designed to support the banker's interests, by controlling the main instrument of social control, the media, the ruling-class gains an absolute control over the language which is used in the media-education system, and it also . . . , allow the monopolies to maintain their dominant social positions.

It should be pointed-out that the Portland/Indy and Boston/Indy have systematically excluded recent posts (eg in 2013) of m concoyle.

That is, the liberal alternative media mind-set is either willing to "define" the words in the Bill of Rights "freedom of the press (whatever)" in a narrow legal manner, so as to legally define "someone's voice" to be excluded from the media, that is, if the phrase "freedom of the press" is not simply an expression for free-speech, but rather it is legally defined to be exactly about "the corporate entity which pays wages to reporters (ie to propaganda people)," ie "the press," then it is clear that law is a "game of words" to be fought-over (or, in reality, controlled) by the high-stake-holders, ie when a society is based on "the letter of the law" then "one can be sure" that very little "justice" is actually being considered by that society (ie justice becomes a morality-circus), or

They (the intellectual-left) are expressing the, quite erroneous, idea that "all" the posts "on the media" must be "verified truths," [but one can simply look at "how the media actually operates," and one can see that "the media is all about lies and mis-representations," which are used to help business interests], so that if one talks about science, in a media, where only ideas verified by the ruling authorities are allowed . . . , then the validity of a scientific idea is to be determined by the experts, eg the ideas of Copernicus would never be allowed, oh "but then the claim would then be" that now we have the category of science, which is, supposedly, independent of business, but this is not true, since the category of science has been sub-categorized by business interests, so that science reflects not the search for "scientific truth," but rather scientific language categories are identified as dogmas which are of use to the business interests,

ie nuclear-bomb engineering has claimed (to be) "the main purpose of all of physics," especially in regard to all the physics-departments at all universities, (because of the great influence which the media can cause on the administration of big institutions), and then in regard to the control of the "language categories" of learning (eg a university catalog of classes) by the big-business interests (so as to be consistent with big-business interests), which is facilitated by institutional managers, the academics are artificially "protected" by peer-review, where it needs to be noted that, if there is a new idea, then it cannot be peer-reviewed, since "it is a new idea," and thus "it" does not fit into any category which "are being" peer-reviewed.

That is, the liberal-left (the intellectual-left) are a bunch of authoritarian dogmatists, and they are dogmatists since the society considers them to be superior intellects (ie they get high-wages, or they are followers of those "intellectuals" who do get high-wages) since they (the superior-intellectuals) make their way through both the media and society's institutions so well (so easily), yeah, and this is because they are expressing their beliefs in the dogmas which support the investment interests of bankers.

Ie The, so called, superior-intellects must agree to the dogma (otherwise they are not allowed to express ideas).

Thus, though they are different from "the right," "since they are intellectuals," but they are intellectuals which support the "intellectual-dogmas provided to them by business-interests," just as the right supports those few people who determine "what the dogmas should be," which are followed by the intellectual-left, ie the dogmas which support business interests.

That is, one only becomes a reporter if one accepts the beliefs upon which . . . working-for-wages as-a-reporter . . . depend, (or as to, how a alternative media, which is funded, needs to behave), while to be a scientist, one must only work on the dogmas which are accepted as being related to the business-interests of the ruling-class, where this narrowing (of the descriptive range of math and science) is done by peer-review, and supported by the managers of universities, who follow-suit.

Appendix I

Notes: Double spaces can mean a sudden new direction of the discussion without a new paragraph title. The *'s represent either favorites (of the author) or (just as likely) indecision and questions about (logical) consistency. Information and discussion about ideas is not a monolithic endeavor pointing toward any absolute truth, the wide ranging usefulness, in regard to practical creativity, might be the best measure of an idea's truth, it is full of inconsistencies and decisions about which path to follow (between one or the other competing ideas) are either eventually made or the entire viewpoint is dropped, but this can occur over time intervals of various lengths.

The marks, ^, associated to letters, eg a^2, indicates an exponent.

The marks, *, in math expressions can have various math meanings, such as a pull-back in regard to general maps which can, in turn, be related to differential-forms, defined on the map's domain and co-domain (or range), but in this book it usually denotes the "dual" differential-form in a metric-space of a particular dimension, eg in a 4-dimensional metric-space the 1-forms are dual to the 3-forms and the 2-forms are self-dual, etc.

The main idea of thought (or of ideas) is that it is about either sufficiently general and sufficiently precise descriptions based on simple patterns, or it is about developing patterns (of description) which lead to particular practical creativity, or to new interpretations of observed patterns, or to directions for new perceptions.

00. Diagrams

This set of diagrams represents a symbolic-map which can be used to help identify a set of analogous higher-dimensional diagrams (or an analogous set of higher-dimensional constructs), where in lower-dimensions these diagrams are consistent with the observed material patterns, though now the ideas of either materialism . . . . , or existence which is contained within a greater set of higher-dimension so that the higher-dimensional analogous constructs possess the properties of macroscopic geometries . . . . , is given a new interpretive context.

(in the Diagram section) The diagrams provide a succinct outline of the simple math, which is based on stable discrete (hyperbolic) shapes, and which is the basis upon which the stability of measurable description . . . . , of the observed stable, definitive properties of physical systems . . . , depends.

The diagrams provide a clear picture (or clear analogy, or clear map in which to think about moving into the higher-dimensions) of the context within which these stable discrete shapes (of non-positive constant curvature) are organized, so as to form a many-dimensional context (whose higher-dimensional properties should be thought of as being macroscopic) of both component containment, and component interaction.

The many-dimensional containment set, possesses a macroscopic and stable geometric context, which is composed primarily of "discrete hyperbolic shapes," which, in turn, are contained in hyperbolic metric-spaces, (wherein it is true that each dimensional level, except the top dimensional level, has a discrete hyperbolic shape associated to itself).

The finite set of stable discrete hyperbolic shapes, which model both the different dimensional levels, as well as the different subspaces of the same dimension, is a geometric foundation upon which the construct of a finite spectral set depends (a finite spectral set for all existence, contained within a high-dimension containing metric-space, ie an 11-dimensional hyperbolic metric-space).

Each dimensional level (and each subspace of any dimensional level) is associated to a very stable "discrete hyperbolic shape," and each subspace (within the many-dimensional set) is characterized by a size-scale (determined in relation to the finite spectral set), where the size-scale of a dimensional level of a particular subspace is also determined by a set of constant multiplicative factors, which are defined both between dimensional levels and between different subspaces of the same dimension.

The fundamental properties of the high-dimension containing space are determined within an 11-dimensional hyperbolic metric-space.

Hyperbolic space is analogous (or isomorphic) to a general model of space-time defined for various dimensions, eg $R(3,1)$=[space-time], while generally, $R(n,1)$, is a "general space-time."

However, there are also various other "metric-function signature" "types of metric-spaces" of the various dimensions and metric-function signatures, $R(s,t)$, which are involved in the description {where s=space dimension, and t=time dimension, where s must be less than or equal to 11 (it seems (?)), and s+t=n}, most notably the "discrete Euclidean shapes," in $R(s,0)$, which possess properties of continuity (of size), which is needed in the interaction process.

These diagrams identify the context in which "material" components exist within each dimensional level, and they identify the context of both "free" components (associated to both parabolic and hyperbolic second order partial differential equations in regard to inertial properties), as well as orbital components (associated to both elliptic and parabolic second order partial differential equations in regard to inertial properties).

These diagrams also (pictorially) show the basis for "material" interactions, and the relation that a new material system, which is emerging from a material interaction, has to being resonant with values of the "finite spectral set" which is defined for the total containment space.

These diagrams provide a context for the emergence of new stable systems from material interactions.

These diagrams of "'material' component interactions" can be identified at any dimensional level (dimension-2 and above).

Note: Interactions are constrained by:

    1. the process itself,
    2. dimension,
    3. size,
    4. subspace, and
    5. a finite spectral set,

where the basic form for such "material" interactions has an analogous structure (or is "the same") for each dimensional level, though there are differences, in regard to the properties of material interaction, between the different dimensional levels, which can be due to dimension, subspace, and size.

The diagrams give low-dimension pictures of the very simple, quantitatively consistent, geometric shapes which are stable, ie most notably the discrete hyperbolic shapes, and it is these stable shapes upon which stable mathematical patterns can be described in a context where measuring is reliable, and because the description is geometric this means that the description can be very useful.

Note: Following "the diagrams themselves" there is a section in which the descriptions associated to the diagrams are re-written in reference to the number of the diagram, eg 1, 2, etc. given on each page of the diagrams.

000. Other things about diagrams

If one cannot read the words on the diagrams, they are here (below), where they are associated to the numbers of the diagrams.

1. If the circle is rigid in its shape then the complex plane defines a commuting number field, as does the real-line.

   Linear measuring directions are perpendicular (or they are independent of both one another's measured values as well as measuring directions) a complex number, z, is represented as, $z=x+iy=r(\cos W, \sin W)=re^{(iW)}$, (W is a measured angle) Line segments and circles are quantitatively consistent shapes.

2. The following (above) shapes are quantitatively consistent shapes, and locally their directions are independent, and form commutative algebraic constructs at each point over the entire shape, ie global commutative algebraic constructs, locally linear and invertible [one-to-one and onto] and this is true everywhere on the shape.

3. Cubical (or rectangular) simplexes are related to circle-spaces by means of "equivalence-relation topologies," or equivalently, by a "moding-out" process.
   On such shapes local geometric measures are based on either measuring rectangular shapes or (equivalently) by a measuring process based on tangents to the circle, which is used as a basis for measuring along a circle's curve, eg $rdW=dx+dy$ along a circle.

4. * Lattice in the "hyperbolic circle"
   This lattice is more restricted than would be rectangles attached at vertices

5. Discrete hyperbolic shapes are composed of toral components

The number of holes in a discrete hyperbolic shape is called the shape's genus

2- holes are surrounded and caught by 1-curves, and defined by 2-dimensional discrete hyperbolic shapes
3- holes are surrounded and caught by 2-surfaces, and defined by 3-dimensional discrete hyperbolic shapes etc

The faces on the fundamental domain (which result from the faces of a hyperbolic shape's rectangular (or "cubical") simplex) form very stable spectral measures on the very stable shapes of the hyperbolic space-forms.

Discrete hyperbolic shapes, or equivalently, hyperbolic space-forms, have open-closed metric-space topological properties, and they may be bounded or unbounded shapes, but all existing hyperbolic space-forms which are 6-dimensional or greater are unbounded shapes, and the dimension of the last known hyperbolic space-form is hyperbolic 10-dimensions (Coxeter).

Orbits on discrete Euclidean shapes and discrete hyperbolic shapes
Subsystems (or sub-metric-spaces) either occupy spectral orbits or they are "free"

6. The shapes obtained from this rectangular simplex for a 3-dimensional hyperbolic space-form . . . , which contains a 'free" 2-dimensional hyperbolic space-form . . . , are contained in a 4-dimensional hyperbolic metric-space.

7. The separation of two hyperbolic material components is, r, where, r, is defined between the two vertices. Take smallest toral component of each hyperbolic space-form, average the sizes of the two toral components, as is also done in center-of-mass coordinates. Represent the average value as a pair of equal oppositely positioned (rectangular) 2-faces, so as to define a right rectangular volume whose separation, r, is the distance between the vertices of the original interacting hyperbolic space-forms.

Then define an interaction differential 2-form on the geometry of this 3-dimensional Euclidean torus, which is contained in Euclidean 4-space. This determines the force-field, defined between the interacting material components.

The local vector geometry of the differential 2-form is relatable to the local geometry of the fiber SO(4) Lie group, since the 2-forms in Euclidean 4-space have the same dimension as SO(4) the geometry of the spatial displacement is determined in SO(4) by it geometric relation to Euclidean 4-space which is given by the 2-forms so that a local spatial displacement occurs due to a local coordinate transformation with SO(4) acting on the positions of the vertices [(in relation to center-of-mass coordinates) of the original pair of interacting hyperbolic space-forms] in Euclidean 4-space.

If the force is attractive and if the interacting (charged) material (the interaction structure) is contained within either 2-dimensional or 3-dimensional, or 4-dimensional Euclidean space then the force is radial and attractive, and r is made smaller. {Note: If the material is of a new type (oscillatory) and contained within Euclidean 4-space then the force-field has a new geometric structure contained in a higher-dimensional Euclidean space.}

Then the same type of process repeats, for time intervals determined by the spin-rotation period of the spin-rotations of opposite metric-space states (about 10^-18 sec).

In this process the Euclidean torus which forms for each discrete time interval, forms in the context of action-at-a-distance.

Classical partial differential equations are defined within very confining and very rigid sets of both discrete hyperbolic shapes and action-at-a-distance material interaction Euclidean toral components which link the hyperbolic material together, so that the force-field differential 2-form is defined on the torus. (7)

The above interaction for material contained in 3-space results in a 4-dimensional descriptive context, but it can be symbolically represented in 3-space.

8. It should be noted that in this new descriptive context eigenfunctions would also be both discrete hyperbolic shapes and discrete Euclidean shapes (tori).

Forming new stable hyperbolic space-forms from a material interaction. Assume an attractive (or repulsive) interaction in 3-space, then the interaction of "free" material components would be similar to a collision of components, if the material components get very close during the (collision) interaction then if the energy of the over-all interaction is within (certain) energy ranges and the closeness allows resonances (with the spectral set of the over-all high-dimension containing metric-space) to begin to form, thus forming a new state of resonance for the interaction simplex, so that the over-all energy, as well as the resonances, allow a "new" stable "discrete hyperbolic shape" (in the proper dimension of the interaction) to form, so a new hyperbolic space-form emerges.

9. Weyl-transformations between two maximal tori within a Lie group (rank-k compact Lie group) Two intersecting circles of the two maximal tori may be angularly related to one another by Weyl group transformations, where the Weyl group defines the conjugation classes of the maximal tori which "cover" the compact Lie group.

10. Forming angular changes between toral components of a discrete hyperbolic shape by using Weyl-transformations, which change the angular relations between circles which compose a toral component of a hyperbolic space-form. These Weyl-transformations allow Envelopes of orbital stability for "free" subsystems (or sub-metric-spaces) to be defined.

11. * But rectangles attached at vertices and then moded-out, "without expanding the vertex," shows a model of a discrete hyperbolic shape's toral components, represented as separate tori attached at separate vertices. (This is to emphasize an apparent toral component

structure of discrete hyperbolic shapes, which is an important aspect of these discrete shapes.)

12. Various types of unbounded 2-dimensional discrete hyperbolic shapes

13. The figure titled "The mathematical structures of stable physical systems," represents information similar to figures 9. and figure 10.

14. The figure titled "Partitioning a many-dimensional containment space" represents two different dimensional levels, where the 2-dimensional level is identified as an un-deformed rectangular lattice, where a deformed lattice shape (in 2-hyperbolic-dimensions) is given in figure 4. Whereas in this 2-dimensional figure there are contained representations of 1-dimensional shapes. The other (larger) representation of the 3-dimensional partitioning structure are the un-deformed "cubical," or right-rectangular, 3-dimensional shapes, which contain within itself the stable 2-dimensional discrete hyperbolic shapes, this process of partitioning space can continue up into higher-dimensions, where a deformed 3-dimensional lattice shape can be moded-out to form into a geometric-shape, which would exist in 4-dimensional hyperbolic metric-space.

15. The figure titled "Perturbing material-components on stable shapes:" shows an atomic orbital structure which is a stable geometric structure with electrons in the outer-orbits of concentric toral-components (folded into their stable shape) and the nucleus in the center small orbital shape, where the electron's orbit is mostly held stable by it (the electron) following the geodesic path, which is defined on its toral component, and the electron is also interacting as if in a 2-body interaction with the nucleus, so that this 2-body interaction (most noticeably) perturbs the orbit of the electron, eg perhaps causing the electron to possess an elliptical path, where these orbital deformation may result in the variations in the details of the atom's discrete energy structure, where the stable orbital shapes, perhaps related to various Weyl-angle shapes, are the basis for the atom's stable discrete energy structure.

16. The figure titled "Describing the dynamics of 'free' material components in higher-dimensions" is a diagram quite similar to figure 8.

Chart of the face structure for rectangular simplex geometry

2-rectangular-simplex
1-face
vertices

3-rectangular-simplex
2-faces
1-face
Vertices

Diagrams

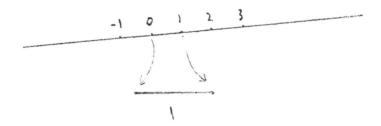

Uniform unit of measurement
(modeled on real-number line),
This unit must remain stable and consistent

If the circle is rigid in its shape
then the complex plane defines
a commuting number field, as
does the real-line

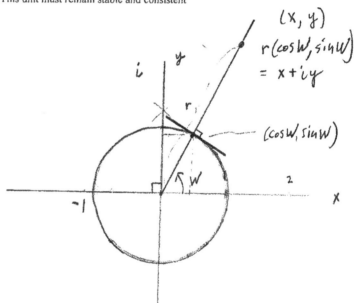

Representation of rectangular coordinates; (x,y)
as well as the complex-number plane,
Linear measuring directions are perpendicular
(or they are independent of both one another's measured values as well as measuring directions)
a complex number, z, is represented as, $z=x+iy=r(\cos W, \sin W)=re^{\wedge}iW$

Line segments and circles are quantitatively consistent shapes.

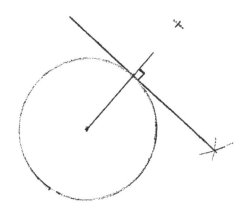

2-dimensions

The following (above) shapes are quantitatively consistent shapes, and locally their directions are independent, and form commutative algebraic constructs at each point over the entire shape, ie global commutative algebraic constructs, locally linear and invertible [one-to-one and onto] and this is true everywhere on the shape

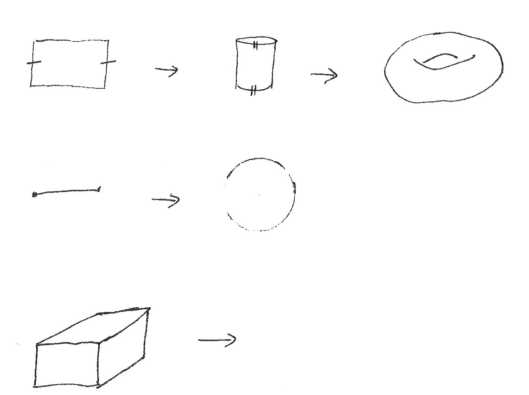

Cubical (or rectangular) simplexes are related to circle-spaces by means of "equivalence-relation topologies," or equivalently, by a "moding-out" process.

On such shapes local geometric measures are based on either measuring rectangular shapes or (equivalently) by a measuring process based on tangents to the circle, which is used as a basis for measuring along a circle's curve, eg rdW=dx+dy along a circle.

Euclidean shapes

Euclidean lattice

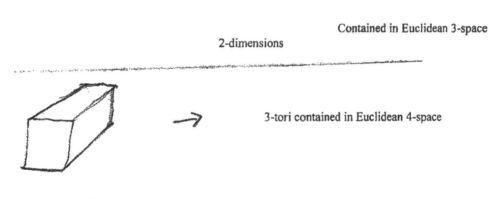

Rectangles
Fundamnetal domains   moding-out   tori

Contained in Euclidean 3-space

2-dimensions

3-tori contained in Euclidean 4-space

3-dimensions

Hyperbolic shapes

Lattices (in hyperbolic 2-space)

infinite

Fundamental domains   moding-out   discrete hyperbolic shapes
                                                                      (contained in hyperbolic 3-space)

Vertices

vertex

Rectangular simplexes   fundamental domains   discrete hyperbolic space-forms

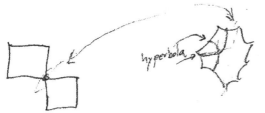

hyperbola

Edges of fundamental domain

The vertex is pulled apart   orthogonal pairs of hyperbolae

Discrete hyperbolic shapes are composed of toral components

The number of holes in a discrete hyperbolic shape is called the shape's genus
2-holes are surrounded and caught by 1-curves, and defined by 2-dimensional discrete hyperbolic shapes
3-holes are surrounded and caught by 2-surfaces, and defined by 3-dimensional discrete hyperbolic shapes
etc

The faces on the fundamental domain (which result from the faces of a hyperbolic shape's rectangular (or "cubical") simplex) form very stable spectral measures on the very stable shapes of the hyperbolic space-forms.
Discrete hyperbolic shapes, or equivalently, hyperbolic space-forms, have open-closed metric-space topological properties, and they may be bounded or unbounded shapes, but all existing hyperbolic space-forms which are 6-dimensional or greater are unbounded shapes, and the dimension of the last known hyperbolic space-form is hyperbolic 10-dimensions (Coxeter).

Orbits on discrete Euclidean shapes and discrete hyperbolic shapes
Subsystems (or sub-metric-spaces) either occupy spectral orbits or they are "free"

Euclidean case

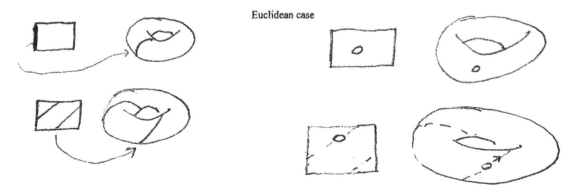

Stable orbits   "free" subsystems

Hyperbolic case

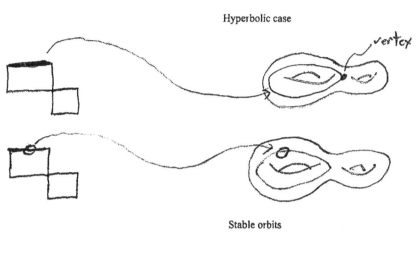

Stable orbits

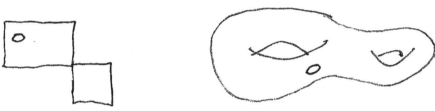

"free" subspaces (or "free" subsystrems)

The shapes obtained from this rectangular simplex for a 3-dimensional hyperbolic space-form , which contains a 'free' 2-dimensional hyperbolic space-form , are contained in a 4-dimensional hyperbolic metric-space.

2-dimensional "free" components (or subsystems) are contained in hyperbolic 3-space

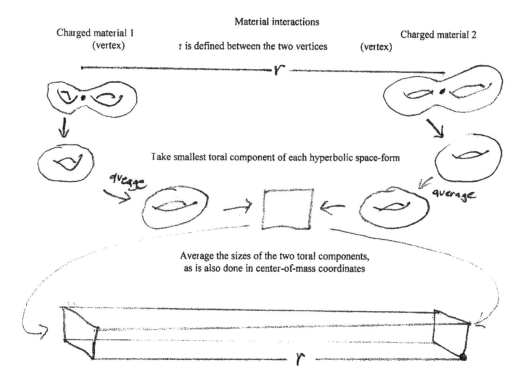

Material interactions

Charged material 1
(vertex)

r is defined between the two vertices

Charged material 2
(vertex)

Take smallest toral component of each hyperbolic space-form

Average the sizes of the two toral components,
as is also done in center-of-mass coordinates

Represent the average value as a pair of equal oppositely positioned (rectangular) 2-faces
So as to define a right rectangular volume whose separation, r, is the distance between the
Vertices of the original interacting hyperbolic space-forms

Then define an interaction differential 2-form on the geometry of this 3-dimensional Euclidean torus, which is contained in Euclidean 4-space This determines the force-field, defined between the interacting material components.

The local vector geometry of the differential 2-form is relatable to the local geometry of the fiber SO(4) Lie group, since the 2-forms in Euclidean 4-space have the same dimension as SO(4) the geometry of the spatial displacement is determined in SO(4) by it geometric relation to Euclidean 4-space which is given by the 2-forms so that a local spatial displacement occurs due to a local coordinate transformation with SO(4) acting on the positions of the vertices [(in relation to center-of-mass coordinates) of the original pair of interacting hyperbolic space-forms] in Euclidean 4-space.

If the force is attractive and if the interacting (charged) material (the interaction structure) is contained within either 2-dimensional or 3-dimensional, or 4-dimensional Euclidean space then the force is radial and attractive, and r is made smaller {Note: If the material is of a new type (oscillatory) and contained within Euclidean 4-space then the force-field has a new geometric structure contained in a higher-dimensional Euclidean space }

Then the same type of process repeats, for time intervals determined by the spin-rotation period of the spin-rotations of opposite metric-space states (about 10^{-18} sec).

In this process the Euclidean torus which forms for each discrete time interval, forms in the context of action-at-a-distance.

Classical partial differential equations are defined within very confining and very rigid sets of both discrete hyperbolic shapes and action-at-a-distance material interaction Euclidean toral components which link the hyperbolic material together, so that the force-field differential 2-form is defined on the torus.

The above interaction for material contained in 3-space results in a 4-dimensional descriptive context, but it can be symbolically represented in 3-space.

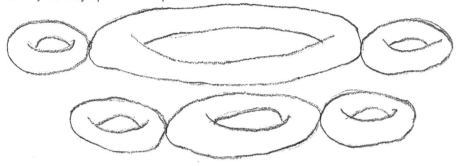

For an attractive interaction in 3-space

It should also be noted that in this new descriptive context eigenfunctions would also be both discrete hyperbolic shapes and discrete Euclidean shapes (tori)

Forming new stable hyperbolic space-forms from a material interaction

Assume an attractive (or repulsive) interaction in 3-space, then the interaction of "free" material components would be similar to a collision of components,

State of resonance   new hyperbolic space-form

if the material components get very close during the (collision) interaction then if the energy of the over-all interaction is within (certain) energy ranges and the closeness allows resonances (with the spectral set of the over-all high-dimension containing metric-space) to begin to form, so that the over-all energy, as well as the resonances, allow a "new" stable "discrete hyperbolic shape" (in the proper dimension of the interaction) to form.

Weyl-transformations
Representing two maximal tori within a Lie group
(rank-2 compact Lie group)

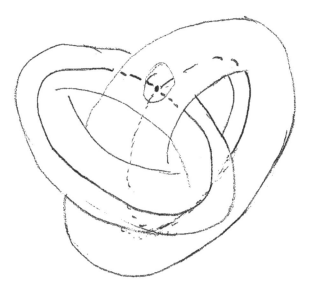

These two intersecting circles of the two maximal tori may be angularly related to one another by Weyl group transformations, where the Weyl group defines the conjugation classes of the maximal tori which "cover" the compact Lie group

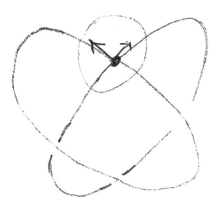

Forming angular changes between toral components of a discrete hyperbolic shape by using Weyl-transformations, which change the angular relations between circles which compose a toral component of a hyperbolic space-form.

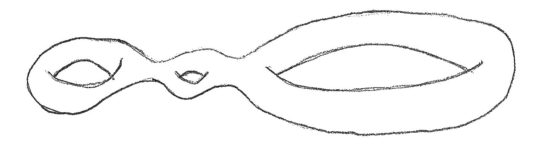

Envelopes of orbital stability for "free" subsystems (or sub-metric-spaces)

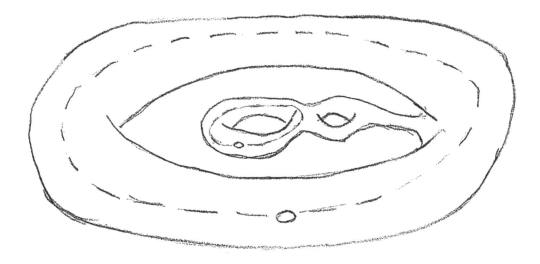

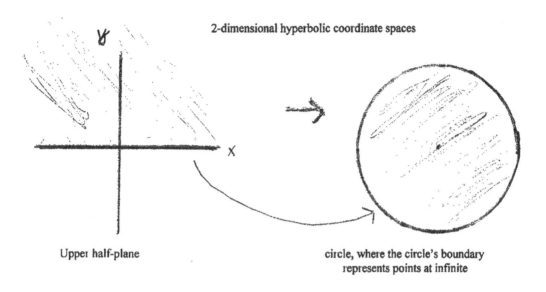

2-dimensional hyperbolic coordinate spaces

Upper half-plane

circle, where the circle's boundary
represents points at infinite

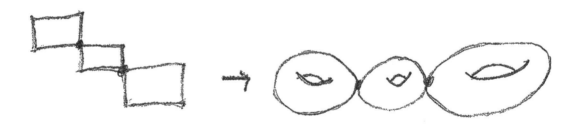

Rectangles attached at vertices and then moded-out, without expanding the vertex, shows a model of a
discrete hyperbolic shape's toral components, represented as separate tori attached at separate vertices

Various types of unbounded 2-dimensional discrete hyperbolic shapes

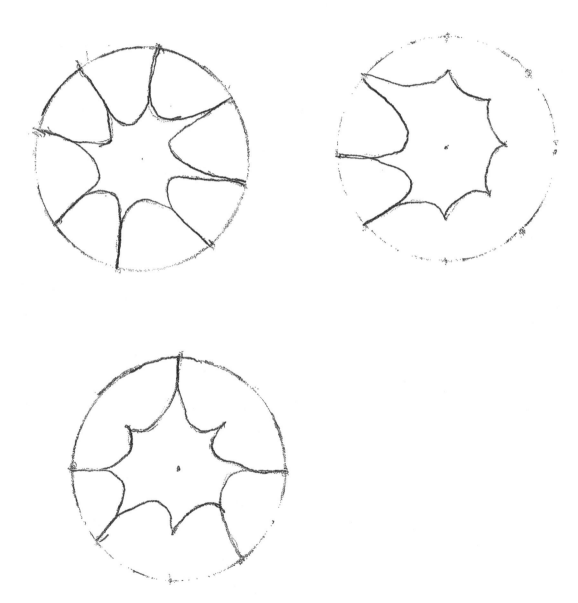

The Mathematical Structure of Stable Physical Systems

The stable geometric shapes

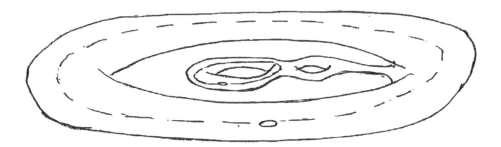

The same shape "folded" by Weyl-angles

Partitioning a Many-Dimensional Containment Space

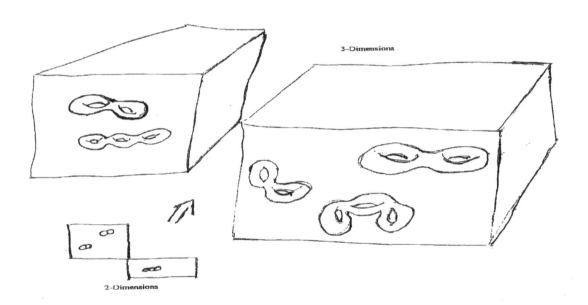

3-Dimensions

2-Dimensions

Perturbing Material-Components on Stable Shapes: How Partial Differential Equations Fit into the Descriptions of Stable Physical Systems

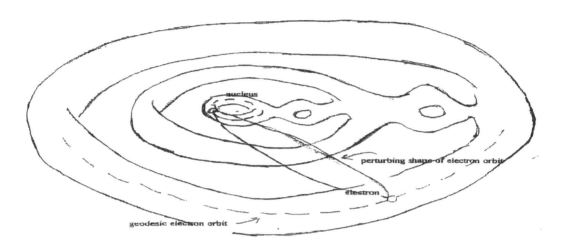

Describing the Dynamics of "Free" Material Components in Higher-Dimensions

Interaction components and the interaction shape

The approximation of the interaction

Appendix II
(San Diego math conference talk
by m concoyle, 2013)

G eometrization and the mathematical context for the solution of physical stability: Nuclei, general atoms, molecules, and the solar system etc.

M Concoyle's talk at San Diego math conference 2013.

Abstracts, Introductions (to the talk), References and books, Speech 1, & re-iterations, Speech 2, Speech 3,
Dimensions* Blurbs
Empty of content (Apparently, no stable patterns exist in professional mathematics)

Abstract I

This relatively new (since 2002) and relatively simple context of math containment provides the setting for a solution to the problem of finding the math structure for the observed stable material systems which are so fundamental and so prevalent. It also provides a basis for a quantitative structure which is defined on a finite set. However, these stable physical systems go without any valid math structure (for these systems) in a currently accepted math context of indefinable randomness (eg improperly defined elementary event spaces), non-linearity, (global) non-commutativity, or only locally commutative, (eg quantitative inconsistency, eg chaos) all defined by a contrived descriptive structure of convergence and divergence onto a continuum. Where these math structure are together used to explain (or identify) the (stable) properties of physical systems. But such a math context really only applies to physical systems in a chaotic transitioning process (eg reactions in weapons) and for feedback systems (eg guided missiles) whose range of applicability is difficult to define and it is a context which applies to quantitative complexity (eg secret codes). But it is also used by the media to create an illusion of expert "mastery" and "expert complexity."

Though many difficult problems now have solutions: nuclei, general atoms, molecules, a new way in which to analyze crystals, and the stable solar system, due to these new ideas, this means that these relatively new ideas should be dominating the attention of the professional mathematicians and physicists, but they are not. Note: Apparently, waves which possess physical properties can be successfully related to solutions by function-space techniques.

Apparently there are stronger social forces involved in an inability of a public, or of an expert-class, "to discern truth."

The US propaganda system is the sole authoritative voice for all of society and it is the propaganda system which directs the attention of the researchers. These researchers are dependent on a funding process. However, these same researchers claim to be the personifications of the highest cultural attainments in the society, nonetheless they have social positions of being both wage-slaves and society's, so called, top intellects in regard to a religious personality-cult, expressed through the media, so public-worship consolidates their belief in their far too authoritative mathematics and physics dogmas, which has failed to solve the problem of the cause of physical stability for nearly 100 years (ie it is a failed dogma), ie the media turns "top intellectualism" and the dogmas upon which such a "measure" of intellect rests (the "intellectual winners" of the competition whose rules, in the education system, are defined by an, essentially, absolute authority) into a religion, this deep religious belief in what the media labels as science [Copernicus would have a more difficult time persuading others to consider an alternative way in which to organize and fashion language within such a current religion (2013) of expert authority, than the difficulties he had in regard to the authoritative religion of his time].

The professionals are following their "deep beliefs" as dictated to them by the propaganda system. Apparently these professionals can rigorously prove properties which are contained in a world of illusion, eg where a description based on randomness also possesses well defined geometric properties, eg particle-collisions.

It should be noted that the best interpretation of the Godel's incompleteness theorem is that precise languages can be very changeable when reduced to the elementary levels of assumption, context, containment, organization etc. Yet the failure to describe the stable underpinnings of physical existence has not been seen as a crisis of the knowledge which is being derived from the currently accepted authoritative dogmas of math and physical description.

There is other social organizational properties which manage society, and with which one must deal with, there is a vast social organization in regard to management which manage the math and physics (or science) communities, eg managing personality types similar to the management of personality types in politics and the justice system.

In the new context of containment one uses the most prevalent of the stable geometric patterns identified in the Thurston-Perelman geometrization, namely, the discrete hyperbolic shapes and the properties that these shapes possess as identified by Coxeter.

Furthermore the ability to "surround" a "hole" by a closed shape, so that a continuous deformation is limited, ie the "holes" introduce stable properties into the context of the continuity of shape.

The discrete hyperbolic shapes [with component interactions mediated by discrete Euclidean shapes (tori)] are also very rigid shapes with very stable spectral properties.

That the solar system is stable is evidence, which can be interpreted, to prove this new context for mathematical descriptions of the physical world is true, especially, since the professionals have no valid model of stability for these stable systems.

Abstract II

A new context in which to apply geometry to: math, quantum physics, and the solar system, etc

Quantum physics assumes the global and descends to the local (ie random particle-spectral measures).

Is geometry a better vehicle to define the stability of quantum systems rather than function spaces?

Is the stable construct to be the very stable discrete hyperbolic shapes, in a many-dimensional context?

A geometrically stable and spectrally finite math construct, where, in adjacent dimensional levels, the bounding discrete hyperbolic and Euclidean shapes are defined, and then mixed as "metric-space states" in a Hermitian (or unitary) context, can provide a structure for stable properties.

Assume that math be consistent with (local) geometric-measures of stable shapes, which define finite spectral sets, contained in higher-dimensions.

The stable shapes in the different dimensional levels are con-formally similar, and resonate with a finite geometric-spectral set contained in a high-dimension space.

A new interaction type consists of a combination of hyperbolic and Euclidean components, but when in an "energy-size range" the system can resonate with the spectra of the containing space, and thus it can change to a new stable, discrete shape.

Abstract III

There are (moded-out) "cubical" simplexes in a many-dimensional context, whose structure is determined by hyperbolic metric-spaces, which can, themselves, be modeled as moded-out "cubical" simplexes. Transition between the different dimensions determine physical constants, and the value of these physical constants can imply that:

1. the different dimensional levels can be hidden from one another, ie the size of the interacting materials change from dimensional level to dimensional level and the geometry of the interaction can also change, ie material interactions are not usually spherically symmetric.
2. The over-all high-dimension containing space can be defined as having a finite spectral set.
3. The descriptions of both mathematical and the physical systems are (or can be) stable, because the "cubical" simplexes can define discrete hyperbolic shapes.

Metric-spaces have properties and subsequently an associated metric-space state, and this determines the dimensional distinction between Fermions and Bosons, as well as determining the unitary (invariant) mixing of (metric-space) states in subsets of complex-coordinates.
Unitary invariance implies continuity, or the conservation laws, eg the conservation of energy and material, etc.
Each dimensional level is a discrete hyperbolic shape, and this implies such a set can define a finite spectral set for the entire space.
A new interaction type consists of a combination of hyperbolic and Euclidean components which are one dimension less that the dimension of their containing metric-space. A 2-form construct emerges from this geometric context which is the same dimension as the adjacent (higher) dimension Euclidean base space of its fiber group which determines discrete spatial displacements. This interaction construct is either chaotic or it could begin to resonate during the interaction and, subsequently, to become a new stable spectral-orbital (discrete hyperbolic shape) structure, by means of its resonance with the spectra of the many-dimension containing space.

Introductions:

Math flyer (1), San Diego joint math meeting (2013)
(come to talk 1-12-13, rm 6E (main bldg) 3:30 pm, #1086-VR-413, Assorted Topics II)
A new context in which to apply Geometry to: Math, Quantum Physics, and the Solar System, etc. By M Concoyle

I.   This new "math construct" addresses unsolved problems, as well as unrealized models of quantitative containment (within a finite quantitative construct). This is done by using the simplest of math structures, and elementary ideas about quantity and shape.

II. The unsolved physical problems are about finding the context within which one can describe very stable fundamental physical systems: relatively stable nuclei, general atoms (atomic number greater than five), molecules and their shapes, crystals (BCS predicted a critical temperature which has been exceeded), and envelopes of orbital stability for orbital planetary systems, [and dark matter, dark energy] etc.

These problems cannot be solved in the math structures (dogmas) based on a non-linear, and an indefinably random context. Nonetheless, this is the context which is used by today's math and physics professions. Such a context leads to the statement, that "the stable properties of fundamental (physical) systems are 'too complicated' to describe," and the focus of the description is on describing fleeting, unstable patterns, in a quantitatively inconsistent manner, and this is done within sets which are "too big," eg the continuum, so as to possess the capacity to be logically inconsistent (eg geometry defined in a random context, allowed by defining various convergences), ie they focus on non-descriptions and irrelevant issues, which they try to describe in too great detail.

III. The new math context is ultimately about, "How to guarantee: stability, quantitative consistency, and to define finitely generated quantitative sets," and about "How to guarantee the stability of a uniform unit of measurement, within descriptions of fundamental systems, which possess stable measurable properties."

IV. The new math context is about basing physical description, as well as stable quantitative structure, on stable geometry. Namely, the geometry based on the discrete hyperbolic shapes, in conjunction with discrete Euclidean shapes, as well as with other metric-space "discrete shapes" associated to non-positive constant curvature spaces, with metric-functions with constant coefficients, eg the R(s,t) metric-spaces and C(s,t) Hermitian-spaces (of finite dimension), (C(s,t) is a result of (4) below), where, s, is the dimension of the spatial subspace, and, t, is the dimension of the temporal subspace, and s+t=n is the dimension of the metric-space. [and where R = real, C = complex numbers]

These are the circle-spaces ie spaces related to "cubical" simplexes or rectangular simplexes (where cubical simplexes are related to circle-spaces by equivalence topologies, ie a moding-out processes).

V. Just as Copernicus and Kepler provided the correct quantitative-geometric context for the properties of the solar system, which could be fitted by Galileo's law and Newton's global solutions to (Newton's) differential equation models of the 2-body, "center-of-mass coordinate" model of the solar system . . . . , which is a more useful context, in regard to feedback systems in gravitating systems, than is the "practically useless," 1-body, spherically-symmetric model of gravity, given by an unrealistic (1-body) non-linear, general relativity theory,

. . . . , this new math context provides the answers (the types of shapes for a solution function) for stable material systems (as well as metric-spaces) modeled in a stable, geometric context, where the assumptions provide the context for the existence of solutions to stable systems, and by modeling metric-spaces as stable shapes, this allows for a finite spectral set to be defined.

But the new context also organizes math patterns, and provides new types of quantitative processes through which fundamental stability can be placed into a general descriptive context, in relation to wide ranging applications, which can lead to sufficiently precise "stable spectral-orbital" descriptions (constructs), which are geometric and thus the information provided can be used in a practically creative manner.

VI. The stable spectral properties of general atoms and nuclei (as well as the other stable material systems) are related to finite (integer) values, ie number of charged components (atomic number), and a number of (uniform) nuclear components (atomic weight), respectively, but they do not have valid descriptions based on physical law, and it has (also) not been understood "how envelopes of orbital-stability form in a macroscopic solar system," but all of these systems can now be modeled, based on stable geometry, ie similar stable geometric constructs hold for both stable microscopic spectral systems, and macroscopic orbital-envelopes of stability.

VII. There are 6 fundamental aspects to the description:

1. The metric-spaces and their associated isometry groups fit into a principle fiber bundle (one ends-up using this construct more later, and it is just as easy to introduce it first).

2. The fundamental shapes of existence: both metric-spaces, and material systems, are to be modeled in the context of discrete hyperbolic shapes and discrete Euclidean shapes (tori), or in non-positive constant curvature spaces for metric-functions with constant coefficients. This is related to the classical Lie groups of both SO(s,t), and spin groups, and SU(s,t) [see 4. below]

3. The containing space is many-dimensional (11-dimensional hyperbolic metric-space) so each dimensional level, as well as each subspace of the same dimension, is to be identified with, either a macroscopic or a microscopic, stable discrete hyperbolic shape.

4. There are (physical) properties associated to metric-spaces of various dimensions, and they are associated to various metric-spaces of the type R(s,t). This leads to metric-space states which come in opposite-pairs. These pairs of opposite metric-space states can be fit into both the real and i(real) [or pure imaginary] subsets of complex-coordinates. Thus the description is (can be) related to SU (unitary) fiber groups. Such descriptions, of opposite metric-space states, also relate to spin-groups, and Dirac operators.

   These properties of metric-spaces can be related to both math and physical patterns since the properties deal fundamentally with "position in space" (Euclidean space), and "a stable existing pattern" (continuity of stable patterns in time, an implied assumption in mathematics), eg energy, mass, and charge conservation assumptions.

5. The derivative as a discrete operator which can be related to: (1) material interactions as discrete operations, (2) dimensional levels involved in dynamics (also physical constants can be modeled as discrete operators defined between dimensional levels) (3) Weyl-transformations are discrete angular transformations, and deal with the shapes of discrete hyperbolic geometries of a physical system's related orbital-spectral properties.

6. A new way in which to model both life and mind.

Introduction 2

Describe the properties of the stable spectral-orbital systems, How does one describe the very stable, many-body, spectral-orbital physical systems which exist at all size scales from nuclei to solar systems, as well as life and mind?

The subject of stability is mostly about using very confining, but very simple set of geometric shapes, and some simple calculus, as well as a few other (local) geometric properties defined in a context of a principle fiber bundle with a many-dimensional base-space, where the base-space is (dimensionally) partitioned into sub-metric-spaces which are identified by discrete shapes, which are open-closed (when observed within the metric-space which possesses the discrete shape), but these metric-spaces (with shapes) form a boundary when viewed from an adjacent higher-dimensional metric-space which contains the discrete shape.

The math properties for stable solvable physical systems, The simple math properties which allow partial differential equations (or differential equations) . . . , which are used to model physical systems (either geometric-inertial or spectral) . . . , to be solved can be listed.

That is, solvability is related to the set of properties: linear, metric-invariant, separable (locally linear and commutative (diagonal matrices at each point in the global coordinate system)), where the metric-functions can only have constant coefficients, (for the various R(s,t)-metric-spaces {where s-space, and t-time; dimensions}, and associated metric-function signatures, where these different metric-spaces are, in turn, related to new types of materials).

[These new material-types are characterized by their properties of being odd-dimensional spatial subspaces, with an odd genus-number, (analogous to discrete hyperbolic shapes) so as to be charge-unbalanced, and thus, naturally oscillating and energy-generating material systems, which model life. Life is defined as new material-types, which are contained in either: R(4,0), R(6,0), R(8,0), or R(10,0) Euclidean spaces.]

The properties of stability are properties which are possessed by the circle-spaces. One can note that, circle-spaces can be modeled as metric-spaces of non-positive constant curvature, whose discrete shapes (or discrete isometry subgroups) are determined from lattices (with an associated fundamental domain, which is related to right-rectangular (or "cubical") simplexes, by a moding-out process (or a process of defining an equivalence topology on the cubical shapes of the fundamental domains).

[Note: The quasi-spectral-geometric properties used to describe particle-physics, ie the non-linear-random-geometric model of particle-physics . . . ., which deals primarily with probabilities of particle-collisions {which, in turn, only model (nuclear) reactions}, and which is a completely irrelevant and practically useless descriptive construct . . . ., are excluded from consideration, since these descriptive constructs only provide . . . , in a chaotic or random fashion . . . , brief and fleeting descriptions of unstable patterns. It is a model which is only applied by means of its relation to reaction-rates.]

Holes-in-space, spectra, force, and the circle-spaces, are all defined within a many-dimensional space. The simple math structures of circle-spaces, as the basis for the stable (geometric) properties of existence, causes both spectral properties and force-fields, to have an analogous (or parallel) math structure . . . , [in relation to holes in space caused by either the shape of a metric-space or by rigid material shapes, eg (usually) 1-dimensional currents defined by rigid material defining a closed curve] . . . , than is usually believed to be true (ie the material-shapes are equivalent to spatial [or metric-space] shapes).

The very stable "discrete hyperbolic shapes" are used as models for both material components and metric-spaces, so as to construct a very rigid geometric structure, in a many-dimensional context, so that Newton's laws still define inertial dynamics. Discrete hyperbolic shapes are stable, with stable spectral properties, and they provide a constrained and stable and very rigid set of boundary conditions for both material containment (confinement) and material interactions, so that differential equations, at first, appear to have a very restrictive containment structure, but this is needed to model stable systems, and it can also be used to model living systems in new ways.

Most material components, as well as metric-spaces, which are contained in this new many-dimensional context, have the shapes of circles-spaces, in particular the "discrete hyperbolic shapes" define a very rigid set of both constraining contexts and boundaries for measurable descriptions. In this new context, descriptions which are related to (partial) differential equations, require that material interactions be mediated by discrete Euclidean shapes, which in turn, are related to the differential equations associated to metric-invariant differential-forms of the force-fields, ie the descriptive structure is based on the geometry of circle-spaces (and holes in the shape of space). Force-fields are applied to material shapes by Newton's law of inertia, which is defined in absolute Euclidean space (which also allows for action-at-a-distance), in this context of very rigid, and confining, geometry (but Newton's universal law of gravitation is modified).

Stable spectral shapes, and material's orbital properties, and the properties of (condensed) "free" material components in space, There are "free" material systems, and orbital material systems, where "free" material components can condense into limited orbital structures, or condensed material can be guided by highly confining orbital constructs, because these (condensed material) components do not have the "correct" size to be stable material components, in their (particular containing) dimensional level, as well as particular subspace, so that the resonances . . . , which allows the existence of material components . . . , are either from another subspace (of the same dimension), or are determined to be from resonances which are defined on the facial structure of the stable simplex structure of the condensed material, in order to have a stable spectral structure for the condensed system. If a higher-dimensional material component is not "big enough" to be a material component, in its particular subspace (of some particular dimension), but nonetheless this higher-dimensional material component does (can) interact. However, the property of "the faces of the material component," which is interacting on a higher-dimensional context, are contained in a closed metric-space. This causes the dynamics of the interaction (in a higher-dimensional level) to be defined on the entire material-containing metric-space. The entire, rigid, metric-space being pushed by the interaction is not noticeable within the rigid geometry defined within the metric-space, ie a perfectly rigid-rod can transmit a "push" instantaneously across the space it occupies.

An example of a true manifestation of inertial affects defined in general relativity, "Free" material components can also be related to orbital structures, where most often condensed material constrained by the orbital structures of discrete hyperbolic shapes, wherein geodesics can (now) affect inertial properties of these "free" material components, and can be described in an explicit manner in a stable (linear) context, ie general relativity is being defined beyond the 1-body problem with spherical symmetry, so that the description is stable, since the geometry in the new descriptive structure is linear, metric-invariant, and separable, so as to define a stable orbit.

[Note: An orbit defined on a discrete hyperbolic shape is pushed . . . , to the limited and rigid structure associated with the shape's geodesics . . . , by the coordinate structure of hyperbolas, which exists away from the geodesic paths (where geodesics will be contained on the faces of the simplex of the discrete hyperbolic shape's fundamental domain)].

The interaction "discrete Euclidean shape," or the interaction torus, Euclidean space-forms (a synonym for a "discrete Euclidean shape") mediate material interactions within the rigid constraints of the "discrete hyperbolic shapes," which model both metric-space existence and material existence. During a material interaction a differential-form is defined upon the geometry of an interaction torus, ie a differential 2-form model of a force-field is defined on the interaction torus. Yet, a "discrete Euclidean shape" of an interaction can also transform, so as to become a toral component of a newly formed discrete hyperbolic shape, where this can happen if both (1) resonances exist for the interaction, and (2) the energy and size of the interaction is within the "correct" energy and size ranges. The geometry of the 2-form is related to the geometry of the fiber SO(n) group in order to determine the direction of the force-field's push [in the base-space containing the interaction torus]. Then the spatial positions of the interacting material's vertices are locally transformed, in a (local) context of opposite metric-space states.

A finite spectral-orbital set, defined upon a many-dimensional containing space, A finite spectral set can be defined on an 11-dimensional hyperbolic metric-space, which is an over-all high-dimension containing space for (of) a model of existence (where the existence is defined by the finite spectral set). This is possible because in each dimensional level (and for each subspace of the same dimension) there is a maximally-sized discrete hyperbolic shape in regard to the finite set of discrete hyperbolic shapes which are used to model the subspace metric-spaces (subspaces of the 11-dimensional space), so that the spectra defined for all the dimensional-levels and subspaces of each of those different dimensional-levels is finite, and thus, it can determine a finite spectral set, upon which all material properties . . . , which are allowed to exist in this 11-dimensional hyperbolic metric-space . . . , depend.

Defining angles between toral components, Weyl-transformations of angles between a discrete hyperbolic shape's toral components, (or a set of folds allowed on the base-space lattice structure), so that there are a finite set of angular relationships which can be defined between the toral components of a discrete hyperbolic shape. These folds between toral components allow envelopes of orbital stability to be defined, based on the orbital (or metric-space shape) structure which a discrete hyperbolic shape can have after its toral components are transformed by certain angular values.

The operation of multiplying by a constant factor, Constant multiplicative factors can be defined . . . , so as to affect the properties of: shapes, sizes, orbits, and the stability of "discrete hyperbolic shapes" :

. . . , between dimensional levels,
. . . , between subspaces of the same dimension, and
. . . , between toral components of a discrete hyperbolic shape.

Physical properties (and math properties) attributed to metric-spaces, There are physical properties associated to metric-spaces. Two examples of physical properties are: (1) position in space of a system's vertex, in regard to the distant stars, (2) the stability of a system's (or a mathematical) pattern.

This results in the definition of "metric-space states" of opposite physical properties, eg (+t) and (-t). These opposite metric-space states are a part of the dynamic processes of material interactions, eg fixed stars, rotating stars (Euclidean); forward time, backward time (hyperbolic space) etc.

In turn, this implies unitary containment, in regard to the containment (of opposite metric-space states), within both real and pure imaginary subsets of finite-dimension Hermitian containing set of coordinates, as well as allowing the definition of the spin-rotation of metric-space states, so that this spin-rotation of metric-space states is defined on opposite metric-spaces states so that these opposite states are a part of the dynamic process. The (time interval of the) period of the spin-rotation of opposite metric-space states is a property used in the dynamic (or inertial) material interaction process.

In this description there is no need of a continuum, instead rigid geometric stability is used rather than using (indefinable) randomness and non-linearity as a basis for (physical, or mathematical) description.

Indefinable randomness and non-linearity seem to possess the properties which are needed to briefly describe patterns which are unstable and fleeting in duration, which, at best, are relatable to feedback constructs, ie it is a description which depends on the validity of the fleeting pattern of a system modeled as a partial differential equation, which has limited descriptive value, and which is used in a relatively unimportant contexts, ie it is a flawed viewpoint which cannot be used describe the observed stable spectral-orbital properties of physical systems at all size scales. (2) (over)

References and Books

San Diego Joint Math Meeting (2013)
A new context in which to apply geometry to: math, quantum physics, and the solar system, etc.
1086-VR-413,
Saturday, 1-12-13, 3:30 Rm 6E (main building), Assorted Topics II, Martin Concoyle PhD.

Just as Copernicus, Kepler and Galileo provided a quantitative-geometric context for the properties of the solar system, which were then precisely identified by the solutions to (the) solvable differential equations of Newton; Martin Concoyle now provides the stable geometric structures which fit . . . , both macroscopically and microscopically . . . , into a many-dimension containment set (hyperbolic 11-dimensional), so that these shapes are the solutions, ie the geometries of the stable spectral-orbital properties, of all the fundamental stable systems which have stable spectra and orbits: nuclei, molecules, solar-systems, etc, and it is the basis for a quantitative system (the spectral set of a measurable existence) which is finite, and These ideas are discussed in the following books: (available at math conference, 2013)

1. A New Copernican Revolution (p286), B Bash & P Coatimundi, Trafford Publishing, 2004.
2. The Authority of Material vs. The Spirit (p483), D D Hunter, Trafford Publishing, 2006.
3. Introduction to the Stability of Math Constructs; and a Subsequent General, and Accurate, and Practically Useful Description of Stable Material Systems (p262), Scribd. com, 2012, Martin Concoyle, and G P Coatimundi.
4. A Book of Essays I: Material Interactions and Weyl-Transformations; (p234), Scribd.com, 2012, Martin Concoyle.
5. A Book of Essays II; Science History, and the Shapes which Are Stable, and the Subspaces, and Finite Spectra, of a High-Dimension Containment Space (p240), Scribd.com, 2012, Martin Concoyle.
6. A Book of Essays III: Elementary Topics; (p303), Scribd.com, 2012, Martin Concoyle.
7. Physical description based on the properties of stability, geometry, and quantitative consistency: Presented to the Joint math meeting San Diego (2013), (p211), Scribd.com, 2013, M Concoyle

The books 3-7 are incorporated into the four new Trafford books, by m concoyle.

San Diego-2013 Math Conference

(talk) Use geometrization (in particular, that the most stable geometric shapes are the discrete hyperbolic shapes) to turn physical description into a geometric exercise as opposed to an exercise about indefinable randomness and (the quantitatively inconsistent) non-linearity

Should physical description be about:

1. Unstable patterns and fleeting contexts (hopefully) relatable to feedback systems, eg guidance systems, and
2. To chaotic contexts of systems briefly transitioning between relatively stable states, eg nuclear reactions,
3. That is, descriptions of marginal, unstable contexts related to the fine-tuned interests of big business, such as improving a complicated instrument's capacities (eg guiding a missile), defined within a fixed descriptive context, so as to help big business. Note:

Computing is about speeding up a computers rate of switching, but a computer is limited to operating on numbers or on symbols strictly related to a process (which might be a symbolic goal). That is, new contexts can establish new ways in which creativity or new ways in which to achieve a goal (or a process). Furthermore, the increase of switching rate for a computer is approaching the obstacle of the limitation of relevant knowledge, ie traditional knowledge has great limitations, it now is being related to many failings, eg the calculations of business risks has failed because the math is failing.

Is math and science about serving commercial interests, and developing complicated instruments (for business interests) but there are limits to an instrument's performance, and/or (in regard to mathematics) does a fixed context and a (fixed) set of authoritative traditions, which are associated to business interests, also have a limit to its performance (or such a context's capabilities, such as in regard to its descriptive range)?

or

Is the range of a measurable description to be determined by a very large set of equal independent free-inquirers, all of whom are to be regarded as equal creators (not simply creators in regard to narrowly defined business interests, ie why should "big oil" or "big banks" be determining the structure of knowledge and hence the structure of creativity within society)? This allows the society . . . , ie big business who in turn has controlled big government so that now big government serves the needs of big business . . . , to identify a small set of pompous intellectual aristocrats (university academics), who are given great authority in society, but who possess a very limited and narrow range of thought, or who have limited intellectual range concerning the possibilities of precise descriptive knowledge, and its possible relation to new creative contexts.) one can only establish oneself as being correct in a fixed context, but fixed contexts are not conducive to the new "development" of precisely described knowledge, nor conducive to the developing new contexts for creativity. (big business is most profitable when the context within which it makes the most money stays fixed). Equality supports development, inequality is defined within a fixed context.

What are stable math patterns? [They are simple math patterns]

Nuclei, atoms with more than 5-charged-components, molecules, crystals, the solar system (etc), are all stable physical systems (which implies that they are formed under controlled linear conditions) but none of these (general) systems have valid descriptions (beginning with the laws of physics and then deriving the spectral [or orbital] properties of these systems).

Is one to try to improve the descriptions of these systems by means of more complications?
or

Should one consider new contexts, and new interpretations upon which to base description?

Gödel's incompleteness theorem can be interpreted to mean "math and science are trying to do the same thing"

{Hilbert wanted to place physics on an axiomatic basis, but when it was shown that math cannot be completely developed by it having an axiomatic basis, and that both disciplines are about developing measurable descriptions of observed patterns, then this leads to this interpretation}

In this context, truth is finding "simple" patterns and processes which are:

1. Widely applicable to general systems (observed properties or observed patterns)
2. Can describe observed patterns accurately and to sufficient precision, can describe definitive, stable patterns, and
3. Strongly relates to practical creativity (measuring, fitting, selecting material or subsystem structure, building, or putting together, and controlling "physical" systems).

Geometry can be both stable and controllable, especially the geometric systems which are contained within a linear, metric-invariant, and separable (or solvable or commutative everywhere) context for their descriptions, (descriptions of physical systems (or shapes))

Geometry is more related to practical use than either randomness or non-linearity (etc).

Note: Random description of systems with only a few components have no practical value (other than perhaps to be used to identify a betting game).

The new idea:

Use the stable circle-spaces, in particular, the discrete hyperbolic shapes, to identify the stable spectral-orbital properties of observed physical systems, or to identify a finite spectral set to be used as a basis for a description's quantitative structure. Note: "Discrete Euclidean shapes" are also needed in regard to describing material interactions.

The new context:

Partition an 11-dimensional hyperbolic metric-space by means of a set of different dimension discrete hyperbolic shapes so that each subspace of each dimensional level is assigned both a discrete hyperbolic shape (of that dimension) and a multiplicative constant (equivalent to physical constants, eg c h etc). One can think of these discrete hyperbolic shapes as being "uniform" shapes, where each face of the shape's fundamental domain has the same geometric measure, ie the same spectral value. Thus the multiplicative constants are the mechanism through which the spectral sizes change between either dimensional levels or between subspaces of the same dimension.

Properties of discrete hyperbolic shapes [Coxeter]

1. The 10-dimensional discrete hyperbolic shapes are the highest dimensional discrete hyperbolic shapes which exist. Thus, the containment set contains all of the discrete hyperbolic shapes.
2. The highest dimension discrete hyperbolic shape, which is bounded, has dimension-five. Thus the 5-dimensional spectra, defined by the 5-dimensional discrete hyperbolic shapes, are the last hyperbolic metric-spaces which are related to spectra with finite values. Note: the 5-dimensional discrete hyperbolic shapes are contained within 6-dimensional discrete hyperbolic shapes, which are unbounded shapes. (Question: Are some of the faces . . . . of the fundamental domain of an unbounded 6-dimensional discrete hyperbolic shape [actually] . . . . bounded 5-dimensional discrete hyperbolic shapes?)

The spectra of an "n-dimensional discrete hyperbolic shape" are determined by the geometric measures of the (n-1)-dimensional faces of the "n-dimensional discrete hyperbolic shape's" fundamental domain.

For example, the spectra for "3-dimensional discrete hyperbolic shapes" would be the 2-dimensional spectral set, but the existence "3-dimensional discrete hyperbolic shapes" in a "4-dimensional hyperbolic metric-space" are determined by the 3-dimensional spectra of the 3-faces of the set of "4-dimensional discrete hyperbolic shapes." Thus, in order to fit into the "4-dimensional hyperbolic metric-space" a "3-dimensional discrete hyperbolic shape" must be in resonance with a set of 3-faces which are smaller than the metric-space within which it is contained, ie it must be in resonance with the spectra defined within a different 4-dimensional subspace (which is itself also modeled as a "4-dimensional discrete hyperbolic shape," but of a "smaller size" than the "4-dimensional discrete hyperbolic shape" within which the (given) "3-dimensional discrete hyperbolic shape" is contained).

The 1-dimensional spectra are "additive" since each higher-dimensional discrete hyperbolic shape also contains a subset of 1-faces, which would also be a part of the set of 1-dimensional spectra. Thus, the set of 1-dimensional spectra is greater than the combinations, 11C1=11, and in the dimensional level and subspace partition by discrete hyperbolic shapes, wherein, if one assumes a uniform shape for each element of the partition, then one ends-up with about 1000 different-sized 1-dimensional spectra for the entire containing space.

That is, it is simplest to model this idea in relation to requiring that each discrete hyperbolic shape be a uniform shape, ie each face has the same spectral value, and then the different sizes are determined by the multiplicative constants defined between both "the different subspaces of the same dimension" and "the different dimensional levels." This simple idea of course can be adjusted.

Results:

The new context identifies a finite spectral set.

There are (can be) other "components=material" of various sizes (but contained in some one of the hyperbolic metric-space) but which must resonate with some value (with the correct dimension) of the given finite spectral-set defined by the new (descriptive or containment) context [ranging over all the various subspaces of the given dimension of the given spectral component, which is to be in resonance with the identified finite spectral set {of the 11-dimensional hyperbolic metric-space}]

Physical description of stable systems is about the set of discrete hyperbolic shapes identified by this finite spectral set in the 11-dimensional hyperbolic metric-space.

Now it is better to model functions . . . , used to describe stable physical systems and which define a function space . . . . , to be discrete hyperbolic shapes. By doing this the idea of "global commutative math structures" defined on such a function space will be more directly related to the properties of physical systems.

Sorting-out our containing 3-space

The spectral values of charges, nuclei, atoms, molecules are "small sized" and some of these shapes are, apparently, close to the "same size," and are either of dimension-2 or dimension-3, while our 3-dimensional containing (spatial) metric-space, ie our 3-dimensional discrete hyperbolic shape, is the size of the solar system, but nonetheless many "small" either 3-dimensional components or 3-dimensionally contained components resonate with the spectra of different 3-subspaces, and/or different 2-subspaces.

In such a large shape, an observer's local probing of the metric-space's shape would not detect such a large hole structure. Furthermore, the observer would "see" that which is "far away," in relation to the lattice of the fundamental domain upon which the "metric-space's shape" is identified, where the lattice is defined on an unbounded hyperbolic metric-space.

It is the hole-structure of the very stable discrete hyperbolic shapes which allows very stable spectral and stable orbital properties to be defined, and it is the stable properties of the solar system, where the orbits are filled with condensed material which takes on a spherical shape because the new structure for interaction has spherical symmetry for (free) condensed material in Euclidean 3-space, and the existence of the stable spectral-orbital properties provides evidence to show that this descriptive structure is the correct model for the containment space of stable "material" systems. These shapes solve both the stability question for the solar system, if the authorities have a better solution please present it to the world, and with the "discrete angular transformations" related to the Weyl-transformations, so that a discrete hyperbolic shape can take-on the shape of concentric orbits, thus it is a descriptive structure which also "solves" the quantum-radial equation, where charged components fit into the exact shapes of the spectral flows of the discrete hyperbolic shapes of the nuclear and atomic system's.

The real point is; that the current descriptive context fails to describe the stable properties of these many fundamental systems, and because it has failed to provide these descriptions, the new model carries more authority than does the currently accepted fixed structure for math and science.

That is, this method of both choosing and organizing a containment set is providing a measurable description for the stable spectra and orbital properties of the observed stable physical systems for the fundamental systems of existence for the physical systems of the sizes ranging from atoms to the solar system. That is, this new descriptive language is already a descriptive structure which is verified by the observed properties of stability for these systems. This is stronger evidence for the verification of this new model than that verifying evidence actually exists for the current authoritative dogmas of physics, as well as math.

The new context also allows for second order parabolic equations related to angular momentum, but the new context has more geometry within circle spaces, in regard to the set of possible angular momentum properties of the new descriptions of physical systems, where this angular momentum can link between different dimensional levels.

Note: If infinite-extent discrete hyperbolic shapes (6-dimensional and higher dimensions) have bounded faces in their fundamental domains, then the size of these bounded faces is determined by the sizes of equivalent lower-dimension discrete hyperbolic shapes which are a part of the partition. The existence of bounded faces on a 2-dimensional infinite-extent discrete hyperbolic shape would be a way of modeling electrons and neutrinos which compose an electron-cloud in an atom or a molecule.

Take Notice!

This solves the most fundamental problem in physical description, the stability of the most fundamental physical systems. This problems has been ignored since it was believed that it was too difficult that these fundamental systems are too complicated to describe, but now it is solved and its solution results in a quantitative construct based (generated) from a finite set (or stable spectral values).

In the new context the descriptive structure is "bounded" by the very stable discrete hyperbolic shapes which model both the "material component" containing hyperbolic metric-space as well as the "material components" whose dimension is one-less that the dimension of their containing metric-space. This means that the very simple but stable geometry has a dominating influence on physical descriptions (not differential equations). However, material interactions are defined between the material components . . . , though discrete [newly determined in a discrete manner every about 10^{-18} sec] . . . , and these interactions are mediated in a continuous manner, in regard to space, by the discrete Euclidean shapes [or Euclidean tori] (which allow spatial continuity for the dynamic processes) so that the interactions between components are generally non-linear and they are defined in a metric-space so in 3-space there are the usual 2nd order elliptic, parabolic, and hyperbolic partial differential equations which are part of the continuous descriptions of locally measurable properties, though now there is a new geometric context for angular momentum.

Note: The solutions to the linear, metric-invariant, and separable partial differential equations related to the discrete hyperbolic shapes has been discussed by L Eisenhart in a chapter in his "about 1930's" book Riemannian Geometry.

However, for interactions between micro-components contained within, say, a thermal reservoir, it should be first noted that these micro-components are always colliding and thus the components are constantly changing from neutrally charged to being ionized, so the interactions identify a Brownian motion for each component. E Nelson (Princeton, 1967) has shown this Brownian motion for micro-components is equivalent to quantum randomness. Furthermore, the vertices of the fundamental-domain of the discrete hyperbolic shapes identify a distinguished point for each shape, about which micro-material interactions are centered. Thus, this accounts for the random motions about an apparent point-like structure for micro-material.

Outline of ideas

Describe the properties of the stable spectral-orbital systems, How does one describe the very stable, many-body, spectral-orbital physical systems which exist at all size scales from nuclei to solar systems, as well as life and mind?

The subject of stability is mostly about using very confining, but very simple set of geometric shapes, and some simple calculus, as well as a few other (local) geometric properties defined in a context of a principle fiber bundle with a many-dimensional base-space, where the base-space is (dimensionally) partitioned into sub-metric-spaces which are identified by discrete shapes, which are open-closed (when observed within the metric-space which possesses the discrete shape), but these metric-spaces (with shapes) form a boundary when viewed from an adjacent higher-dimensional metric-space which contains the discrete shape.

The math properties for stable solvable physical systems, The simple math properties which allow partial differential equations (or differential equations) . . . , which are used to model physical systems (either geometric-inertial or spectral) . . . , to be solved can be listed.

That is, solvability is related to the set of properties: linear, metric-invariant, separable (locally linear and commutative (diagonal matrices at each point in the global coordinate system)), where the metric-functions can only have constant coefficients, (for the various $R(s,t)$-metric-spaces {where s-space, and t-time; dimensions}, and associated metric-function signatures, where these different metric-spaces are, in turn, related to new types of materials).

[These new material-types are characterized by their properties of being odd-dimensional spatial subspaces, with an odd genus-number, (analogous to discrete hyperbolic shapes) so as to be charge-unbalanced, and thus, naturally oscillating and energy-generating material systems, which model life. Life is defined as new material-types, which are contained in either: $R(4,0)$, $R(6,0)$, $R(8,0)$, or $R(10,0)$ Euclidean spaces.]

The properties of stability are properties which are possessed by the circle-spaces. One can note that, circle-spaces can be modeled as metric-spaces of non-positive constant curvature, whose discrete shapes (or discrete isometry subgroups) are determined from lattices (with an associated fundamental domain, which is related to right-rectangular (or "cubical") simplexes, by a moding-out process (or a process of defining an equivalence topology on the cubical shapes of the fundamental domains).

[Note: The quasi-spectral-geometric properties used to describe particle-physics, ie the non-linear-random-geometric model of particle-physics . . . . , which deals primarily with probabilities of particle-collisions {which, in turn, only model (nuclear) reactions}, and which is a completely irrelevant and practically useless descriptive construct . . . . , are excluded from consideration, since these descriptive constructs only provide . . . , in a chaotic or random fashion . . . , brief

726

and fleeting descriptions of unstable patterns. It is a model which is only applied by means of its relation to reaction-rates.]

Holes-in-space, spectra, force, and the circle-spaces, are all defined within a many-dimensional space. The simple math structures of circle-spaces, as the basis for the stable (geometric) properties of existence, causes both spectral properties and force-fields, to have an analogous (or parallel) math structure . . . , [in relation to holes in space caused by either the shape of a metric-space or by rigid material shapes, eg (usually) 1-dimensional currents defined by rigid material defining a closed curve] . . . , than is usually believed to be true (ie the material-shapes are equivalent to spatial [or metric-space] shapes).

The very stable "discrete hyperbolic shapes" are used as models for both material components and metric-spaces, so as to construct a very rigid geometric structure, in a many-dimensional context, so that Newton's laws still define inertial dynamics. Discrete hyperbolic shapes are stable, with stable spectral properties, and they provide a constrained and stable and very rigid set of boundary conditions for both material containment (confinement) and material interactions, so that differential equations, at first, appear to have a very restrictive containment structure, but this is needed to model stable systems, and it can also be used to model living systems in new ways.

Most material components, as well as metric-spaces, which are contained in this new many-dimensional context, have the shapes of circles-spaces, in particular the "discrete hyperbolic shapes" define a very rigid set of both constraining contexts and boundaries for measurable descriptions. In this new context, descriptions which are related to (partial) differential equations, require that material interactions be mediated by discrete Euclidean shapes, which in turn, are related to the differential equations associated to metric-invariant differential-forms of the force-fields, ie the descriptive structure is based on the geometry of circle-spaces (and holes in the shape of space). Force-fields are applied to material shapes by Newton's law of inertia, which is defined in absolute Euclidean space (which also allows for action-at-a-distance), in this context of very rigid, and confining, geometry (but Newton's universal law of gravitation is modified).

Stable spectral shapes, and material's orbital properties, and the properties of (condensed) "free" material components in space, There are "free" material systems, and orbital material systems, where "free" material components can condense into limited orbital structures, or condensed material can be guided by highly confining orbital constructs, because these (condensed material) components do not have the "correct" size to be stable material components, in their (particular containing) dimensional level, as well as particular subspace, so that the resonances . . . , which allows the existence of material components . . . , are either from another subspace (of the same dimension), or are determined to be from resonances which are defined on the facial structure of the stable simplex structure of the condensed material, in order to have a stable spectral structure for the condensed system. If a higher-dimensional material component is not "big enough" to be a material component, in its particular subspace (of some particular dimension), but nonetheless this higher-dimensional material component does (can) interact. However, the property of "the faces of the material component," which is interacting on a higher-dimensional context, are contained in a closed metric-space. This causes the dynamics of the interaction (in a higher-dimensional level) to be defined on the entire material-containing metric-space. The entire, rigid, metric-space being pushed by the interaction is not noticeable within the rigid geometry defined within the metric-space, ie a perfectly rigid-rod can transmit a "push" instantaneously across the space it occupies.

An example of a true manifestation of inertial affects defined in general relativity, "Free" material components can also be related to orbital structures, where most often condensed material constrained by the orbital structures of discrete hyperbolic shapes, wherein geodesics can (now) affect inertial properties of these "free" material components, and can be described in an explicit manner in a stable (linear) context, ie general relativity is being defined beyond the 1-body problem with spherical symmetry, so that the description is stable, since the geometry in the new descriptive structure is linear, metric-invariant, and separable, so as to define a stable orbit.

[Note: An orbit defined on a discrete hyperbolic shape is pushed . . . , to the limited and rigid structure associated with the shape's geodesics . . . , by the coordinate structure of hyperbolas, which exists away from the geodesic paths (where geodesics will be contained on the faces of the simplex of the discrete hyperbolic shape's fundamental domain)].

The interaction "discrete Euclidean shape," or the interaction torus, Euclidean space-forms (a synonym for a "discrete Euclidean shape") mediate material interactions within the rigid constraints of the "discrete hyperbolic shapes," which model both metric-space existence and material existence. During a material interaction a differential-form is defined upon the geometry of an interaction torus, ie a differential 2-form model of a force-field is defined on the interaction torus. Yet, a "discrete Euclidean shape" of an interaction can also transform, so as to become a toral component of a newly formed discrete hyperbolic shape, where this can happen if both (1) resonances exist for the interaction, and (2) the energy and size of the interaction is within the "correct" energy and size ranges. The geometry of the 2-form is related to the geometry of the fiber SO(n) group in order to determine the direction of the force-field's push [in the base-space containing the interaction torus]. Then the spatial positions of the interacting material's vertices are locally transformed, in a (local) context of opposite metric-space states.

A finite spectral-orbital set, defined upon a many-dimensional containing space, A finite spectral set can be defined on an 11-dimensional hyperbolic metric-space, which is an over-all high-dimension containing space for (of) a model of existence (where the existence is defined by the finite spectral set). This is possible because in each dimensional level (and for each subspace of the same dimension) there is a maximally-sized discrete hyperbolic shape in regard to the finite set of discrete hyperbolic shapes which are used to model the subspace metric-spaces (subspaces of the 11-dimensional space), so that the spectra defined for all the dimensional-levels and subspaces of each of those different dimensional-levels is finite, and thus, it can determine a finite spectral set, upon which all material properties . . . , which are allowed to exist in this 11-dimensional hyperbolic metric-space . . . , depend.

Defining angles between toral components, Weyl-transformations of angles between a discrete hyperbolic shape's toral components, (or a set of folds allowed on the base-space lattice structure), so that there are a finite set of angular relationships which can be defined between the toral components of a discrete hyperbolic shape. These folds between toral components allow envelopes of orbital stability to be defined, based on the orbital (or metric-space shape) structure which a discrete hyperbolic shape can have after its toral components are transformed by certain angular values.

The operation of multiplying by a constant factor, Constant multiplicative factors can be defined . . . , so as to affect the properties of: shapes, sizes, orbits, and the stability of "discrete hyperbolic shapes" :

. . . , between dimensional levels,
. . . , between subspaces of the same dimension, and
. . . , between toral components of a discrete hyperbolic shape.

Physical properties (and math properties) attributed to metric-spaces, There are physical properties associated to metric-spaces. Two examples of physical properties are: (1) position in space of a system's vertex, in regard to the distant stars, (2) the stability of a system's (or a mathematical) pattern.

This results in the definition of "metric-space states" of opposite physical properties, eg (+t) and (-t). These opposite metric-space states are a part of the dynamic processes of material interactions, eg fixed stars, rotating stars (Euclidean); forward time, backward time (hyperbolic space) etc.

In turn, this implies unitary containment, in regard to the containment (of opposite metric-space states), within both real and pure imaginary subsets of finite-dimension Hermitian containing set of coordinates, as well as allowing the definition of the spin-rotation of metric-space states, so that this spin-rotation of metric-space states is defined on opposite metric-spaces states so that these opposite states are a part of the dynamic process. The (time interval of the) period of the spin-rotation of opposite metric-space states is a property used in the dynamic (or inertial) material interaction process.

In this description there is no need of a continuum, instead rigid geometric stability is used rather than using (indefinable) randomness and non-linearity as a basis for (physical, or mathematical) description.

Indefinable randomness and non-linearity seem to possess the properties which are needed to briefly describe patterns which are unstable and fleeting in duration, which, at best, are relatable to feedback constructs, ie it is a description which depends on the validity of the fleeting pattern of a system modeled as a partial differential equation, which has limited descriptive value, and which is used in a relatively unimportant contexts, ie it is a flawed viewpoint which cannot be used describe the observed stable spectral-orbital properties of physical systems at all size scales. (2) (over)

San Diego Joint Math Meeting (2013)
A new context in which to apply geometry to: math, quantum physics, and the solar system, etc.
1086-VR-413,
Saturday, 1-12-13, 3:30 Rm 6E (main building), Assorted Topics II,
Martin Concoyle PhD.

Just as Copernicus, Kepler and Galileo provided a quantitative-geometric context for the properties of the solar system, which were then precisely identified by the solutions to (the) solvable differential equations of Newton; Martin Concoyle now provides the stable geometric structures which fit . . . , both macroscopically and microscopically . . . , into a many-dimension containment set (hyperbolic 11-dimensional), so that these shapes are the solutions, ie the geometries of the stable spectral-orbital properties, of all the fundamental stable systems which have stable spectra and orbits: nuclei, molecules, solar-systems, etc, and it is the basis for a quantitative system (the spectral set of a measurable existence) which is finite, and These ideas are discussed in the following books:

1. A New Copernican Revolution (p286), B Bash & P Coatimundi, Trafford, 2004.
2. The Authority of Material vs. The Spirit (p483), D D Hunter, Trafford, 2006.

Concerning the media and professional math and scientists:
Succinctly-put the media makes bigger-suckers out of the "successful" professional intellects, than it makes suckers out-of the public, since it traps the professional intellectuals into a fixed context upon which the social value of these professionals depends, and it is a dogmatic context within which all valid authoritative ideas are "to be" expressed, eg peer review.

However, Gödel's incompleteness theorem implies that the professionals should also seek to consider new contexts within which to express math and physics ideas, ie professional publication should not be peer reviewed but rather the assumptions upon which ideas are expressed should be made clear and then the expression of ideas should be placed into categories

Where one category needs to be marked the context which most supports (big monopolistic) businesses.

These new ideas are expressed in a new context, in a similar way as Copernicus expressed a new context which was different from the authoritative context of Ptolemy.

It is the professionals who need to enter the new context so as to discuss the new ideas, new ideas which solve the very difficult problem of the stability of fundamental physical systems and the solution in the new context is very simple (the hallmark of a superior context (or a new paradigm)) the new ideas do not need to accept their context, even though it is their context which allows them to play the roles of wage-slave professionals who serve monopolistic business interests, the new way of expressing ideas only needs to provide an interpretation of the observed data.

Addenda:

It is strange that when confronted with such a simple solution the current authorities find the pattern interesting, but do not comprehend the significance of the true authority (its truth and great capacity for wide ranging usefulness) of this new context.

They are so caught-up in their own dogmatic authority, that they do not realize that they have been dethroned, and that their authority has already been lost. This is because they are so self-important, but it is the social structure which has created their sense of being so superior, and so fixed and traditional in their sense of possessing authority.

The overly authoritative fixed structures of containment and interpretation are used to define the "aristocracy of intellect," those chosen few, who are so keen to be intellectual aristocrats, but who, in fact, so weakly . . . . , (in relation to mental awareness about the society and about either what knowledge is, or what knowledge does (is supposed to do) within society, and how big business corrals and uses knowledge for its selfish interests) . . . . serve the interests of big business, with their business monopolies, which allow them to be such very domineering social forces, where these dominant monopolies are allowed to exist in a society, because the laws of our society are set-up and enforced, so that it is a society which values property more than both life and creativity.

Pompous, self-important math and physics professionals "believe the hype" and they believe that they are the "culturally superior people" of the society, but who are, in reality, people with personality flaws; they are manipulative, narrowly obsessive, authoritarian, and selfish people who are themselves easily manipulated by social forces (eg mainly by-means-of the media) which is why they were chosen to have these social positions, where these superior people dutifully are servile to their own image of being uppity where they promote within society the social traits of domination and fixed-ness of knowledge and its uses, due to their servile social positions, which are, in fact, anti-knowledge and anti-creativity social positions, so as to assure the rich owners of society that the façade of knowledge is all about complicating descriptions to filter the public out of social positions which would allow them greater capacity to create, and thus causing a competitive structure for products within the so called free-market.

These professional authorities are presented to the public by the media as those people who have attained the highest cultural achievements within society, but they do this by serving the few owners of society. Because they depend on their pay-master, this makes them lesser people rather than greater people. They retreat behind the big bully (the owners of society, and these bullies allies in both the government and in the justice system) who create the social images of these, so called, experts, and this is done so as to serve the interests of the owners of society.

San Diego-2013, Math Conference talk-2

The problem is the propaganda system, which is the society's sole authoritative, reliably-truthful vehicle of social expression (but which only promotes monopolistic interests) where it continuously spews-out mis-information which, when placed in the "context" of honest reporting, the public always interprets to be an absolute truth.

Thus when monopolistic, unregulated, but almost completely controlled "market-place" fails in a complete collapse, due to criminal fraud aided by the justice system and the congress the propaganda system continues to "sing the praises" of (and need for) the "magical" unregulated "market-place", where it should be clear that "unregulated", now especially, means license to steal along with a complicit justice system and political system, since these institutions have been manipulated by the propaganda system so as to simply to have become a part of the propaganda system, themselves.

The authorities (or technical experts) which serve this system by adjusting the complicated instruments for the monopolistic ruling interests (the interests of the owners of society) are also controlled by the absolute-truth espoused (in the context of "honest" reporting) by the media related to "technical-development" (but really only small adjustments to fixed traditional technologies, since the context of math and science is not allowed to seek new creative contexts within the context of stable math patterns, ie the monopolies depend on society continuing to use products and resources in a fixed way which allows the monopolies to continue to make money based on their products), where for the intellectual class, the academics, the authoritative experts, are provided with a set of "prized" problems whose context is:

Indefinable randomness
Non-linearity
Non-commutativeness, or

At best locally commutative in a context of a general metric but non-linear in regard to the containing coordinates of geodesic coordinates or set of functions

So that the propaganda system, validated by certain narrowly interested experts insists in "big bangs" particle-physics, string-theory geometry which are needed to understand the singular points of a black hole's gravitational field, and it is claimed by some possibly charges related to black holes and in regard to singularities within nuclei, so as to realize a "grand-dream" (truly a pipe-dream) of the control of worm-holes in space, or control over a "many-world" context of existence, though the math structures through which these ideas are to be described, based on probability and non-linearity, does not allow any control, . . .

since there are not any stable patterns in their descriptive context

Nonetheless these math structures do apply to the business interests:

Is math staying too traditional, fixed, formal, complicated, irrelevant?
Is math only about serving business interests in regard to:

1. Unstable patterns to be used in fleeting contexts (feedback systems),
2. Chaotic transitory systems, eg nuclear reactions, transiting between two stable states,
3. Manufacturing complications, eg formulating security codes, etc,
4. Pulling the wool over the eyes of the public, so as to provide irrelevant and inadequate descriptive structures for physical systems?

. . . , of such constructs even if they are mathematically modeled (as it is so claimed that quantum systems are correctly mathematically modeled) due to the properties of the math constructs and due to the over-whelming complexity of such models these models (of controlling a "worm-hole") cannot possibly be controlled, as the propaganda system is suggesting that they can be controlled. Consider, if the current descriptive context cannot describe the stable, definitive fundamental systems, systems so stable that it implies that these systems form in a linear controlled context, then how can their exotic models of physical systems (changing between worlds, in a many-world model) ever be realized, if they are models which are not consistent with the actual structure of the (external) world? or of these ideas

That is, these prized problems are delusional-ly based, yet the propaganda system promotes them as do the experts themselves so they become the basis for identifying an elite in-crowd of "knowledgeable" experts.

This, in-crowd, of so called superior intellectual elites, obsessed with complicated math patterns (one must note the autistic connection, the manipulation of personality types by institutional managers, used for the purpose of deceiving the public) who are led to believe, by the media itself, that they are zeroing-in on the wonderful goal of their great intellectual prowess and intellectual creativity.

Yet it is clearly a failed intellectual exercise since there are basic stable definitive physical systems which exist at all size scales but which go without valid description (based on physical law).

Nonetheless (however) when these stable systems are actually solved the intellectual community and society has a great mental inertia (almost entirely caused by the propaganda system, it unwillingness to publish the new solutions, since they define a completely new math and physics context) to realize just what they have heard, but nonetheless assured that it is they the intellectual elites who will be the ones to forge new roads into new technical landscapes and thus the elites will not (will refuse to) listen to the new ideas which must originate from an inferior mentality and thus must be wrong and surely wrong within the dogmatic authority which defines their truth for them. The elites only find these new ideas somewhat interesting but from an intellect inferior to their own since they have memorized their contexts and their always correct interpretations and their always intelligent evaluations of the state of knowledge, they are not tricked by prized problems, no not them.

That is, the propaganda system is a communication system (vehicle) which expresses a dogmatic authority (an absolute truth) which is followed by the experts which in fact determines the faith of the high-valued institutions which serve the ruling class and the experts have an absolute faith in the authoritative truth of that dogma. The propaganda system defines the true religion of society and it is a religion of personality-cult (not unlike Roman emperors or Egyptian pharaohs) and a deep belief in inequality and a manufactured property of a society dominated by selfish monopolistic interests, ie it is a society opposed to life and opposed to adapting to change

Wake-up you popes of the religion of science and math, eg general relativity, particle-physics, string-theory, indefinable randomness, non-linearity non-commutativity or only locally commutative.

Gödel's incompleteness theorem has a simple interpretation:

Because precise language has sever limitations the axiomatic basis, the containment set, the organization of this containment set, the context of the description, and the interpretations of the observed properties (of the observed measurements must be fully considered and when an alternative, well defined example of new ways in which to present axioms and contexts is provided, especially if it solves the most difficult problems in math and physics then the math physics community of experts should listen and take it seriously.

Please pay-attention (ironically) it is, me, your superior, talking to you, and it truly is, the irony is, that, in fact, we are all equal, but you have tried to achieve in the eyes of the paymaster and you have lost your way and you have made yourselves lesser (not superior), so now by artificial measures I am smart and you old experts are now stupid and disposable which is the context which your over-reaching superiority has caused the public to be, ie the public is disposable since we have "the great experts"

The crux of the problem with knowledge and education in society is its capture by corporate and private interests capture by the owners of society even though it is most often "public" education institutions, nonetheless the professors are trying to adjust the complicated instruments for the corporate and private interests and they are not concerned with descriptive knowledge in its most general and most powerful sense where new creative context get developed by new contexts for descriptive knowledge which result when assumptions, contexts, containment, organization of pattern use are considered at their most elementary levels, as Faraday developed the language of electromagnetic description while he also developed a new context in relation to the instruments related to electromagnetic properties. Though such a dramatic chain of developing events is not a necessary attribute for developing a new descriptive context at its most elementary level it means that new languages can be related to new creative contexts.

Since the professors of public universities have been captured by corporate and private interests through the mechanism of funding research and identifying prize problems in math so as to keep math traditional and under the control of peer review the talks at conferences, such as at the joint math meeting in San Diego 2013, are either about developing even more complicated theories and more complicated formal professional math language where these formal math languages have very limited if any relation to practical development they may only be marginally related to corporate interests, essentially related to bomb technology,

or

About applying technical complicated math so as to be able to adjust rather complicated instruments of interests to corporate and private interests, eg feedback systems, imaging systems, recognition systems and improperly defined statistical constructs which often appear to be valid, since the propaganda system is capable of making all of society to continue to use language and product in very certain narrowly defined ways so as to place an artificial stability on the statistics where such stability of the statistical context does not really exist.

Propaganda, in regard to science and math is provided in the context of great breakthroughs and, supposedly, new things but which has little bearing on the uselessness of the descriptive language except in regard to weapons technology

Science and math are often about making adjustments to systems which will reduce labor if things remain in their fixed social context in regard to the corporation's products

That is developing new knowledge in regard to new contexts and new ways in which to organize descriptive language, in regard to solving fundamental mysteries, is effectively stopped by peer review prize problems traditional authority and mostly by the process of funding which is controlled by the corporations and private businesses.

Nonetheless, there are many unsolved fundamental problems in physics which are ignored due to dogmatic authority of math and funding traditions within public educational institutions an authority which is essentially related to military development and banking investment interests

The sad thing is that now (2013) these fundamental problems have been solved but the structure of both propaganda and the "knowledge institutions" keeps-out ideas which are different from the inter-related interests of business and traditional academic authority.

Academic science was invented by the mercantile class in the 1600's (after Newton) to support their investment and productively-creative interests, in the 1500's science development based on measuring was centered around schools and about literate people, so that investment in science was centered around the schools (or universities).

However, public schools should take notice of Godel's incompleteness theorem and the logical positivists who proclaimed to limitations of precise language (or the limitations of measurable descriptions) and to consider the example of Faraday wherein he both invented a new math language to describe electric and magnetic properties but he also created the instrumental context through which these properties could be used and controlled

That is, descriptive knowledge best leads to new contexts for creativity at the elementary level of language assumption context interpretation etc not at the complicated formal level, eg no one can use the principles of particle-physics to describe the stable spectral properties of general nuclei.

That is, one wants new contexts for creativity to come from precise descriptive languages one does not simply want adjustments to complicated instruments since instruments also have limitations as to their capabilities or if there is not a better instrument to do the same thing, though digital electronic seems to be able to deal with arithmetic and math patterns well but this has led to an attempt to deal with non-linear systems but non-linear systems can only be related to feedback systems and only for limited ranges of time or distance in regard to arithmetically determined solutions.

There are many mysteries in regard to fundamental physical systems: why do there exist stable physical systems with definitive measurable properties, eg nuclei, general atoms, molecules and their shapes, crystals, life stable solar systems, there are many galaxies with planar spiral structures, etc.

These fundamental physical systems go without valid descriptions based on what is considered to be "physical law" instead one hears about all the hyped-up ideas through the propaganda system and the professional mathematicians and physicists about "big bangs," Black holes, worm-holes, Higg's particles, transforming neutrinos, all related to general relativity, particle-physics grand unification and string-theory etc, are all expressions which are wild speculations if they cannot describe the stable properties of fundamental systems whose stability implies that they come into being in a controlled context, while believing the wild speculations is mostly driven by (or caused by) the propaganda system and a traditional context of math and physics authority and a personality cult which forms around these academics who mostly interfere with the development of new contexts for knowledge and creativity but the monopolistic business interests do not want new contexts for creativity. Where it might be noted that general relativity was shown to be untrue in regard to the non-local properties of material interactions in Euclidean space since non-localness was demonstrated by A Aspect's experiments.

The correct answer as to why there are stable fundamental systems require s that the context of containment be changed in a drastic manner from 3-space and time (quantum, Newton) or space-time (electromagnetism, particle-physics (?)) and materialism where the measurable descriptions are either Classical often leading to non-linearity or Quantum indefinable randomness, spectra supposedly derived from $1/r$ potentials, function spaces usually non-commutative and Lie groups all used in the context of a continuum and a loose idea about convergence to this continuum, eg renormalization (something Dirac rejected) but apparently personality cult and propaganda was able to establish as an authoritative technique.

Instead consider a new context:

An 11-dimensional hyperbolic metric-space is partitioned . . . into its different dimensional levels and the different subspaces of the same dimensions . . . by discrete hyperbolic shapes which exist up to hyperbolic dimension-10. This can be used to define a finite spectra on the over-all 11-dimensional hyperbolic metric-space.

The set of all resonating discrete hyperbolic shapes which are contained within this high-dimension containing space form the bounding stable structures of the more usual physical description of material defined as one-lower dimensional shapes in each dimensional level, where the usual metric-invariant, second-order elliptic, parabolic, hyperbolic, differential equations . . . associated to material interactions or material properties . . . are defined, but the stable elliptic structures are defined on the discrete hyperbolic shapes. The elliptic case is mostly about describing condensed material contained within a (higher-dimension) discrete hyperbolic shape, or the orbital path of a material component is (becomes) resonant with the discrete hyperbolic shape upon which the component is contained (or can become "so contained," due to resonance).

The stable physical systems are the discrete hyperbolic shapes of the various dimensional levels which are in resonance with the finite spectra of the over-all 11-dimensional hyperbolic metric-space which has been so partitioned.

Each n-dimensional level "sees" the bounding geometry of the (n-1)-dimensional material "surfaces" but the open-closed topology of these shapes allows light to be observed from outside the metric-space shape's fundamental domain out to the unbounded lattice, this is especially true for metric-spaces whose fundamental domains are large eg as large as the solar system, where it should be noted that lower-dimensional shapes than (n-1)-dimension tend to condense onto the (n-1)-material's shape.

That is, the shapes imply the discontinuity of a metric-space experience between dimensional levels.

Interactions between micro-components imply Brownian motions which implies (due to E Nelson 1967) quantum randomness. Furthermore the distinguished points on discrete hyperbolic shapes implies that the interactions appear point-like, but, nonetheless, mediated by discrete shapes.

It might also be noted that this descriptive context provides a definitive spectral relation between different 11-dimensional containment sets, ie there are many different worlds where each world is well defined by a definitive spectral set, and the best instrument to realize transitions between these world might very well be a (human) life-form.

Speech 3

The observed stable, precise, patterns of physical systems are associated to finite properties, eg bounded-ness and/or the finite number of a physical system's components, eg atomic-number, and these stable physical system properties are fundamental and observed features of a reliably measurable context associated to the observers of physical patterns. This implies both the existence of stable patterns which allow reliable measuring, (or which are associated to the context of measuring (for an observer)), and the existence of stable-controllable patterns associated to a set of fundamental physical systems which possess stable features of "what may, or may-not be" "material" systems, eg nuclei, general-atoms, molecules and their shapes, crystals, solar systems, dark-matter (ie orbital properties of solar-system's in galaxies) etc, the physical patterns upon which the relatively stable aspects of our life experiences depend, and upon which our mental constructs also depend.

Thus, science and math are about identifying stable, quantitatively-consistent, math patterns which are generally applicable to these stable, measurable, and apparently controllable, physical properties so as to result in descriptions of these patterns which are accurate (to sufficient precision), and general so as to be able to describe the observed stable physical patterns of existence, so as to provide a context for practical usefulness, ie measurable and controllable, so that one can: measure, fit together (or couple), and interact with these various patterns (using the natural structures of these patterns, eg life-forms and its coordinated chemical properties (but, apparently, coordinated by an unknown structure), ie not feedback mechanisms nor carefully prepared structures so as to cause reactions), so that this descriptive knowledge can be related to "practical" creativity (as opposed to literary creativity, essentially associated to a world of illusion, ie a world without stable features).

What are (math) patterns?
Patterns are:

1. consistent relationships, or
2. operators acting on quantitative sets so as to have fixed "consistent properties" related to the application of an operator on a quantitative set, and these consistent patterns are related to the "meaning" of the quantitative-set's elements (where quantities represent properties of: type and [measurable] size), or
3. stable shapes, etc.

Can the current descriptive language of mathematics and physics describe stable patterns? [Apparently not.]

There are essentially the three ways in which to try to describe stable math-physical patterns . . . ,

I.   stable geometry, which strongly limits both a descriptive context and the patterns it is trying to describe (the new context for physical description, the circle-spaces, or the very stable discrete hyperbolic shapes),

II. differential equations in a geometric context (unfortunately, this method most often leads to non-linear patterns),

III. differential equations in an operator context (this methods seems to only work for harmonic properties which possess actual physical attributes))

. . . , so as to try to use quantitative descriptions so as to try to identify stable patterns which provide valid information, as well as control, over relatively stable (physical) system properties.

A new interaction-construct can be constructed which is general, but its stable properties are determined from a context defined by a many-dimensional set of discrete metric-space shapes, which, in turn, define existence.

The professional mathematicians and scientists in regard to descriptions of fundamental stable physical systems express symbolic nonsense, ie they provide a set of nonsense symbols which result in descriptions which are neither general, nor accurate (to sufficient precision), nor do they provide a practical context for useful creativity.

Physical systems which are very stable and definitive, but which are many-(but relatively few)- body systems, nonetheless, because these systems are so stable and definitive, it is clear that they are forming within a very controlled context, so that the descriptions (of the professionals) which are based on:

1. (vague) randomness (which is an uncontrollable description for a system which is composed of only a few components),

2. non-linearity (quantitatively inconsistent, and chaotic), and

3. non-commutative (not invertible, or equivalently, not solvable, eg non-linear or spectrally- un-resolvable), context, which is

4. contained in a continuum (a containing set which is far "too big" allowing logically inconsistent descriptive constructs to be put-together as if they belong to the same containment set), and

5. it is a description (when based on randomness) which begins from a global viewpoint (a function space) but the methods of the description focus on local spectral-particle events

in space, ie it is a description which gives-up information leaving one in an inaccurate and non-useful context in regard to information.

It is a description which "in general" is not accurate, yet it also is a description which is "intent on" losing information about the stable definitive properties of the [assumed to be random] system.

That is the descriptive structure of the "dogmatically pure" set of experts of math and science is simply a bunch of nonsense.

Yet one must list the places and contexts within which it is a valid descriptive context:

1. It is a description which is relatable to a system whose initial conditions, and initial properties are carefully put-together so as to be a system which is easily broke-apart, so as to form a transitioning system which is chaotic, so that the rates of reactions (in this context, based on component-collision probabilities) are determined by cross-sections of the broken-apart components, where these cross-sections determine the rates of certain aspects of the (a) reaction, and

2. They are descriptive contexts which relate a limited set of metrically measurable (observable) properties to a feedback structure, which is mostly associated to the critical-points and limit-cycles of a non-linear (usually classical) partial differential equation, where the range of relevance of the differential equation is difficult to determine or to control. Furthermore, the initial or boundary conditions of this type of a system relate to the properties of the descriptive context (or properties associated to the solution) of the system's differential equation in a chaotic manner.

That is, difficult math methods . . . , which are related to fleeting, unstable math patterns . . . , are descriptive constructs which have no content (and possess no useful information), they are patterns which apply only to unstable contexts, where control emanates from a higher abstract, and manipulative, context imposed on properties which are only definable in a metric-space, and which requires a lot of preparation (in regard to sensing and reacting in the desired way to the detected properties), a context which is at-odds with the system's natural properties, ie rather than controlling a system by simple adjustments to affect the system's properties in regard to affecting the properties of several system-components being coupled together.

So we have the tradition of "western hypocrisy," where failure is rewarded if those who perpetuate it, are in the high social classes.

What is wanted, by the owners of society, is that the social structures through which the powerful derive their power are kept in place.

That is, it is a social structure which is opposed to new, creative changes and thus it is also opposed to equality and the creativity associated to equality. However, the traditional social structure which upholds dominant interests so violently, and it expresses its interest in "lyrical creativity" in regard to the science and math experts . . . . , where these authoritative experts define the "literary" creative development of science and math, which is authoritative, but unrelated to practical creative-development, and the owners of society support the "creativity" of the elite artists, those who also compete in a "narrow context of authoritative cultural value," as well as those journalists and intellects whose ideas are judged (by the owners of society) to possess "cultural value," so that the ideas expressed are consistent with the ideas of (or can be used by) the owners of society, so as to be distributed by the material-instruments of the media, which are owned and controlled by the owners of society . . . , then even the failures of the experts can become part of the social structure which allows the powerful to remain powerful.

The top-intellects and top-artists are defined as a social class, along with artists and journalists, so that the intellectuals can dogmatically dominate those many-others who question the authority of assumptions, or who have different ideas.

The main tool used to maintain the power of the owners of society is the single voice of authority which the media has become (most clearly controlled by ownership, or by a set of funding processes).

That is, it is violence and domination (intellectual domination) which is fundamental to social power, not knowledge.

Knowledge is relevant, within today's social structure, only in regard to the creativity which is a part of the organization of society (ie business productivity) which, in turn, maintains the power of the few. However, the organization of society, and the use of resources and the ownership of technology within society, essentially, remains fixed and traditional.

For example, the many-purpose phone, eg an i-phone, is about developing 19th century ideas of electromagnetism, and the micro-chip circuit boards in these devices depend on 19th century optics.

Whereas identifying stability "as a needed property" in both math and physics, in regard to the useful descriptions of controlled (or controllable) physical systems, is a focus (in regard to the valid descriptions of math patterns) which the math professionals, apparently, have not considered.

Furthermore, very simple math patterns can be used to create new math patterns, which can be used to describe the stable material properties, so that these descriptions are based on a finite quantitative set, within which the descriptive containment of physical properties depends, ie the containment set is not a continuum and the derivative and its integral-inverse function-operators become discrete operators (the continuum can, instead, be the set of rational numbers).

In fact, the math patterns of stability are very simple, and relating these simple structures (which are best characterized by the stable discrete shapes, or circle-spaces) to many-dimensions, can be done by a simple process of partitioning the dimensional-levels of a hyperbolic 11-dimensional containment metric-space (base-space) by means of stable shapes, ie partitioning the dimensional levels by means of the discrete hyperbolic shapes (or circle-spaces), so as to form a finite spectral-orbital set, where the sequences of spectral-size are defined (either increasing or decreasing) as the dimensional level increases, so that these size-sequences of spectra are fundamental, in regard to how the description is organized, so that a finite spectral set is the basis for physical descriptions of the observed spectral-orbital-material order which the stable (material and containing metric-space) structures of existence possess.

Speech 4

Stability

In order to describe stable, "measurably consistent," and precise patterns of material systems one needs stable, "quantitatively consistent" math patterns.

In math such patterns are (quite often related to) linear, metric-invariant, separable partial differential equations, ie non-linearity does not work. Stable and quantitatively consistent math patterns also deal with the geometric models used for measurable quantities of the line (or line segment) and the circle, where these two geometric (or quantitative) structures can be easily organized so as to be quantitatively consistent with one another, eg the real-line and the complex-number-plane.

Circle-spaces

The circle-spaces fit both categories which identify reliable and stable quantitative descriptive (or measurable) constructs. The circle-spaces are the tori (the dough-nut shape) and those shapes with toral components, these include the discrete Euclidean shapes (single torus) and the discrete hyperbolic shapes (composed of toral components). The patterns of the circle-spaces are stable and quantitatively consistent. The circle-spaces are characterized by the properties of being related to non-positive constant curvature metric-spaces where the metric-functions have constant coefficients. The geometric properties of circle-spaces fit into the geometric structure of analytic complex-functions (where an analytic function is supposed to be consistent with the algebraic structure of quantitative sets, but the series must be put into the context of finitely defined polynomials to ensure quantitative consistency)

Lie groups

These are also the discrete isometry subgroups of the classical Lie groups (associated to metric-invariance). These spaces (shapes) include the discrete Euclidean shapes of R(n,0), the discrete hyperbolic shapes of hyperbolic n-space, which in turn, is associated to the general (n+1)-space-time spaces of R(n,1), as well as more general R(s,t) spaces and their associated discrete shapes related to circle-spaces.

Properties associated to metric-spaces

These metric-spaces, which may be modeled as discrete shapes, have associated to themselves both math and physical properties, so their discrete shapes can identify both metric-spaces and stable material components within the metric-spaces.

That is:

1. The property of position in space (in relation to the distant stars), these are related to the Euclidean spaces, which includes the property of action-at-a-distance, when the shapes have distinguished points and the context is the very rigid properties of hyperbolic spaces filled with sets of very rigid discrete hyperbolic shapes.
2. A stable well-defined pattern or shape, ie properties which are continuous in time, or conserved properties (or conserved patterns, or shapes), this property is related to the hyperbolic spaces [and the R(s,t)-spaces]. Time states are defined by the properties of the opposite flow of time advanced and retarded potentials, as well as the opposite pair of wave-equation solution functions.

Opposite time states and wave-propagation

How are these properties related to the propagation of wave-functions in either odd-spatial dimensions of a metric-space (distinct directions of time associated to wave-propagation [surface]), or even-spatial dimensions (always a mixture of time states, waves fill space after propagation "surface" distinguishes a wave property of a solution wave-function)?

Spin and unitary fiber groups

The assignment of properties to metric-spaces leads to the existence of pairs of opposite metric-space states and this, in turn, leads to the two: real and pure-imaginary subsets, of the complex coordinates, and the unitary fiber groups, as well as the spin-groups, where the spin-groups spin-rotate between the pairs of opposite metric-space states, where pairs of opposite time-states (associated to hyperbolic-space) are a part of the local dynamic process, and they are a part of the stable spectral-flow structure, which exists on the discrete hyperbolic shapes, and is related to the sub-face structure of a discrete hyperbolic shape's fundamental ["cubical"] domain.

Note: For new (3-dimensional) material associated to (or contained in) R(4,2), there are two time-dimensions so that each direction of time can be associated to a different domain space for the spatial positions of distinguished points (of the discrete hyperbolic shape models of material components) being transformed by SO(4)=SO(3) x SO(3) in (x,y,z,w)-space.

Size (Cardinality) of quantitative sets

[problems in description which are introduced by using sets which are too big]

There is another problem in regard to the size (cardinality) of a containing set, which is used in descriptions. The continuum is "too big of a set" allowing the inconsistent properties of both stable geometry (of particle-collisions) to be defined by means of convergence to the domain (or containing) space so that this explicit geometry (of particle-collisions) is defined in a context of assumed fundamental randomness, in regard to quantum physics, and (non-linear) particle-physic's random math structures do not allow microscopic material geometries (such as the geometry of a particle-collision). The continuum is "too big of a set" since it allows inconsistent constructs to be defined through convergence into the same containing (or domain) space, which is a continuum (containing both particle-collisions and fundamental random math structures).

To create a finite quantitative set

Thus, one needs to base quantitative descriptions, especially of physical systems, on a quantitative set built upon a finite spectral set. That is, one needs a stable, quantitatively consistent, and finite based quantitative set upon which to base a measurable description of physical systems.

Properties of discrete hyperbolic shapes

By following (or using) the patterns associated to the stable discrete hyperbolic shapes:

1. the last type of discrete hyperbolic shapes which is compact is 5-dimensional,
2. the dimension of the last type of discrete hyperbolic shapes to exist is 10-dimensional (hyperbolic-dimension) and it is an unbounded shape.

Partitioning a many-dimensional containing-space with shapes, so that the partition depends on dimension and the set of separate subspaces (of the same dimension, as well as of different dimension) which are defined on a many-dimensional space.

The number of n-dimensional subspaces "for n less that 11" is given by the "combinations" 11Cn, eg 11C2=55.

The properties of the dimensional partition

The idea is to partition the different dimensional levels . . . , and the various subspaces of each dimensional level, of the 11-dimensional hyperbolic metric-space . . . , by a (finite) set of (bounded when possible) discrete hyperbolic shapes, ie circle-spaces. Thus an 11-dimensional hyperbolic containing metric-space which is so partitioned is only continuous within each dimension, and subspace shape, of n-dimensions, and the continuity defined in such a n-dimension subspace is mostly defined (identified) in regard to the set of (n-1)-dimension boundaries of the (n-1)-dimension material component shapes contained in the containing n-metric-space, which, in turn, is contained in an 11-dimensional space. Inside a discrete hyperbolic shape, one does not see holes in the shape, but instead sees an open-closed topological space, ie one sees the space within which the lattice [("cubical" partition) of the discrete hyperbolic shape] is defined.

The nature of the spectral sequence based on the partition

That is, one is defining (on most of the "many sets of subspaces," defined within the 11-dimensional space) an increasing spectral sequence (as the dimension increases), and where the spectra are defined on the bounded, discrete hyperbolic shapes, which are a part of the (finite) partition, though on some subset of "the set of subspaces" there may be some decreasing spectral sequence defined (as the dimension increases).

This can be used to define a finite spectral set upon which the stable shapes and components must be resonant in order to be contained within such a containing set (or metric-space) structure. The sets of subspaces within which an increasing spectral set is defined (as the dimension increases) (can) have material-components contained within themselves (within the discrete hyperbolic shapes which model hyperbolic metric-spaces) so the contained material components (within the partitioning shapes) will have stable geometric-spectral properties which resonate with the finite spectral set of the partitioned 11-dimensional over-all containment space. Furthermore, condensed matter . . . . , ie material components whose size is too small, in regard to both the dimension and the subspace (of that same dimension) within which the material component is contained . . . . , will still be contained within a discrete hyperbolic shape, within which the condensed material can be in an orbit (orbital structure defined by the discrete hyperbolic shapes, within which the condensed material is contained). That is, the stable solar system can be interpreted to be evidence which proves that these ideas are correct.

Thus, the idea that "geometry dominates (or is more important than) the traditional authority of the partial differential equation of a physical system, defined in a context of materialism, non-linearity, and (undefined) randomness."

The decreasing spectral sequence, as dimension increases (the current, fixed, overly authoritative viewpoint)

On-the-other-hand a decreasing spectral sequence (as dimension increases) will not have any stable properties "to speak of," except for the material components themselves, and in such a decreasing spectral sequence (as the dimension increases), the high-dimension containing metric-space should be continuous for all the material components, but since this is not observed (that is, we do not experience the fact that we are in an 11-dimensional space, when we confine our observations to a low-dimension subspace of this 11-dimensional containing space) so these small spectra are curled-up, eg string-theory, so that the idea of materialism is maintained . . . ,

where the existence of material, ie materialism, is related to the fact that both the 1-dimensional discrete hyperbolic shapes and 2-dimensional discrete hyperbolic shapes are close to the same size, while the size of 3-material-components, in our metric-space, are the size of the solar system, thus, in our containing metric-space, the atoms and molecules (which are 3-dimensional discrete hyperbolic shapes) are condensed material, which have relatively smaller energy ranges of stability, than do nuclei,

. . . , and with a decreasing spectral sequence (as dimension increases) (in a metric-space whose subspace structure of a higher-dimension containing space is not detectable), one is left with descriptive structures which are unstable, indefinably random, and quantitatively inconsistent, where logic becomes irrelevant, as is now the "current way" in which quantitative language is now being organized to describe the observed material properties. Furthermore, this organization of precise language is held onto with stifling authority, and this is because the probabilities of particle-collisions are used in the randomly-directed transitioning system (of a nuclear reaction), so as to identify properties of rate and energy release of such a reaction (where a reaction is randomly transitioning system, wherein material interactions are modeled as collisions).

That is, the current descriptive context assumes that the spectral sequence of the containing structures decreases as the dimension increases, so all of the "many-component systems" (but still containing relatively few components), become "too complicated to describe," if one is trying to describe their stable patterns based on material interactions, which, in turn, are based on random, non-linear, (but nonetheless geometric) particle-collisions.

Other new attributes of description

In the new description there are new sets of operators, or properties, but the geometry of the description becomes the dominant attribute of the description, often this is because the new geometry describes (identifies) the new context, in a most dramatic way.

1. There exist conformal factors defined between dimensional levels (ie physical constants), as well as between different subspaces of the same dimension,

2. (perhaps) conformal factors can be defined between toral components of a discrete hyperbolic shape (though this might simply be relatable to the existence of particular varied discrete hyperbolic shapes which can exist based on the reflection group structure of the lattices at vertices which can also be related to the various possible sizes of the faces of the fundamental doamin)
3. Discrete Weyl-transformations of angles can be defined between toral components of discrete hyperbolic shapes (these Weyl-transformations define "allowable folds on the lattice" of a discrete hyperbolic shape).

The derivative becomes a discrete operator in regard to:

1. Time intervals defined by the periods of the spin-rotations of opposite metric-space states
2. Dimensional levels
3. (possibly) Between toral components of a discrete hyperbolic shape

This new descriptive structure defines material interactions using many aspects of Newton's law of inertia. Thus, there are also the "usual types" of 2nd order material interaction math (or equation) patterns:

1. elliptic (or orbital)
2. parabolic (or free, or angular momentum)
3. hyperbolic (waves with physical properties, or collisions of material components)

But these types of interactions depend on the spectral values, and they apply in a more restricted geometric context, but now it is within a more diverse many-dimensional construct where quantitative descriptions are to be guided by the geometry (both shape and size) of any of the dimensional levels.

Note:

It should be noted that with both materialism and the belief that the spectra of higher-dimensions need to identify a decreasing spectral sequence as the dimension increases, means that conventional science tries to sort-out the spectral properties, of quantum systems composed of "five, or more," charged components, by attaching a 1/r potential (also associated with spherical symmetry), for each charged component, where the assumed spherical symmetry, of each 1/r term, is "deduced" from the assured-ness that the random-particle model of material interactions is to be spherically symmetric in each dimension (if the dimensions are not curled into small shapes), because of fundamental randomness, so if particles are being emitted from a "force-field source," then the field-particles will emanate in any random direction, and then this will define a spherically symmetric force-field.

But such a model . . . , (or such an assumption that a "physical description of material interaction is to be based on fundamental, indefinable-randomness, associated to random particle-collisions," so that particles are emitted in random directions from a "force-field source" so as to cause a spherically-symmetric force-field, in all dimensions) . . . , is far from the truth.

In fact, material interactions are mediated by a toral shape, associated to action-at-a-distance toral shape defined for each (small) time interval ($\sim 10^{-18}$ sec), where this time interval is defined by the period of the spin-rotation of metric-space states, so that the tangent structure of each interaction-torus (defined for each time interval) is related to a 2-form force-field, which, in turn, is related to the geometry of the fiber group (of the containing space of the interaction torus), and this geometry (of material interaction) only results in a spherically-symmetric force-field for SO(3), ie the interaction torus which is contained in R(3,0).

Quantum randomness of point-particles

The new model of material interactions, defined for small components whose properties of "being neutral" or "being charged" changes rapidly, results in these small material-components defining Brownian motions, where these Brownian motions, in turn, determine an appearance of quantum randomness. Furthermore, the vertices of the fundamental domains of the circle-space shapes, define a distinguished point on the circle-space shapes, about which material interactions are defined. This, in turn, creates the illusion that material-interactions are interactions between point-particles.

The commercial world is related to a fixed stationary way of behaving or acting, a commercial structure is a very narrow context, based on a limited range of creativity and a fixed way in which to use material resources. The power of business monopolies depend on society not changing how it uses the material resources a business monopoly supplies to a society. The law is supporting this type of narrowness, essentially based on property rights and minority rule (creditor vs. debtor, smart vs. stupid, etc), and it supports such selfish actions with great violence. In fact, the economy is tied to a fixed narrow way in which to live and create, and this model of monopolistic economies is being used as a means to conquer ever larger populations, but it is being put into-place by means of extreme violence and coercion (often an economic coercion).

Does one want a society to be based on a fixed way to use material, and a fixed way in which one is to serve the material based, and fixed structure of society, and a fixed overly authoritative organization of descriptive knowledge, so that this type of power, and associated narrowly defined knowledge, depends on expansion in the form of an ever greater exploitation of particular types of material (usage)?

Dimensions*
Dimensions, shape (holes, stability), size, measurable description, and spectra

The dimensions of the set of 2-forms defined on an n-metric-space is also the dimension of SO(n), where dim(SO(n))=dim(spin group of SO(n)).

In the new descriptive construct the geometry of the 2-forms in an n-space is related to the local (tangent) geometry of an (n-1)-dimension "discrete Euclidean shape," ie an (n-1)-torus, which in turn are related to the geometry of the SO(n) fiber group of the n-base-space, in regard to the local coordinate changes of the positions of the interacting materials (determined by both the 2-form force-fields and the local coordinate transformations) where these local coordinate changes are associated to each discrete time interval, in turn, defined by the spin-rotation of metric-space states ($\sim 10^{-18}$ sec).

Thus dimensional relations can be found between the various possible spaces related to a material interaction and the associated spectral properties (ie properties of spectral-size) of the different containment levels.

The list of the dimensions of the 2-form spaces of the different dimensions up to Euclidean 6-space . . . , since the spectral sizes of discrete hyperbolic shapes of dimension 6 and up to dimension-10 are infinite extent, so that shape loses its intuitive sense of bounded-ness, thus the geometry of the interaction-shapes are difficult to identify . . . , are as follows:

$$2C2=1, \ 3C2=3, \ 4C2=6, \ 5C2=10, \ 6C2=15$$

From this list . . . ., and along with some information about the geometric structure of the SO(n) fiber group, such as SO(4) = SO(3) x SO(3) . . . ., one can make some determinations (or guesses) about the nature of force-field interaction geometry. The relation of the fiber group geometry to the 2-form geometry to the geometry of the containing n-dimensional metric-space, or possibly the geometry of an (n-1)-dimensional metric-space associated to the: subspace, material, and dimensional structure of our containment set.

That the 2-form on n-space has the same dimension as SO(n) means that there needs to be a geometric-vector relation to force-fields acting on (n-1)-shapes contained in n-space. The "charges" on (n-1)-shapes are the (n-2)-flows (or (n-2)-faces of an (n-1)-shape) so the geometry of SO(n) is related to the geometry of (n-2)-flows on (n-1)-shapes in turn, contained in n-space.

1. For 3-shapes contained in 4-space it is useful to identify the normal to the 3-shape with time instead of the 4th spatial-dimension, so the geometry of the 2-forms is related to 3-space, while

2. for the 4-shape contained in 5-space so the geometry of the 3-flows (or the 2-forms) is related to 5-space, but

3. for 5-shapes contained in 6-space the 4-flow (or 2-form) geometry is again dimensionally more convenient to treat is being related to the toral 5-shapes again letting the normal-direction of the 5-shape to be associated to a time-direction, etc.

In regard to odd-dimensional spatial subspaces, which are the material-component containment spaces, the dimensional properties of the 2-forms are related directly to the local tangent geometric properties of the containment space.

In regard to even-dimensional spatial subspaces, which are the material-component containment spaces, the dimensional properties of the 2-forms are related (directly) to the local tangent geometric properties of the interaction toral shape which is 1-dimension less than the dimension of the containment space, with the normal direction to the toral shapes being associated to a time-dimension (instead of the extra dimension of the containment space).

The most important geometric relation in regard to our "3-space experience" is that the geometries of the 3-space and the 4-space can have a fairly complicated relation with one another, especially since (or if) the spectral-size of 4-space seems to be the size of the solar system, thus the geometry of the 3-tori do not manifest, in regard to our human size-scale on earth, in 4-space, but rather are related to the 3-space in our experience (due to their spectral and subsequent resonance relation with another 4-dimensional subspace, a subspace which possesses "smaller spectral values" than does the 4-space within which we are contained, where our planetary-orbits are defined by the large 4-spectra of our 4-subspace containment). That is, our 4-space material geometries are in fact 2-surfaces contained in a 3-space, due to issues of spectral size in our subspace for 3-shapes, thus the 2-forms defined on a 3-torus (contained in 4-space) which define the interactions of 2-surfaces, which model material components in 3-space, must be related to 3-space. This is possible since SO(4) = SO(3) x SO(3) so for 4-space, (x,y,z,w) can be separated into a pair of 3-spaces (x,y,z) and (x,y,w) subspaces of the 4-space, and thus relatable to 3-space and to a 2-plane in 3-space.

Thus, there is a geometric relation of the condensed material of our 4-subspace to the spectra of our 3-space containment, allowing a 6-dimensional electromagnetic-field, etc.

Whereas for a 2-form defined on a 4-torus contained in 5-space, the 5-space relates to either a 4-field (or a 4-vector) and a 2-form of 4-space structure, or a pair of 5-fields (or a pair of 5-vectors).

Whereas a 5-torus defined in a 6-space can be related to three 5-vectors, or a 5-vector and a 2-form of 5-space, etc. or a pair 2-forms defined on 4-space and a 3-vector defined on 3-space (or a 2-form defined on 3-space), or a 2-vector-field defined on 2-space, the intersection space of a pair of 4-subspaces defined on 6-space.

The geometry of the 2-tori in 3-space, ie the vector-fields defined on the 2-torus which is contained in 3-space is, thus, associated to SO(3) (which has the geometry of a 3-sphere), is the geometry which causes inertial interactions to be spherically symmetric in 3-space, while the geometry of 4-space and SO(4), along with the spectral sizes of 3-dimensional discrete hyperbolic shapes in 4-space seems to also allow for some aspects of spherical symmetry for the inertial properties of material interactions being related to 3-tori, which are contained in 4-space.

The partition

Partitioning the dimensional levels by defining a discrete hyperbolic shape for each subspace of each dimensional level.

On the other hand, the new viewpoint requires that a catalog of spectral values be found for the different dimensional levels, which are related to (or modeled by) bounded "discrete hyperbolic shapes" of which the 5-dimension "discrete hyperbolic shape" is the last such "discrete hyperbolic shape" which can be a bounded shape.

This spectral-catalog would be similar to the periodic table of the elements.

Thus there is:

The spectra related to subspaces as follows:

$[11C1 + 11C2 + 11C3 + 11C4 + 11C5] = [11 + 55 + 165 + 330 + 462]$ (respectively) = 1023,

. . . , where 1023 is the number of subspaces of the various dimensions in regard to the 11-dimensional containment set, where each subspace of each dimensional level can be associated to "discrete hyperbolic shapes," whose shapes might be bounded, so as to define a specific (well-defined) spectral set for each subspace.

This number, 1023, may also be interpreted to be the number of 1-spectra which compose the finite spectral set upon which all material systems (or bounded discrete hyperbolic shapes) depend for their existence by means of resonance with this finite spectral set.
This spectral set will define sequences of spectral size where the sequence is defined as the dimension increases, so that these spectral sequences may be increasing (allowing for stable material structures) or decreasing which would imply that order to material systems would be much more limited, where an increasing spectral sequence allows the lower dimension shapes to be contained in the upper-dimensional shapes.

The time subspaces in higher-dimensions

Consider the metric-spaces whose metric-functions have constant coefficients, R(s,t), such as R(3,0) [Euclidean 3-space] and R(3,1) [space-time], where s is the dimension of the spatial subspace, and t is the dimension of the time subspace of R(s,t).

In such spaces the time-dimension changes when a new material is added to the structure of existence.

For example, one can hypothesize that for R(2,0) there is only the material property of inertia, which exists as a 1-dimensional discrete shape, when charge is added into the descriptive context, it is a 2-dimensional discrete hyperbolic shape, and then it is contained in R(3,1). Thus, one might hypothesize that there is a new type of material to be contained in R(4,2), namely, the odd-dimension 3-shapes which possess an odd-number of holes in their shapes, ie an odd-genus, but nonetheless, the inertial properties of the material interactions, ie changes in spatial position, of the material contained in R(4,2) would be in R(4,0) space, so "in general" inertial changes of interacting materials' positions in an R(s,t) space would be identified in R(s,0) Euclidean space.

Sizes of discrete hyperbolic shapes

The spectral size sequence, defined as the dimensional levels increase, can be increasing, decreasing or neither increasing nor decreasing, but the sequence of increasing spectral sizes allows the associated material systems to be ordered and stable, while decreasing spectral sequences do not allow for the order of material systems, or decreasing spectral-orbital sequences only allow for a limited (stable) spectral-orbital order to exist.

There is also the issue of infinite-extent "discrete hyperbolic shapes," where the last existence of bounded discrete hyperbolic shapes existing in a 5-dimensional discrete hyperbolic shapes contained in a 6-dimensional hyperbolic metric-space, where the idea that lower-dimensional shapes are contained in a larger higher-dimensional metric-space which (also) posses shapes would only exist (or be possible) for an increasing spectral sequence.

However, there are discrete hyperbolic shapes which have the property of being "infinite-extent" (or unbounded shapes) within (or for) all hyperbolic metric-spaces.

The neutrino is best modeled as an (semi) infinite-extent discrete hyperbolic shapes, due to its, apparent, zero-mass, where its property of being semi-infinite-extent allows a neutrino to be both infinite-extent and to possess a spatial position (associated to the atom which the neutrino is a neutrally charged component), but such an infinite-extent model of a discrete hyperbolic shape (at low dimensions) allows the "infinite-extent property" to, subsequently, be contained in a higher-dimensional metric-space shape, which is bounded, and this allows the low-dimension material systems, wherein these low-dimension systems possess components like neutrinos, whose unbounded geometric property comes to be contained in a bounded metric-space shape.

The infinite-extent properties of neutrinos can be contained in a bounded metric-space. Thus, the systems which contain neutrinos can still be contained in a bounded spectral (or metric-space) set, thus metric-spaces of higher-dimensions can carry within themselves the finite spectral-set which defines a "world of experience."

In turn, this would allow other higher-dimensional infinite-extent discrete hyperbolic shapes to determine (by containment) an "arbitrary" bounded, (relatively) low-dimension spectral set, so that this spectral set can include atomic-type systems, whose components are neutrinos.

However, such arbitrary bounded spectral sets require a higher dimensional experience, ie higher-dimensional containment.

Blurbs

I

If a fiber group's (primarily isometry, or unitary) local matrices are always diagonal (or commutative) on the global coordinates of a shape's locally-identified vector-field then the geometry is stable, quantitatively consistent, and simple enough to be considered to be a good candidate to be an element in a set of shapes which is to be used to model existence in a many-dimensional, macroscopically-geometric context so that stable spectral-orbital properties of physical systems [which exist at all size scales, eg nuclei to solar systems] can: generally, accurately (with sufficient precision), and practically usefully; be described (where a stable geometric description is a practically useful description).

Basically these simple commutative shapes are the stable circle-spaces, defined within metric-spaces of non-positive constant curvature, where (but) they are also primarily the "discrete hyperbolic shapes," though metric-spaces which are different from R(n-1,1) [which is either a general space-time or an (n-1)-hyperbolic metric-space] can be considered, such as R(s,t), wherein new higher-dimensional material can be (newly) defined, and a higher-dimensional model of a life-form can be defined.

Furthermore, the metric-spaces possess intrinsic properties, which exist as opposite pairs of metric-space states. This leads to complex coordinates and unitary fiber groups, as well as fiber spin-groups and the spin-rotations of opposite metric-space states.

The different dimensional levels are to be modeled as discrete hyperbolic shapes, where up to and including hyperbolic dimension-5, these shapes may be assumed to be bounded, and assume there is a shape which is maximal for each dimensional level and for each subspace of that same dimensional value. This is the basis for defining a finite spectral set, associated to the over-all high-dimension containing space, ie an 11-dimensional hyperbolic metric-space, since the last existing infinite-extent discrete hyperbolic shape has hyperbolic dimension-10.

Thus, the different dimensional levels of the higher-dimensional (over-all) containment space are partitioned by open-closed shapes, which are associated to "rectangular-like" fundamental domains in hyperbolic space.

The "rectangular" simplex fundamental domains for the metric-space within which we and our solar system are (both) contained are 4-dimensions, upon which our metric-space is a 3-flow, and this fundamental domain would be the size of the solar-system, these block-like fundamental domains (or equivalently circle-space metric-spaces) would appear open, and thus the incoming light (from outside the solar system), where light is modeled as an infinite-extent discrete hyperbolic shape, would pass through the blocks from far away, while on-the-other-hand the apparently infinite extent neutrino discrete hyperbolic shapes which define the 2-faces of the "discrete hyperbolic 3-shapes," which model electron clouds, would (could) have its infinite extent defined by being bounded by the bounded discrete hyperbolic 4-shape metric-space which is also a model of the discrete hyperbolic 4-shape of the solar system. Thus, one has that when one looks away from the rectangular 4-simplex of the solar system (or looking out to the universe) one would see light coming in from the distant places, while looking within the rectangular 4-simplex one would see the closed boundaries of bounded discrete hyperbolic 2-shapes (and/or possibly bounded discrete hyperbolic 3-shapes) modeling the material components contained in the 3-flow model of our containing metric-space

In this context, the way to use higher-dimensions is to model the different dimensional-levels as the stable circle-spaces, and the way in which to get a (math) pattern which is associated to the existence of stable spectral-orbits is to multiply the different dimensional levels by constants, ie the nature of physical constants, so that the spectra . . . , of the adjacent next higher dimensional level . . . . , increases in value, where this allows the lower-dimensional shapes to be contained within a stable shape, where, in turn, this shape can define a relatively stable orbital geometry, for the material which is contained within the shape (ie the metric-space is a shape).

The idea of materialism as well as of quantum physics and particle-physics (both consistent with the idea of materialism) assumes that the higher dimensions are either continuous, in an open context, and that material reduces to point-particles and the spectra (associated to the physical systems which these point-particles occupy) decrease in value (spectral size) as the dimensions of containment increase, where in both cases (continuous and discrete particle context) the force-fields are assumed to be spherically symmetric (though perhaps not always inverse square), but this spectral structure implies that no stable spectral-orbits exist (that is, where does one now (2013) find valid descriptions of the general and spectrally relatively stable: nuclei, atoms, molecules, crystals, and solar-systems?).

The derivative operator can be re-defined as a discrete operator, defined between: (1) dimensional levels (2) time intervals, and (3) toral components, wherein Weyl-transformations can be used to identify stable orbital shapes associated to a set of angularly-deformed "discrete hyperbolic shapes," which in turn, define envelopes of orbital stability for the condensed material components, which these metric-space shapes might contain.

Blurb II

Topic list

The stable properties of general sets are related to precisely identifiable properties of physical systems where these precisely identifiable properties of physical systems exist at all size scales; from nuclei to solar systems, and these fundamental systems have no valid descriptions, eg Hartree-Fock etc.

List of fundamental topics concerning these new math-science ideas:

From
1. precise language (build new languages at an elementary level of assumption, and context, and interpretation, in order to broaden the capacity to create; Godel's incompleteness theorem can be interpreted to mean add more assumptions or it can be interpreted to mean review and alter one's precise language at the level of assumptions),
    Does language fit into a "fixed scene" (which is implicitly assumed to be moving toward some absolute truth) which is similar to the idea that an authoritative math language is always relatable to ever more complicated instruments, but in an industrial society all the instruments are built so as to be based on the same principles (electromagnetism or other classical theories which allow stable patterns which are controllable), so that when the instruments are adjusted in their "complicated context," they either improve or they reach their limits of performance, likewise physical theory and math patterns are carefully adjusted within the realm of fixed principles, but the math and the physics do not work, and the careful adjustments either do not work or they have already reached their limits of performance (and thus, they have lost their relevance), to
2. quantitative structure
    (continuum, quantitative consistency, comparisons [length, time, material {particle-number or density}] or spectral values [momentum, energy {single-valued in regard to holes in the shape of the domain}, as well as an assumption of fundamental randomness [which apparently, for physics, cannot be defined as a valid elementary set of random events]]), to

3. failing descriptions (for valid descriptions of the stable spectral-orbital physical systems which exist at all size scales, rather the descriptions are of fleeting and unstable patterns), to

4. the stability of mathematical patterns (how can their stability be established), to

5. the proper role of geometry (stability of patterns, a measuring context, a useable context), to

6. interpreting observed patterns (Are the incessant examination of the properties of elementary-particles best interpreted to mean that existence is higher-dimensional and unitary? Furthermore, the existence of high-energy cosmic-rays, as well as an apparent property of dark-matter, are best interpreted to mean the existence of a large-scale spectral structure which exists in higher-dimensions), to

7. fundamental structure of math, functions vs. numbers, (algebraic equations and partial differential equations, is math really about finding the stable geometric confines related to the existence of stable describable math patterns or stable measurable properties).

For example, How to determine the structure of the derivative?: {From a derivative interpreted as

1. an operator on a function space, to

2. a model of locally measured properties within a sufficiently determined containing space, to

3. a discrete operator in a finite math structure determined by (discrete) stable geometry in a newly organized, many-dimensional, containing space with new interpretations, which can lead to many more possibilities (the functions in the function space are [now] the discrete shapes).}

Math is about quantity and shape, . . . , but when shape is considered primarily in terms of measurable quantities and functions (or functions and their coordinate domain spaces), it (the shape which is being described) is most often non-linear and unsolvable, except locally (the fiber diffeomorphism group is locally invertible), but non-linear quantitative properties are chaotic, so the local pieces of shapes cannot be put together in a quantitatively consistent manner, and . . . , thus it (shape) fits into an indefinable random structure, in turn, to be fit into function spaces which, in turn, depend on indefinable sets of spectral functions (whose associated operator structures, or measurable properties, usually do not commute).

However, Thurston's geometrization finds that the variety of stable geometric systems depends most strongly on the discrete hyperbolic shapes, while Coxeter found that the last bounded discrete hyperbolic shapes are 5-dimensional, and the last set of infinite-extent discrete hyperbolic shapes are 10-dimensional.

Consider:

1. The stable properties of general sets of precisely identifiable physical systems which exist at all size scales; from nuclei to solar-systems, and these fundamental systems have no valid descriptions, eg Hartree-Fock, general relativity, etc.
2. Basing measurable descriptions of systems which possess stable properties on a quantitative set which is determined by a finite spectral set, put into a context of stable (linear, solvable) geometries, contained in a higher-dimensional set which is also organized around stable shapes, ie organized around stable circle-spaces. Namely, a finite number of discrete hyperbolic shapes contained in an 11-dimensional hyperbolic metric-space.

Where as consider the following patterns of spectral values associated to a containment space, as the dimension increases there is either:

(a) an increasing finite sequence of spectral values with an upper bound, or
(b) a decreasing finite sequence of spectral values with a lower bound.

That is, (a) is the new proposal, while (b) is essentially what is assumed today (2012) when guided by the principle of materialism and the principle that material reduces to a set of fundamentally random elementary-particles, where (b) is helpful in regard to building nuclear weapons.

The consideration of (a) is about circle-spaces, which in turn, are associated to complex-numbers and subsequently complex-coordinates, and, in turn, the relation of circle-spaces to: quantitative consistency, stability of measured patterns, and the relation of the shape of a circle-space to both spectral values and the geometric properties of spectral constructs (multiplicative constants, and Weyl-transformations [or allowable folds] on the lattice of a discrete shape, as well as action-at-a-distance {or non-local} structures of material interactions) and the subsequent rigidity of measurable structure (eg analytic [complex] function structure), and its relation to the stability of pattern, and the existence of either analytic continuation, or the controllability of linear solvable systems.

Though the apparent randomness of point-particles, may appear to dominate observed material phenomenon, it is really the confinement to a set structure which allows for stable patterns to: exist, be measured, and used, ie controlled (and/or formed). This is about how material components relate to either a stable math structure of their own, or a stable math context needed for reliable measurements (existing in stable orbits or existing as free components in a metric-space, which in turn is about the math structure's, eg spectral values of the different dimensional levels, as well as the energies of an interacting context, which determine condensation

758

vs. resonances with the existing constructs of the stable circle-spaces, where resonances allow for a system's stable math structure to form within itself).

3. Life: The odd-dimensional discrete hyperbolic shapes which also possess an odd-genus, when their faces (or spectral-flows) are occupied then they are charge unbalanced, and would naturally oscillate and generate energy, ie a simple model of life.
4. Mind (related to the spectra which can be contained within a maximal torus of the fiber group)
5. Intent (directing the flow of energy within a cognizant system)
6. Creativity (creating and expanding the possibilities of existence, itself, in a direct manner, or how the instrument of life can be used) Etc.

The truth of a precisely identified pattern should be determined by the relation that the pattern has to practical usefulness, generality of application, so as to provide a wide range of accurate descriptions, made to a sufficient level of precision, based on simple easily applicable laws, or based on the "correct" context which can be seen to limit possibilities, so the information is accurate and the context is practically useful. That is, math truth, not necessarily a context about so many independent variables defining containment so that the description needs the same number of independent equations. That is understanding the context of existence in the context of its stable shapes which organize the different dimensional levels of existence.

There are all types of social issues concerning the expression, and consideration of new ideas, where these issues get mixed up in the way in which tradition and authority dominate the published expressions of a society, but more strikingly the structures of investment and the condition of wage-slavery which afflicts the development to of knowledge, and the subsequent set of lies about personal value and worth as well as competition, where both worth and the competitive game are narrowly defined by the investors, and these forces constitute the social context, and it is a context which opposes new ideas and new expressions, the investors require that knowledge serve the creative interests of the investors, thus such a system grinds itself to a halt, because of the highly enforced narrow viewpoints. It should be noted that, these social structures, which are based on inequality, are created and maintained by means of extreme violence, this emanates from the justice system, and from a militarized management system.

One needs the math properties of stability:

1. Linear (partial differential equations)
2. Metric-invariant, with non-positive constant curvature, where the metric-function has constant coefficients,

3. The shape must be parallelizable and orthogonal at each point of the global shape of the containing coordinate system, ie the partial differential equation is separable so that the locally linear matrix structure associated to derivatives and differential equations is always commutative, or diagonal.

These properties of stability are about the properties which the discrete Euclidean shapes and the discrete hyperbolic shapes, possess, as well as being possessed by the (discrete) shapes in R(s,t), where space-time is R(3,1).

Other properties:

I.   principle fiber bundle with metric-space base spaces and isometry and unitary fiber groups
II. the metric-spaces possess properties, eg properties of position and the property of a stable pattern. This leads beyond the isometry Lie groups to both spin-groups and unitary groups
III. Both the metric-spaces and the material components are discrete hyperbolic shapes, essentially modeling adjacent dimensional metric-spaces, ie a higher-dimensional context is not based on continuity of the lower dimensions until one reaches the discrete shape which defines a particular dimensional level, a 3-shape does not see a 7-shape as part of its containment context.
IV. The containing space is an 11-dimension hyperbolic metric-space, a hyperbolic metric-space is chosen since the discrete hyperbolic shapes are so stable in both their shapes and their spectral properties. There are constant factors defined between dimensional levels and between toral components of discrete hyperbolic shapes
V.   The derivative can be defined as a discrete operator between

     (a) dimensional levels, between
     (b) time intervals, and between
     (c) toral components of discrete hyperbolic shapes by way of the Weyl-transformations.

VI. Life, mind, etc
     As well as:
          Holes in metric-spaces; material either resonates with a shape so as to occupy a hole, or spectral-flow, defined within a discrete hyperbolic shape, or it condenses and orbits around the holes (spectral flows) defined by a discrete hyperbolic shape which is much larger than the size of the condensed material.

Increasing or decreasing spectral-size sequences

[The properties of a physical/mathematical construct of an increasing spectral sequence, as the dimension increases, can be organized, based on: shapes, sizes, and formation processes (or formation structures) so as to cause an observer within a particular dimensional level to not perceive the higher dimensions. Furthermore, in an n-metric-space the observer mostly sees only the (n-1)-material components. In a decreasing spectral sequence, as the dimension increases, which is the assumption of both particle-physics and string-theory, it is assumed that there exists the property of continuity between all dimensional levels, and all subspaces, so the spectral sequence must decrease, as the dimension increases, so the observer cannot see the affects of the higher dimensions, thus the higher-dimensions are curled into small shapes which possess small spectral values, but such an assumption of continuity between dimensional levels is not needed, and it is certainly not necessary.]

Material condensation (often due to the material sizes which are defined in a particular subspace of a particular dimensional level, where the condensed material is smaller than the material-component sizes, ie discrete hyperbolic shapes, of the particular dimension and subspace [of that same dimension])

Only valid model expressing the principle of inertia, as identified in general relativity, is "orbits of condensed material on the discrete hyperbolic shapes of the condensed material's containing metric-space."

A main issue is: "finding one of the finite spectral sets, which identify the stable context of an 11-dimension hyperbolic metric-space."

Empty of content (Apparently, No stable patterns exist)

The content, or focus, or motivation of today's science and math languages used in professional (peer reviewed) journals are the elaborate and complicated techniques that are either unrelated to the observed patterns, ie the observed stable patterns of material systems, or these techniques are unrelated to reliable descriptions of stable, well-defined, measurable patterns, (where well-defined patterns are: shapes, reliable quantitative relationships, observed stable and precise properties, laws which are truly applicable to a wide range of different contexts so as to provide relatively accurate solution functions (or spectral sets), and the math-physical conditions which allow for reliable measurements). The dogmatically authoritative literature of the science and math communities has become devoid of content. It has become elaborate complicated descriptions of a world of illusions. Its main social function is to define an authority which identifies inequality, yet its descriptive context is primarily formulated to develop weapons and to allow ever more control over communication (channels).

Today's professional math-scientists are not describing stable patterns, since the context of the authoritative descriptive precise language, and its associated techniques, are unrelated to describing stable identifiable patterns, rather the intellectual content of their descriptive focus is only about describing complicated elaborate calculating techniques which are unrelated to the observable order of the world.

The context of the professional dogmatists is defined by:

1. indefinable randomness (the elementary random events are not stable and they are not well-defined),
2. non-linearity (quantitatively inconsistent), and
3. contained within a set of measurable properties (or measurable coordinates) which have the properties of a continuum (a very large set, high cardinality), and either
3a. geometries or
3b. functions spaces

both of which whose properties are non-commutative, where

3b1. the functions-space spectral techniques associated to random descriptions are focused on local spectral properties, as opposed to
3b2. solution functions which provide global system information, and

4. the description has the property of being about random descriptions of the relatively few components which compose a stable system, the logic of this construct is inconsistent.

Furthermore, it is a description which is logically inconsistent where convergences . . . of geometric properties which are based on random descriptive structures . . . where such geometry and randomness are both defined upon (into) the domain space's continuum structure.

Furthermore, there is an improper focus on "as to what constitutes" a valid frame of descriptive reference for a physical system:

I. A valid "frame" for the containment of a physical system's properties is not about coordinate frames associated to motions as in general relativity, this unduly narrows the context within which a physical system obtains its ordered-form, but rather

II. Physical systems are determined by, and contained within, the shapes (both macroscopic and microscopic shapes, which are defined for all dimensional levels) of metric-spaces, where these discrete (isometric) shapes are models of metric-spaces, and where these metric-space shapes are needed in relation to a finite spectral-orbital set observed for the observed material systems at all size scales, where a finite spectral-orbital set is to be defined on "an over-all" high-dimension containment set (an 11-dimensional hyperbolic metric-space), in which, the partition of the dimensional levels by shapes, defines, either "increasing or decreasing sequences of spectral-sizes as the dimension increases" within the high-dimension containment set . . . ,

. . . , so that, within the partition of the dimensional levels by spaces with stable shapes, there is an associated, and prevalent, stable set of holes on these dimension-partitioning shapes.

Note: The genus of such a shape is the number of holes defined on the "discrete isometric shape," upon which the "holes in space" are prevalent, but these holes are not seen by the observer who is contained within a (particular) dimensional level.

Furthermore, there is an inability (of a learner) [in a society dominated by monopolistic businesses] to question the traditions and authoritative structure of "what has come to be thought of as a discipline," but dogmatic authority, ie religion, is a fallacious mental context "in which to develop both knowledge and a descriptive language," where the descriptive language should not be fixed, but rather a change should (always) be considered in regard to a precise descriptive language. That is changes in a precise descriptive language "should be the main focus" in regard to the intellectual context (or condition) of causing changes in descriptive knowledge for the better.

The both overly-authoritative and fixed intellectual state (of our US society) is a result of the experts "need to be subservient to the process of 'peer review,'" which protects, or ensures that, the knowledge fits into the interests of the monopolistic businesses which dominate the society.

That is, one needs to question:

1. the continuum,
2. indefinable randomness,
3. non-linearity,
4. materialism,
5. what are the fundamentals of a differential equation, and
6. the context (of knowledge associated to business interests) of newly forming systems . . . . , after a stable system has become unstable and has broken apart so as to transition to a new stable state by a series of many-component collisions (where business concerns are interested in the probabilities of collisions and its relation to rates of reactions).

These dogmas need to be questioned if one wishes knowledge to develop and change.

Yet one must list the places and contexts within which it is a valid descriptive context: (see below for more details)

1. It is a description which is relatable to a system whose initial conditions, and initial properties are carefully put-together so as to be a system which is easily broke-apart, so as to form a transitioning system which is chaotic, so that the rates of reactions (in this context, based on component-collision probabilities) are determined by cross-sections of the broken-apart components, where these cross-sections determine the rates of certain aspects of the (a) reaction, and

2. They are descriptive contexts which relate a limited set of metrically measurable (observable) properties to a feedback structure, which is mostly associated to the critical-points and limit-cycles of a non-linear (usually classical) partial differential equation, where the range of relevance of the differential equation is difficult to determine or to control. Furthermore, the initial or boundary conditions of this type of a system relate to the properties of the descriptive context (or properties associated to the solution) of the system's differential equation in a chaotic manner.

These contexts identify structure related to (1) nuclear weapons and (2) guiding missiles and drones.

Furthermore, there are the overly general contexts, wherein the experts consider holes in shapes, but they view holes in shapes as arbitrary structures, which most often, the experts, relate to complicated shapes and distortions of very general, but unstable, geometries, and these overly general contexts also need o be questioned, though seeking generality can have great value, it should not be a dogmatic command, since it is the limitations which allow useful information about patterns to be found.

Though there is a great imagination, by the experts, for great generality, but nonetheless there is an unimaginative viewpoint about "how holes in shapes" can be (might be) related to physical descriptions.

The nature of shapes, and shapes with holes in themselves, identifying valid descriptive frames (or frames of containment) through which the spectral-orbital properties which characterize physical systems can be modeled.

Holes in space affect (or interfere) with single-valued-ness of values determined from integral operators, yet the relation that holes have with stable spectral values is seldom considered. Namely, the discrete hyperbolic shapes, placed into a new construct, which is to be used for a new precise descriptive language.

One might note that:

The observed stable, precise, patterns of physical systems are associated to finite properties, eg bounded-ness and/or the finite number of a physical system's components, eg atomic-number, and these stable physical system properties are fundamental and observed features of a reliably measurable context associated to the observers of physical patterns. This implies both the existence of stable patterns which allow reliable measuring, (or which are associated to the context of measuring (for an observer)), and the existence of stable-controllable patterns associated to a set of fundamental physical systems which possess stable features of "what may, or may-not be" "material" systems, eg nuclei, general-atoms, molecules and their shapes, crystals, solar systems, dark-matter (ie orbital properties of solar-system's in galaxies) etc, the physical patterns upon which the relatively stable aspects of our life experiences depend, and upon which our mental constructs also depend.

Thus, science and math are about identifying stable, quantitatively-consistent, math patterns which are generally applicable to these stable, measurable, and apparently controllable, physical properties so as to result in descriptions of these patterns which are accurate (to sufficient precision), and general so as to be able to describe the observed stable physical patterns of existence, so as to provide a context for practical usefulness, ie measurable and controllable, so that one can: measure, fit together (or couple), and interact with these various patterns (using the natural structures of these patterns, eg life-forms and its coordinated chemical properties (but, apparently, coordinated by an unknown structure), ie not feedback mechanisms nor carefully prepared structures so as to cause reactions), so that this descriptive knowledge can be related to "practical" creativity (as opposed to literary creativity, essentially associated to a world of illusion, ie a world without stable features).

The violent nature of today's society

The entry into science and math of a set of overly authoritative dogmas (essentially, defined by the authority of the peer-review process), which are overly protected (dogmas), where the protection is accomplished by means of extreme mental, social violence, and the extremely violent-intellectual demands of the dogma, which are required for a person to be admitted into the realm of being a "valid authoritative person within society." It is authoritative dogmas which also define the image of "true science and math" and this adherence to narrow dogma turns the wage-slave scientists and mathematician into a protector of a fixed viewpoint of high-valued knowledge. However, this overly demanding authority comes to be knowledge which serves only the narrow, monopolistic, dominating, business interests (the interests of the owners of society) within society.

The social-instrumental structure with defines both high-value and an "authoritative truth" for all of society is the media, and the ideas which are authoritative are the ideas which are to be expressed on the media, all of the media (including the alternative media).

Most of what the professional (peer-reviewed) science and math communities do, is marginal at best, and it is essentially irrelevant, where its focus is on creating (in a literary sense) [not in a practical sense of creativity] elaborate methods which are associated to unstable contexts which possess only fleeting patterns (measurably distinguishable, but in an unstable context) which are only useful within a larger global context (than the structure of observed stable measurable patterns) of measuring abstracted components of an (unstable) arbitrary context (a context within which stable measurable patterns [associated to the partitioning-components of the measurable attributes] do not exist).

That is, distinguishing features in an arbitrary and abstract context is not evidence that there exist stable patterns within such a context.

An idea appears to be about a context, but one also needs to identify properties, and one needs to possess an ability to measure these properties in a reliable context, and furthermore, these properties also need to be associated to (valid) measurements which are described within a context, wherein, the properties and the contexts are (should be) related to the capacities of human capabilities to create in a practical manner, to measure and to control the patterns. This is about the relation that a measurable descriptive knowledge has to creativity.

The unstable contexts, which nonetheless possess short-lived, identifiable (or measurable) [but unstable] patterns can be used . . . .
. . . . either in a context of feedback-systems where the "relevance of the measurable context" is difficult to define (determine) or in the random context of a transitioning system, [from a broken-apart system, so as to transition to its final (relatively) stable state], where this transition takes place under a context of component-collisions, and where the probability of these collisions is related to the rate of the reaction, ie the context of nuclear reactions . . . ,
. . . , this is one of the main business interests associated to our society's fixed way of organizing society, so that this fixed structure is upheld and maintained by extreme violence, in which the violence and coercion is needed to in order for the society to remain fixed, so that the monopolistic business structure can continue to exist.

In the context of feedback of locally measurable (and non-linear) properties the observed properties (observed in the context of a metric-space) seem to be "relatively stable," but in fact, are fleeting and unstable patterns, which are built upon abstract interpretations of contexts, [in turn, built upon an underlying set of fundamental stable material properties].

The social context of the descriptive structure is built upon an overly fixed social-structure, and the symbolic structure, which the owners of society impose on material constructs which define the products of businesses within society. If knowledge is structured primarily to be used to build weapons and to control knowledge and information then this is what people will be best suited to create.

However, the context of un-identifiable spectra, or the context of unstable events placed in a context of physical attributes which are measurable in a metric-space, is a descriptive context which possesses virtually no relation to:

1. the stable patterns [of "stable spectral-orbital material systems" which exist at all size-scales], nor to
2. valid models of chemistry,
3. models of life, or
4. mind, and even at
5. the higher abstract level of human experience, often labeled "religion," where one can perceive the "world as it really is," or as it could possibly "appear to be" in regard to the "true nature" of a living observer (What is the complete context of life?).

The dogmas of science and math are expressed in the social context of "intellectual exclusion," and it is formalized (or defined) by peer-review, where these authoritative dogmas are used to protect business interests.

This is possible because of the fact that the dogmas and the "elite structure of science and math" exclude the development of knowledge, in regard to creativity, which is not under the control of the monopolistic business interests (which the justice system so violently upholds), where business can control science by, controlling (1) an authoritative media (2) the laboratories, and (3) educational institutions, which are owned and/or controlled by the business interests, and thus these institutions serve these (same) business interests.

It is these institutions, and the associated authoritative dogma so presented by the media, which is used by the media to express both the identity (and absolute authority) of science and the business interests of those few people who dominate the society and its organization.

What are (math) patterns?
Patterns are:

1. consistent relationships, or

2. operators acting on quantitative sets so as to have fixed "consistent properties" related to the application of an operator on a quantitative set, and these consistent patterns are related to the "meaning" of the quantitative-set's elements (where quantities represent properties of: type and [measurable] size), or

3. stable shapes, etc.

Can the current descriptive language of mathematics and physics describe stable patterns?

Consider:

I. A new context which is identified by the special shapes (circle-spaces) which relate the local quantitative operators more directly to a more stable (and fixed) set of separable solution-functions (which might exist for a system's set of partial differential equations), where the process has a more significant, and more restrictive, geometric dependence (than does the notion of materialism), each dimensional level is given a more restrictive context, but in this new construct there is a many-dimensional context which is highly relevant to the set of observed properties, and how math constructs can be related to these observed stable, definitive patterns of existence.

II. Partial differential equation (a process for finding formulas for measurable properties of a system by relating local measures of the system's measurable properties, to the local measures of the containing space's coordinates), ie sets of locally linear operators which relate function values to domain values, seems to be a construct which is "less important in the new math construct" than are the importance of shapes.

III. But, in the current "descriptive authority," derivatives are being used to identify "local" spectral values "associated to random local particle-spectral events," so that operator-types are believed to be related to various spectral-types. Thus, the descriptive context is about finding (complete) sets of commuting Hermitian operators, which identify (in a unitary-invariant (or energy-invariant) context) a system's set of identifying local particle-spectral set which, supposedly, can be used to identify the system's set of spectral-measurable properties.

That is, the containment set is defined in a context of measurable, local, random spectral-events, so functions represent the randomness of the system's components, and the spectral sets represent the containment set of the system's identifying measurable properties.

After, nearly 100-years this idea has not been successful at a level of generality which is needed to make such an idea valid.

That is,

There are essentially the three ways in which to try to describe stable math-physical patterns . . . ,

I. stable geometry, which strongly limits both a descriptive context and the patterns it is trying to describe (the new context for physical description, the circle-spaces, or the very stable discrete hyperbolic shapes),

II. differential equations in a geometric context (unfortunately, this method most often leads to non-linear patterns),

III. differential equations in an operator context (this methods seems to only work for harmonic properties which possess actual physical attributes))

. . . , so as to try to use quantitative descriptions so as to try to identify stable patterns which provide valid information, as well as control, over relatively stable (physical) system properties.

If a measurable descriptive language is without the properties of stability (ie stable properties of the description do not exist) and the descriptions are also without the property of quantitative consistency, but the descriptions are still associated to relatively distinguishable patterns, which, unfortunately, are unstable and fleeting patterns, then one's math methods end-up being only complicated exercises, which possess no content (or, at best, unstable patterns may allow for control by feedback in a fleeting pattern whose range of stability is even more difficult to identify than is the unstable pattern).

If a descriptive structure is associated to many elaborate techniques for "framing and describing" observed physical phenomenon, but if actual, generally accurate descriptions are not forthcoming from such techniques, and the context has virtually no practical purposes, then such a descriptive structure is without content, and is only the basis for elaborate, but irrelevant, techniques, which are devoid of any content, and such techniques are without the capacity to identify (in an accurate manner) stable patterns.

That is, the descriptive patterns of the current (overly authoritative) beliefs (dogmas) of math and science are (have become) irrelevant, in regard to using these (such) patterns to describe the observed, "relatively stable" and definitive, and (often) discrete, patterns of material systems. Furthermore, they are "descriptive" patterns which have no relation to practical creative developments. That is differential equations based on geometry work for some classical systems, and the operator viewpoint acting on function spaces only works for waves (harmonic functions) whose attributes have physical properties, but not for general, stable, precise quantum systems (whose defining property has been assumed to be the randomness of their spectra-carrying particle-components), so it seems that only the context of a very limiting geometric context which also defines a new context for the derivative as a discrete operator on discrete geometries is available for a valid way in which to represent and quantify descriptions of stable observed patterns.

The question is open: What other ideas are there?

When (If) math procedures are: non-commutative (in regard to both geometry and when used in the context of function spaces), non-linear, and indefinably random . . . . (where indefinably random events are events which are neither stable nor calculable), ... then the math patterns, which these procedures are trying to describe, are fleeting, unstable patterns, which are neither "generally accurate," nor do such unreliable-patterns have any practical use.

That is, difficult math methods . . . , which are related to fleeting, unstable math patterns . . . , are descriptive constructs which have no content (and possess no useful information), in regard to the stable observed patterns which do exist.

That is, we are equal creators, and this needs to be expressed in a context of equal free-inquiry where a precise description needs to be related to some type (preferably new types) of practical creativity, creativity intrinsic to the intent of life, not creativity which identifies and maintains inequality and its associated violence.

Knowledge is not fixed, and knowledge does not need to only be related to some intricate instrument's further development, or to some intricate authoritative viewpoint about "how knowledge should be developed." That is, knowledge is based on elementary properties of language and the patterns realized which are related to these beginning elementary language structures are sets of assumptions, and contexts of a set of descriptive patterns. Furthermore, a particular set of assumptions, contexts, and interpretations is not moving toward an absolute knowledge, where this is because descriptive knowledge is limited as to the patterns which it (the language based on assumptions) can describe, and the limits of the patterns and/or the practical usefulness of a set of patterns (of a precise descriptive language) can be reached, so that further development leads to irrelevance and illusion. This is analogous to the idea that there are limits to the capabilities of instruments (precise descriptive languages built upon sets of assumptions), and other contexts might very-well be related to a better way to do the same (functioning aspects to which the patterns of language are related) capacities of the fixed set of (complicated) instruments.

The set of assumptions about which the current overly narrowly defined authority depends are both "far too general" and also "far too restrictive" based on an (almost) arbitrary narrowness emanating from (social) authority. There is the very narrow idea of materialism (which assumes that no-holes exist in the material-containing coordinate-space modeled as a continuum), continuity of dimension (or the holes of spaces which define higher-dimensions, and the spectral-lengths which fit-into these higher-dimensional constructs, get smaller), and randomly based spherically symmetric force-field geometry (often an inverse square field), which emanate from material-particle components, where a quantitative structure (a measurable pattern) is imposed on a blank-canvass (but consistent with materialism) either by means of a (geometric) solution to a differential equation, or by a set of operators acting on a function-space (which model the randomness of harmonic local point-particle-events), both of these structure-imposing constructs are defined upon what is believed to be a blank structure, wherein a prevalent spherically-symmetric inverse-square force-field is always assumed to be that which imposes a spectral-orbital structure, but wherein holes, twists, and cuts-points (in all generality) are believed to be relevant (valid), apparently for shapes imposed by material properties, or properties of high-dimension space whose spatial regions are assumed to diminish in size, while [and analogously] the high-energy spectra (observed in particle-colliders), whose origin is assumed to be in higher-dimensions, the values of the spectra must descend in size (implying increased energy). The idea is that if one finds either the force-field or the energy structure which applies to the operator structure (associated to a material system) then one can identify, by calculation on a blank canvass, the observed order of the material system.

The construct of "a blank canvass which hosts the "material" of random points of spherically symmetric force-fields, modeled as non-linear random relationships," has no relation to any (general) precise stable pattern which is re-construct-able from the laws of this context. This description has no relation to a stable pattern.

It is a descriptive context which is neither general, nor accurate (so as to have sufficient precision), nor practically useful. It is relatable to a state of free-material components transitioning between stable states, ie the reaction rates (or collision probabilities) of transitioning systems, where the original system has broke-apart.

Instead (alternatively), holes are prevalent at all dimensional levels (and in all subspaces), the dimensional structures are partitioned by an increasing set of spectral-orbital values as the dimension increases (at least on most [or some] subspaces), so that the spatial structure of the high-dimensional containment set is not "continuous between dimensional levels," ie furthermore continuity in n-space is defined by the continuity of (n-1)-faces of the lower-dimension metric-spaces, (or equivalently "the material-components," which the n-space contains, or a system's spectral-functions (or the functions in the function spaces), are tied to the geometric structure of the coordinates, so that: orbits, angular momentum, the state of being free-material-components, and component-collisions, as well as the properties of physical waves are all "closely tied" to a limited set of stable shapes and the usual second-order differential equations, whose context is now (in the new context) limited to the prevalent shapes (of existence), on which both spectra and orbits are analogous constructs. These fundamental spectral-orbital properties are related to a hole structure of the shapes, but the shapes are placed in a many-dimensional context, which allows material-components to be contained on (or to exist on) "linear shapes," which, nonetheless, guide the material to an orbital structure based on the material trying to adhere to the geodesics of the (linear) shape (defining envelopes of orbital stability).

The spectral-orbital properties which determine the organization of material structures are defined by the geometric-measures of the faces of the (difficult to perceive) fundamental domains which determine the shapes of the metric-spaces and the material components which determine existence (though there can also exist condensed material).

The new interaction construct is general, but its stable properties are determined from a context of the metric-space shapes of existence.

Re-iterating

A new interaction-construct can be constructed which is general, but its stable properties are determined from a context defined by a many-dimensional set of discrete metric-space shapes, which, in turn, define existence.

The professional mathematicians and scientists in regard to descriptions of fundamental stable physical systems express symbolic nonsense, ie they provide a set of nonsense symbols which result in descriptions which are neither general, nor accurate (to sufficient precision), nor do they provide a practical context for useful creativity.

Physical systems which are very stable and definitive, but which are many-(but relatively few)-body systems, nonetheless, because these systems are so stable and definitive, it is clear that they are forming within a very controlled context, so that the descriptions (of the professionals) which are based on:

1. (vague) randomness (which is an uncontrollable description for a system which is composed of only a few components),
2. non-linearity (quantitatively inconsistent, and chaotic), and
3. non-commutative (not invertible, or equivalently, not solvable, eg non-linear or spectrally-un-resolvable), context, which is
4. contained in a continuum (a containing set which is far "too big" allowing logically inconsistent descriptive constructs to be put-together as if they belong to the same containment set), and
5. it is a description (when based on randomness) which begins from a global viewpoint (a function space) but the methods of the description focus on local spectral-particle events

in space, ie it is a description which gives-up information leaving one in an inaccurate and non-useful context in regard to information.

It is a description which "in general" is not accurate, yet it also is a description which is "intent on" losing information about the stable definitive properties of the [assumed to be random] system.

That is the descriptive structure of the "dogmatically pure" set of experts of math and science is simply a bunch of nonsense.

Yet one must list the places and contexts within which it is a valid descriptive context:

1. It is a description which is relatable to a system whose initial conditions, and initial properties are carefully put-together so as to be a system which is easily broke-apart, so as to form a transitioning system which is chaotic, so that the rates of reactions (in this context, based on component-collision probabilities) are determined by cross-sections of the broken-apart components, where these cross-sections determine the rates of certain aspects of the (a) reaction, and
2. They are descriptive contexts which relate a limited set of metrically measurable (observable) properties to a feedback structure, which is mostly associated to the critical-points and limit-cycles of a non-linear (usually classical) partial differential equation, where the range of relevance of the differential equation is difficult to determine or to control. Furthermore, the initial or boundary conditions of this type of a system relate to the properties of the descriptive context (or properties associated to the solution) of the system's differential equation in a chaotic manner.

These contexts identify structure related to (1) nuclear weapons and (2) guiding missiles and drones.

That is, difficult math methods . . . , which are related to fleeting, unstable math patterns . . . , are descriptive constructs which have no content (and possess no useful information), they are patterns which apply only to unstable contexts, where control emanates from a higher abstract and manipulative context imposed on properties which are only definable in a metric-space, and which requires a lot of preparation (in regard to sensing and reacting in the desired way to the detected properties), a context which is at-odds with the system's natural properties, rather than controlling a system by simple adjustments to affect the system's properties in regard to affecting the properties of several system-components being coupled together.

These professionals are deemed, by the media, to be the intellectual top-experts of the society.

Yet their failed descriptive context is claimed to be the best descriptive range that they can offer. Namely, a descriptive structure which essentially destroys the context of creative development, by the experts providing a failed descriptive structure.

Nonetheless, these experts proclaim that only "the dogmatically pure" can join in on the discussion.

That is, the professionals are getting high-marks (big salaries) [by the owners of society, ie those few who assign value within society] for playing a role of top-intellect in society. Yet their true goal, which they seem to not be aware of, (which is to develop knowledge, which, in turn, is useful and applicable over a wide range, in regard to developing practically useful physical systems).

All that these, so called, top experts do is to develop contexts which are hopelessly narrow in their application, but which demonstrate elaborate and complicated methods, but they are methods which do not describe stable patterns, they do not (one cannot use the laws of quantum or particle physics to) generally and accurately describe the observed stable patterns of existence (of general but fundamental quantum systems), and these descriptive structures have virtually no relation to practical creative development. That is, the experts can provide patterns which have literary interest to other experts, but these, essentially, unstable patterns . . . , (which are contained in an illusionary world) upon which the experts dwell . . . , have no physical interest (or they have no relation to the stable patterns of the physical world).

Like most aspects of the current society, those on the top tiers of society are held in high social esteem for being total and complete failures (the media and the corruption of institutions allows this).

This is the result of the justice system of the US society, where according to the Declaration of Independence US law is supposed to be based on equality . . . , (the point of "freedom based on equality" is about each person having the right to develop knowledge as they want (by the process of equal free-inquiry), so as to be able to create what they envision, and then give as a gift, which the individual can give to society in a selfless manner, where the society cannot judge their value, and the society is committed to giving everyone the material needs to live, prosper, and, subsequently, to create in a selfless manner [note: only in this context can a truly free-market exist, but the profits of the most successful products should be well below 1%]), ... , but the elites, who opportunistically began administrating "the independent US nation," instead of instituting equality within the law, so that selfishness was to get punished, acted in a selfish manner, so as to base law on property rights and minority rule, [which is the essential law of the emperor of the Holy-Roman-Empire]. That is, the US governance began as a total failure, so as to be run by opportunistic elitists, who instead of instituting equality and "free-inquiry based on equality" so that knowledge and creativity were to be developed by the culture, instead the elitists in charge used to law to steal, coerce, and destroy, those in the lower social classes, in the name of their own selfish advantage, the selfish advantage of the few, ie power and production were based on social domination of the many by the few.

So we have the tradition of "western hypocrisy," where failure is rewarded if those who perpetuate it, are in the high social classes.

What is wanted, by the owners of society, is that the social structures through which the powerful derive their power are kept in place, ie it is a social structure which is opposed to new, creative changes and thus is is also opposed to equality and the creativity associated to equality. However, the traditional social structure which upholds dominant interests so violently, and it expresses its interest in lyrical creativity of the science and math experts . . . . , where these authoritative experts define the "literary" creative development of science and math, which is authoritative, but unrelated to practical creative development, and the owners of society support the "creativity" of the elite artists, those who also competed in a "narrow context of authoritative cultural value," and those journalists and intellects whose ideas are judged to possess cultural value, so that the ideas expressed are consistent with the ideas of (or can be used by) the owners of society, so as to be distributed by the material-instruments of the media which are owned and controlled by the owners of society . . . , then even the failures of the experts can become part of the social structure which allows the powerful to remain powerful. The top-intellects and top-artists are defined as a social class, along with artists and journalists, so that the intellectuals can dogmatically dominate those many-others who question the authority of assumptions, or who have different ideas. The main tool used to maintain the power of the owners of society is the single voice of authority which the media has become (most clearly controlled by ownership, or by a set of funding processes). That is, it is violence and domination (intellectual domination) which is fundamental to social power, not knowledge.

Knowledge is relevant, within today's social structure, only in regard to the creativity which is a part of the organization of society (ie business productivity) which, in turn, maintains the power of the few. However, the organization of society, and the use of resources and the ownership of technology within society, essentially, remains fixed and traditional.

For example, the many-purpose phone, eg an i-phone, is about developing 19th century ideas of electromagnetism, and the micro-chip circuit boards in these devices depend on 19th century optics.

Whereas identifying stability "as a needed property" in both math and physics, in regard to the useful descriptions of controlled (or controllable) physical systems, is a focus (in regard to the valid descriptions of math patterns) which the math professionals, apparently, have not considered.

Furthermore, very simple math patterns can be used to create new math patterns, which can be used to describe the stable material properties, so that these descriptions are based on a finite quantitative set, within which descriptive containment of physical properties depends, ie the containment set is not a continuum and the derivative and its integral-inverse become discrete operators (the continuum can, instead, be the set of rational numbers).

In fact, the math patterns of stability are very simple, and relating these simple structures (which are best characterized by the stable discrete shapes, or circle-spaces) to many-dimensions, can be done by a simple process of partitioning the dimensional levels of a hyperbolic 11-dimensional containment metric-space (base-space) by means of stable shapes, ie discrete hyperbolic shapes (or circle-spaces), so as to form a finite spectral-orbital set, where the sequences of spectral-size is defined (either increasing or decreasing) as the dimensional level increases, so that these size-sequences of spectra are fundamental in regard to how the description is organized, so that a finite spectral set is the basis for physical descriptions of the observed order which the stable (material and containing metric-space) structures of existence possess.

Furthermore, these simple ideas seem to be much better ideas than are the ideas which the experts possess, [ie than are the ideas that the professional "dogmatically pure" intellectual-army of experts (who work for the owners of society) possess], where the "top-intellects" of society (as proclaimed by the media, where the media is the single authoritative voice of the society) allow the ownership (the management) to bend the minds of these so called experts, ie the pay-masters bend the minds of the salaried-help, but those who possess the best resume's get the best jobs (as everyone competes to help develop the high-value of society, but the high-value of society are those ideas which are proclaimed [or expressed] by the media) the minds of the experts who serve the high-value (defined by the media) have their minds bent in any way which the management wants to bend their minds.

That is, demonstrating high-value in regard to an external model of high-value compromises the internal value (and thus the real creative value) of a person, and it destroys (or greatly limits) knowledge and creativity.

That is, the commercial world is related to a fixed stationary way of behaving or acting, a commercial structure is a very narrow context, based on a limited range of creativity and a fixed way in which to use material resources. The power of business monopolies depend on society not changing how it uses the material resources a business monopoly supplies to a society. The law is supporting this type of narrowness, essentially based on property rights and minority rule (creditor vs. debtor, smart vs. stupid, etc), and it supports such selfish actions with great violence. In fact, the economy is tied to a fixed narrow way in which to live and create, and this model of monopolistic economies is being used as a means to conquer ever larger populations, but it is being put into-place by means of extreme violence and coercion (often an economic coercion).

Does one want a society to be based on a fixed way to use material, and a fixed way in which one is to serve the material based, and fixed structure of society, and a fixed overly authoritative organization of descriptive knowledge, so that this type of power, and associated narrowly defined knowledge, depends on expansion in the form of an ever greater exploitation of particular types of material (usage)?

References

1. A New Copernican Revolution, Bill G P H Bash and George P Coatimundi, Trafford Publishing, 2004. www.trafford.com/03-1913,

2. The Authority of Material vs. The Spirit, Douglas D Hunter, Trafford Publishing,2006. www.trafford.com/05-3038

3. Topology and Geometry for Physicists, C. Nash and S. Sen, Academic Press, 1983.

4. The Infamous Boundary, David Wick, Springer-Verlag, 1995.

5. Function Theory, C. L. Siegel,

6. Three-dimensional Geometry and Topology, W. Thurston, Princeton University, 1997.

7. Gauge Theory and Variational Principles, D Bleeker, Addison, 1981.

8. Geometry II, E B Vinberg, Springer, 1993.

9. Spaces of Constant Curvature, J Wolf, Publish or Perish, 1977.

10. Contemporary College Physics, Jones and Childers, Addison-Wesley, 1993 (High School text).

11. I M Benn and R W Tucker, in, An Introduction to Spinors and Geometry with Applications in Physics, 1987, (Chapter 2)

12. Representations of Compact Lie Groups, T Brocker, T tomDieck, Springer-Verlag, 1985.

13. Dynamical Theories of Brownian Motion, E. Nelson, Princeton University Press, 1967 (1957).

14. Quantum Fluctuations, E. Nelson, Princeton University Press, 1985.

15. Algebra, L Grove.

16. Electron magnetic moment from gonium spectra, H Dehmelt (Nobel prize winner) et al, Physical Review D, Vol 34, No. 3, Aug 1, 1986.

17. Newton's Clock, Chaos in the Solar System, I Peterson, W H Freeman and Company, 1993.

18. Quantum Mechanics, J L Powell and B Crasemann, Addison-Wesley Publishing, 1965.

19. The End of Science, J Horgan, Broadway Books, 1996.

20. Riemannian Geometry, L. P. Eisenhart, Princeton University Press, 1925.

21. Reflection Groups and Coxeter Groups, J Humphreys, Cambridge University Press, 1990.

22. Partial Differential Equations, J Rauch, Springer-Verlag, 1991.

23. The Foundation of the General Theory of Relativity, A Einstein, 1916, Annalen der Physik (49).

D Coxeter

Katok

Just as Copernicus, Kepler and Galileo provided a quantitative-geometric context for the properties of the solar system, which were then precisely identified by the solutions to (the) solvable differential equations of Newton; Martin Concoyle now provides the stable geometric structures which fit . . . , both macroscopically and microscopically . . . , into a many-dimension containment set (hyperbolic 11-dimensional), so that these shapes are the solutions, ie the geometries of the stable spectral-orbital properties, of all the fundamental stable systems which have stable spectra and orbits, and it is the basis for a quantitative system (the spectral set of a measurable existence) which is finite, and

These ideas are discussed in the following books: (available at math conference, 2013)

1. A New Copernican Revolution (p286), B Bash & P Coatimundi, Trafford, 2004.

2. The Authority of Material vs. The Spirit (p483), D D Hunter, Trafford, 2006.

3. Introduction to the Stability of Math Constructs; and a Subsequent General, and Accurate, and Practically Useful Description of Stable Material Systems, Concoyle, and G P Coatimundi, (p262), 2012, Scirbd.com.

4. A Book of Essays I: Material Interactions and Weyl-Transformations, Martin Concoyle Ph. D., (p234), 2012, Scribd.com.

5. A Book of Essays II: Science History, and the Shapes which Are Stable, and the Subspaces, and Finite Spectra, of a High-Dimension Containment Space, Martin Concoyle, (p240), 2012, Scribd.com.

6. A Book of Essays III: Elementary Topics, Martin Concoyle (p303), 2012, Scribd.com.

7. Physical description based on the properties of stability, geometry, and quantitative consistency: Short essays which are: simple, "clear," and direct Presented to the Joint math meeting San Diego (2013), (p208), Martin Concoyle, 2013, Scribd.com.

8. Describing physical stability: The differential equation vs. New containment constructs, Martin Concoyle, 2013, (p378), Scribd.com. (also equivalent to, VII 3, at Scribd.com)

Old SD

9. Introduction to the stability of math constructs; and a subsequent: general, and accurate, and practically useful set of descriptions of the observed stable material systems, Martin Concoyle Ph. D., 2013, (p70), Scribd.com,

See scribd.com put m concoyle into web-site's search-bar

As well as in the following (new) books from Trafford:

1. The Mathematical Structure of Stable Physical Systems, Martin Concoyle and G. P. Coatimundi, 2013, (p449) Trafford Publishing (equivalent to 3. And 5. Above, Scribd)
2. Partitioning a Many-Dimensional Containment Space, Martin Concoyle, 2013, (p477) Trafford Publishing, (equivalent to 4. And 6. Above, Scribd)
3. Perturbing Material-Components on Stable Shapes: How Partial Differential Equations Fit into the Descriptions of Stable Physical Systems, Martin Concoyle Ph. D., 2013, (p234) Trafford publishing (Canada) (equivalent to 7. And 9. Above, Scribd, and new material)
4. Describing the Dynamics of "Free" Material Components in Higher-Dimensions, Martin Concoyle, 2013, (p478) Trafford Publishing (equivalent to 8. Above, Scribd, and new material)

Alternative title to any of 1-4 Trafford:
The Unbounded Shape, and the Self-Oscillating, Energy-Generating Construct

Copyrights

These new ideas put existence into a new context, a context for both manipulating and adjusting material properties in new ways, but also a context in which life and creativity (practical creativity, ie intentionally adjusting the properties of existence) are not confined to the traditional context of "material existence," and material manipulations, where materialism has traditionally defined the containment of material-existence in either 3-space or within space-time.

Thus, since copyrights are supposed to give the author of the ideas the rights over the relation of the new ideas to creativity [whereas copyrights have traditionally been about the relation that the owners of society have to the new ideas of others, and the culture itself, namely, the right of the owners to steal these ideas for themselves, often by payment to the "wage-slave authors," so as to gain selfish advantages from the new ideas, for they themselves, the owners, in a society where the economics (flow of money, and the definition of social value) serves the power which the owners of society, unjustly, possess within society].

Thus the relation of these new ideas to creativity is (are) as follows:

These ideas cannot be used to make things (material or otherwise) which destroy or harm the earth or other lives.

These new ideas cannot be used to make things for a person's selfish advantage, ie only a 1% or 2% profit in relation to costs and sales (revenues).

These new ideas can only be used to create helpful, non-destructive things, for both the earth and society, eg resources cannot be exploited to make material things whose creation depends on the use of these new ideas, and the things which are made, based on these new ideas, must be done in a social context of selflessness, wherein people are equal creators, and the condition of either wage-slavery, or oppressive intellectual authority, does not exist, but their creations cannot be used in destructive, or selfish, ways.

Index
(key words)

fundamental domains
Geometrically separable
Hermitian form (finite dimensions)
hyperbola
hyperbolic
hyperbolic metric-space
Independent
infinite extent space-forms
interaction
interaction potentials
Inverse
Invertible
Isometry
Isometry groups
Lattices
Linear
Lie algebra valued connection 1-forms
Lie group
Maximal torus
metric spaces
metric-space states
moding out
non-reducible
Orthogonal
Parallelizable

Physical properties and fundamental invariance's, eg translations and linear momentum,
principle fiber bundles
sectional curvature
signature of a metric,

Solvable
space-forms,
space-time,
unitary groups

Weyl group

The problems which physics and math now (2014) face, and their simple resolution, only to again face the social problems causing society's inability to realize change, and subsequently, to solve its problems

There is the example of Copernicus who challenged the experts in 1550 "an earth centered view of existence" so that Copernicus, instead, provide a "sun-centered basis for the descriptions of the motions of the heavenly bodies."

The two ideas (the Copernican system and the Ptolemaic system) have similar contexts about bodies in space, or descriptions within the context of math patterns which are used to describe "observed properties."

Where, in turn, challenging the language of experts, is also about the issues raised by Gödel's incompleteness theorem.

Namely, about a precise descriptive language not being able to describe certain patterns, due to the implicit limitations of the precise expert language, where the limitations are imposed by the language's assumptions, contexts, and interpretations, etc, etc. So this set of assumptions and contexts, etc, defines the elementary level at which knowledge is really built.

That is, current physics (and subsequently also math, 2014) is assumed to be based on:
1. materialism,
2. the invariance of partial differential equations
(where geometry is, supposed to be, the cause of material-motion, {but how this occurs has only been recently identified in a limited geometric context, but where the new geometric context is fundamental, for there to be a truthful model of both existence and its observed properties}),
3. contained in a material context, which is governed by both
(1) fundamental randomness, and

(2) reduction of all material and field-energy to elementary-particles, where the material elementary-particles identify unstable elementary random events (in space and time, or in space-time), and where

4. material-interactions (which cause material motion) are to be based on (or modeled on) both

(1) an explicit geometry related to the random elementary-particle-collisions, which exist between both the various field-particles and material elementary-particles (and where these types of particle-collisions represent the interaction process), and

(2) an associated non-linear "local geometry," defined in particle-physics by complex-coordinates, which, in turn, is associated to local particle-state transformations (and energy-particle transformations), which, supposedly, occur during the particle-collisions between both the elementary-particles and the field-particles, (but, where such an explicit geometry of "a particle-collision" cannot exist within a context which is fundamentally random), where

5. these particle-collision interactions, are supposed to perturb the wave-functions of a general quantum system, but

6. the wave-functions for general quantum systems cannot be found by calculations, which are based on the laws of (regular) quantum physics, and where

7. A (regular) quantum system's wave-function is, supposed to be based on the existence of a "random wave-function, which represents the random behaviors of the system's 'component and energy structure'" where the wave-function oscillates, or forms, around the energy-operator representation of a quantum system, where this energy structure is basically represented as an eigenvalue equation.

But, in general, the eigenvalue equation cannot be solved, ie it is not a valid context in which to define the valid laws of physics, since these spectral-systems possess very stable and precise spectral-orbital properties, which means that they form under very controlled descriptive contexts.

This (above) set of assumptions defines an inconsistent descriptive context (random but geometric, so as to describe material interactions which perturb non-calculable spectral properties) which cannot be used to describe the observed stable spectral-orbital properties of the very prevalent many-(but-few)-body systems

That is, this is a circular set of relations which define an explicit inability (or incapacity) to describe the fundamental systems of our existence, eg namely, the very stable many(-but-few)-body (material) systems which exist at all size-scales: from the nucleus to general atoms, to molecules, to molecular shapes, and even in regard to the, effectively, classically described solar-system, so that none of these… stable, and precisely (spectrally) identifiable fundamental systems…, can be described by using the laws of physics, as outlined above.

A group of people…

who have failed to solve the most fundamental problem of physics (and math), namely, "how to describe in an accurate and in a practically useful way the properties of the many-(but-few)-body problem which exists at all size-scales?"

….are not the top intellects of society.

Yet these are the people who are deified, as the "top intellects" of society, by the propaganda-education system, Despite the fact that they have failed to provide a valid description (solution) of this many-(but-few)-body problem.

But there are deep risks for the investor-class, in regard to "new knowledge," which would be related to describing the observed stable spectral-systems, in comparison to the "current knowledge" which is being provided to society by today's propaganda-education system, since the current communities of peer-reviewed science and math people are not providing a valid description of physical systems and their stable spectral-properties, except, perhaps, in regard to the explosion of a nuclear weapon (ie where such knowledge is desired by the monopolistic interests of the big military industries).

In a nuclear explosion, a stable system chaotically transitions to another stable state, so that the transition takes place in a chaotic context, where the transition rate is related to probabilities of random particle-collisions, this chaotic descriptive context is what the so called top intellects dwell upon.

The intellectuals need to get out of this descriptive scenario, which they embrace, which puts a high-value on indefinable randomness, and on non-linearity, and where measuring is no longer reliable, and there are no stable patterns upon which to base their descriptions.

The current paradigm (in 2014) of the fundamental pattern of material interactions described in the context of "particle-collisions in particle-physics," which describes a general state of existence, in which there is always "a chaotic transitioning process" which exists and is, supposedly, manifested as random elementary-particle collisions, between particles which are not stable, and which is a chaotic process which is supposed to be perpetually occurring.

However, the descriptions…. of the wide range of the general stable states of the many-(but-few)-body systems into which this "forever chaotically transitioning process" are supposed to settle… do not exist when one tries to use a random context which is assumed to be reducible field-particle colliding with unstable elementary material particles.

Instead the experts say that "such stable, many-(but-few)-body systems are too complicated to describe."

That is, if this chaotically process of transition, which exists between two relatively stable states of material systems, is supposed to occur and to actually cause the stable spectral states of the general many-(but-few)-body systems, then where are the precise descriptions for all of the general systems, which possess such precise stable spectral properties?

That is, the precise descriptions..., of the stable spectral states of the general many-(but-few)-body systems which are a result of this model of material interactions, where material interactions are being modeled so as to be a transitory state of randomness and elementary-particle-collisions...., do not exist.

This is what has happened to physics and math when the focus of these academic disciplines is on the commercial interests of nuclear weapons, where the descriptions of the "nuclear weapon explosions" are based on a chaotically transitioning process, modeled as particle-collisions, and defined between two relatively stable states of material, where the rate of the transition process is related to the probabilities of the particle-collisions.

But this is the fundamental nature of all the institutions, which are the pillars of our (hierarchical) society. The expert languages, associated to all of the fundamental social institutions of our society, are, in fact, arbitrary, which are used to justify any actions which these institutions perform, where math and physics are about nuclear weapons and communication systems, but these institutional actions depend on old knowledge which is limited in its narrow range of descriptive validity (or descriptive capacities).

To rectify this problem

There are choices which can be made, in regard to many well known properties either Physical law as partial differential equations, and/or an invariant partial differential equations, where locally invariant partial differential equations are invariant to local coordinate transformations which only need to be ("general" in the sense that they are) smooth and locally invertible, but this defines a measuring context which is unreliable, in-part because the patterns are non-linear and chaotic, or Physical law as geometry, and the structure of metric-spaces, ie the very stable (linear and continuously commutative) metric-space shapes of non-positive constant curvature metric-spaces (for metric-functions which only have constant coefficients) in a many-dimensional context, where partial differential equations are still relevant, but mainly in regard to perturbing stable spectral-orbital properties

Path invariance in a space (where the space has a shape) implies that the space's shape can be deformed to a point, That is, one needs holes in "the shape of spaces," upon which (or so that) stable measurable properties can be defined,

The shapes of stable quantitative patterns are:

Lines, rays, line-segments, cubes, rectangles, circles, (discs, but discs are deformable to a point), cylinders, tori, shapes with toral-components, etc.

Other shapes are not stable.

The context of stability is quantitative consistency, solvability, metric-invariance, linearity, stable metric-space shapes, and one needs holes in the shape of spaces, upon which (or so that) stable measurable (spectral-orbital) properties can be defined.

But it is incorrect to claim that partial-differential-equation-invariance is a basis for the descriptions of stable physical systems.

Furthermore, one does not need the reduction of all material systems to unstable elementary point-particle components in a descriptive context based on randomness

Solvability, or equivalently, system controllability, in regard to partial differential equations, means linear, (metric-invariant), and continuously commutative local coordinate relations, which are associated to either the shape of the space, within which the description of locally measurable properties are (is) contained, or the shape of the system which is being described. However, the physical systems must, in fact, be the stable (discrete) shapes of (hyperbolic) metric-spaces, ie shapes which possess holes. It is only this math context allows the patterns, which can be described in this math context, to be stable patterns.

So, what is this very controlled descriptive context which is needed for the observed very stable spectral-orbital systems?

It is built from the stable metric-space shapes, in particular the "discrete hyperbolic shapes," which, in turn, can determine a partition of a high-dimensional containment set, so that this partitioned-set (of, effectively, independent subspaces) is associated to a finite spectral set, so that the material and spatial components, which are contained in (or, equivalently, which compose) this high-dimension containment set, are in resonance with the containment set's finite spectral set.

But such an idea will be excluded from the media, it will be excluded from (or ignored by) the communication channels, so that only the "correct dogmas" are allowed to be expressed in the highly controlled communication channels of our banker-controlled society.

It is an idea cannot be peer-reviewed since it is a new idea. Peer-review articles are, essentially, the technical dogmas which support the bankers' investment interests. It is an expression of the experts' own faith in their, overly narrow, indoctrinated viewpoint.

That is, the issue is, that the formalized axiomatic structures which are being used (in math and physics), are consistent with, and apparently helpful in regard to, the interests which the investor class, namely, in regard to the narrowly defined "creative" commercial interests…, eg oil-energy, controlling communication systems, and military advantages (tied to nuclear weapons)…, of the investor-class and

They are not consistent with the mathematical need for "being able to describe stability."

This book is an introduction to the simple math patterns used to describe fundamental, stable spectral-orbital physical systems (represented as discrete hyperbolic shapes, ie hyperbolic space-forms), the containment set has many-dimensions, and these dimensions possess macroscopic geometric properties (which are also discrete hyperbolic shapes). Thus, it is a description which transcends the idea of materialism (ie it is higher-dimensional, so that the higher-dimensions are not small), and it can also be used to model a life-form as a unified, high-dimension, geometric construct, which generates its own energy, and which has a natural structure for memory, where this construct is made in relation to the main property of the description being, in fact, the spectral properties of both (1) material systems, and of (2) the metric-spaces, which contain the material systems, where material is simply a lower dimension metric-space, and where both material-components and metric-spaces are in resonance with (or define) the containing space. Partial differential equations are defined on both (1) the many metric-spaces of this description and (2) the lower-dimensional material-components which these metric-spaces contain, ie the laws of physics, but their main function is to act on either the, usually, unimportant free-material components (so as to most often cause non-linear dynamics) or to perturb the orbits of the, quite often condensed, material which has been trapped by (or within) the stable orbits of a very stable hyperbolic metric-space shape.

It could be said that these new ideas about math's new descriptive context are so simple, that some of the main ideas presented in this book may be presented by the handful of diagrams which show these simple shapes, where these diagrams indicate how these simple shapes are formed and folded, or bent, to form the stable shapes, which can carry the stable spectral properties of the many-(but-few)-body systems . . . . , where these most fundamental-stable-systems have no valid quantitative descriptions within the, so called, currently-accepted "laws of physics," (ie the special set of partial differential equations associated to the, so called, physical laws) . . . . so that the diagrams of these stable geometric shapes are provided at the end of the book.

This new measurable descriptive context is many-dimensional, and thus, it transcends the idea of materialism, but within this new context the 3-dimensional (or 4-dimensional space-time) material-world is a proper subset (in a subspace which has 3-spatial-dimensions),

The, apparent, property of fundamental randomness (in a currently, assumed, absolutely-reducible model of material, and its reducible material-components) is a derived property, but now in a new context in which stable geometric patterns are fundamental,

The property of spherically-symmetric material-interactions is shown to be a special property of material-interactions, which exists (primarily, or only) in 3-spatial-dimensions, of Euclidean space, wherein inertial-properties are to, most naturally, be described,

It is both reductive . . . . , (to some sets of small material-components, but elementary-particles are most likely about components colliding with higher-dimensional lattice-structures, which are a part of the true geometric context of physical description) . . . , and unifying in its discrete descriptive contexts (relationships) which exist, between both a system's components, and the system's (various) dimensional-levels (where these dimensional-levels are particularly relevant, in regard to understanding both (1) the chemistry and (2) the functional organization of living systems),

But most importantly, this new descriptive language (new context) describes the widely observed properties of stable-physical-systems, which are composed of various dimensional-levels and of various types of components and interaction-constructs, so that this new context provides an explanation about both (1) "how these systems form" and (2) "how they remain stable," wherein, partial differential equations, which model material-interactions, are given a new: context, containment-structure, organization-context, interpretation, and with a new discrete character,

It provides a (relatively easy to follow, in that, the containment set-structure for these different-dimensional stable-geometries are simple dimensional relations) 'map' "up into a higher-dimensional context (or containment set) for existence," wherein some surprising new properties of existence can be modeled, in relation to our own living systems also being modeled as higher-dimensional constructs, and this map can shed-light onto our own higher-dimensional structure, and its relation to both existence, and to the types of experiences into which we may enter (or possess as memory) (or within which we might function), where because any idea about higher-dimensions is difficult to consider, and is relatively easy to hide and ignore these higher-dimensions, especially, if we insist on the idea of materialism.